ACS SYMPOSIUM SERIES **602**

Proteins at Interfaces II

Fundamentals and Applications

Thomas A. Horbett, EDITOR
University of Washington

John L. Brash, EDITOR
McMaster University

Developed from a symposium sponsored
by the Division of Colloid and Surface Science
at the 207th National Meeting
of the American Chemical Society,
San Diego, California,
March 13–17, 1994

American Chemical Society, Washington, DC 1995

Library of Congress Cataloging-in-Publication Data

Proteins at interfaces II: fundamentals and applications / Thomas A. Horbett, John L. Brash, editors.

  p.   cm.—(ACS symposium series, ISSN 0097–6156; 602)

"Developed from a symposium sponsored by the Division of Colloid and Surface Chemistry at the 207th National Meeting of the American Chemical Society, San Diego, CA, March 13–17, 1994."

Includes bibliographical references and index.

ISBN 0–8412–3304–7

1. Proteins—Congresses. 2. Surface chemistry—Congresses. 3. Biological interfaces—Congresses.

  I. Horbett, Thomas A, 1943–   . II. Brash, John L., 1937–   . III. American Chemical Society. Division of Colloid and Surface Chemistry. IV. American Chemical Society. Meeting (207th: 1994: San Diego, Calif.) V. Series.

QP551.P697782   1995
574.19′245—dc20
                                                  95–36863
                                                     CIP

This book is printed on acid-free, recycled paper.

Foreword

THE ACS SYMPOSIUM SERIES was first published in 1974 to provide a mechanism for publishing symposia quickly in book form. The purpose of this series is to publish comprehensive books developed from symposia, which are usually "snapshots in time" of the current research being done on a topic, plus some review material on the topic. For this reason, it is necessary that the papers be published as quickly as possible.

Before a symposium-based book is put under contract, the proposed table of contents is reviewed for appropriateness to the topic and for comprehensiveness of the collection. Some papers are excluded at this point, and others are added to round out the scope of the volume. In addition, a draft of each paper is peer-reviewed prior to final acceptance or rejection. This anonymous review process is supervised by the organizer(s) of the symposium, who become the editor(s) of the book. The authors then revise their papers according to the recommendations of both the reviewers and the editors, prepare camera-ready copy, and submit the final papers to the editors, who check that all necessary revisions have been made.

As a rule, only original research papers and original review papers are included in the volumes. Verbatim reproductions of previously published papers are not accepted.

Contents

CONFORMATION AND ORIENTATION OF PROTEINS
AT INTERFACES

ROLE OF ADSORBED PROTEINS IN CELL INTERACTIONS
WITH SOLID SURFACES

PROTEIN BEHAVIOR AT FLUID–FLUID INTERFACES

INDEXES

Preface

THE SURFACE ACTIVITY OF PROTEINS is a fundamental property of these complex macromolecules that derives from their large size, amphipathic nature, and the many types of chemical interactions that can occur between proteins and surfaces. Thus interfaces of almost any type that come into contact with protein solutions tend to become quickly occupied by proteins, leading to profound alterations in the physicochemical and biological properties of the interfaces. Proteins at interfaces are important in many applied areas, including separation and purification, the biocompatibility of biomaterials, mammalian and bacterial cell adhesion, blood coagulation at solid and membrane surfaces, solid-phase immunoassays, biosensor development, the opsonization of particulates used as therapeutic agents, food processing, and biotechnology in general. Despite a fairly long history of study of proteins at interfaces, many of the fundamental mechanisms remain only partly understood, and research on proteins at interfaces remains very active.

Reflecting the diversity of situations affected by proteins at interfaces, investigations in this area are also very diverse with respect to types of interfaces, proteins studied, methodology employed, and practical problems to which the work is directed. Consequently, research on proteins at interfaces is presented at diverse scientific meetings with typically only a few papers at each meeting. Similarly, the research is published in a wide variety of journals and other publications, resulting in very few comprehensive sources of information on this topic. By bringing workers from the many disparate application areas together into one symposium, we believed we could provide a means to foster advances in proteins at interfaces via presentation of the many common concepts and approaches embodied in the diverse systems and applications being studied.

To make participation in the symposium on which this book is based as comprehensive as possible, we began more than a year in advance by inviting many, if not most, of the investigators actively involved in studies of proteins at interfaces to participate, and we subsequently included their contributions in this book. This volume is similar in its intent to ACS Symposium Series 343, *Proteins at Interfaces*, edited by us in 1987, but the approach we have taken to the content is somewhat different. In the previous volume, we invited overview chapters, whereas in this volume, the chapters are similar to journal articles that reflect the current interests and work of each group.

This book provides a broad collection of articles on the behavior of proteins at interfaces, most of which derive from recently completed investigations focusing on topics of current interest. The major themes include molecular mechanisms, competitive adsorption, conformation of proteins at interfaces, surface chemistry effects, protein effects on cell interactions, and the behavior of proteins at fluid–fluid interfaces.

We believe this book can provide a sound introduction for those new to the field but will have its greatest impact as a convenient way for experienced investigators to broaden their understanding of the behavior of proteins at interfaces.

The introduction by Leo Vroman, surely the expert's expert on protein interfacial interactions in "outrageously complex protein mixtures", relates specifically to blood–material interactions, phenomena that have motivated many of the researchers who study proteins at interfaces and whose work is described in this volume. However, its message can be transposed and applied to any complex biofluid. Vroman's introduction indicates, in a way that only a scientist–poet of long experience could, the complex nature of these interactions and warns us, lest we become prematurely smug, that we still have a long way to go in achieving anything resembling a full understanding.

THOMAS A. HORBETT
Department of Chemical Engineering
University of Washington
Seattle, WA 98195–1750

JOHN L. BRASH
Department of Chemical Engineering
McMaster University
Hamilton, Ontario L8S 4L7
Canada

June 28, 1995

Prologue

THE MOST COMPLICATED LANGUAGE we can ever attempt to learn is the one spelled by the proteins within us and rapidly written by them on any blank surface they face. Each of the "words" is long enough to fill an entire page if printed legibly, and as they are being "written" on the blank surface, they may rapidly change their meaning or, in the presence of more protein species, be displaced by a succession of these others. How then can we ever hope to read the significance of such a briefly present and changing text and context?

Perhaps we have been very slow to face this problem, misled since we invented glass test tubes, especially once we learned to put blood in them that had been anticoagulated and could be swung around to separate its deceptively simple-looking plasma. Slowly it dawned on us that the resulting interactions were far from simple, and we now know that on many surfaces, outrageously complex protein mixtures such as blood plasma write their own opinions by means of adsorbing and interacting protein molecules.

Of course our blood's authorship did not evolve to entertain us; it merely tries to protect us. I am sure that any spilled blood is not aware of having left its host forever, nor of the relative immensity of foreign surface it faces. All interactions with this surface are aimed at protecting the host against what this poured blood perceives as an invasion of its host's body by strange and unfathomed matter.

Thus it is the blood itself that is doing the reading. It reads the surface properties with the help of its own proteins and passes the information on to others: the intrinsic clotting system, the complement system, and from there to the platelets and white cells that may arrive just in time to read a few code elements exposed by some of the plasma's "written words"—protein epitopes that these cells and platelets are tuned to by their receptors.

So it is not surprising that we can find a wild paradise of elements left on devices that blood has streamed over. Fixed and stained, it represents a rather slow snapshot of a few physiologically significant and not entirely simultaneous events that were driven by very local conditions. For example, we found that heparinized blood injected between a glass slide and a convex lens resting belly-down on the slide may leave a small ring of platelets surrounded by a fine line of fibrin. What happened, we believe, is that platelets adhered where high molecular weight kininogen (HMK) could not remove fibrinogen before the platelets arrived. Immediately beyond this spot, where HMK did compete successfully for the surface, the clotting system was activated next to the area where heparin had been neutralized by platelet antiheparin activity so that fibrin could form.

A reasonable and yet naive question is what keeps our plasma from writing its graffiti on the surfaces of all cells floating in it? And if we create surfaces resembling those of cells, would proteins not be adsorbed on them? Only the second question allows experimentation and can be answered. There is no logical way of separating or cultivating normal healthy cells in total absence of added or generated proteins in their medium. Our ability to separate cells from their plasma may let us forget that these cells were born in plasma and have been exposed to hundreds of proteins at their interfaces before we put our hands on them. Clean cell surfaces do not exist until they are dead. Even single purified proteins must be suspected of not behaving as they would normally in the environment where they evolved their specific and often unknown functions.

We can create surfaces now that vaguely resemble cell surfaces, e.g., by coating materials with phospholipid bilayers. The mobility of molecules in such a coating lets it respond to very local events, such as the approach of a protein molecule. If it is true that a protein molecule simply cannot attach itself on such a "soft" surface and will not be forced to spread, the complexity of such an interaction has been taken away from this protein molecule and transferred to the substrate. When we attempt to write on oatmeal, the invisible words may still change the oatmeal.

Such interactions between surface and proteins, and between adsorbed and hence modified proteins and formed elements, are echoed in this volume by the interactions among those scientists who study them.

LEO VROMAN
Department of Chemical Engineering,
   Materials Science, and Mining
Columbia University
500 West 120th Street
New York, NY 10027

June 29, 1995

Chapter 1

Proteins at Interfaces

An Overview

John L. Brash[1] and Thomas A. Horbett[2]

[1]Department of Chemical Engineering, McMaster University, Hamilton, Ontario L8S 4L7, Canada
[2]Department of Chemical Engineering, University of Washington, Seattle, WA 98195-1750

Proteins at interfaces are involved in a wide variety of phenomena, including mammalian cell growth in culture, reactions to implanted biomaterials, growth of soil bacteria, and formation of organized layers of proteins at the oil/water and air/water interfaces. This broad range of phenomena has attracted the interest of a correspondingly broad range of scientists and engineers, many with different backgrounds and different perspectives. As a result, the field of proteins at interfaces displays as much breadth as depth, and many of the investigators in this field of research conduct investigations that are unique with respect both to the particular protein/surface system studied and to the methods used. The many different approaches and systems under investigation, and in particular the wide variety of applications towards which the research is directed, means that an overview such as this one which is limited in length, must be selective. In selecting subtopics for discussion we had to be somewhat arbitrary, and therefore tried to focus on broad aspects which are common to most application areas.

The contributions to the San Diego Symposium and to this volume represent a significant fraction of the workers who are actively involved in studying the behavior of proteins at interfaces. We have therefore chosen to organize our overview around six major aspects of protein behavior at interfaces which emerged from the Symposium. These aspects are: (1) theory, including molecular mechanisms; (2) competitive adsorption; (3) conformation and orientation of proteins at interfaces; (4) surface chemistry effects on adsorption; (5) the behavior of proteins at fluid interfaces; and (6) effects of proteins on cell interactions with surfaces. For each topic, current understanding is briefly summarized, and the contributions of the articles in this volume are then discussed. Inevitably this type of subdivision is to some extent arbitrary and artificial, so there is some overlap of material from one section to another. We trust, however, that on balance our approach will facilitate the readers' (as well as the authors') task of understanding the state of the art in this field.

0097–6156/95/0602–0001$12.75/0

Theory and molecular mechanisms of protein adsorption

It is fundamental to recognize at the outset of this discussion that proteins are large, amphipathic molecules. As a result they are intrinsically surface active, and one of their natural habitats may be said to be in the interface between phases. This idea gains in plausibility if one considers that within cells proteins are often membrane associated. Thus it may be argued that to understand the behavior of proteins at interfaces is to understand an important aspect their normal behavior. A good perspective for adoption by the novice approaching this field is that all proteins adsorb to all surfaces. It is rarely a problem how to achieve the adsorption of a protein, but rather how to prevent it. Consequently, protein adsorption is the central event in the biofouling of surfaces.

The interactions recognized as occurring in protein adsorption are mostly noncovalent, ie H-bonding, electrostatic, and hydrophobic interactions. Examples of covalent adsorption (as opposed to intentional covalent immobilization) are rare. These are the same phenomena that occur in small molecule adsorption. Protein adsorption is distinguished by the large size of the adsorbate and by the fact that while adsorbed the protein can undergo various transformations, both physical and chemical. Apart from the theoretical aspects of these transformations, which are of themselves of considerable interest, they often entrain changes in the biological activity of the protein, eg enzyme activity.

Most of the effects mentioned have been recognized for many years and we will not attempt to review the relevant literature in detail. Rather a few examples of recent developments bearing on adsorption theory and mechanistic aspects will be discussed.

Modeling oriented studies of adsorption in single protein systems (as opposed to mixtures, for which see below) have been relatively few in recent years. Many authors continue to fit adsorption data to the Langmuir equation despite its obvious shortcomings in relation to proteins. A major one is that the Langmuir model requires adsorption to be reversible while protein adsorption in most systems examined to date is to all intents and purposes irreversible on a realistic time scale. There seems little doubt that this stems from the multivalent binding interactions which are typical of protein adsorption. However, the data in many experimental studies appear to fit well to the Langmuir equation and some authors have interpreted this as confirmation that the Langmuir mechanism is applicable. Estimates of the affinity constants for binding have been made from such data fits, although there is considerable scepticism about this practice. Such estimates should strictly speaking not be considered as thermodynamic equilibrium constants. They are at best apparent, and at worst pseudo binding constants. However they can be regarded as giving a qualitative indication of binding affinity.

This more conservative approach to the interpretation of isotherm data is illustrated by the work of Norde and Anusiem (1) on BSA and lysozyme adsorption to silica and hematite. In this work the relative values of the initial isotherm slopes for a series of protein-surface systems were interpreted in terms of the relative adsorption affinities. An interesting result from these studies was that BSA that had been adsorbed and then desorbed showed higher affinity than native BSA, suggesting that adsorption caused a significant physical transition in the protein. No such effect was found for lysozyme, a smaller protein also considered to be "harder" than BSA.

The paper by Norde and Haynes in this volume takes yet another look at reversibility of protein adsorption. It shows that significant internally created entropy

$\Delta_i S$ may be generated, leading to the conclusion that in such cases adsorption is indeed irreversible. Their estimates of $\Delta_i S$ are obtained from hysteresis loops in the adsorption-desorption isotherms, and it is argued that the higher adsorption values seen in the descending branches of the isotherms indicate higher affinity for the desorbed than for the native protein. As indicated this is borne out by experimental data on the BSA-silica system (*1*).

An important approach to the modeling of protein adsorption has come from the studies of Schaaf and Talbot (*2*). These groups have investigated the random sequential adsorption (RSA) model in relation to proteins. RSA is essentially surface filling governed by geometry and does not allow for desorption or diffusion of the protein over the surface. It may be considered as the opposite of fully reversible equilibrium adsorption. An important result of this theory is that a so-called jamming limit of surface coverage is reached, beyond which no additional molecules can be accommodated in the layer. For circular discs the jamming limit of surface coverage is 0.547 (*2*). The fact that coverage in protein adsorption is known to reach higher values (close to 1.0) suggests immediately that some desorption and/or surface diffusion can occur as has in fact been demonstrated by experimental means (*3,4*). These authors have also drawn attention to a perceived weakness of the Langmuir model which gives the available surface function (related to the rate of adsorption) as $(1-\theta)$, where θ is the fractional surface site coverage. This is a consequence of the Langmuirian specification that one molecule occupies only one site. Clearly for proteins, which are generally big enough to cover several sites, this is inappropriate. More appropriate available surface functions have been suggested on theoretical grounds. In general such functions are nonlinear implying that in contrast to the Langmuir mechanism, the rate of adsorption should show a nonlinear dependence on coverage. A possible flaw in these arguments is that the definition of a site may have to be different for a large heterogeneous molecule like a protein than for the small gas molecules that were considered by Langmuir. For physical adsorption, for example, it may be that a site is defined simply by the protein footprint, and not by any recognizable chemical entity.

The RSA model is relevant to protein adsorption in that it stipulates irreversibility. However the other major requirement that surface diffusion should not occur is not necessarily met. There have been a number of reports showing that protein molecules can diffuse over surfaces even though they cannot desorb. Recently Rabe and Tilton (*5*) have shown using fluorescence methods that adsorbed BSA diffuses over the surface of acrylic polymers with diffusivities of the order of 10^{-8} cm^2/s, ie approximately two orders of magnitude smaller than the bulk diffusivity. Moreover the diffusivity showed a sharp increase in the region of the glass transition temperature of the polymer indicating that motions in the surface itself can augment surface diffusion of adsorbed protein.

The past several years have witnessed considerable activity in the search for protein resistant/repellent surfaces. Such materials are expected to be bioinert and should resist cell adhesion and biofouling generally, with applications in biocompatibility, biosensors, liposomes (*6*) and other devices. Based on an understanding of protein adsorption mechanisms it should in principle be possible to design protein resistant surfaces, but the approaches taken to this problem have been more empirically than mechanistically based.

It has been found by a number of groups that surfaces with polyethylene oxide (PEO) grafts show greatly reduced protein adsorption (7). The mechanism of resistance to protein adsorption is not well understood but appears to be related to the high hydrophilicity, high flexibility and therefore high excluded volume of PEO chains. These properties allow steric repulsion interactions to dominate over van der Waals attraction. There is no general agreement on the optimal structure of such a surface, but the important variables are believed to be the PEO chain length and spacing on the surface (8,9). In the present volume the paper by McPherson, Lee and Park shows that PEO chains of length 3 to 128 immobilized on glass (through the use of PEO-PPO-PEO block copolymers) are effective in reducing the adsorption of fibrinogen and lysozyme. The most important property of such materials is the surface concentration of grafted PEO.

The paper by Ivanchenko et al in this volume also shows that shorter PEO chain lengths (in the range 4 to 24) are more effective in reducing protein adsorption. However the relevance of these data to the question of the optimum PEO graft length is somewhat obscured by the fact that the PEO chains are not grafted directly to the surface, but are in fact side chains on methacrylate polymers which are grafted to the surface. In addition the length of the methacrylate chains, which was controlled by a chain transfer agent, was also found to affect adsorption, with the short chains giving the best results.

The idea that protein resistance may be achieved through the use of phosphoryl choline (PC) moieties has also been pursued by a number of groups. PC is the head group of phosphatidyl choline (lecithin) which is a major component of the exterior surface of cell membranes. The expectation of protein resistance is based on the hypothesis that these groups should mimic the lipid component of the cell surface (10) which is supposed to be essentially protein resistant. Ishihara et al (11) have suggested that arrays of such PC moieties attached to a surface in contact with blood will bind phospholipids and organize them into structures that are lipid bilayer-like, and thus protein resistant. The paper by Nakabayashi et al in this volume provides additional data in support of this concept.

It is not clear whether the expectation that the outer surface of lipid bilayers should be protein resistant is entirely justified. In fact there have been few investigations of protein interactions with such surfaces. The tendency of liposomes and foreign cells to become opsonized and then phagocytosed (6) presumably results from adsorption of immunoglobulins.

A highly significant recent development is the use of protein mutants as a way to investigate the effects of structure on protein adsorption to solids. Previous studies of the behavior of protein mutants at the air-water interface showed a linear dependence of several indicators of surface activity with the thermodynamic stability of mutants of tryptophan synthase alpha subunit (12). Although just beginning, this approach appears to hold much potential to elucidate the details of the adsorption process. Pioneering work in this area has been done by McGuire et al (13) using mutants of bacteriophage T4 lysozyme. In their paper in this volume they show that minimal changes in amino acid composition can change adsorption behavior significantly. For example substitution of the isoleucine residue at position three of lysozyme gave proteins of altered stability and correspondingly altered adsorption kinetics and resistance to elution.

Strength of attachment to the surface appeared to be greater for the less stable mutants. It may be anticipated that exploitation of this approach will increase in the near future.

Also relevant to the mechanism of protein adsorption are the studies of Nygren and Alaeddine reported in this volume, which address the question of the distribution of protein over the surface as this relates to the kinetics of the adsorption process. This is a little explored aspect and indeed it is generally assumed that for a homogeneous surface the adsorption is random (see discussion of random sequential adsorption above). Nygren and Alaeddine find that adsorption kinetics for several systems shows an initial lag phase followed by autoacceleration and finally a phase of logarithmically decreasing growth rate. This behavior is described as fractal kinetics and may be due to nonrandom distribution of the components in the surface regions. The initial nucleation and acceleration phases produce surface clustering of protein, while the later phase (in which desorption plays an important role) leads to re-randomization. There are some novel and stimulating ideas in this approach and it will be interesting to see if such results are more generally observed.

Mechanistic aspects of protein adsorption should in principle benefit greatly from the recent advances in scanning probe microscopy. Atomic force microscopy (AFM) in particular with its potential to yield images of adsorbed protein molecules while still in contact with the mother solution seems well adapted to give new insights. The promise of this method, however, has not yet been fulfilled, and technical problems remain. In particular it has been found that the AFM probe tends to rearrange as well as image the protein molecules (*14*). Also limitations on resolution due to the finite radius of curvature of the probe tip have been noted (*15*). Nonetheless, some useful studies have been carried out. For example Marchant et al have obtained images of von Willebrand factor adsorbed on mica showing a domainal structure in agreement with earlier TEM images (*15*).

An interesting development in AFM is the use of modified probes containing ligands that can interact with adsorbed proteins (*16,17*). In principle this approach can be used to measure the forces between ligand and protein and to provide images of the distribution of the specific binding sites.

Competitive adsorption of proteins

The competitive adsorption of proteins is important in the many interfacial phenomena that occur in the presence of mixtures of proteins. These phenomena include the blood and tissue compatibility of biomaterials, cell culture on solid supports, bacterial adhesion to teeth and to implanted materials, soil bacterial growth, fouling of contact lenses in tear fluids, and the protein fouling phenomena that occur during food processing. Thus, for example, the mechanisms of cell adhesion are affected by competitive protein adsorption processes because the outcome of competition between fibronectin and vitronectin in serum are different on different surfaces (see article by Steele et al in this volume). Similarly, many investigators have tried to develop biomaterials that would adsorb layers enriched in passivating proteins like albumin and depleted of activating proteins like fibrinogen. The well known tendency of glass-like surfaces to activate the intrinsic clotting enzymes much more than hydrophobic surfaces like polyethylene is another example of the importance of competitive processes. Surface induced clotting is believed to arise from the higher relative affinity of the clotting proteins, especially

HMWK, for glass-like surfaces, whereas on hydrophobic surfaces other proteins compete too well and reduce the uptake of the clotting proteins (see papers by Elwing et al, Arnebrant et al, and Turbill et al in this volume).

The irreversibility of protein adsorption results in a fundamental difference between the competitive adsorption of proteins and other competitive adsorption processes. The adsorption behavior of reversibly adsorbed, low molecular weight detergents in mixtures is predictable from the adsorption behavior of the individual species because the species can exchange readily until an equilibrium mixed layer is achieved that reflects the relative affinity and concentration of the competing species (18). In contrast, a previously adsorbed protein molecule at a given site on the surface may or may not be displaceable by a later arriving molecule of another protein, depending on how long a time elapsed between the two events. The removability of the protein depends on residence time because the adsorbed protein molecules tend to undergo a molecular relaxation or spreading process on the surface.

The effect of these relaxation processes on competitive adsorption cannot be predicted from data taken with single protein solutions. In competitive, multiprotein systems one often sees time, concentration, and spatially dependent maxima in adsorption. The best known example of this behavior is the Vroman effect observed for fibrinogen adsorption from plasma (see Slack and Horbett, this volume). The existence of peaks in adsorption at intermediate dilutions of plasma or serum is expected on the basis of differences in transport rates and binding affinities. Thus abundant proteins of low affinity are expected to be adsorbed initially and later replaced by scarcer proteins of high affinity. However when adsorbed proteins relax to irreversibly adsorbed states in a time dependent manner, the situation is complicated since displacement is then more difficult. Thus the time or plasma dilution at which adsorption peaks occur will be altered by relaxation effects, and in some cases the peaks may be eliminated altogether if relaxation is very rapid, and the protein becomes irreversibly adsorbed before concentrations of competing proteins reach effective levels.

The finite time required to achieve the irreversible state is a key point, because it allows for multiple hits of the adjacent surface sites by incoming molecules to occur. This molecular race for the surface will in some cases prevent the initial adsorbate from ever achieving the irreversible state. Instead, later arriving molecules will occupy the sites which the initial adsorbate is trying to occupy as it undergoes a surface relaxation process. The outcome of such a race between reversibly adsorbed species would normally be an equilibrium mixture on the surface reflecting the relative binding affinity of the competing species, but the irreversibility of some of the adsorbing events prevents the equilibrium mixture from being achieved. Consequently, competitive adsorption events involving proteins are affected by factors in addition to affinity.

Such additional factors include the rate of transport (diffusion and convection) and the rate of relaxation or unfolding of the protein in the adsorbed state. Because of the complex and progressively irreversible nature of protein adsorption, the only reliable way to characterize the competitive adsorption effectiveness of different proteins is to compare them under identical conditions on a given surface, e.g. in binary (or more complex) mixtures. Even then, however, the apparent competitive adsorption effectiveness of a protein will depend on exactly how the experiments are run, because of the kinetic factors mentioned above. Thus, for example, Horbett has shown that the amount of competing protein required to inhibit the adsorption of fibrinogen to solid surfaces from binary mixtures depends on the absolute concentration of the species involved, suggesting that the apparent competitive effectiveness is occupancy dependent (*19*). Such complexities notwithstanding, the extreme surface activity of a few proteins seems indisputable. Among these should be mentioned hemoglobin (*19*) and high molecular weight kininogen (*20*). Convincing, structurally based explanations of the high surface activity of these proteins have not so far been proposed, although the positively charged histidine-rich domain of HMWK (*21*) may account for its adsorption to negatively charged surfaces.

While irreversibility causes unique features in competitive protein adsorption to surfaces, other aspects are generally similar to competitive adsorption processes of all types. A major factor determining the outcome of any competitive process is the relative concentrations of the competing species. Thus, for example, increasing the concentration of fibrinogen in plasma results in a roughly proportional increase in the amount of fibrinogen adsorbed to surfaces (*22,23*). Similarly, the adsorption of fibrinogen from binary mixtures decreases as the concentration of the competing species increases, and the sigmoid shapes of fibrinogen adsorption versus the competing protein/fibrinogen ratio are similar to those observed for other competitive binding processes (*24*). A rigorous description of the process of competitive adsorption of proteins will therefore include unique aspects deriving from irreversibility as well as more general features characteristic of any competitive adsorption process. A complete, molecularly based model of competitive adsorption will probably require inclusion of the rate of transport of the various species, the rate of adsorption, and the rate of conversion to an irreversible state for each species.

Recent studies by Norde et al, and by McGuire et al reported in this volume, stress the concept of the stability of the molecule in determining its relative adsorption affinity (e.g., "soft" or more denatureable proteins are expected to have greater surface affinity than "hard" or more stable proteins). However, these ideas derive from measurements of equilibrium unfolding energies of proteins, whereas it seems likely that the rate of unfolding of the proteins at the surface may also be important in determining the outcome of competitive adsorption of proteins. To date, actual measurements of the rates of unfolding of proteins have not been used in models of competitive protein adsorption, and would appear to be a fruitful area for investigation in view of the importance of irreversibility in the competitive adsorption process.

A number of the chapters in the present volume address the issues of competitive protein adsorption, and most of them support and confirm the general ideas and trends discussed above. The paper by Baszkin and Boissonade reports on competition between albumin and fibrinogen. A rather unusual observation is that the presence of albumin can cause an increase in the adsorption of fibrinogen beyond its value in the corresponding single protein system. The notion of adsorption driven by spreading pressure, akin to the relaxation phenomena already discussed, is advanced in this work. Of interest also is the Vroman type peak in the kinetics of fibrinogen adsorption to polyethylene from mixtures with albumin. The absence of such peaks at the air-water interface is attributed to denaturation of initially adsorbed albumin which can then not be displaced.

The chapter by Le et al emphasizes the role of transport in protein interfacial exchange in plasma and shows that the displacement of adsorbed fibrinogen is more rapid under flow than static conditions. A similar theme is taken up by Nadarajah et al who present a model for competitive adsorption which includes differential transport as well as differential adsorption behavior of proteins. Such models predict the classical maxima in the adsorption of fibrinogen from plasma. Although the models include a surface relaxation step leading to irreversibility, the authors conclude that this step is unimportant for initial adsorption and exchange in multiprotein systems. It is clear that the latter conclusion depends on the relative magnitudes of the exchange and relaxation rates as has been pointed out by others in analyses of similar models (25). If relaxation to an irreversibly adsorbed state is rapid then exchange cannot occur regardless of the intrinsic relative affinities of the native proteins.

Similar ideas are discussed by Sevastianov et al in relation to competition between IgG and HSA in binary mixtures. Interestingly it is shown that albumin is able to displace IgG from quartz and polydimethyl siloxane surfaces, but the reverse is not true. Addition of IgG to albumin solutions simply suppresses albumin adsorption, presumably by direct competition for surface "space", with no evidence of ability to displace albumin once adsorbed. The displacement of IgG by albumin is explained in terms of a re-orientation of IgG by interactions with HSA.

The chapter by Arnebrant and Wahlgren discusses protein-surfactant-surface interactions, which have strong relevance to competitive protein adsorption. It is recognized that one mechanism by which a surface may be "cleaned" of a protein is via solubilization involving formation of complexes between adsorbed protein and surfactant. Analogous mechanisms may be involved in the exchange of one protein for another in adsorbed layers.

Extensive data in the chapter by Warkentin et al emphasize the role of protein-protein interactions, both in solution and at the surface, in multiprotein systems. The detergent like action of a solution protein on an adsorbed one is pointed out. Also related aspects of protein antigen-antibody interactions at surfaces are discussed which bear on the technology of immunoassays involving immobilized reactants. As has been pointed out elsewhere, the quantification of adsorbed antigens by reaction with antibody depends on knowledge of epitope availability which can be greatly affected by the manner in which the proteins are adsorbed.

Conformation and orientation of proteins at interfaces

The conformational and orientational state of proteins at interfaces has been the subject of continuing interest and investigation for both theoretical and practical reasons. Many of the theories attempting to explain the behavior of proteins at interfaces postulate some contribution from the entropy gain upon unfolding of the protein, while variations in the degree of retention of native structure upon adsorption of enzymes, clotting factors, antibodies, and cell adhesion proteins is thought to be an important way in which surfaces affect a variety of biological processes. For example, variations in the ability of adsorbed adhesion proteins to influence cell adhesion, depending on the substrate to which the protein is adsorbed, are thought to arise from conformational or orientational changes in the adsorbed proteins that modulate the availability and potency of their cell binding domains (see below).

It is reasonable to suppose that proteins adsorbed to solid surfaces may undergo some conformational change because of the relatively low structural stability of proteins and their tendency to unfold to allow formation of additional contacts with the surface

(26). Consequently, several models of protein adsorption include a transition from a reversibly adsorbed state to a more tightly held state, the latter brought about by a molecular restructuring or relaxation of the protein on the surface *(27-29)*. However, since enzymes and antibodies retain at least some of their biological activity in the adsorbed state, and biologic activity is exquisitely dependent on maintenance of a native structure, it would seem that at least on some surfaces, conformational changes in adsorbed proteins are limited in nature.

The competitive aspects of unfolding on surfaces may result in complete access to the surface for some molecules, while others are prevented from unfolding because the surface sites adjacent to them are already occupied by other molecules. In this case, one would expect selected parts of the adsorbed molecules to be altered, while others remain unperturbed. This model should be attractive to those who believe in unfolding as well as to others who observe retention of structure since it allows both to be correct. However, this model also implies a critique of both points of view since typically neither deals with the fact that knowledge regarding the fraction of the surface molecules that are native or denatured is incomplete. It has often been stated simply that denaturation occurs without attempting to assess whether all molecules on the surface are similarly denatured or instead whether there is a distribution of states. Many earlier studies are consistent with the idea that proteins at the solid/liquid interface range from native to, at most, partially unfolded (see *27,28,30* for reviews). On the other hand, recent studies have indicated that at least for some proteins and surfaces, there are major changes in the structure of the adsorbed protein. In this brief overview selected studies will be used as examples to indicate that the conformation of adsorbed proteins can vary from native or nearly so to extensively unfolded, depending greatly on the particular protein and surface.

Before beginning that discussion, however, two caveats should be indicated. First it should be pointed out that most of the studies to be discussed use techniques that provide only a global indication of the structure of the protein. Physicochemical methods do not usually give information about the state of specified local regions of adsorbed proteins. Instead some overall measure is obtained such as change in alpha helix content from circular dichroism (CD) measurements or from infrared spectra. In some cases a change in the signal from an extrinsic label (e.g. a fluorescent label) whose location is specific but typically not known may be measured. An exception to this is the labeling of human serum albumin at cysteine 34 with a spin labeling reagent (see Nicholov et al, this volume). This label allows detection, with electron spin resonance spectroscopy, of changes in a specific region of the adsorbed protein due to the adsorption process. Electron spin resonance spectroscopy has also been used to detect conformational changes upon adsorption of fibronectin to polystyrene beads *(31)* and Cytodex microcarrier beads *(32)*. Another method to detect localized changes in adsorbed proteins is the binding of monoclonal antibodies to known regions of an adsorbed protein. This method has been used to detect changes in regions of the fibrinogen molecule involved in platelet adhesion (see below for a brief review).

Secondly, there is at present little direct evidence on the orientational arrangement of proteins at interfaces, despite its likely importance in many phenomena, including the biological activity of adsorbed cell adhesion proteins. One cannot tell, for example, if reduced binding of monoclonal antibodies specific to cell binding domains reflects changes in orientation that sterically hinder access to the epitope or if instead the reduction in antibody binding is to due to conformational changes rendering the array of amino acid side chains in the binding region ineffective for proper docking to the antibody. However, the theoretical studies of Park et al predict that there should be energetically favored orientations (33,34). Also, studies of lysozyme adsorbed to mica using the surface force apparatus have shown that at low concentrations, the contact thickness corresponds to a side-on monolayer, while at higher concentrations, the contact thickness corresponds to end-on molecules, suggesting that proteins may assume preferred orientations, depending on conditions (Blomberg and Claesson, this volume). In addition, Lee and Saavedra (this volume) provide evidence that cytochrome c adsorbed to hydrophilic and hydrophobic wave guide surfaces is not randomly oriented because the average tilt angle of the porphyrin is respectively 17±2 and 40±7 degrees, while the average tilt angle expected for a completely isotropic distribution of molecular orientations is 54.7 degrees. The tilt angle of 17 degrees for cytochrome c on the hydrophilic surface indicates that the heme plane is oriented essentially vertical to the waveguide plane. In contrast, Haynes and Norde recently cited some unpublished data taken with the internal reflection fluorescence method suggesting that cytochrome c adsorbed to charged silica is randomly oriented (35).

As already indicated, there is substantial evidence suggesting that in some systems native or near native structure of adsorbed proteins is retained. For example, enzyme activity is frequently retained after adsorption although the most thorough studies of this phenomenon have shown that the degree of retention varies with enzyme type and the degree of loading of the surface (36). Similarly, the widespread success of the solid phase immunoassay method presumably depends on the retention of structure in at least some of the adsorbed antibody or antigen molecules because the binding of the antibody and antigen depends on a complex, sterically sensitive interaction of an array of interacting groups in the two molecules. Furthermore, the thickness of adsorbed protein films has frequently been found to be close to the dimensions of the native molecule, as previously discussed by Norde (37). As reported in this volume (Blomberg and Claesson), studies with the surface force apparatus have shown that the thickness of a lysozyme layer adsorbed to mica is similar to the dimensions of the native molecule. In the case of beta casein at the air/water interface studied with the novel technique of neutron reflectivity (Atkinson et al, this volume), the data are consistent with the presence of both a thin, dense inner layer of protein (1 nm thick) and a much less dense, more tenuous outer layer (4-5 nm) which may indicate the presence of both the denatured and native states.

Despite the indications of retention of native structure in adsorbed protein obtained from studies of enzyme activity or layer thickness, several recent studies with new differential scanning microcalorimetry methods indicate that some adsorbed proteins may lose much of their structure. Since the release of heat occurs for native proteins in solution due to unfolding at the transition temperature, an absence or reduction of this effect for an adsorbed protein suggests that it has already undergone the transition, i.e. that it has already unfolded upon adsorption, and thus has released the heat of unfolding prior to the calorimetry measurement. Haynes and Norde observed that the transition enthalpy of lysozyme adsorbed to negatively charge polystyrene was much less than for the protein in solution (0-170 kJ/mol for the adsorbed protein depending on the pH (35) versus about 600 kJ/mol for the native protein). However, for lysozyme adsorbed on hematite, the unfolding enthalpy was only about 20% less than for the native protein, indicating that changes in the enthalpy of unfolding depend on the adsorbing surface. Furthermore, for lactalbumin the heat released was nearly zero when adsorbed to either the polystyrene or the hematite surface, suggesting complete unfolding of lactalbumin on both surfaces. It was suggested that these observations could be explained in terms of the lower stability of lactalbumin in comparison to lysozyme. Recently, Feng and Andrade have shown that several proteins adsorbed to pyrolytic carbon no longer show any release of heat at the expected transition temperature, suggesting that pyrolytic carbon induces complete unfolding, a result that is consistent with the tenacious binding of proteins to this surface (*38*). Using the same calorimetric methods, Yan et al (this volume) have shown that albumin and lysozyme adsorbed to polystyrene exhibit no unfolding enthalpy, while lysozyme adsorbed to a hydrophilic contact lens still exhibits about 50% of the heat released by the native protein. Yan et al also showed that streptavidin adsorbed to polystyrene displays an unfolding enthalpy that is very similar to that for the native protein in solution, attributing this to the greater stability of streptavidin in comparison to lysozyme or albumin.

An important new experimental development in the study of the structure of adsorbed proteins was reported by Kondo (*39*). The method uses CD spectra to provide a measure of secondary structure, but instead of measuring the spectra of eluted proteins as others have done (*40*), the proteins in the adsorbed state are examined directly. This is made possible by the use of small silica particles (of the order of 15 nm) as adsorbent. Since these particles do not cause scattering at the wavelengths of interest, the particle suspension can be examined directly. Results from this work demonstrate the range of behavior which has been referred to above. Thus it was shown that "soft" proteins like albumin and hemoglobin undergo extensive loss of α-helix on adsorption, while smaller "hard" proteins like ribonuclease do not. Interestingly it was also shown that upon desorption into solution, albumin "recovered" to its native structure, presumably via refolding and re-formation of α-helices. The same principle has been used by Elwing et al (*41*) to measure the fluorescence of proteins adsorbed on silica nanoparticles, thus eliminating the need for total internal reflectance methods which are experimentally demanding.

The influence of surface properties on protein adsorption

Surface properties have an enormous effect on the rate, the extent, and the mechanism of adsorption. Perhaps the broadest, most widely accepted generalization regarding surface properties concerns hydrophobicity and holds that the more hydrophobic the surface the greater the extent of adsorption. Recent confirmation of this "rule" or "principle" is provided by the work of Prime and Whitesides (*42*) using surfaces consisting of self assembled monolayers of long chain alkanes with terminal groups of differing hydrophobicity.

The hydrophobicity "rule" derives its strongest support from studies of hydrophobicity gradient surfaces. The concept and implementation of the gradient surface was first reported by Elwing et al (*43*). The gradient surface provides an excellent tool with which to study the effects of surface properties, and several research groups have made use of it (*44-46*). A single surface specimen spans a wide range of hydrophobicity (or other property) so that adsorption properties over the entire range can be examined in the same experiment. Much of the data obtained with this method have shown that adsorption is greater at the hydrophobic end of the gradient for a variety of proteins including fibrinogen, IgG and complement proteins.

It should be pointed out that gradient surfaces must be heterogeneous at some level, and analyses which take this into account have yet to appear. By the nature of the gradient forming process, the substrate is increasingly covered with an "overlayer" of the modifying substance so that macroscopic properties such as contact angle change smoothly along the gradient. However at the level of an adsorbing protein molecule the properties may not change smoothly depending on the size of the hydrophobic and hydrophilic domains or islands which form the gradient. If the domains are large compared to the protein molecule then the protein adsorption response will effectively be the sum of the responses for the two types of domain, reflecting merely the relative quantities of each. For domains that are small compared to the protein, the protein molecules will straddle the two domain types and the response will be potentially quite different.

Three articles in the present volume describe results using the gradient method. The work of Warkentin et al is concerned mostly with competitive adsorption and has been mentioned above in that context. It is worth pointing out here that the effect of hydrophobicity can be very different for a given protein depending on whether it is present alone or in admixture with other proteins. Thus Warkentin et al (this volume) have shown that when alone HMWK is adsorbed more at the hydrophobic end of a methyl-hydroxyl gradient, whereas in serum or plasma it predominates at the hydrophilic end.

The paper by Ho and Hlady reports on the adsorption of low density lipoprotein to a gradient surface having hydrophilic silica (advancing water contact angle 0°) at one end and a self assembled monolayer of C-18 hydrocarbon chains (contact angle 104°) at the other. It was found that the adsorption capacity was higher at the hydrophilic end. The affinity was also higher due to a more rapid adsorption rate at the hydrophilic end. These results are in contrast to the general rule of thumb regarding hydrophobicity and protein adsorption. Interactions at the hydrophilic part of the gradient in this work may be influenced by negative charges on the silica since LDL binding is known to be sensitive to charge. The responses observed may thus not be determined entirely by hydrophobicity.

The chapter by Elwing et al is concerned with competitive adsorption from plasma and shows that hydrophobicity plays a role in these phenomena also. The gradients in this work were created by methylation of oxidized silicon and the water contact angle ranged from 10 to 90°. Both albumin and fibrinogen adsorption were favored at the hydrophobic end of the gradient, but the accumulation of high molecular weight kininogen occurred at both ends of the gradient with lower adsorption at intermediate hydrophobicities.

It seems important to point out that some of the older evidence for the hydrophobicity rule may be suspect. Adsorption experiments where measurements are made after separating the surface from the solution and then washing to remove residual solution may underestimate adsorption on hydrophilic surfaces which tends to be more easily reversible. Thus the most convincing evidence comes from in situ methods, particularly ellipsometry on which much of the more recent work cited above is based (42,43).

More recently gradient surfaces based on properties other than hydrophobicity have been fabricated. The fabrication methods are generally based on the counter diffusion principle established by Elwing et al. Thus Liedberg and Tengvall (*45*) have described a method of making chemical gradients on gold substrates by reaction with thiols of different structure. Yueda-Yukoshi and Matsuda (*46*) have produced gradients in carbonate and hydroxyl groups by slowly immersing a polyvinylene carbonate surface in a solution of hydrolyzing agent. These and similar materials will be immensely useful for the study of surface chemistry effects in protein adsorption.

The search for surfaces that do not adsorb proteins at all, or at least minimize adsorption, has been discussed above. The use of polyethylene oxide in this connection is in a sense related to the hydrophobicity rule. However it appears that factors more subtle than global hydrophilicity (eg chain flexibility, effective excluded volume) are involved. It may be that hydrophilicity of the grafted chains is a necessary but not sufficient condition for protein resistance, suggesting that a more systematic investigation of the broad class of hydrophilic polymers for this application may be warranted. In this regard a comparison between PEO and dextran has recently been published by Osterberg et al (*8*). They suggest that thickness of the attached polymer layer (beyond the requirement for a certain minimum value) is less important than packing density. Achievement of a packing density sufficient to prevent access of the protein to the bare substrate material is crucial. This work also confirms the unique protein rejecting properties of grafted PEO.

Electrical charge is another general surface property that has been much investigated in relation to protein adsorption, although no clear consensus has developed as to its effects. It seems obvious that attractive interactions should dominate between opposite charges and repulsive interactions between like charges. Although this is often the case, charge effects can be confounded by "lurking" factors such as small multivalent counterions bridging between a protein and surface having the same charge, which would normally be expected to repel each other. Also proteins are large molecular entities which usually contain many charged groups, some negative and some positive. The question of whether they behave as particles with a net "point" charge or as assemblages of individual charges thus arises.

Perhaps the most extensive studies in this area have been from the group of Norde et al (see reference *35* for a review, and the chapter by Norde and Haynes in this volume). Along with several other groups they have observed maxima in adsorption capacity at the isoelectric point of the protein, suggesting a strong effect of global charge. They have also observed greater adsorption of net positive proteins on negative polystyrene surfaces and vice versa. Other examples indicate that while electrostatic interactions are important they do not necessarily dominate protein adsorption. Indeed it is suggested that the dominant effects are "structural rearrangements in the protein molecule, dehydration of the sorbent surface, redistribution of charged groups and protein surface polarity". The latter effects, which can be probed by proton titration, are discussed by Norde and Haynes in this volume. They have found that carboxyl groups in some proteins which are dissociated in solution, are protonated in the adsorbed state.

Considerable evidence has accumulated showing that sulfonate-containing surfaces have high affinity for protein adsorption. The high affinity has been observed for a wide variety of proteins and sulfonated surfaces. Thus several reports have noted the extensive and strong adsorption of fibrinogen to sulfonated polyurethanes from buffer and from plasma or blood (47,48). Thrombin behaves in a like manner on similar materials (49). Similarly plasminogen has been shown to adsorb strongly to sulfonated silica surfaces (50) and sulfonated polystyrenes have been found to adsorb large amounts of proteins from plasma (51). It may be inferred that these interactions are nonspecific and indiscriminate with respect to protein type, ie all such surfaces appear to adsorb all proteins extensively and with high affinity. The nature of the sulfonate-protein interaction is unknown. A possibility is that sulfonamide bonds could be formed by reaction of sulfonate with amine groups in the proteins, but this seems unlikely under the the usually mild conditions of protein adsorption experiments.

The extensive adsorption of proteins to sulfonated surfaces is quite similar to the binding of many proteins to sulfoethylated chromatography matrices (eg SE Sephadex). The separation of proteins in mixtures using these matrices can occur over a wide range of pH and ionic strength because the sulfoethyl group retains its full charge except at very low pH and because this group interacts strongly with positively charged groups in proteins.

The work of Han et al (52,53) is of interest with respect to sulfonate groups and protein adsorption. They have developed polyurethane surfaces grafted with PEO chains which terminate in sulfonate groups at the free end. These are described as negative cilia surfaces (52), and appear to combine the protein repellent properties of PEO with the protein binding properties of sulfonate groups. Interestingly it has been shown that compared to control surfaces grafted with unmodified PEO, the sulfonated materials adsorb increased amounts of all proteins from plasma (53). By far the greatest increase is in albumin adsorption, perhaps explaining the relatively low platelet adhesion observed on these materials. From these various observations on protein adsorption to sulfonated surfaces it appears that more detailed investigations of sulfonate-protein interactions at a fundamental level are warranted.

Over the past several years many reports have appeared in which surfaces have been designed for the selective adsorption of specific proteins from mixtures. The entire class of substrates for immunosorbent assays falls into this category. These substrates are based on the immobilization of an appropriate antigen which then captures the corresponding specific antibody on exposure to the test fluid. An example of this approach is provided by the chapter of Melzac and Brooks in this volume in which a ferrichrome A antigen is immobilized on silylated silica. Affinity chromatography matrices also provide many examples of design of materials for binding of a specific protein.

Examples of ligand-derivatized surfaces from the biomaterials area are lysinized materials to bind plasminogen from plasma (50,54), heparinized materials to capture antithrombin III, and tyrosine-containing materials for the removal of anti-factor VIII antibodies from hemophiliac plasma (55). By and large these surfaces do not adsorb the target protein exclusively, but bind other proteins as well, presumably in a nonspecific manner. Affinity chromatography matrices which combine a protein resistant base material (eg Sepharose) with a specific protein binding ligand appear to give the best performance with respect to suppression of nonspecific binding.

Proteins at fluid interfaces

Fluid-fluid interfaces have traditionally been considered separately from solid-fluid interfaces in discussions of protein interfacial behavior. Fundamentally there is no difference between these two interfacial types insofar as protein interactions are concerned, and the same phenomena occur at both, including exchange between adsorbed and dissolved protein, surface diffusion, and conformational change. However the applications are different, and interest in the adsorption of proteins at liquid-gas and liquid-liquid interfaces relates mostly to food colloids including emulsions and foams. As will be discussed below, the interactions of proteins at cell membranes can also be considered to be a liquid interface phenomenon. Adsorption will of course occur at any protein solution-air interface leading to possible conformational change, loss of biological function and creation of stable foams. These phenomena can therefore be important simply in the handling of protein solutions, especially at very low concentrations.

Another difference is the fact that adsorbed protein can move more freely on the fluid-fluid interface than on the solid-fluid interface and can therefore undergo reorientation, diffusion and conformational change more readily, probably due to the greater relative mobility of the interface itself. In this regard Baszkin and Boissonade point out (this volume) that the air-water interface may have unique properties in that accommodation of additional protein at higher coverage can be made through facilitated penetration of adsorbing protein and rearrangement of already adsorbed protein.

Different experimental techniques have been developed for the study of fluid-fluid interfaces, including surface pressure (Langmuir trough) and surface viscosity measurements. These methods reflect the different properties of interest at the fluid-fluid interface compared to the solid-fluid interface.

The reader is referred to a comprehensive earlier review of this topic by MacRitchie (56). An excellent recent review has been given by Dickinson and Matsumura (57). The latter article introduces the interesting idea that the structure of globular proteins adsorbed at liquid interfaces resembles the so-called molten globule state, considered to be intermediate between the native and completely unfolded states (secondary structure intact, tertiary structure destroyed). They suggest that proteins adsorbed at liquid interfaces may retain normal secondary, but not tertiary structure.

In the present volume there are four chapters which focus on various aspects of proteins at fluid interfaces. The paper by Vogel examines the role of surface interactions in the transformation of fibronectin from its soluble form in plasma to its multimeric/fibrillar form in connective tissue. Data are presented on the behavior of plasma fibronectin at the air-water interface, and on phospholipid layers spread at the air-water interface. The latter is used as a model for the cell membrane which does not contain specific receptor molecules. The data obtained suggest that fibronectin can self assemble into fibrillar structures at the phospholipid interface.

The theme of protein interactions at the cell surface is also the subject of the articles by Maloney et al and by Lecompte and Duplan. In the work of Maloney et al the hydrolysis of phospholipid monolayers by the adsorbed enzyme phospholipase A2 was studied, and evidence is presented suggesting the formation of interfacial microphases consisting of reaction products (fatty acids and lyso-lipids) and the adsorbed enzyme.

This work makes use of a broad range of experimental methods including surface pressure, surface potential and fluorescence microscopy.

The paper by Lecompte and Duplan reports an investigation of the adsorption of prothrombin to monolayers of phospholipids (phosphatidyl ethanolamine and phosphatidyl choline) at the air-water interface. These experiments constitute a model for the interactions of clotting factor proteins at the platelet surface during blood coagulation. Using alternating current polarography methods these authors were able to demonstrate that at higher solution concentrations, prothrombin can penetrate the mixed phospholipid layers as well as adsorb to them. Penetration is probably driven by hydrophobic interactions whereas adsorption is believed to be electrostatic in origin. It is of interest to speculate on whether such phenomena may occur in the binding of proteins to cell membranes.

The paper of Magdassi et al, while reporting data on the adsorption of IgG at the oil-water interface, is primarily directed to the development of methods to modify IgG antibodies so that they become more surface active. Such antibodies would be better adapted for use as targeting moieties on the surface of drug delivery vehicles such as liposomes. Hydrocarbon chains were attached to IgG by reaction with the amine groups of lysine residues. It was found that such molecules are much more surface active than unmodified IgG in the sense that the adsorption capacity of oil-in-water emulsions is considerably higher for the hydrophobized IgG. At low degrees of modification it was demonstrated that the IgG retained its antigen recognition properties. Thus the approach of promoting attachment of targeting antibodies to therapeutic devices by increasing their surface activity appears to hold promise.

The development of protein repellent surfaces has been discussed above. However it is appropriate to mention this aspect again under the liquid interface heading. A major problem in the development of liposomes as intravenous injectable drug carriers is their rapid clearance in the reticuloendothelial system, probably mediated by the adsorption of opsonins (IgG, complement proteins etc). The adsorption of these proteins to liposomes also suggests that the putative protein resistance of lipid bilayers may be somewhat more limited than has been supposed by some investigators. As indicated above the incorporation of water soluble polymers, particulary PEO, into the surface of the liposomes appears to be a promising approach to this problem (6).

Finally, it is of interest to note that some of the newer techniques which were applied originally to investigations of proteins at the solid-liquid interface are being used to study the liquid-liquid interface also. Thus Corredig and Dalgleish (58) have used DSC to examine the structure of whey proteins at the oil-water interface, and have concluded that these proteins are at least partly denatured. This work complements that of Caldwell and of Norde and Haynes in the present volume.

The role of adsorbed proteins in cell interactions with solid surfaces

The adsorption of proteins to solid surfaces plays an important role in the adhesion, spreading, and growth of cells, primarily because of the existence of cell surface receptors that bind specifically to certain proteins that act as adhesive agents. In this section, three mechanisms by which adsorbed adhesion proteins affect cells will be discussed. These mechanisms are the differential affinity of the adhesion proteins for different surfaces, the modulation of the biological activity of the adsorbed adhesion protein by the surface, and substrate activation of adhesion proteins.

The differential affinity of adhesion proteins. In this mechanism, differences in the relative affinity of the adhesion protein for various surfaces lead to variable degrees of enrichment of the adsorbed protein layer in the adhesion protein, which then cause differences in the behavior of cells. For example, changes in the ability of a surface to adsorb fibronectin would result in different degrees of enrichment of this protein in comparison to the many competing proteins, most of which inhibit cell adhesion.

The differential affinity model is supported by studies that showed large differences in the adsorption of fibronectin onto various surfaces (59-62). In addition, the initial spreading of 3T3 cells on HEMA-EMA copolymers is linearly correlated with the amount of fibronectin adsorbed to the surfaces from serum (61). Similarly, plasma deposited polymers which enhanced 3T3 and muscle myoblast cell growth also enhanced fibronectin adsorption from serum in comparison to untreated surfaces (63). Depletion studies, in which the fibronectin or vitronectin are selectively removed from the serum, have shown that vitronectin, rather than fibronectin, appears to be the primary adhesion factor for certain cells and surfaces (62,64-66). However, the most recent studies of this type show that the enhanced affinity of fibronectin for the nitrogen rich Primaria tissue culture surface in comparison to ordinary oxygen rich tissue culture surfaces is sufficient to provide for the attachment of human vein endothelial cells to Primaria from serum depleted of vitronectin. In contrast, cell attachment to tissue culture polystyrene in the absence of vitronectin is greatly reduced because there is not enough fibronectin adsorbed to this surface (Steele et al, this volume).

The differential affinity model is also supported by the effect of serum dilution on cell behavior. Fibronectin adsorption displays a maximum in adsorption at intermediate serum dilutions, a phenomenon similar to the maximum in fibrinogen adsorption at intermediate plasma dilutions (19,67). A corresponding maximum in cell spreading or attachment has been observed in several studies, using BHK cells (68), platelets (69), 3T3 cells (61), and endothelial cells (70). The magnitude and position of the peak in adsorption of fibronectin depends somewhat on the surface composition of the substrate (60).

Vitronectin adsorption from serum does not appear to exhibit a peak at intermediate serum dilution, at least on the surfaces studied to date (62,71) so surfaces tend to have much more adsorbed vitronectin than fibronectin, especially at higher serum concentrations. Binary competitive adsorption studies have indicated the ranking of surface activity of vitronectin relative to other plasma proteins to be in the approximate order: vitronectin=fibrinogen»albumin=IgG (72).

Modulation of the biological activity of adsorbed adhesion proteins. Surfaces with similar amounts of adsorbed adhesion proteins sometimes exhibit substantial differences in cell attachment or spreading, suggesting that the substrate properties somehow modulate the biologic activity of the adhesion protein (see 73 for a more detailed review). Modulation of the biological activity of adsorbed fibrinogen by the substrate is indicated by differences in platelet retention on a series of poly (alkylmethacrylates) despite similar amounts of adsorbed fibrinogen (74). Since the binding of anti-fibrinogen antibody to fibrinogen adsorbed on these surfaces did vary, the adsorbed fibrinogen was thought to have different conformations on the various poly(alkyl methacrylates) (74). Variations in monoclonal antibody binding to various epitopes on fibrinogen adsorbed to the different poly(alkyl-methacrylate) polymers also suggested differences in the accessibility of epitopes on adsorbed fibrinogen molecules that would likely also influence the reactivity of the fibrinogen with the platelets (75). Fibrinogen adsorbs in high amounts to PTMO-based sulfonated polyurethanes but nonetheless these surfaces exhibit very low platelet adhesiveness in vivo (48,76). Kiaei et al (this volume) have shown that when adsorbed to PTFE fibrinogen is much more supportive of platelet adhesion than when adsorbed to CF_3-rich plasma-deposited TFE polymers.

Modulation of the biologic activity of adsorbed fibronectin is indicated by several studies showing that fibronectin adsorbed to various surfaces is not equivalent in its interactions with cells, despite the presence of similar amounts of adsorbed fibronectin (77-82). For example, fibronectin adsorbed to tissue culture grade polystyrene supports BHK cell attachment and spreading, whereas fibronectin adsorbed to ordinary polystyrene does not support spreading unless some albumin is added to the fibronectin solution (77). Similarly, substrata of varying chemical composition to which fibronectin was pre- adsorbed varied considerably in their ability to induce cell spreading and intracellular stress fiber formation in fibroblasts and neural cells, in their ability to induce neurite formation in neuroblastoma cells, and in their degree of inhibition of stress fiber formation by an RGDS peptide (78). Finally, the ability of fibronectin adsorbed to various surfaces to promote the outgrowth of corneal cells was found to vary considerably, even when compared at similar amounts of adsorbed fibronectin. The changes in outgrowth were better correlated with the strength of fibronectin binding onto the substrate (83) than with changes in availability of the RGD sequence measured with a monoclonal antibody (80).

Substrate activation of the adhesion proteins. It appears that adsorption of fibronectin and fibrinogen to surfaces potentiates their adhesive properties, a phenomenon termed "substrate activation". Early studies suggested that fibronectin was somehow "activated" by adsorbing to surfaces because the binding of the cells to the adsorbed form appeared to be much stronger than to the soluble form (59,84-86). Substrate activation of fibrinogen is indicated by the observation that unstimulated platelets bind and spread readily on adsorbed fibrinogen and other adsorbed adhesion proteins (87-89) while platelets do not bind to fluid phase fibrinogen or other adhesion proteins unless the platelets have first been exposed to an agonist such as ADP or thrombin. Inhibition of platelet adhesion to adsorbed fibrinogen required much higher concentrations of RGD peptides (16 µM) than was required to inhibit aggregation of the platelets in suspension (1.5 µM), suggesting that platelets demonstrate a higher affinity for the adsorbed form of fibrinogen (90).

Observations of this type have led to a two step model for platelet interactions with fibrinogen, in which either the platelet GPIIb/IIIa receptor or fibrinogen must first become activated to allow an initial "recognition" between these agents (91). The recognition step is then followed by the induction of additional high affinity ligand-receptor interactions. Three mechanisms for the substrate activation of fibrinogen have been proposed: multivalency; conformation/orientation; and tightness of binding.

Multivalency and substrate activation of adhesion proteins.   In this model, adsorption of proteins to surfaces appears to accentuate the adhesion receptor- adhesion protein interaction, probably because of the concentrating and localizing effect of immobilizing the proteins at the interface.   Thus, the mere presence of enough fibrinogen may be sufficient to cause platelet activation because it allows simultaneous cooperative interactions between the platelet and many fibrinogen molecules (*92-94*).

Conformational or orientational changes.   A second and more prominent mechanism for substrate activation involves conformational and/or orientational changes in adsorbed fibrinogen.   Changes are detectable with monoclonal antibodies, some of which bind to adsorbed but not soluble fibrinogen.   In addition, certain of the antibodies that bind to adsorbed but not soluble fibrinogen have also been shown to bind to receptor-bound fibrinogen but not to fibrinogen in solution.   Fibrinogen binding to the receptor evidently causes the exposure of so-called receptor-induced binding sites (RIBS) on the fibrinogen molecule.   Three RIBS are now known: RIBS-I lies near the C-terminus of the gamma chain (residues 373-385 are involved); RIBS-II is in the middle of the gamma chain (involving 112-119); and RIBS-III lies near the N-terminus of the Aa chain involving the RGDF (95-98) sequence (*95*).   Since soluble fibrinogen is not activated and will only bind to a pre-activated receptor, it is thought that the changes in receptor bound fibrinogen are induced by binding to the activated receptor.   Thus, when it was found that adsorption also induced changes in fibrinogen that allow these same RIBS antibodies to bind, it suggested that surface activation involves conformational changes in fibrinogen that allow recognition by unactivated platelet receptors.   More recently, it was proposed that the degree of exposure of the RIBS-III, RGD site induced by adsorption to different surfaces may be a key factor determining the thrombogenicity of foreign surfaces (*95*).

Tightness of binding.   Fibrinogen adsorbed from plasma or pure solutions undergoes transitions in its physicochemical and biological state that depend on the postadsorptive residence time and surface properties.   For example, fibrinogen adsorbed from plasma quickly becomes non-displaceable by plasma but at very different rates on different substrates (*29,96*).   Transitions in adsorbed fibrinogen have been detected with several physical methods, including decreases in SDS elutability (*96*) and changes in the FTIR spectra (*97*).   Transitions in adsorbed fibrinogen have also been detected using biological assays, including reductions in polyclonal antibody binding and in platelet adhesion to adsorbed fibrinogen with increasing residence time of the protein on the substrate (*98*).   The reduction in displaceability of adsorbed fibrinogen that occurs with residence time is thought to involve further contact between the fibrinogen molecule and the substrate, resulting in a more tightly held molecule that is less subject to competitive displacement by other proteins in plasma.   Changes in the binding of antibodies specific for three known platelet binding regions of fibrinogen did not correlate with changes in platelet adhesion as a function of residence time (*99*).

The morphology of adherent platelets and the ability of the platelets to remove substrate bound fibrinogen are affected by the residence time of the adsorbed fibrinogen (*100*).   Placing surfaces pre-adsorbed with fibrinogen in albumin containing buffers prevents residence time dependent decreases in platelet (*98*) and polyclonal anti-fibrinogen binding to adsorbed fibrinogen (*101*), presumably because albumin occupies some of the empty sites on the surface and prevents further contact formation between the fibrinogen molecule and the surface. The effect of albumin in preventing decreases in fibrinogen elutability and platelet adhesion, together with the fact that the binding of antibodies to the platelet binding regions of fibrinogen does not correlate with changes in platelet adhesion, are all consistent with the possible role of tightness of binding of fibrinogen to the substrate in platelet adhesion.

Acknowledgements

Financial support of the authors' research by the following agencies is gratefully acknowledged: Medical Research Council of Canada; Natural Sciences and Engineering Research Council Of Canada; Heart and Stroke Foundation of Ontario; Ontario Center for Materials Research (JLB); National Heart, Lung and Blood Institute, NIH (TAH).

Literature cited

1   Norde, W.; Anusiem, A.C.I. *Colloids Surfaces* **1992**, *66*, 73-80.
2   Schaaf, P.; Talbot, J. *J Chem Phys* **1989**, *91*, 4401-4409.
3   Burghardt, T.P.; Axelrod, D. *Biophys J* **1981**, *33*, 455-467.
4   Tilton, R.D.; Robertson, C.R.; Gast, A.P. *J Colloid Interface Sci* **1990**, *137*, 192-203.
5   Rabe, T.E.; Tilton, R. *J. Colloid Interface Sci* **1993**, *159*, 243-245.
6   Lasic, D.D.; Papahadjopoulos, P. *Science* **1995**, *267*, 1275-1276.
7   Jeon, S.I.; Lee, J.H.; Andrade, J.D.; de Gennes, P.G. *J. Colloid Interface Sci* **1991**, *142*, 149-158.
8   Osterberg, E.; Bergstrom, K.; Holmberg, K.; Schuman, T.P.; Riggs, J.A.; Burns, N.L.; Van Alstine, J.M.; Harris, J.M. *J Biomed Mater Res* **1995**, *29*, 741-747.
9   Prime, K.L.; Whitesides, G.M. *J Am Chem Soc* **1993**, *115*, 10714-10721.
10   Chapman, D.; Charles, S.A. *Chemistry in Britain* **1992**, *28*, 253-256.
11   Ishihara, K.; Oshida, H; Endo, Y.; Ueda, T.; Watanabe, A.; Nakabayashi, N. *J Biomed Mater Res* **1992**, *26*, 1543-1552.
12   Kato, A.; Yutani, K. *Protein Eng* **1988**, *2*, 153-156.
13   McGuire, J.E.; Wahlgren, M.C.; Arnebrant, T. *J Colloid Interface Sci* **1995**, *170*, 182-192.
14   Marchant, R.E.; Lea, S.A.; Andrade, J.D.; Bockenstedt, P. *J Colloid Interface Sci* **1991**, *148*, 261-272.
15   Eppell, S.J.; Zypman, F.R.; Marchant, R.E. *Langmuir* **1993**, *9*, 2281-2288.
16   Lee, G.U.; Kidwell, D.A.; Colton, R.J. *Langmuir* **1994**, *10*, 354-357.
17   Stuart, J.K.; Hlady, V. *Langmuir* **1995**, *11*, 1368-1374.
18   Soriaga, M.P.; Song, D.; Zapien, D.C.; Hubbard, A.T. *Langmuir* **1985**, *1*, 123-127.
19   Horbett, T.A. *Thromb Haemostas* **1984**, *51*, 174-181.
20   Brash,J.L.; Scott, C.F.; ten Hove, P.; Wojchiechowski, P.; Colman, R.W. *Blood* **1988**, *71*, 932-939.

21   Muller-Esterl, W. *Sem Thromb Haemostas* **1987**, *13*, 115.
22   Slack, S.M.; Horbett, T.A. *J Colloid Interface Sci* **1988**, *124*, 535-551.
23   Wojchiechowski, P.; ten Hove, P.; Brash,J.L. *J Colloid Interface Sci* **1986**, *111*, 455-465.
24   Horbett, T.A.; Weathersby, P.K.; Hoffman, A.S. *J Bioeng* **1977**, **1**, 61-78.
25   Dejardin, P.; ten Hove, P.; Yu, X.J.; Brash, J.L. Submitted for publication.
26   Horbett, T.A. *Cardiovasc Pathol* **1993**, *2*, 137S-148S.
27   Andrade, J.D. In Surface and Interfacial Aspects of Biomedical Polymers; 2: Protein Adsorption; Andrade, J.D., Ed.; Plenum Press: New York, **1985**, pp. 1-80.
28   Lundstrom, I. *Prog Coll Polymer Sci* **1985**, *70*, 76-82.
29   Slack, S.M.; Horbett, T.A. *J Colloid Interface Sci* **1989**, *133*, 148-165.
30   Horbett, T.A.; Brash, J.L. In Proteins at Interfaces: Physicochemical and Biochemical Studies, Brash, J.L.; Horbett, T.A., Eds.; *ACS Symposium Series 343*; American Chemical Society: Washington, D.C., **1987**, pp. 1-33.
31   Narsimhan, C.; Lai, C.-S. *Biochemistry* **1989**, *28*, 5041-5046.
32   Wolff, C.E.; Lai, C.-S. *Biochemistry* **1990**, *29*, 3354-3361.
33   Lu, D.R.; Park, K. *J Biomat Sci Polymer Edn* **1990**, *1*, 243-260.
34   Lu, D.R.; Lee, S.J.; Park, K. *J Biomat Sci Polymer Edn* **1991**, *3*, 127-147.
35   Haynes, C.A.; Norde, W. *Colloids Surfaces B: Biointerfaces* **1994**, *2*, 517-566.
36   Sandwick, R.K.; Schray, K.J. *J Colloid Interface Sci* **1988**, *121*, 1-12.
37   Norde, W. *Adv Colloid Interface Sci* **1986**, *25*, 267-340.
38   Feng, L.; Andrade, J.D. *J Biomed Mater Res* **1994**, *28*, 735-743.
39   Kondo, A.; Oku, S.; Higashitani, K. *J Colloid Interface Sci* **1991**, *143*, 214-221.
40   Chan, B.; Brash, J.L. *J Colloid Interface Sci* **1981**, *84*, 263-265.
41   Clark, S.R.; Billsten, P.; Elwing, H. *Colloids Surfaces B: Biointerfaces* **1994**, *2*, 457-461.
42   Prime, K.; Whitesides, G.M. *Science* **1991**, *252*, 1164-1167.
43   Elwing, H.; Askendal, A.; Ivarsson, B.; Nilsson, U.; Welin, S.; Lundstrom, I. *ACS Symposium Series* **1987**, *343*, 468-488.
44   Lin, Y.S.; Hlady, V.; Janatova, J. *Biomaterials* **1992**, *13*, 497-504.
45   Liedberg, B.; Tengvall, P. *Langmuir*, in press.
46   Ueda-Yukoshi, T.; Matsuda, T. Personal communication.
47   Santerre, J.P.; ten Hove, P.; Vanderkamp, N.H.; Brash, J.L. *J Biomed Mater Res* **1992**, *26*, 39-57.
48   Grasel, T.G.; Cooper, S.L. *J Biomed Mater Res* **1989**, *23*, 311-338.
49   Tian, Y.; Weitz, J.I.; Brash, J.L., unpublished observations.
50   Woodhouse, K.A.; Weitz, J.I.; Brash, J.L. *J Biomed Mater Res* **1994**, *28*, 407-415.

51 Boisson-Vidal, C.; Jozefonvicz, J.; Brash, J.L. *J Biomed Mater Res* **1991**, *25*, 67-84.
52 Han, D.K.; Jeong, S.Y.; Kim, Y.H.; Min, B.G.; Cho, H.I. *J Biomed Mater Res* **1991**, *25*, 561-575.
53 Han D.K.; Ryu, G.H.; Park, K.D.; Kim, U.Y.; Min, B.G.; Kim, Y.H. *J Biomed Mater Res* **1995**, in press.
54 Deutsch, D.G.; Mertz, E.T. *Science* **1970**, *170*, 1095-1096.
55 Dahri, L.; Boisson-Vidal, C.; Muller, D.; Jozefonvicz, J. *J Biomat Sci Polymer Edn* **1994**, *6*, 695-705.
56 MacRitchie, F. *Adv Protein Chem* **1978**, *32*, 283-326.
57 Dickinson, E.; Matsumura, Y. *Colloids Surfaces B: Biointerfaces* **1994**, *3*, 1-17.
58 Corredig, M.; Dalgleish, D.G. *Colloids Surfaces B: Biointerfaces* **1995**, in press.
59 Klebe, R.J.; Bentley, K.L.; Schoen, R.C. *J Cell Physiol* **1981**, *109*, 481-488.
60 Bentley, K.L.; Klebe, R.J. *J Biomed Mater Res* **1985**, *19*, 757-769.
61 Horbett, T.A.; Schway, M.B. *J Biomed Mater Res* **1988**, *22*, 763-793.
62 Steele, J.G.; Johnson, G.; Underwood, P.A. *J Biomed Mater Res* **1992**, *26*, 861-884.
63 Chinn, J.A.; Horbett, T.A.; Ratner, B.D.; Schway, M.B.; Haque, Y.; Hauschka, S.D. *J Colloid Interface Sci* **1989**, *127*, 67-87.
64 Knox, P. *J Cell Sci* **1984**, *71*, 51-59.
65 Steele, J.G.; Johnson, G.; Norris, W.D.; Underwood, P.A. *Biomaterials* **1991**, *12*, 531-539.
66 Underwood, P.A.; Bennett, F.A. *J Cell Sci* **1989**, *93*, 641-649.
67 Brash, J. L.; ten Hove, P. *Thromb Haemostas* **1984**, *51*, 326-330.
68 Grinnell, F.; Feld, M.K. *J Biol Chem* **1982**, *257*, 4888-4893.
69 Grinnell, F.; Phan, T. *Thromb Res* **1985**, *39*, 165-171.
70 van Wachem, P.B.; Vreriks, C.M.; Beugeling, T.; Feijen, J.; Bantjes, A.; Detmers, J.P.; vanAken, W.G. *J Biomed Mater Res* **1987**, *21*, 701-718.
71 Bale, M.D.; Wohlfahrt, L.A.; Mosher, D.F.; Tomasini, B.; Sutton, R.C. *Blood* **1989**, *74*, 2698-2706.
72 Fabrizius-Homan, D.J.; Cooper, S.L. *J Biomed Mater Res* **1991**, *25*, 953-971.
73 Horbett, T.A. *Colloids Surfaces B: Biointerfaces* **1994**, *2*, 225-240.
74 Lindon, J.N.; McManama, G.; Kushner, L.; Merrill, E.W.; Salzman, E.W. *Blood* **1986**, *68*, 355-362.
75 Shiba, E.; Lindon, J.N.; Kushner, L.; Matsueda, G.R.; Hawiger, J.; Kloczewiak, M.; Kudryk, B.; Salzman, E.W. *Am J Physiol* **1991**, *260*, C965-C974.
76 Silver, J.H.; Lin, H.-B.; Cooper, S.L. *Biomaterials* **1993**, *14*, 834-844.
77 Grinnell, F.; Feld, M.K. *J Biomed Mater Res* **1981**, *15*, 363-381.
78 Lewandowska, K.; Pergament, E.; Sukenik, C.N.; Culp, L.A. *J Biomed Mater Res* **1992**, *26*, 1343-1363.
79 Lewandowska, K.; Balachander, N.; Sukenik, C.N.; Culp, L.A. *J Cell Physiol* **1989**, *141*, 334-345.

80   Pettit, D.K.; Horbett, T.A.; Hoffman, A.S.   *J Biomed Mater Res* **1994**, *28*, 685-691.
81   Grinnell, F.; Feld, M.K. *J Biol Chem* **1982**, *257*, 4888-4893.
82   Juliano, D.J.; Saavedra, S.S.; Truskey, G.A. *J Biomed Mater Res* **1993**, *27*, 1103-1113.
83   Pettit, D.K.; Horbett, T.A.; Hoffman, A.S. *J Biomed Mater Res* **1992**, *26*, 1259-1275.
84   Grinnell, F. *Int Rev Cytol* **1978**, *53*, 65-141.
85   Akiyama, S.K.; Yamada, K.M. *Advances in Enzymology* **1987**, *59*, 1-57.
86   Schwartz, M.A.; Juliano, R.L. *Exp Cell Res* **1984**, *153*, 550-555.
87   Coller, B.S. *Blood* **1980**, *55*, 169-178.
88   Chinn, J.A.; Horbett, T.A.; Ratner, B.D. *Thromb Haemostas* **1991**, *65*, 608-617.
89   Savage, B.; Ruggeri, Z.M. *J Biol Chem* **1991**, *266*, 11227-11233.
90   Hantgan, R.R.; Endenburg, S.C.; Cavero, I.; Marguerie, G.; Uzan, A.; Sixma, J.J.; de Groot, P.G. *Thromb Haemostas* **1992**, *68*, 694-700.
91   Ginsberg, M.H.; Xiaoping, D.; O'Toole, T.E.; Loftus, J.C.; Plow, E.F. *Thromb Haemostas* **1993**, *70*, 87-93.
92   McManama, G.; Lindon, J.N.; Kloczewiak, M.; Smith, M.A.; Ware, J.A.; Hawiger, J.; Merrill, E.W.; Salzman, E.W. *Blood* **1986**, *68*, 363-371.
93   Sixma, J.J.; Hindriks, G.; Van Breugel,H.; Hantgan, R.; de Groot, P.G. *J Biomat Sci Polymer Edn* **1991**, *3*, 17-26.
94   Ruoslahti, E. *Ann Rev Biochem* **1988**, *57*, 375-413.
95   Ugarova, T.P.; Budzynski, A.Z.; Shattil, S.M.; Ruggeri, Z.M.; Ginsberg, M.H.; Plow, E.F. *J Biol Chem* **1993**, *268*, 21080-21087.
96   Rapoza, R.J.; Horbett, T.A. *J Colloid Interface Sci* **1990**, *136*, 480-493.
97   Lenk, T.J.; Horbett, T.A.; Ratner, B.D.; Chittur, K.K. *Langmuir* **1991**, *7*, 1755-1764.
98   Chinn, J.A.; Posso, S.E.; Horbett, T.A.; Ratner, B.D. *J Biomed Mater Res* **1991**, *25*, 535-555.
99   Horbett, T.A.; Lew, K.R. *J Biomat Sci Polymer Edn* **1994**, *6*, 15-33.
100   Sheppard, J.I.; McClung, W.G.; Feuerstein, I.A. *J Biomed Mater Res* **1994**, *28*, 1175-1186.
101   Chinn, J.A.; Posso, S.E.; Horbett, T.A.; Ratner, B.D. *J Biomed Mater Res* **1992**, *26*, 757-778.

RECEIVED July 12, 1995

THEORY AND MOLECULAR MECHANISMS OF PROTEIN ADSORPTION

Chapter 2

Reversibility and the Mechanism of Protein Adsorption

Willem Norde[1] and Charles A. Haynes[2]

[1]Department of Physical and Colloid Chemistry, Wageningen Agricultural University, P.O. Box 8038, 6700 Wageningen, Netherlands
[2]Biotechnology Laboratory, University of British Columbia, 6174 University Boulevard, 237 Wesbrook Building, Vancouver, British Columbia V6T 1Z3, Canada

Detailed adsorption isotherm data are combined with thermodynamic arguments in an effort to determine whether protein adsorption to solids is a reversible or irreversible process. We then examine the dominant driving forces for protein adsorption and present a thermodynamic framework for understanding the protein-adsorption process which is consistent with results from our study of process (ir)reversibility.

Phenomenologically, a system is in equilibrium if no further changes take place at constant surroundings. At constant pressure P and temperature T, the equilibrium state of a system is characterized by a minimum value of the total Gibbs energy G. Any other state, away from this minimum, is nonequilibrium and there will be a spontaneous transition (*i.e.* a process) towards the equilibrium state provided the energy barriers along this transition are not prohibitively large. By definition, a process is reversible if, during the whole trajectory of the process, the departure from equilibrium is infinitesimally small, so that in the reverse process the variables characterizing the state of the system return through the same values but in the reverse order. Since a finite amount of time is required for the system to relax to its equilibrium state upon changing the conditions, investigations of the reversibility of a process must be designed such that the time of observation exceeds the time required for the system relaxation.

In this paper, we address the question whether adsorption of proteins from aqueous solutions to solids is a reversible process with respect to variations in the bulk-solution protein concentration. The answer to this question, which is too often ignored in literature, determines what thermodynamic criteria apply to the protein adsorption process and also provides information about affinities between proteins and sorbent surfaces.

We then examine the dominant driving forces for protein adsorption and present a thermodynamic framework for understanding the protein-adsorption process which is consistent with results from our study of process (ir)reversibility. A complementary set of adsorption-isotherm, isothermal-titration-microcalorimetry, potentiometric-titration, and differential-scanning-calorimetry data are used to argue that three effects, namely, structural rearrangements in the protein molecule, dehydration of the sorbent and protein surfaces, and redistribution of charged groups in the interfacial layer, usually make the primary contributions to the overall driving force for adsorption.

Adsorption Isotherms and Reversibility

Adsorption data are often presented as adsorption isotherms, where, at constant T, the amount adsorbed Γ is plotted against the concentration of sorbate c_p in solution (see Figure 1). For adsorption to be reversible with respect to variation of c_p, an increase in c_p from a value corresponding to point A in Fig. 1 to that corresponding to point C should result in a change in the adsorbed amount that is independent of the manner in which c_p has been changed. In this case, Γ should increase to its value at $c_p = c_p(C)$ regardless of whether the change in concentration is made in one step [*i.e.*, $c_p = c_p(A)$ $\rightarrow c_p = c_p(C)$] or in multiple steps [*e.g.*, via path AB'BC''C in Fig. 1]. Reversibility also requires that a decrease in c_p from its value at point C to its value at point A" result in a reduction in Γ to the value corresponding to point A, again independent of path. Therefore, in a reversible adsorption process, the ascending (increasing concentration in the bulk) and descending (decreasing concentration in the bulk) branches of the isotherm must overlap at all c_p. Only for such reversible processes can the adsorption isotherm be used to determine the (equilibrium) binding constant K, from which the thermodynamic functions of state (*i.e.*, Gibbs energy of adsorption $\Delta_{ads}G$, enthalpy of adsorption $\Delta_{ads}H$, and entropy of adsorption $\Delta_{ads}S$) can be derived by applying reversible thermodynamics.

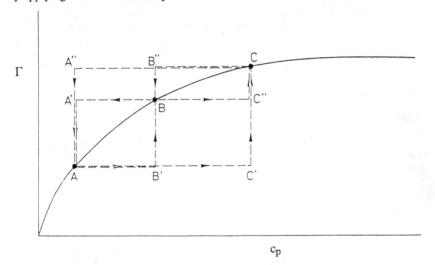

Figure 1: Adsorption isotherm distinguishing reversible and irreversible pathways.

Protein adsorption from aqueous solution often results in high-affinity isotherms where the initial slope of the ascending branch merges with the Γ-axis as depicted in Fig. 2a. Plateau values are reached at very low bulk protein concentrations and the region in which Γ depends on c_p is limited to values very near the Γ-axis. Verification of reversibility in such systems is difficult because precise measurement of the ascending and descending branches of the isotherm requires a method for determining bulk protein concentrations in very dilute solutions. Occasionally, low-affinity isotherms are observed where the isotherm is distinguishable from the ordinate at low c_p (see Fig. 2b). In such systems, the form of the ascending branch is usually but not always (see refs. 1,2) independent of the number and size of c_p steps used in measuring the curve. However, when diluting such systems, Γ rarely if ever follows the same path backwards, thereby making the descending and ascending branches of the isotherm distinguishable. As a rule it is found that, when shear is excluded, dilution does not lead to detectable desorption of proteins from solid sorbents, particularly hydrophobic sorbents, even when the observation time is extended to several days and is therefore much longer than the relaxation time of the protein at the surface [2,3]. Such a deviation between the ascending and descending branches of the isotherm is defined as hysteresis. The occurrence of hysteresis indicates that at a given c_p the system has two equilibrium/meta-stable states: one on the ascending branch and the other on the descending branch. These two states are characterized by *local* minima in G which are separated by a Gibbs energy barrier that prevents the transition from the one state to the other and, hence, prevents the adsorption process from following a reversible path. The fact that the ascending and descending isotherms represent different equilibrium states implies that during the transition from adsorption to desorption a physical change has occurred in the system.

Irreversible Protein Adsorption

In spite of the irreversible nature generally observed for protein adsorption, many authors erroneously interpret their experimental data using theories that are based on reversible thermodynamics. The most common example is the determination of $\Delta_{ads}G$ by fitting the ascending adsorption isotherm to the Langmuir or Scatchard equation [*e.g.*, 4,5]. Another common approach involves calculation of the Gibbs energy of adhesion $\Delta_{adh}G$ using the reversible thermodynamic result known as the Dupré equation

$$\Delta_{adh}G = \gamma_{sp} - \gamma_{sw} - \gamma_{pw} \qquad (1)$$

where γ is the interfacial tension of the interface indicated by the subscript and $\Delta_{adh}G$ is the reversible work (at constant T and P) of forming a protein (p)/sorbent (s) interface at the expense of sorbent (s)/solution (w) and protein (p)/solution (w) interfaces [*e.g.*, 6]. It is not clear how $\Delta_{adh}G$ relates to $\Delta_{ads}G$ since the latter quantity reflects an irreversible process which, as shown below, often includes contributions from structural rearrangements in the protein molecule during adsorption; however, $\Delta_{adh}G$ and the quantity of interest $\Delta_{ads}G$ are not equivalent.

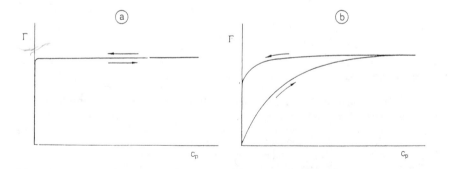

Figure 2: Schematic representation of ascending and descending adsorption isotherms: (a) high-affinity isotherm, and (b) isotherms for which the descending branch shows a higher affinity between the protein and the sorbent surface.

A persistent argument by those who continue to interpret protein-adsorption phenomena with reversible thermodynamics theories involves the assertion that the kinetics of protein adsorption are such that structural perturbations in the protein do not contribute to the driving force for adsorption because they occur after initial attachment of the protein to the interface [*e.g.*, 7,8]. However, the Gibbs energy change $\Delta_{ads}G$ driving the protein-adsorption process refers to the global free energy required or released when taking a mole of protein in its native conformation in solution to its perturbed steady-state structure(s) on the sorbent surface. Thus, when they occur, protein structural rearrangements are an integral part of the adsorption process and cannot be ignored in any meaningful adsorption theory.

The minimum error involved in treating protein adsorption as a reversible process can be estimated by calculating the entropy *production* due to the irreversibility of the process. In a closed system, the entropy change associated with any internal process can be written as

$$\Delta S = \Delta_e S + \Delta_i S \tag{2}$$

where $\Delta_e S$ is the reversible entropy exchange between the system and the surroundings and $\Delta_i S$ is the internally created entropy in the system. For a reversible process $\Delta_i S = 0$ and for an irreversible process $\Delta_i S > 0$. According to Everett [9], $\Delta_{ads,i}S$ can be calculated from the hysteresis loop (*i.e.*, the closed-loop integral) between the ascending and descending branches of the adsorption isotherm

$$\Delta_{ads,i}S = R \int \frac{\Gamma(c_p)}{\Gamma^*} \, d \ln c_p \tag{3}$$

where Γ^* is the adsorbed amount at the upper closure point of the hysteresis loop and R is the universal gas constant. Accurate solution of Eq. 3 requires detailed knowledge of $\Gamma(c_p)$ for both the ascending and descending isotherms, especially in the very-dilute c_p region. Regrettably, such data are rarely available. However, a minimum value for $\Delta_{ads,i}S$ can be determined by letting the hysteresis loop close at the lowest experimentally detectable c_p. As an example, outlines of the ascending and descending isotherms for bovine serum albumin (BSA) on silica particles are shown in Fig. 3a [data can be found in ref. 10]. Replotting the data as shown in Fig. 3b and subsequent application of Eq. 3 above the lowest detectable c_p provides a minimum value for $\Delta_{ads,i}S$ of 37 J K^{-1} mol^{-1}, indicating that irreversible entropy changes will lower the overall driving force for adsorption $\Delta_{ads}G$ (at 298 K) by more than 11 kJ mol^{-1}. It should be realized that if the ascending and descending branches of the Γ/Γ^* versus ln c_p plot do not coincide at concentrations below the detectable c_p limit (which in view of the shapes of the isotherms in Fig. 3b is almost certain), the true value of $\Delta_{ads,i}S$ will be far greater than the minimum values calculated above.

The hysteresis loop reflects a higher adsorption affinity for the descending branch of the isotherm as compared to the ascending branch. Assuming the sorbent surfaces before and after adsorption are identical, the following identity for the molar Gibbs energy of the protein g_p must therefore hold. Thus,

$$g_p(\text{adsorbed}) - g_p(\text{desorbed}) < g_p(\text{adsorbed}) - g_p(\text{native}) \tag{4}$$

must therefore hold. Thus,

$$g_p(\text{desorbed}) > g_p(\text{native}) \tag{5}$$

indicating that the protein undergoes a physical change during the adsorption process. The nature of this physical change is illustrated by the transmission circular dichroism data of Norde et al. [11] for BSA adsorbed to and desorbed from silica. In these experiments, Norde et al. used morpholine to displace the adsorbed BSA after verifying that dilution did not lead to significant desorption. The average α-helix contents of BSA in its native, adsorbed, and desorbed (displaced) states are shown in Table 1. Adsorption to silica involves a severe reduction in the helical content of BSA, particularly when the surface coverage is low enough to allow for the increase in molecular volume which follows from a loss in ordered intra-atomic packing. Upon desorption from silica, BSA regains only a fraction of the helix content lost during adsorption and does not return to its native-state conformation (at least within the two-day observation time of the experiment).

Table 1: α-helix content of bovine serum albumin in native, adsorbed to silica, and desorbed from silica states as measured by transmission circular dichroism (θ is the fractional surface coverage, i.e. $\theta = \Gamma/\Gamma^{pl}$, where Γ^{pl} is the plateau adsorption value)

pH	percentage α-helix content			
	Native State	Adsorbed State		Desorbed State
		$\theta = 0.24$	$q = 1.00$	
4.0	69	-	-	50
4.7	70	-	-	51
7.0	74	28	38	55

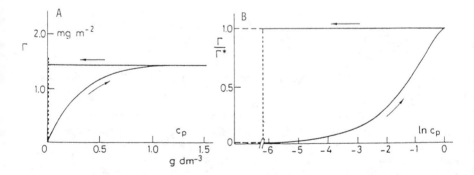

Figure 3: Adsorption and desorption data for bovine serum albumin (BSA) on silica particles: (A) conventional representation of ascending and descending isotherms, and (B) replot of data according to Eq. 3, where the vertical dashed line indicates the lowest BSA detection level ($c_p = 0.002$ g dm^{-3}; 0.05-M PBS; 25 °C).

According to Eq. 5, re-incubation of desorbed BSA with fresh silica should show an adsorption affinity for silica which is stronger than that of native BSA. This is indeed the case. Time dependent reflectometry data [10] for adsorption of native and desorbed BSA to oxidized silicon wafers are shown in Figure 4a. Initial slopes of the reflectometry curves indicate that rates of adsorption are substantially higher for the previously (pre-)desorbed protein and, thus, that a higher fraction of the pre-desorbed protein molecules attach to the surface upon contact. As shown in Figure 4b, initial slopes of adsorption isotherms for native and pre-desorbed BSA also indicate that the latter conformation has a higher affinity for the silica surface.

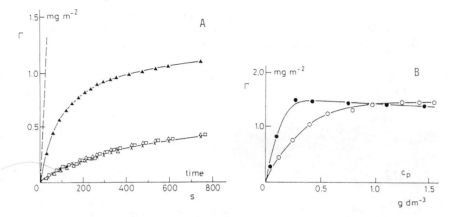

Figure 4: Adsorption of bovine serum albumin (BSA) on silica: (A) adsorbed amounts from stagnation-point-flow reflectometry data ($c_p = 0.01$ g dm^{-3}) where (x) is native BSA, (▲) is BSA previously desorbed from silica by morpholine, and (Δ) and (□) are native BSA pre-exposed but not adsorbed (*i.e.*, that remaining in the supernatant after equilibration) to silica and to morpholine, respectively; (B) adsorption isotherms for (○) native BSA and for (●) BSA previously desorbed from silica by morpholine (all experiments at 0.05-M PBS, pH 7, 25 °C).

Dominant Driving Forces for Protein Adsorption

The simplest realistic chronology of irreversible protein adsorption to a solid nonporous surface involves the three steps shown in Figure 5: (1) protein transport to the energetic boundary layer where the potential at the sorbent surface influences the rate of approach of the protein (including diffusion through the stagnant boundary layer), (2) interaction and attachment of the protein with the surface which may involve perturbations in protein structure, and (3) relaxation of the adsorbed protein to its steady-state conformation(s). As demonstrated by Northrup [12] and others [13], in the absence of convection, step (1) is a stochastic process which can be accurately described by the Langevin equation and Brownian-dynamics simulations. We will not concern ourselves further with this step of the adsorption process. Instead, we focus on steps (2) and (3), which are primarily controlled by direct forces (*e.g.*, coulombic and hydration forces) between the protein, the sorbent surface, solvent (water) molecules, other adsorbed protein molecules in close proximity, and low-molecular-weight ions in the interfacial region. The ability of water to hydrogen bond, the heterogeneous, usually charged surface chemistry of solid sorbents, and the amphipolar, amphoteric, compact nature of native proteins combined with their marginal structural stabilities suggest that no type of molecular interaction is unimportant in the adsorption process. However, our aim is to identify the nature and magnitudes of the dominant driving forces for globular protein adsorption, bearing in mind the irreversibility of the adsorption process.

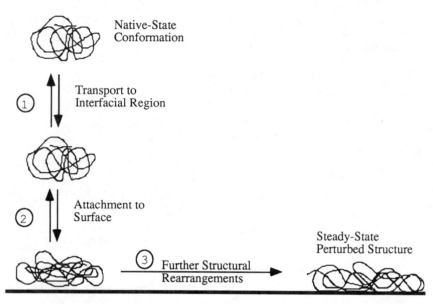

Figure 5: Simplified chronology of irreversible protein adsorption to a charged solid.

Table 2: Physico-chemical Properties of Model Proteins and Sorbent Surfaces

Property	LSZ	αLA	PS-	αFe$_2$O$_3$ (pH 9.5)
Molar Mass (D)	14,600	14,200		
Dimensions (Å^3)	46x30x30	37x32x25		
Isoelectric Point	11.1	4.2		
Total Hydrophobicity (J g^{-1})	- 7.6	- 5.8		
% Apolar Surface Area	53	61		
$\Delta_{N\text{-}D}G$ (J g^{-1})	4.1	1.7		
% a-Helix Content	42	26		
Surface Charge Density (μC m^{-2})			- 23	-
Electrophoretic Mobility (10^{-8} m^2 V^{-1} s^{-1})			- 4.9	- 2.9
Electrokinetic Potential (mV)			- 69	- 47
Hydrophobicity			82°	≈ 0°
(contact angle of a sessile drop of 0.05-M PBS)				

Regardless of the mechanism and kinetics of the process, protein adsorption at constant T and p can only occur if the Gibbs energy of the system decreases:

$$\Delta_{ads}G = \Delta_{ads}H - T\Delta_{ads}S \qquad (6)$$

The irreversible nature of the protein-adsorption process eliminates the possibility of direct measurement of the overall driving force for adsorption, $\Delta_{ads}G$, as well as determination of $\Delta_{ads}S$. This leaves $\Delta_{ads}H$ and the heat capacity change upon adsorption, $\Delta_{ads}C_p$, as the only directly measurable thermodynamic parameters describing the irreversible adsorption process. Regrettably, such data are severely limited for protein adsorption systems [14-16]. Here we report $\Delta_{ads}H$ and $\Delta_{ads}C_p$ data for adsorption of two model proteins, hen egg-white lysozyme (LSZ) and bovine milk α-lactalbumin (αLA), to a negatively-charged polystyrene latex (PS-). As shown in Table 2, these proteins are of similar size, shape, and primary structure (40% sequence homology), but differ in native-state structural stabilities, hydrophobicities, and electrical properties. Electrokinetic, proton-titration, and differential-scanning-microcalorimetry data are also reported for adsorption of these two model proteins to PS- and used to elucidate further those subprocesses involved in the adsorption process.

Heats of Adsorption. $\Delta_{ads}H$ represents the total enthalpy required or released when taking a mole of protein in its native conformation in solution to its perturbed steady-state structure(s) on the sorbent surface. The sign and magnitude of $\Delta_{ads}H$ are therefore governed by a competition between the energetic subprocesses occurring within the protein molecule and between the protein and the sorbent surface. For instance, the total contribution from the electric field overlap, which includes the enthalpy change associated with net protein-protein and protein-sorbent coulombic interactions $\Delta_{ads}H_{el}$, with low-molecular-weight ion (including proton) coadsorption in the interfacial layer $\Delta_{ads}H_{ion\text{-}coad}$, and with formation of ion pairs between adjacent oppositely-charged residues on the protein and sorbent surfaces $\Delta_{ads}H_{ion\text{-}pair}$, can be endothermic or exothermic depending on the solution pH, the ionic strength, and the nature and density of charges on the protein and sorbent.

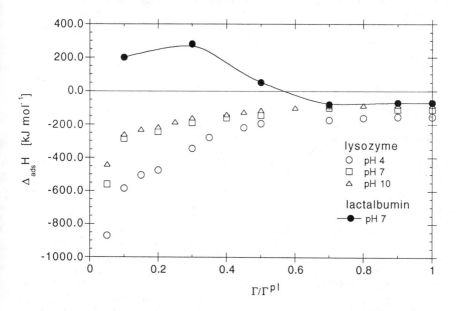

Figure 6: Enthalpy of adsorption data $\Delta_{ads}H$ at 25°C for lysozyme and α-lactalbumin on negatively-charged polystyrene microspheres in 50-mM KCl solution. All data from isothermal titration microcalorimetry measurements.

Figure 6 shows $\Delta_{ads}H$ data at 25° C measured by isothermal titration calorimetry for adsorption of LSZ and αLA on PS- as a function of surface coverage and, for LSZ, as a function of adsorption pH. In the LSZ system, increasing the adsorption pH and, thus, decreasing the attractive charge differential between the protein (pI 11.1) and the negatively-charged sorbent leads to a progressive reduction in the enthalpic driving force for adsorption. Less exothermic $\Delta_{ads}H$ values are also observed at high surface coverages, where the influence of repulsive lateral electrostatic interactions between adsorbed LSZ molecules becomes significant. The $\Delta_{ads}H$ data for LSZ therefore provide a first indication of the importance of electrostatic interaction in the overall driving force for globular protein adsorption.

Although important, coulombic interactions in general do not dominate protein adsorption to solid surfaces, as illustrated by the strong adsorption of αLA to PS- at conditions where the protein carries a substantial net negative charge. The complex dependence of $\Delta_{ads}H$ on fractional surface coverage for αLA adsorption to PS- at pH 7 (see Fig. 6) suggests that a number of subprocesses in addition to electrostatic effects make substantial contributions to the overall driving force for adsorption. For instance, $\Delta_{ads}H$ is endothermic at surface coverages less than 0.55, indicating that adsorption is driven by an increase in entropy at these conditions. Thus, entropically driven subprocesses such as sorbent and protein dehydration and protein denaturation must contribute to the overall driving force of adsorption.

Dehydration Effects. Essentially all globular proteins, regardless of their native-state stabilities and electrokinetic charges, adsorb to some extent on hydrophobic surfaces. The signature of a dehydration process is a large decrease in the heat capacity of the system. The heat capacity change upon adsorption $\Delta_{ads}C_p$ can be determined from the temperature derivative of $\Delta_{ads}H$ at constant pressure, composition and pH:

$$\Delta_{ads}C_p = \left(\frac{\partial \Delta_{ads}H}{\partial T}\right)_{P,\, c_p,\, pH} \tag{7}$$

As discussed by Brandts [17], two subprocesses in addition to sorbent and protein dehydration are known to influence the sign and magnitude of $\Delta_{ads}C_p$. The transfer of ions from aqueous solution to an apolar environment causes an increase in heat capacity. A loss of ordered secondary structure (e.g., α-helices and β-sheets) in a protein molecule also leads to an increase in heat capacity which is proportional to the increase rotational mobility along the polypeptide chain.

Table 3 shows $\Delta_{ads}C_p$ values calculated from $\Delta_{ads}H$ data at 15° C and 25° C for the adsorption of LSZ and αLS on PS-. The large negative $\Delta_{ads}C_p$ values observed make substantial contributions to the overall driving force for protein adsorption to PS-. In both adsorption systems, $\Delta_{ads}C_p$ becomes more positive with increasing charge (either positive or negative) on the protein molecule. This trend is most pronounced in the αLA adsorption system, where $\Delta_{ads}C_p$ eventually takes on positive values at high pH. A pH 10, αLA carries a large net negative charge and, consequently, a relatively low native-state stability and a strong electrostatic repulsion for the PS- surface. Adsorption of αLA to PS- at pH 10 will therefore involve a large transfer of positive counterions (including protons) to the interfacial layer to preserve electroneutrality; it may also involve a relatively large change in protein structure. The sensitivity of $\Delta_{ads}C_p$ to these subprocesses indicates that, although clearly important, dehydration effects alone do not dominate the overall driving force for protein adsorption at all conditions.

Table 3. $\Delta_{ads}C_p$ values for lysozyme and α-lactalbumin adsorbed to a negatively-charged polystyrene latex in 50-mM KCl. Values determined using Eq. (7) and $\Delta_{ads}H$ data at 15° C and 25° C and $\Gamma = \Gamma^{pl}$.

Protein	Sample pH	$\Delta_{ads}C_p$ (μJ m^{-2} K^{-1})
Lysozyme	4.0	- 110
	7.0	- 305
	10.0	- 420
α-Lactalbumin	3.0	- 310
	7.0	- 150
	10.0	+ 120

Protein Structural Changes. Protein folding involves a considerable loss in the conformational entropy of the polypeptide chain. Creighton [18] estimates that this loss in conformational entropy destabilizes the native state relative to the fully denatured state of a 100-residue protein by *ca.* 2500 ($\pm$ 1200) kJ mol^{-1}. Under certain solution conditions, other effects, particularly dehydration of hydrophobic residues, outweigh this entropic opposition to folding and the native state is marginally preferred. However, there is now substantial evidence that solid/water interfaces upset this delicate balance by providing a region on which the polypeptide backbone can unfold without exposing hydrophobic residues to water molecules [19]. The extraordinary intra-atomic packing densities of native-state globular proteins suggest that protein unfolding at a surface leads to an increase in the conformational entropy of the polypeptide chain and, thus, a second entropic driving force for adsorption. Evidence for rearrangements in protein structure upon adsorption have come from transmission circular dichroism, NMR and fluorescence spectropscopy, FTIR, and proton titrations [19]. Unfortunately, few of these studies have provided a quantitative understanding of the extent of protein unfolding at solid-liquid interfaces. As a result, structures of adsorbed protein, their dependence on sorbent and solution properties, and the contribution of protein structural rearrangements to the driving force for adsorption remain the most poorly understood aspects of the protein adsorption process.

The differential-scanning-microcalorimetry (micro-DSC) method pioneered by Privalov has provided much of the direct thermodynamic data characterizing the stabilities and structures of globular proteins [20]. Recently, we demonstrated that micro-DSC can also be used to quantify losses in ordered secondary structure of globular proteins upon adsorption to solids [15]. The direct thermodynamic observables in the micro-DSC experiment are the enthalpy change $\Delta_{P-D}H$, the denaturation temperature T_d, and the heat capacity change $\Delta_{P-D}C_p$ accompanying the temperature-induced denaturation of the protein in its adsorbed perturbed-state (P) structure. The magnitude of $\Delta_{P-D}H$ relative to $\Delta_{N-D}H$, where $\Delta_{N-D}H$ is the denaturation enthalpy of the native-state protein at the same solution pH, provides an unambiguous gauge of the extent of ordered secondary structure loss in proteins as a result of adsorption. A high T_d is indicative of a highly stable adsorbed-state structure. Figure 7 shows enthalpy of denaturation data, measured by micro-DSC, for LSZ dissolved in 50-mM KCl at 25° C and adsorbed to PS- at the same solution conditions [15]. In all cases, the micro-DSC data corresponds to a total protein concentration set at 95% of the Γ value where the adsorption isotherm deviates from the ordinate.

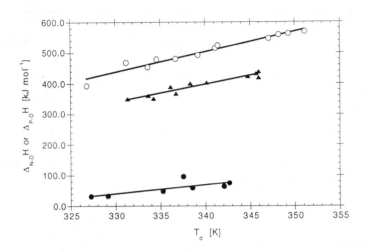

Figure 7: Micro-DSC data for lysozyme (○) dissolved in 50-mM KCl, (●) irreversibly adsorbed to PS- in 50-mM KCl, and (▲) irreversibly adsorbed to hematite in 50-mM KCl. (See text for addition information.)

Figure 7 also shows micro-DSC data for LSZ adsorbed to an aqueous dispersion of hematite (i.e., αFe_2O_3). Comparison of the micro-DSC data for the P-state and N-state protein indicates that LSZ retains most of its native-state structure when adsorbed to this hydrophilic surface. Thus, for structurally stable proteins such as LSZ, perturbations in protein structure can make a substantial contribution to the driving force for adsorption, particularly when the sorbent is hydrophobic, but they do not appear to dominate adsorption under all conditions.

In contrast, when the adsorbing protein has a low native-state stability, $\Delta_{ads}G$ is strongly influenced by rearrangements in protein structure irrespective of the nature of the sorbent surface. For instance, Figure 8 compares micro-DSC data at 25 °C for αLA dissolved in 50-mM KCl, and adsorbed from a 50-mM KCl solution to both PS- and hematite. Adsorption of αLA to either surface involves a near complete loss of ordered secondary structure.

Sequential micro-DSC scans of adsorbed protein samples indicate that thermal denaturation of P-state proteins is an irreversible process. Moreover, the magnitude and location (i.e., T_d) of $\Delta_{P-D}H$ recorded in the initial micro-DSC thermogram are dependent on the scan rate, indicating that the protein unfolding reaction on the sorbent surface is kinetically controlled. A general theory for such scan-rate dependencies has been provided by Sanchez-Ruiz et al. [21], who assumed that irreversible protein denaturation reactions can be represented by a two-step mechanism where I is the final state of the protein irreversibly arrived at from a reversible unfolded state D. In this model, k_1 and k_2 represent the forward and reverse rate constants, respectively, for the reversible transition from the N-state to the D-state and k_3 is the rate constant for the irreversible transition from D to I; all of the kinetic constants are first order and change with temperature according to the Arrhenius equation. If $k_3 \gg k_2$, all of the D-state molecules formed are converted to I and the reaction mechanism reduces to

$$N \xrightarrow{\ k\ } I \qquad\qquad (8)$$

where k equals k_1 or k_3 depending on the rate-determining step. Assuming the thermogram is initiated at a low enough temperature to assure that all protein present is in its adsorbed perturbed state, Eq. (8) and the kinetic model associated with it predict that the transition temperature T_d should vary with scan rate v (K min^{-1}) according to

$$\ln\!\left(\frac{n}{T_d^2}\right) = \ln\!\left(\frac{AR}{E_a}\right) - \frac{E_a}{R}\frac{1}{T_d} \qquad\qquad (9)$$

where A (min^{-1}) is the frequency factor, E_a (kJ mol^{-1}) is the activation energy for the surface denaturation process, and R is the universal gas constant.

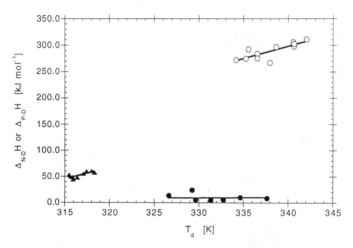

Figure 8: Micro-DSC data for α-lactalbumin (O) dissolved in 50-mM KCl, ($\bullet$) irreversibly adsorbed to PS- in 50-mM KCl, and ($\blacktriangle$) irreversibly adsorbed to hematite in 50-mM KCl. (See text for additional information.)

Figure 9 shows a plot of $\ln(v/T_d^2)$ as a function of T_d^{-1} for LSZ and αLA adsorbed to PS- at pH 7. The activation energy for the irreversible surface denaturation reaction (*i.e.*, P-state to I-state), determined from the slope of each line, is 243 ± 13 kJ mol^{-1} for LSZ and 132 ± 16 kJ mol^{-1} for αLA. Thus, although the amounts of ordered secondary structure retained in the P-states are low and similar for the two proteins, the activation energy which must be overcome to reach the I state on the surface is substantially higher for LSZ. This suggests that the conformational dynamics of P-state LSZ remain fairly limited compared with adsorbed αLA, providing further evidence that the contribution of protein structural changes to the overall driving force for adsorption is dependent on the structural stability of the native-state protein.

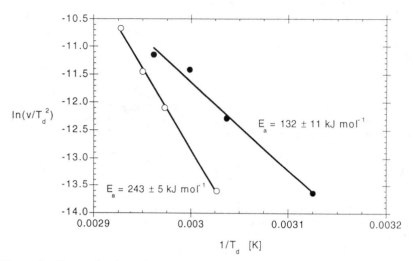

Figure 9: Determination of activation energies for the irreversible denaturation of (O) lysozyme and (●) α-lactalbumin adsorbed at pH 7 and 25 °C to PS- in 50-mM KCl.

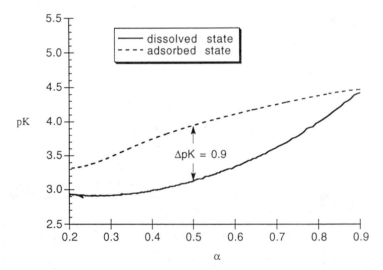

Figure 10: Carboxyl-residue dissociation data at 25 °C for lysozyme dissolved in 50-mM KCl and adsorbed to PS- in 50-mM KCl. Ordinate was calculated from proton titration data and the fundamental relation $pK = pH + \log_{10}[(1+\alpha)/\alpha]$ where α is the degree of dissociation.

Redistribution of Charged Groups. The fact that electrostatic forces must contribute to protein adsorption at solid surfaces has long been recognized. For instance, many researchers have noted a bell-shaped dependence of Γ^{pl} on adsorption pH with a maximum centered on the isoelectric point of the protein-sorbent complex [19]. However, the relative magnitude of coulombic contributions to the overall driving force for protein adsorption remains unknown.

As shown by Haynes *et al.* [22], proton titration experiments provide an effective (but underutilized) method for probing charged-group redistribution in protein-adsorption processes. Figure 10 plots pK as a function of the degree of proton dissociation (α) for the carboxyl groups on LSZ when it is adsorbed to PS- in 50-mM KCl and when it is dissolved in its native-state conformation in 50-mM KCl. The 0.9 positive shift in the average pK of the carboxyl groups reveals the strong influence of the negatively charged apolar sorbent on the electrostatic properties of the adsorbed protein. This shift in pK at pH 4 is consistent with the protonation of 3 carboxy groups in the adsorbed state which remain deprotonated in the dissolved state. For αLA, a positive pK shift of 1.1, which at pH 4 requires titration of an additional 8 carboxyl groups, was observed upon adsorbing the protein to PS- [22]. Clearly, charge redistribution effects contribute to protein adsorption.

However, as discussed above, electrostatic effects do not dominate protein adsorption at all conditions. For instance, αLA adsorbs to hydrophilic sorbents such as αFe_2O_3 and glass at all solution conditions, including those where the net electrostatic force between the protein and sorbent is large and repulsive.

Summary

The complex irreversible nature of protein adsorption to solids suggests that in most systems the overall driving force for adsorption $\Delta_{ads}G$ is not controlled by a single force or subprocess, but rather by an interplay between several adsorption subprocesses. Dehydration of a sorbent surface requires close contact between the sorbent and the adsorbed protein which, for structurally rigid globular proteins, can only be achieved through sincere changes in protein conformation. If those conformational changes involve a breakdown in ordered secondary structure, the concomitant increase in rotational freedom of the polypeptide backbone contributes further to the driving force for adsorption. Moreover, the increased backbone flexibility improves the ability of the protein to form strong ion pairs with oppositely charged residues on the sorbent and to reduce lateral electrostatic repulsions with proximal proteins. This interplay between major adsorption processes points to a synergistic adsorption mechanism; in addition to its inherent contribution, each subprocess amplifies the other contributions to the overall driving force for adsorption.

1. Jonsson *et al.*, *J. Colloid and Interface Sci.* **1982**, *90*, 148.

2. Jennissen, H.P. In *Surface and Interfacial Aspects of Biomedical Polymers*; volume 2; Andrade, J.D., Ed.; Plenum Press: New York, 1985; p. 295.

3. Barbucci, R.; Casolaro, A.; Magnani, A. *Clinical Materials* **1992**, *11*, 37.

4. Mizutani, T.; Brash, J.L. *Chem. Pharm. Bull.* **1988**, *36*, 2711.

5. Moreno, E.C.; Kresak, M.; Hay, D.I. *Biofouling* **1991**, *4*, 3.

6. van Oss, C.J. *Biofouling* **1991**, *4*, 25.

7. Shastri, R.; Roe, R.J. *Org. Coat. Plast. Chem.* **1970**, *40*, 820.

8. Andrade, J.D. In *Surface and Interfacial Aspects of Biomedical Polymers*; volume 2; Andrade, J.D., Ed.; Plenum Press: New York, 1985; ch. 1.

9. Everett, D.H. *Trans. Faraday Soc.* **1954**, *50*, 1077.

10.   Norde, W.; Anusiem, A.C.I. *Colloids & Surfaces* **1992**, *66*, 73.

11.   Norde, W.; Favier, J.P. *Colloids & Surfaces* **1992**, *64*, 87.

12.   Northrup, S.N. *J. Phys. Chem.* **1988**, *92*, 5847.

13.   McCammon, J.A.; Harvey, S.C. *Dynamics of Proteins and Nucleic Acids*; Cambridge Press: Cambridge, England, 1987.

14.   Norde, W. In *Surface and Interfacial Aspects of Biomedical Polymers*; volume 2; Andrade, J.D., Ed.; Plenum Press: New York, 1985, p. 263.

15.   Haynes, C.A.; Norde, W. "Structures and Stabilities of Adsorbed Proteins", *J. Colloid Interface Sci.* **1994**, in press.

16.   Nyilas, E.; Chiu, T.-H.; Lederman, D.M. In *Colloid and Interface Science*; volume 5; Kerker, M., Ed.; Academic Press: New York, 1976, p. 77.

17.   Brandts, J.F. In *Biological Macromolecules*; Fasman, G.; Timesheff, S.N., Eds.; Marcel-Dekker: New York, 1969, p. 213.

18.   Creighton, T.E. *Biochem. J.* **1990**, *270*, 1.

19.   Haynes, C.A.; Norde, W. *Colloids and Surfaces B* **1994**, *2*, 517.

20.   Privalov, P.L. *Adv. Protein Chem.* **1979**, *33*, 167.

21.   Sanchez-Ruiz, J.M.; Lopez-Lacomba, J.L.; Cortijo, M.; Mateo, P.L. *Biochemistry* **1988**, *27*, 1648.

22.   Haynes, C.A.; Sliwinski, E.; Norde, W. *J. Colloid Interface Sci.* **1994**, *164*, 394.

RECEIVED May 31, 1995

Chapter 3

Logarithmic Growth of Protein Films

Simon Alaeddine and Håkan Nygren

Department of Anatomy and Cell Biology, University of Göteborg, Medicinaregatan 5, S–413 90 Göteborg, Sweden

The kinetics of protein adsorption and antibody binding to surface-immobilised antigen was measured with off-null ellipsometry under non-diffusion limited conditions. An initial lag-phase was seen, followed by an accelerating reaction rate (auto catalysis). The reaction rate then decreased at a surface concentration far below monolayer coverage and a continuously decreasing rate of binding was seen. The kinetics can be described by a logistic law of limited growth function. The theoretical description of surface reactions with fractal kinetics is discussed.

The reaction rate of macromolecular reactions at interfaces decreases logarithmically over long periods of time as shown for protein adsorption (*1, 2*) and antigen-antibody reactions (*3, 4*). The general phenomenon of logarithmically decreasing reaction rates has been collectively named fractal kinetics (*5*) and has been demonstrated experimentally in a number of situations (6). Spatial and/or energetic heterogeneity of the medium or non randomness of the reactant distribution in low dimensions have been suggested as mechanisms behind the phenomenon of fractal kinetics. Experimental studies of ferritin adsorption, a suitable model system with fractal kinetics, have revealed that the logarithmic growth of the protein film is preceded by an initial acceleration-phase of adsorption (2).

The acceleration of the initial adsorption can be described by nucleation-and-growth- like kinetics assuming attraction between adsorbed molecules and molecules in solution. The initial cooperative adsorption can be described theoretically by an exponential growth (*7*), which rapidly leads to depletion of reactants in the reaction zone making the reaction mass-transport limited.

It has also been shown that ferritin clusters are restructured during adsorption with fractal kinetics. Orderly structured aggregates seen in previous phases of adsorption disappear and a more random distribution is seen. This process of rearrangement of dense clusters is not necessarily a continuous one but may take place through critical dissociation of orderly structured aggregates (*2, 8*). An intellectual model of this phenomenon is the continuous build up and discontinuous fall down of sand on sand piles in an hour-glass. Theoretical models of such processes have been elaborated (*9, 10*) and may serve to explain the fractal kinetics of macromolecular reactions at surfaces. The use of such models have been suggested for intermolecular protein dynamics (*11*).

The present study was undertaken in order to further describe the molecular mechanism behind the kinetics of macromolecular reactions at interfaces.

0097–6156/95/0602–0041$12.00/0
© 1995 American Chemical Society

Theory

1. Nucleation. From previous studies with TEM it is known that there is a limited number of sites available for monomolecular binding. The number of nucleation sites differ between surfaces but is typically of the order of 10^9-10^{10} molecules/cm^2 (7, 12). It is not easy to measure the kinetics of this initial binding, but a probability for independent binding is relevant, at least for short incubation time (< 5 sec). Thus the initial nucleation rate can be written as (7):

$$\frac{\partial S}{\partial t} = R_s \, k(t) \left(N_{max} - N \right)$$

$$k(t) \propto exp\left[-\frac{\Delta\mu_{des}}{kT} - \lambda t\right]$$

(1)

where the function, k(t), corresponds to the sticking probability of independent binding at the nucleation sites (N_{max}); N is the number of occupied nucleation sites; λ is the frequency of collision with the nucleation sites; $\Delta\mu_{des}$ is the mean activation energy per molecule; R_s is the molecular flux towards the surface (number of molecules striking 1cm^2 of the surface per second) that is dependent on the boundary conditions of diffusion and the concentration of reactant in the bulk. The subscript "s" represents the concentration close to the surface.

2. Growth. The cluster can grow in an arbitrary shape, favouring the interaction energy, and may be described in different ways. One, two, and three-dimensional aggregates will grow if the mean interaction free energy, $\Delta\mu$, decreases with the number of adsorbed molecules, S. Thus the interaction free energy of the molecules in the bulk phase can be expressed as (13)

$$\mu_m^0 = \mu_\infty^0 + \frac{\alpha_{m-m} kT}{S^q}$$

(2)

where α_{m-m} is a positive constant dependent on the strength of the intermolecular interactions and q is a number that depends only on the shape or the dimensions of the aggregates.

For clusters adsorbed to the surface it is reasonable to assume that the interaction free energy is a superposition of the strength of the intermolecular interaction, α_{m-m}, and the strength of the surface-molecules interaction, α_{s-m}

$$\mu_{sm}^0 = \mu_\infty^0 + \frac{\alpha_{m-m} kT}{S^q} - \alpha_{m-s} kT$$

(3)

The adsorbed protein film grows from initial nucleation sites, dependent on binding growth (14, 15) until critical clusters of various size and shape reach equilibrium with the bulk molecules. In the initial stages of adsorption, clusters of various sizes are in metastable equilibrium with adsorbed monomers. As these clusters grow, they deplete the surrounding region of adsorbed monomers so that further nucleation (or cluster formation) is not possible in this region (called a capture zone). Taking such a capture zone to be of radius r_c, one gets

$$r_c \propto \sqrt{D\tau_a}$$

(4)

where D is the surface-diffusion coefficient of adsorbed monomers on the substrate, τ_a is the mean free residence time before adsorbed monomer desorption. Hence the rate of addition of new molecules per cm^2 per second is proportional to the number of occupied nucleation sites, N, the flux, the sticking probability, and the free residence time

$$\frac{\partial S}{\partial t} \cong R_s f(t) N$$

$$f(t) \propto \frac{\lambda^p}{(p-1)!} t^{p-1} exp\left[-\left(\frac{\Delta\mu_{sd}}{kT} + \frac{\Delta\mu_{des}}{kT}\right) - \lambda t\right] \tag{5}$$

where $\Delta\mu_{sd}$ is the Gibbs free energy of activation for surface diffusion of adsorbed monomers, $\Delta\mu_{des}$ is the Gibbs free energy of activation for monomer desorption, and p is the density of independent binding sites per nucleation site. It is of interest to note that the theory behind the derivation of the first factor in f(t) is based on the gamma distribution (p, λ) where p represents the density of independent binding sites and λ is the frequency of collision with the nucleation sites.

Another process of growth around a nucleation site, where limiting factors are taken into account, has been derived in reference (*16*):

$$\frac{\partial S}{\partial t} \cong g(t)S\left(1 - \frac{S}{S_{max}}\right)e^{-\frac{\Delta\mu_{des}}{kT}} \tag{6}$$

where S_{max} is the maximum number of population per nucleation site, and g(t) is a time dependence function which corresponds to the probability of favourable interaction between bound and surface diffusing molecules. This growth model is usually called the *logistic law of limited growth*. A third possibility of describing cluster growth is by using a statistical model

$$S = R_s(1 - e^{-\alpha t}) \tag{7}$$

where α relates to the probability of favourable interaction between adsorbed and adsorbing molecules.

Any of the equations mentioned above predicts a rapid growth of protein films with an exponential time dependence in the initial stage. This is also seen experimentally but only for short periods of time before growth is limited by external constraints (*17*), e.g., (a) the mass transport, which is often found to limit the rate of macromolecular reactions at surfaces (*18*), and (b) due to the geometry of growing clusters (Nygren, H. Biophys. Chem. in press.)
3. Desorption. An obvious reason for a decreasing rate of growth is desorption of bound molecules. Spontaneous desorption of single molecules is described by:

$$\frac{\partial S}{\partial t} \propto -Se^{-\frac{\Delta\mu_{des}}{kT}} \tag{8}$$

or when limiting factors are taken into account

$$\frac{\partial S}{\partial t} \cong -h(\lambda)(S - S_0)e^{-\frac{\Delta\mu_{des}}{kT}} \tag{9}$$

$$0 \leq S_0 \leq S$$

where S_0 is the number of molecules at which the surface-rate density becomes constant, and $h(\lambda)$ is desorption rate depending on the collision frequency and the adsorption time . Reference (19) suggested that the desorption rate of aggregates containing i monomers is given by

$$\frac{h(\lambda)}{exp\left[-\frac{\Delta\mu_{des}}{kT}\right]} \cong \frac{1}{1 + \left[\frac{a}{\lambda_{1i} R_s \tau_a^{(i+1)}}\right]} \tag{10}$$

where a is a constant.

Associative desorption due to collisions between bound and colliding particles, or due to association by surface diffusion is described by (20):

$$\frac{\partial S}{\partial t} \propto -S^n \exp\left[-\frac{\Delta\mu_{des}}{kT}\right] \tag{11}$$

where n is the number of molecules taking part in the reaction. Associative desorption may also be described in a statistical model by the term

$$S(1 - e^{-\beta t}) \tag{12}$$

where β relates to the probability of desorption.

In analogy with the discussion in paragraph 2, the time dependence of desorption is limited by the geometry of the adsorbing clusters, the number of collisions of monomers with the aggregate and the fluctuation density of molecules close to the surface. The rate of adsorption, taking into account nucleation, growth and limited desorption can be written as:

$$\frac{\partial S}{\partial t} \cong R_s\left((N_{max} - N)k(t) + Nf(t)\right) - h(\lambda)(S - S_0) \tag{13}$$

where equation (13) is the result of combining equations (1), (5), and (9). A statistical description of the adsorption process is obtained by combining equations (7) and (12):

$$S = R_s \frac{(1 - e^{-\alpha t})}{(2 - e^{-\beta t})} \tag{14}$$

Experimental

Measurements of the kinetics of protein adsorption are often hampered by the effect of mass transport limitation (18). The mass transport limitation is due to depletion of protein in the reaction zone close to the surface and no conclusions can be made from such measurements regarding the sticking probability of protein molecules at the surface.

In the present study precautions were taken to avoid mass transport limitations by using a flow cuvette and fast ellipsometric detection. The kinetics of protein adsorption was measured by off-null ellipsometry in situ allowing a time resolution of 0.1 seconds which makes it possible to measure the initial adsorption taking place before the diffusion layer in the solution is depleted of protein. The flow cuvette had a small volume (50µl) and a high flow rate was used to give a thin unstirred layer and a continuous exchange of the bulk phase to ensure constant concentration of protein in the solution close to the surface. Under these boundary conditions, mass transport limitation is described by the relationship (*18*):

$$\frac{\partial S_0}{\partial t} = \frac{DC_0}{R} \qquad (15)$$

where S_0 is the amount of protein reaching the surface, D is the diffusion constant of the protein, C_0 is the concentration of protein in the solution, R is the thickness of the unstirred water film and t is time.

The experimental set-up makes it possible to use high concentrations of protein in the solution which further decreases the risk for depletion of protein in the diffusion layer. The benefit of reducing the risk of depletion of protein near the surface was obtained at the cost of losing the defined boundary conditions for diffusion since the thickness of the unstirred layer is unknown and almost impossible to estimate.

Chemicals: Fibrinogen (Kabi, Stockholm, Sweden) was dissolved in PBS. Monoclonal antibody directed against dinitrophenol (DNP) was a generous gift from professor M. Steward, London School of Hygiene and Tropical Medicine. Hexamethyldisilazane (E. Merck, Darmstadt, Germany) was used for methylation of SiO_2 surfaces (oxidized silicon wafers) as described (*21*). Hydrophobic quartz was coated with antigen as described elsewhere (*22*).

Ellipsometry: A null ellipsometer (Rudolph Research model 436) was used as described previously (*17*). The film thickness during adsorption is given by:

$$d = k' \sqrt{I - I_0} \qquad (16)$$

The thickness d is an equivalent optical thickness. The relation between this film thickness and the amount of bound protein can be found in different ways depending on what properties of the protein used are best characterised experimentally. With a well known density of the protein, the relation

$$\Gamma = d\,\zeta * 100 \qquad (17)$$

can be used, where Γ is the surface concentration (ng/cm^2), d is film thickness (nm) and ζ is the density of the protein (g/cm^3). The density of the proteins used was determined pycnometrically as described (*23*).

Results and Discussion

It is easy to realize that equation (13) has no analytical solution, especially when the integrand contains a power of exponential functions. However, there are two possible ways of expressing this equation in a simple form: a) N = N(t), i. e, N as a function of time, or b) N = N(S), i. e, N as a function of S. Thus using the first case equation (13) takes the form

$$\frac{\partial S}{\partial t} = z_1 t^{\rho_1 - 2} e^{-\lambda_1 t} (1 - e^{-kQ(\rho, t_0)}) + z_2 t^{\rho_2 - 2} e^{-\lambda_2 t} e^{-kQ(\rho, t_0)} - \left(ct + z_3 t^{\rho_1 - 1} e^{-\lambda_1 t} (1 - e^{-kQ(\rho, t_0)}) - \right.$$

with

$$\lambda \propto \lambda_1, \qquad \lambda_2 = (\lambda_1 + a\lambda^{-1}), \qquad \rho \propto \rho_1, \qquad \rho_2 = \rho + \rho_1 \qquad (18)$$

Where z_1, z_2, z_3, k, c, and B are proportionality constants. $Q(\rho, t_0)$ is called the gammaregularized function (*24*). Integrating equation (18) results in the following

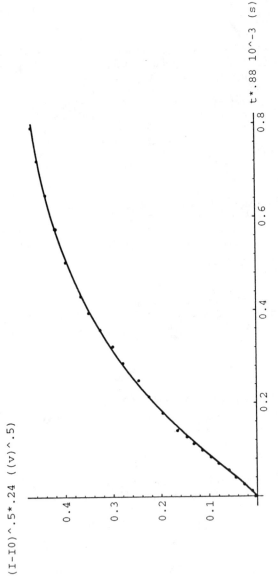

Figure 1. Ellipsometry data showing the surface concentration, S, of fibrinogen adsorbed onto hydrophobic surface from a bulk concentration of 0.1 mg/ml versus the adsorption time, t. Simulation of equation (19) fitted to the experimental data of fibrinogen adsorbed onto the hydrophobic surface.

$$S = 0.6089 \, t^{0.26} \, e^{-0.000047 \, t} \left(1 - e^{-2Q(1.01, 0, 2.049t)}\right) + 0.1259 \, t - 0.1439 \, t^2$$

$$S(t) = z\, t^{\rho_1 - 1}\left(1 - e^{-k\, CDF[\Gamma(\rho,\lambda),at]}\right)e^{-\lambda_1 t} + Bt - \frac{c}{2}t^2 \tag{19}$$

where CDF is the cumulative distribution function (*24*).

The kinetics of adsorption of fibrinogen onto a hydrophobic or hydrophilic quartz surface, measured by ellipsometry, is shown in Figures 1 and 2 respectively together with a fitted theoretical model (equation 19). Antibody binding to immobilised hapten (DNP) is shown in Figures 3 and 4, together with a fitted theoretical model (equation 19).

The initial accelerated kinetics of the reaction suggests favourable interactions between adsorbed molecules and molecules in the bulk solution. This initial phase of adsorption can be described by an exponential growth which is limited with time. A statistical model (equation (7)) of the initial kinetics of adsorption is shown in Figure 5. The parameter a will critically determine the kinetics of growth from linear at low a-values to strongly auto catalytic growth at high a-values. In model experiments with adsorption of ferritin, showing fractal kinetics, it has been found that the limitation of growth is due to discontinuous desorption of large molecular clusters that are formed. A statistical model of auto catalytic growth and critical desorption (equation (14)) is shown in Figure 6. Note that two exponential terms with different sign will result in logarithmic growth. In order to explain the solubility of protein in solution and the discontinuous desorption from surfaces some mechanism has to be assumed counteracting the intermolecular attractive forces. Thermal motion is an obvious candidate. A useful formalism to describe this duality of attractive forces and thermal motion was suggested by Brönsted (*25*)

$$\frac{C_1}{C_2} \equiv \exp\left[-\frac{\lambda A}{kT}\right] \tag{20}$$

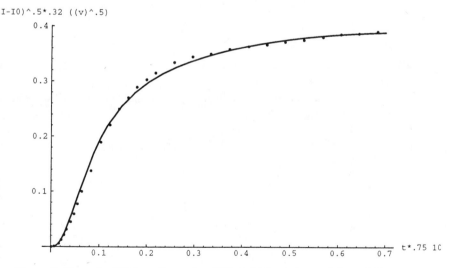

Figure 2. Numerical fitting of equation (19) to the experimental data of fibrinogen binding onto a hydrophilic surface. Surface concentration (S) adsorbed from a bulk concentration of 0.1 mg/ml as a function of time (t).

$$S = 0.5561\, t^{0.28} e^{-0.000047 t}\left(1 - e^{-2.1 Q(2.5, 0, 22.222 t)}\right) + 0.00847 t - 0.1196 t^2$$

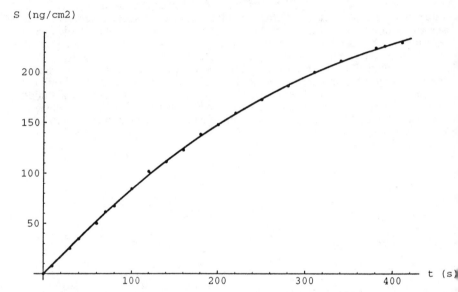

Figure 3. Surface concentration of antibody bound to immobilised DNP-hapten versus the time, t. (concentration of Mab = 30 μg/ml). The experimental data are fitted using equation (19).

$$S = 15.292\, t^{0.25} e^{-0.0000357\,t} \left(1 - e^{-2Q(1.1,0,0.005288\,t)}\right) + 0.64317\,t - 0.0005210\,t^{2}$$

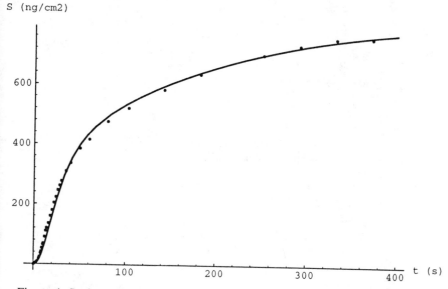

Figure 4. Surface concentration of antibody bound to immobilised DNP-hapten versus the adsorption time (concentration of Mab = 100 µg/ml). The experimental data are fitted using equation (19).

$$S = 178.192\, t^{0.25} e^{-0.0000533\, t}\left(1 - e^{-2Q(2.5, 0, 0.067568\, t)}\right) + 0.57814\, t - 0.00084138\, t^2$$

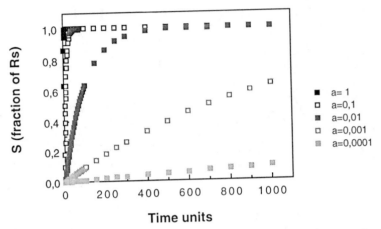

Figure 5. A plot of the fraction of the flux of molecules adsorbed to the surface as a function of the parameter α of equation (7).

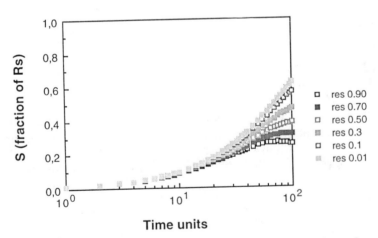

Figure 6. A plot of the fraction of the flux of molecules adsorbed to the surface as a function of the parameter β of equation (14). $\alpha = 0.01$. Resorption (β) = 1% to 90% of α.

where λ denotes molecular properties, e.g., hydropathy; A is the molecular area and C_1/C_2 are the concentration of the molecules in two phases (bulk and surface). This simple relationship, reflecting the duality of protein molecules, may explain the existence of short-lived clusters as the mechanism behind the complex kinetics of protein adsorption.

Acknowledgments

The present study was supported by growth from the Swedish Medical Research Council (12x-06235) and the Research Council of Engineering Sciences.

Literature Cited

1   Cuypers, P.A.; G.M. Willems.; J.M. Kop.; J.W. Corsel.; M.P. Jansen.; W.T. Hermens. In: *Proteins at interfaces*. Physicochemical and biochemical studies. Eds.; J.L. Brash.; T.A. Horbett. ACS Symposium Series 343; American Chemical Society, Washington, **1987**; pp. 208-221.
2   Nygren, H. *Biophys. J.* **1993**, 65, 1508-1512.
3   Nygren, H.; M. Stenberg. *Immunology.* **1989**, 66, 321-327.
4   Werthén, M.; M. Stenberg.; H. Nygren. *Progr. Colloid Polymer Sci.* **1990**, 82, 349-352.
5   Kopelman, R. *Science.* **1988**, 241, 1620-1625.
6   Kopelman, R. *J. Stat. Phys.* **1986**, 42, 185-200.
7   Nygren, H.; Alaeddin. S.; Lundström. I.; Magnusson. K-E. *Biophys.Chem.* **1994**, 49, 263-272.
8   Nygren, H. *Progr. Colloid Polymer Sci.* **1992**, 88, 86-89.
9   Bak, P.; C. Tang.; K. Wiesenfeld. *Phys. Rev.* A. **1988**, 38, 364-374.
10   Bak, P.; K. Chen. *Physica D.* **1989**, 38, 5-12.
     Bak, P.; K. Chen.; M. Creutz. *Nature.* **1989**, 342, 780-782.
11   Dewey, T.G.; J.G. Bann. *Biophys. J.* **1992**, 63, 594-598.
12   Nygren, H.; Stenberg, M. *Biophys. Chem.* **1990**, 38, 67-76.
13   Jacob N, Israelachvili. *Intermolecular and surface forces.* Academic Press Limited San Diego. CA. **1985**.
14   Nygren, H. *J. Immunol. Methods.* **1988**, 114, 107-111.
15   Nygren, H.; Arwin, H. *Progr. Colloid Polymer Sci.* **1993**, 93, 321-323.
16   Smith, C. A. B. Biomathematics C. Griffin & Co London. **1954**.
17   Nygren, H.; Arwin, H.; Welin, S. *Colloids and Surfaces.* **1993**, 76, 87-93.
18   Stenberg, M.; Nygren, H. *J. Immunol. Meth.* **1988**, 113, 3-15.
19   Zinsmeister, G. *Proc. Intern. Symp. Basic Problems in Thin Film Physics*, Van denhoeck and Rupprecht, Goettingen. **1965**, p. 33
20   Zhdanov, V.P. *Elementary physicochemical processes on solid surfaces.* Plenum Press N.Y.**1992**.
21   Jönsson, U.; G. Olofsson.; M. Malmqvist.; I. Rönnberg. *Thin Solid Films.* **1985**, 124, 117-123.
22   Werthén, M.; Nygren, H. *Biochim. Biophys. Acta.* **1993**, 1162, 326-332.
23   Bernhardt, J.; H. Pauly. *J. Phys. Chem.* **1980**, 84, 158-162.
24   Wolfram, S. *Mathematica: a system for doing mathematics by computer.* Addison-Welsley Publishing Co., Redwood, California. **1991**.
25   Brönsted, J. N. *Z. Phys. Chem.* **1931**, A (Bodenstein-Festband), 257.

RECEIVED April 25, 1995

Chapter 4

Comparative Adsorption Studies with Synthetic, Structural Stability and Charge Mutants of Bacteriophage T4 Lysozyme

J. McGuire[1], V. Krisdhasima[1], Marie C. Wahlgren[2], and
Thomas Arnebrant[2]

[1]Department of Bioresource Engineering, Oregon State University,
Gilmore Hall 116, Corvallis, OR 97331–3906
[2]Department of Food Technology, University of Lund, P.O. Box 124,
S–221 00 Lund, Sweden

We have purified wild type, three structural stability mutants and four charge mutants of bacteriophage T4 lysozyme from *E. coli* strains harboring desired expression vectors. Structural stability mutants were produced by substitution of the isoleucine at amino acid position three, yielding a set of proteins with stabilities ranging from 1.2 kcal/mol greater, to 2.8 kcal/mol less, than that of the wild type. Charge mutants were produced by replacement of positively charged lysine residues with glutamic acid, yielding a set of molecules with formal charges ranging from +5 to +9 units. Adsorption kinetic data, along with the dodecyltrimethylammonium bromide-mediated elutability of each protein, has been monitored with *in situ* ellipsometry at hydrophobic and hydrophilic silica surfaces. A simple mechanism that allows adsorbing protein to adopt one of two states, each associated with a different resistance to elution and a different interfacial area occupied per molecule, has been used to assist interpretation of the adsorption data. Conditions implicit in the model have been used to estimate the fraction of molecules present on the surface just prior to surfactant addition that had adopted the more resistant state, and this fraction has been observed to correlate positively with resistance to elution. For the stability mutants, these properties were clearly related to protein stability as well. Concerning the charge mutants, results have not been clearly explainable in terms of protein net charge, but rather in terms of the probable influence of the location of each substitution relative to other mobile, solvent-exposed, charged side chains of the molecule.

Study of molecular influences on protein adsorption has received much attention due in part to the relevance of this matter to better understanding of the nature of adsorption competition in complex mixtures. Important contributions to current understanding of molecular influences on protein adsorption have evolved from several comparative studies of protein interfacial behavior, in which similar or otherwise very well-characterized proteins (*1-4*), genetic variants (*5-7*) or site-directed mutants (*8*) of a single protein had been selected for study. A number of factors are known to affect protein adsorption, and these studies have stressed the importance of protein charge, hydrophobicity and structural stability in interfacial behavior.

Comparative studies with genetic variants and site-directed mutants of single proteins have particularly demostrated the importance of structural stability in protein interfacial behavior, beginning about 20 years ago with investigation of the interfacial behavior of several single-point mutants of human hemoglobin (5). Study of human hemoglobin variants was motivated by the finding that the oxy-form of the abnormal hemoglobin involved in sickle cell disease precipitated very quickly upon mechanical treatment relative to normal hemoglobin (9,10). A number of similar approaches to better understand protein adsorption have taken place since then. For example, Horsley et al. (6) made a comparison of isotherms constructed for hen and human lysozyme at derivitized silica surfaces, Xu and Damodaran (7) compared adsorption kinetic data measured for native and denatured hen, human, and bacteriophage T4 lysozymes at the air-water interface, and Kato and Yutani (8) evaluated the interfacial behavior of six mutants of tryptophan synthase α-subunits, produced by amino acid substitution at a single position in the protein's interior. In general, less stable mutants have been observed to be more surface active, i.e., they more rapidly adsorb and/or more readily unfold or otherwise rearrange at an interface. Similarly, adsorption from single-component solutions of model globular proteins of similar size, but differing charge and structural stability, has shown that at a given surface adsorption is related to structural stability; i.e., proteins of high stability behave like "hard" particles at a surface, with the interactions governed by hydrophobicity and electrostatics, while adsorption of proteins of low stability ("soft" proteins) may be influenced by structural rearrangement, allowing adsorption to occur even under conditions of electrostatic repulsion (1,2). Still, although the importance of protein charge and structural stability on adsorptive behavior is well-accepted, it remains incompletely quantified.

The elutability of adsorbed protein by surfactant has been used to provide an index of protein binding strength (11-15), and Wahlgren and coworkers (16-19) have used *in situ* ellipsometry to continously monitor the effects of different surfactants on the elutability of selected proteins, as well as adsorption from protein/surfactant mixtures, at a number of interfaces. We recently reported on the adsorption and dodecyltrimethylammonium bromide-mediated elutability of wild type, stability mutants and charge mutants of bacteriophage T4 lysozyme (McGuire et al., *J. Colloid Interface Sci.*, in press). Here we review the most significant conclusions from that work, and provide a brief treatment of results from longer-term, single-component adsorption kinetic experiments, as well as circular dichroism experiments performed with the stability mutants following their adsorption to ultrafine silica particles.

Bacteriophage T4 Lysozyme

Serious study of the molecular basis for any aspect of protein behavior would require a set of very similar proteins, and the use of mutant proteins with single amino acid substitutions is arguably the best way to achieve this. T4 lysozyme was selected for study as numerous variants of this protein have been synthesized and characterized with respect to their deviations in crystal structure and thermodynamic stability from the wild type.

A schematic of the α-carbon backbone of T4 lysozyme is shown in Figure 1. Figure 1 shows that the molecule is comprised of two distinct domains: the C-terminal and N-terminal lobes, joined by an α-helix (residues 60-80) that traverses the length of the molecule (20). The two-domain structure of T4 lysozyme is more distinct than is that of hen lysozyme, and in general several important differences between these two variants exist. There is very little homology between the two with respect to primary structure, T4 lysozyme has no disulfide linkages (21), and although some homology exists at the tertiary level, most of the C-terminal domain of T4 lysozyme has no counterpart in hen lysozyme (22). T4 lysozyme has 164 amino acid residues with a molecular weight of about 18,700 daltons (23). Crystallographic data would support more prolate molecular dimensions for T4 lysozyme in solution than those of hen

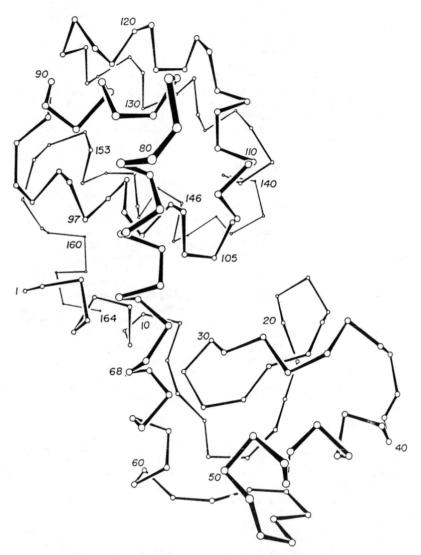

Figure 1. The α-carbon backbone of the wild type lysozyme from bacteriophage T4. (Reproduced with permission from McGuire et al., *J. Colloid Interface Sci.*, in press. Copyright 1995 Academic Press.)

lysozyme, allowing estimation as an ellipsoid about 54 Å long, with the diameter of the C-terminal lobe being about 24 Å, and that of the N-terminal lobe about 28 Å (22-24). Similar to hen lysozyme, however, T4 lysozyme is a basic molecule with isoelectric point above 9.0, and an excess of nine positive charges at neutral pH. In addition to the terminal amino and carboxylate groups, the wild type molecule has 27 positively-charged side chains and 18 negatively-charged side chains at pH 7, with nearly all of the out-of-balance charge located on the C-terminal lobe (20,21).

Stability Mutants. The isoleucine at position three (Ile 3) has been replaced with 13 different amino acid residues by site-directed mutagenesis (25). In that work, thermodynamic measurements of stability for each mutant, along with high resolution X-ray structural analyses were used to unambiguously illustrate the contribution of hydrophobic interactions at the site of Ile 3 to the overall structural stability of the protein. The structural stability of each mutant was quantified by $\Delta\Delta G$: the difference between the free energy of unfolding of the mutant protein and that of the wild type at the melting temperature of the wild type (20). Ile 3 contributes to the major hydrophobic core of the C-terminal lobe, and helps to link the C- and N-terminal domains. The side chain of Ile 3 contacts the side chains of methionine at position six, leucine at position seven, and isoleucine at position 100; it also contacts the main chain of cysteine at position 97. These residues are buried within the protein interior (25).

In addition to the wild type lysozyme, three stability mutants have been studied thus far. A mutant with cysteine substituted for Ile 3 (Ile 3 → Cys (S-S)), within which a disulphide link is formed with Cys 97, was selected as that mutation yields a more stable protein than the wild type ($\Delta\Delta G$ = +1.2 kcal/mol at pH 6.5). A mutant with tryptophan substituted for Ile 3 (Ile 3 → Trp) was selected as it is one of the least stable lysozymes characterized to date ($\Delta\Delta G$ = -2.8 kcal/mol at pH 6.5). The third mutant selected was one in which Ile 3 was replaced with serine (Ile 3 → Ser), as its structural stability falls near the middle of the range bounded by wild type and Ile 3 → Trp ($\Delta\Delta G$ = -1.2 kcal/mol at pH 6.5).

Charge Mutants. Five positively charged surface residues of this molecule have been individually and collectively replaced with glutamic acid, yielding mutants with formal charges ranging from +1 to +9 units (26). Individual substitutions were made for the lysine at position 16 (Lys 16), the arginine at position 119 (Arg 119), Lys 135, Lys 147, and Arg 154. High resolution X-ray analysis of these five molecular structures showed each to be very similar to the wild type. Eight additional mutants were produced, in which two, three or four replacements were made at the sites indicated above. Small differences in stability among the variants were attributed to local interactions at the site of the substitution: these workers stated there was no suggestion that the substitutions altered the stability of the folded relative to the unfolded form. With reference to their catalytic activity, like the stability mutants of the same molecule described by Matsumura et al. (25), each of the charge mutants produced were fully functional lysozymes.

In addition to the wild type lysozyme, four charge mutants have been studied thus far, each involving substitution of a lysine with a glutamic acid residue (Lys → Glu). Two mutants were used where the substitutions were made only on the C-terminal lobe: the single-point mutant Lys 135 → Glu, and the double-point mutant Lys 135 → Glu + Lys 147 → Glu. These mutants will be abbreviated 135 and 135-147, respectively. Another mutant involved a substitution on only the N-terminal lobe: Lys 16 → Glu, or simply 16. The fourth mutant involved a substitution at both the C-terminal and N-terminal lobes: Lys 135 → Glu + Lys 16 → Glu, or 135-16. Inspection of the α-carbon backbone shows that all of the substitutions relevant to the present study were made on the "back" of the molecule, defined here as the opposite face of the molecule from that presented in Figure 1.

Experimental Methods

Protein Isolation. Synthetic mutants of T4 lysozyme are produced from transformed cultures of *E. coli* strain RR1. Individual bacterium strains, containing the desired mutant lysozyme expression vectors are kindly provided by Professor Brian Matthews and co-workers at the Institute of Molecular Biology, University of Oregon, Eugene. Expression and purification of the mutant lysozymes from this point are performed at Oregon State University, Corvallis, and generally follow established procedures as described with some detail earlier (McGuire et al., *J. Colloid Interface Sci.*, in press). In summary, cells bearing a desired expression vector, which carries an ampicillin resistant gene, are grown overnight in media containing ampicillin. This is allowed to ferment (total volume about 5 L) for further growth at 35°C. The temperature is then lowered to 30°C and lysozyme expression induced by addition of isopropyl-β-thiogalactoside. Fermentation continues for an additional period after which time the cells are harvested and centrifuged. From this point, all purification procedures are performed at 4°C. Mutant proteins are purified from both the pellet and supernatant fractions. The pellets are combined, resuspended then centrifuged again, and this supernatant combined with that from the original centrifugation. This is dialyzed against about 4 L of deionized, distilled water, until its conductivity is between 2 and 3 μmho/cm for stability mutants, or below 2 μmho/cm for charge mutants. The pH is adjusted and the solution loaded onto a CM Sepharose ion exchange column; lysozyme is eluted with a salt gradient from 0.05 M to 0.30 M NaCl. Eluted fractions are combined and dialyzed against 50 mM sodium phosphate buffer, pH 5.8, containing 0.02% sodium azide, then concentrated in a SP Sephadex column. Mutant proteins are eluted with 0.10 M sodium phosphate, pH 6.5, containing 0.55 M NaCl and 0.02% azide. The yield of lysozyme is usually between about 40 and 150 mg, in concentrations anywhere from about 20 to 100 mg/mL. Preparations are stored without further treatment at 4°C. On the average, SDS-gel electrophoresis shows the isolated proteins to be over 95% pure.

Surface Preparation, Adsorption and Elution. Preparation of hydrophilic silica and silanized, hydrophobic silica surfaces, and *in situ* monitoring of adsorption and dodecyltrimethyl-ammonium bromide (DTAB)-mediated elution of each protein with ellipsometry, have been described earlier (McGuire et al., *J. Colloid Interface Sci.*, in press). For convenience, the experimental procedure used to monitor adsorption and elution with ellipsometry is restated here. Experiments were performed in 0.01 M sodium phosphate buffer, pH 7.0. Solutions were stirred with a magnetic stirrer at 325 rpm, and all tests were conducted at 25°C. The pseudo-refractive index of the bare surface was determined prior to addition of protein. A plot illustrating the course of a typical adsorption-elution experiment is shown in Figure 2. An experiment began with addition of 0.5 mL of protein solution to the cuvette containing 4.5 mL of buffer, prepared to yield a final protein concentration of 1.0 mg/mL. Adsorption was monitored for 30 min. The cuvette was then rinsed with buffer for 5 min at a flow rate of 20 mL/min, and film properties were monitored for an additional 25 min. This was followed by addition of 0.5 mL of DTAB (Sigma Chemical Co.), prepared to yield a final solution concentration in the cuvette of 0.03 M. After 15 min the cuvette was rinsed for 5 min and the sample monitored an additional 25 min as before. Each experiment was performed at least twice, with an average deviation from the mean of about 0.02 mg/m^2 for stability mutants, and about 0.03 mg/m^2 for charge mutants.

Other Methods. Kinetic data have been monitored over a longer period for wild type, Ile 3 → Cys (S-S) and Ile 3 → Trp with ellipsometry (Singla et al., Oregon State University, unpublished data). In those experiments, silica surfaces were silanized to exhibit a low or high hydrophobicity by reaction with 0.01 or 0.10% dichlorodimethylsilane in xylene. Adsorbed mass was monitored for 8 h under static

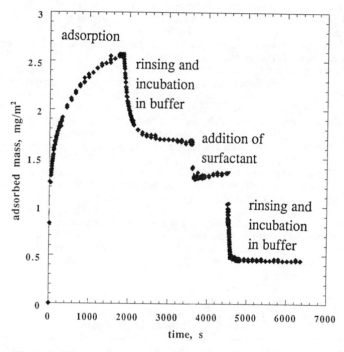

Figure 2. The pattern of a typical adsorption-elution experiment.

conditions, i.e., no stirring and no flow. The change in secondary structure exhibited by the same proteins upon adsorption to silica nanoparticles (particle size about 9 nm) has been measured as well, using circular dichroism (Billsten et al., Linköping Institute of Technology, Sweden, unpublished data). In each case experiments were performed in 0.01 M sodium phosphate buffer, pH 7.0.

Results and Discussion

Results obtained for the adsorption-elution experiments at hydrophilic and hydrophobic silica surfaces for both the stability mutants and the charge mutants were discussed earlier (McGuire et al., *J. Colloid Interface Sci.*, in press). The goal here is to propose adsorption mechanistic features apparently shared by all of the variants, then to support that proposal with reference to what we can learn from simple kinetic models for adsorption from single-component solutions. Although data recorded for each of the lysozyme variants can be interpreted in a manner consistent with the following hypotheses, here we will specifically refer only to wild type, Ile 3 → Cys (S-S), Ile 3 → Trp, 135, and 135-147. Adsorption of the stability mutant Ile 3 → Ser, and of charge mutants involving substitution in the N-terminal lobe (16 and 135-16), is apparently accompanied by molecular associations that complicate the simple analysis to be applied to the five former proteins. Adsorption of 16 was apparently affected greatly by intermolecular associations inhibiting the general process experienced by wild type, 135, and 135-147, while adsorption of 135-16 and of Ile 3 → Ser is thought to occur with associated monomers as the dominant surface active species.

Table I shows, for each adsorption-elution test, the average values of mass of protein remaining on the surface after 60 min, in addition to the average amount remaining following contact with DTAB, rinsing and incubation in buffer, i.e., after 105 min. The average resistance to elution, also tabulated, was calculated as the mass of protein remaining after 105 min divided by that remaining after 60 min. Considering "end-on" adsorption of a T4 lysozyme molecule to require about 784 Å^2 (28 x 28 Å), and "side-on" adsorption to require about 1512 Å^2 (28 x 54 Å), one can estimate limits in surface coverage expected for a monolayer of molecules adsorbed end-on and side-on as 3.96 and 2.05 mg/m^2, respectively. The amounts of protein remaining on the surface after rinsing and incubation in buffer (i.e., after 60 min) would therefore correspond to monolayer or sub-monolayer coverages in each case.

Table I. The Mass of Each Protein at Selected Times in the
Adsorption-Elution Tests, and the Fraction of Protein that
Resisted Elution in Each Case

Protein	Surface	Adsorbed Mass (mg/m^2)		Resistance to Elution
		60 min	105 min	
wild type	hydrophilic	2.38	0.84	0.35
	hydrophobic	2.94	0.65	0.22
Ile 3 → Cys	hydrophilic	1.63	0.45	0.28
	hydrophobic	1.32	0.22	0.17
Ile 3 → Trp	hydrophilic	1.90	1.11	0.58
	hydrophobic	2.16	0.62	0.29
135	hydrophilic	2.24	0.96	0.43
	hydrophobic	2.24	0.73	0.33
135-147	hydrophilic	2.14	0.98	0.46
	hydrophobic	2.35	0.70	0.30

SOURCE: Adapted from McGuire et al., *J. Colloid Interface Sci.*, in press.

T4 Lysozyme Behavior at Hydrophilic Silica Surfaces. Nearly all of the out-of-balance charge on these molecules resides in the C-terminal lobe, and it is probably fair to hypothesize that the C-terminal lobe would first be oriented toward the negatively-charged, hydrophilic surface during adsorption. Molecular rearrangement may take place whereby the most mobile regions of positive charge are brought near the interface. The mobility of a residue can be estimated from the refined X-ray crystal structure of the molecule, by calculating its average thermal displacement. Alber et al. (27) calculated the average thermal displacement separately for each of the side chain and main chain atoms in the wild type molecule. This property of an atom is quantified by the crystallographic thermal factor, B ($Å^2$), where B is related to the unidirectional mean square thermal displacement (27). Considering only the side chains of each residue in T4 lysozyme, the average value of B (calculated as the mean of the individual B values of each atom of the side chain) is 26.5 $Å^2$. Generally, only side chains of low mobility and low solvent accessibility contribute substantially to the thermal stability of a protein (27). A residue side chain characterized by a B value greater than 40 $Å^2$ is considered "very mobile" (26,27).

So each charged group is not equally capable of interacting with a surface, as the mobility of some may be inhibited by interactions with side chains of opposite charge, with helix dipoles, or involved in hydrogen bonding interactions. Using data found in (27), a total of 18 positively-charged groups and eight negatively-charged groups can be identified as solvent exposed and very mobile (these include side chains of residues 16, 135 and 147). These are distributed on the molecule such that the front of the C-terminal and N-terminal lobes has an excess of very mobile, positively charged side chains, while there is an excess of very mobile, negative charges on the back of the C-terminal domain for 135 and 135-147. Again, the front and back of the molecule are defined as shown in Figure 1.

Thus, molecular features that would facilitate rearrangement or orientation of adsorbed molecules such that mobile, positively-charged side chains are brought to the hydrophilic silica should enable that molecule to bind relatively tightly to the surface. Concerning facilitated rearrangement, this is consistent with the high resistance to DTAB-mediated elution shown by the least stable variant, Ile 3 → Trp, relative to wild type, and the low resistance shown by Ile 3 → Cys (S-S). Concerning CD spectra recorded before and after adsorption to colloidal silica, changes in spectra for Ile 3 → Trp corresponded to a 29% loss in α-helix content upon adsorption, while the loss in α-helix was calculated to be 13 and 9% for wild type and Ile 3 → Cys (S-S), respectively (Billsten et al., Linköping Institute of Technology, Sweden, unpublished data).

Orientation of the face of the C-terminal lobe toward the silica surface would orient the back of the C-terminal lobe to solution. The greater resistance to elution exhibited by 135 and 135-147 would be consistent with each of these protein molecules being more readily anchored to the surface in this favorable orientation, presumably associated with a greater number of noncovalent bonds with the surface. Such rearrangement or orientation may also facilitate formation of a less tightly bound second layer, expected to more readily undergo exchange reactions with protein in solution, and be more readily removed by rinsing.

T4 Lysozyme Behavior at Hydrophobic Surfaces. At hydrophobic surfaces, we may expect adsorption to be mediated first by the N-terminal lobe, as it has approximately the same numbers of positively- and negatively-charged residues at pH 7. Molecular graphic depiction of space-filled models of hen and human lysozymes have shown these molecules to possess a hydrophobic patch, opposite the active-site cleft (6). Similar techniques applied to T4 lysozyme show it can be considered as possessing a hydrophobic patch opposite the active site cleft, surrounding the region of Ile 3. This patch may serve to facilitate packing of molecules if in an end-on orientation, or side-on, hydrophobic association with the surface. Molecular features that would favor either local rearrangement toward optimization of such association, or simply orientation of the hydrophobic patch to the interface, should enable that molecule to bind relatively tightly to the surface. Again, concerning facilitated rearrangement, this is consistent with the high resistance to DTAB-mediated elution shown by the least stable variant, Ile 3 → Trp, relative to wild type, and the low resistance shown by Ile 3 → Cys (S-S). And, as the back of the C-terminal lobe is made increasingly negative, orientation of the molecule with its hydrophobic patch toward the surface would allow association between the back of the C-terminal lobe of one molecule and the front of a neighboring molecule; indeed, the definition of front and back used for the purpose of this discussion defines the region of the patch as one "side" of the molecule, such that hydrophobic association with the surface would avail the front and back to interaction with neighboring molecules. Thus relative to wild type we would expect 135 and 135-147 to rapidly attain a more or less side-on orientation at the hydrophobic surface, with closest packing in that orientation being predicted for 135-147, and in any event less electrostatic repulsion experienced among neighboring 135-147 molecules. This was consistent with the behavior observed prior to DTAB addition, as it is with the high resistance to elution observed for the mutants relative to wild type (McGuire et al., *J. Colloid Interface Sci.*, in press). Formation of a second layer can be fairly expected at hydrophobic surfaces as well, but molecular rearrangements allowing for a strong hydrophobic association with the surface may render regions of the molecule exposed to solution more positively-charged and hydrophilic than their counterparts at hydrophilic surfaces, inhibiting formation of an outer layer. In any event, greater amounts of each protein were removable by rinsing at hydrophilic as opposed to hydrophobic surfaces (McGuire et al., *J. Colloid Interface Sci.*, in press).

Interpretation with Reference to Kinetic Models. With reference to their model for bulk-surface exchange reactions at an interface involving two different types of protein, Lundström and Elwing (*28*) considered an experimental situation in which adsorption to a solid surface is allowed to occur from a single-component protein solution, followed by incubation in buffer, after which time a second, dissimilar protein is added. They showed that their model would lead to an experimentally verifiable expression relating the fraction of originally adsorbed protein that is nonexchangeable to rate constants governing conversion of the originally adsorbed protein to an irreversibly adsorbed form, and exchange of adsorbed protein by the dissimilar protein introduced to the solution. Krisdhasima et al. (*29*) made use of that development by adapting it to the sodium dodecylsulfate-mediated removal of selected milk proteins from silanized silica surfaces, where the tests were conducted in a manner similar to the present experiments. Considering adsorbed protein to exist in either a removable or nonremovable state, where θ_1 is the fractional surface coverage of removable protein, and θ_2 the fractional surface coverage that is nonremovable, they showed the consequence of the original kinetic model (*28*) would be, after a sufficiently long time,

$$\theta_2/\theta_{1,ti} = 1 - [k_s/(s_1 + k_s)] \cdot \exp(-s_1 t_s), \qquad (1)$$

where t_i is the time at which protein is contacted with pure buffer and t_s the time of surfactant addition, k_s is a concentration-independent rate constant for surfactant-mediated removal of protein, and s_1 the rate constant governing conversion from the removable to the nonremovable state. The left side of equation 1 is a measure of adsorbed protein binding strength, and increases with increasing value of s_1 associated with the adsorption.

Concerning the stability mutants, one could assume that k_s is independent of protein type, i.e., adsorbed mutants would differ only in their relative populations of removable and nonremovable forms. Approximating the left side of equation 1 as resistance to elution (given in Table I), differences in binding strength among the proteins could be considered a function of s_1 alone, in which case s_1 would correlate well with $\Delta\Delta G$ at each surface. Thus, we would conclude that less stable proteins more readily make the conversion from a removable to a nonremovable form. We would reach the same conclusion using a mechanism consistent with that described in (28), but for irreversible adsorption of protein from single-component solutions. Such a mechanism has been used by Krisdhasima et al. (29,30), and is shown in Figure 3a. In step 1, corresponding to short contact time, the protein molecule reversibly adsorbs to the surface, with its adopted surface conformation closely approximating its native form. In step 2, a conformational change takes place in which the reversibly adsorbed molecule is changed to an irreversibly adsorbed form. Solving equations describing the time-dependent fractional surface coverage of protein in each of the two states, one reversibly adsorbed (θ_1) and one irreversibly adsorbed (θ_2), yielded an expression for total surface coverage (θ) as a function of time:

$$\theta = \theta_1 + \theta_2 = A_1 \exp(-r_1 t) + A_2 \exp(-r_2 t) + A_3, \tag{2}$$

where A_1, A_2 and A_3 are constants, the roots (r_1 and r_2) are known functions of the three rate constants defined in Figure 3a, and t is time. An expression for total adsorbed mass as a function of time can be obtained from equation 2 as:

$$\Gamma = a_1 \exp(-r_1 t) + a_2 \exp(-r_2 t) + a_3. \tag{3}$$

The parameters a_1, a_2 and a_3 are the products of Γ_{max}, the "equilibrium" adsorbed mass, with A_1, A_2 and A_3, respectively.

Nonlinear regression performed on adsorption kinetic data fit to equation 3 would yield estimates of the parameters a_1, a_2 and a_3, r_1 and r_2. Parameters r_1 and r_2 are known functions of k_1, k_{-1} and s_1. In particular (30),

$$(r_1 + r_2) = (k_1 C + k_{-1} + s_1) \tag{4}$$

and

$$r_1 r_2 = s_1 k_1 C, \tag{5}$$

where C is protein concentration. Assuming k_1 and k_{-1} are similar among the T4 lysozyme stability mutants, kinetic experiments for any two stability mutants allow equations 4 and 5 to be written for each kinetic curve yielding four equations and four unknown variables, since k_1 and k_{-1} are assumed not to change among these mutants, with only s_1 being affected by stability. This analysis was performed with each pair permutation allowable for the three stability variants wild type, Ile 3 → Cys (S-S) and Ile 3 → Trp. Values of rate constant s_1 for each mutant relative to that for wild type are shown in Table II, and agree fairly well with both $\Delta\Delta G$ and the elution data of Table I.

Table II. The s_1 Value for Each Stability Mutant Relative to that of the Wild Type

Mutant	$\Delta\Delta G$ (kcal/mol)	Surface	$\dfrac{s_{1,mutant}}{s_{1,wild\ type}}$
Ile 3 → Cys (S-S)	+1.2	hydrophobic	0.93
		hydrophilic	0.64
Ile 3 → Trp	- 2.8	hydrophobic	1.63
		hydrophilic	2.19

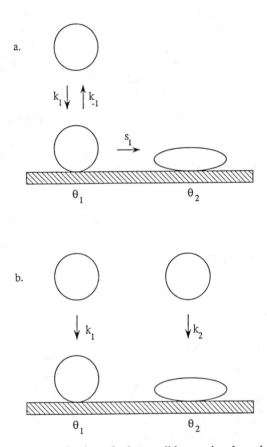

Figure 3. Two simple mechanisms for irreversible protein adsorption from single-component solutions: (a) initial, reversible adsorption into one state followed by conversion to an irreversibly adsorbed form (adapted from ref. 30); and (b) adsorption into one of two states exhibiting different resistances to elution and occupying different interfacial areas (adapted from McGuire et al., *J. Colloid Interface Sci.*, in press).

The developments leading to equations 1-5 assumed that generation of irreversibly adsorbed molecules occurs only after adsorption in the removable form, that s_1 is much smaller than the adsorption rate constant, and that the conversion involves no change in interfacial area occupied by the molecule (28-30). But results to-date indicate that adoption of a more tightly bound state may occur rather quickly, and molecules in that state may occupy a greater interfacial area. A more suitable mechanism (for adsorption of T4 lysozyme monomers) is shown in Figure 3b. In this case, protein may again adopt one of two states, where state 1 molecules are less tightly bound (less resistant to elution) than those in state 2, but where molecules are considered partially elutable in either state. Additionally, a state 2 molecule is defined as occupying a greater interfacial area (A_2) than it would occupy in state 1 (A_1). Although drawn to depict the possibility of attainment of state 2 immediately upon adsorption, the parallel reaction characterized by rate constant k_2 should be taken only as a convenient means to account for the apparent observation that some population of adsorbing molecules rapidly attains a more tightly bound state.

It is most instructive here to interpret the elutability data in terms of certain conditions implicit in the mechanism of Figure 3b. We will define the maximum mass of molecules that could be adsorbed in a monolayer as Γ_{max} (mg/m²). We will define θ_1 as the mass of state 1 molecules adsorbed at any time (mg/m²) divided by Γ_{max}, and θ_2 as the mass of state 2 molecules adsorbed at any time divided by Γ_{max}. For cases where the surface can be considered as covered following incubation in buffer, and in the absence of desorption at that time, the definitions of θ_1 and θ_2 require that

$$\theta_1 + a\theta_2 = 1 \tag{6}$$
and
$$\Gamma_{60} = \Gamma_{max}(\theta_1 + \theta_2), \tag{7}$$

where a is A_2/A_1, and Γ_{60} (mg/m²) is the value of adsorbed mass recorded after 60 min (Table I). Equations 6 and 7 allow expressions for θ_1 and θ_2 to be written as follows:

$$\theta_1 = 1 - a(\Gamma_{60} - \Gamma_{max})/[\Gamma_{max}(1 - a)] \tag{8}$$
and
$$\theta_2 = (1 - \theta_1)/a. \tag{9}$$

Γ_{max} can be approximated as 3.96 mg/m², corresponding to a monolayer of molecules adsorbed end-on. Equations 8 and 9 state that knowledge of A_1 and A_2 for adsorption to a given surface, along with Γ_{60} recorded for any protein, would allow estimation of the mass of protein adsorbed in each of the states 1 and 2. In particular, θ_1 and θ_2 are calculable, and can be used to estimate the fraction of protein that was adsorbed in state 2 immediately prior to DTAB addition in each test:

$$\Gamma_{60, \text{ state } 2}/\Gamma_{60} = \theta_2/(\theta_1 + \theta_2). \tag{10}$$

Table III shows the results of this analysis. At the hydrophilic surface, parameter a was set equal to 2.08 (= 3.96/1.90), where 1/1.90 was taken as the specific interfacial area occupied by side-on molecules, based on the data for Ile 3 → Trp, and 1/3.96 the specific interfacial area occupied by end-on molecules. At the hydrophobic surface, A_1 and A_2 were approximated as the specific interfacial areas occupied by end-on and side-on molecules, respectively, or a = 3.96/2.05 = 1.93.

Stability Mutants. As the surface coverage of Ile 3 → Cys (S-S) was below that corresponding to a monolayer in any orientation on either surface, equations 8 and 9 could not be used, and the different populations were estimated as follows.

Considering data for wild type and Ile 3 → Trp shown in Tables I and III, and defining x_1 and x_2 as the fractions of state 1 and state 2 molecules that resist surfactant

Table III. The Fraction of Protein at Each Surface Adsorbed in State 2 Immediately Preceding Contact with DTAB

| Protein | $\Gamma_{60,state\ 2}/\Gamma_{60}$ | |
	Hydrophilic	Hydrophobic
wild type	0.61	0.37
Ile 3 → Cys (S-S)	0.48	0
Ile 3 → Trp	1	0.90
135	0.71	0.83
135-147	0.79	0.74

SOURCE: Adapted from McGuire et al., *J. Colloid Interface Sci.*, in press.

elution (i.e., the fraction of each remaining after 105 min), for the hydrophobic surface we can write $1.84\ x_1 + 1.10\ x_2 = 0.65$ and $0.22\ x_1 + 1.94\ x_2 = 0.62$. Solution of this pair of equations would set $x_1 = 0.17$ and $x_2 = 0.30$. At the hydrophilic surface, by inspection $x_1 = 0$ and $x_2 = 0.58$. As expected, $x_2 > x_1$ in each case. At the hydrophobic surface, the fraction of Ile 3 → Cys (S-S) that resisted elution was 0.17, and as that number is equal to x_1, all of the molecules were assumed to be adsorbed in state 1. At the hydrophilic surface, taking $x_1 = 0$ and $x_2 = 0.58$ yields the result shown for Ile 3 → Cys (S-S) in Table III. Thus, resistance to elution correlates well with the fraction of molecules present on the surface that were in state 2 at the time of surfactant addition, while each correlates well with $\Delta\Delta G$.

Charge Mutants. For the hydrophobic surface, performing the calculation with each of the three allowable pair permutations for wild type, 135 and 135-147, $x_1 = 0.12$ (standard deviation, s, = 0.04) and $x_2 = 0.37$ (s = 0.02). At the hydrophilic surface $x_1 = 0$ and $x_2 = 0.59$ (s = 0.02). At the hydrophobic surface, although the mechanism of adsorption has been described as generally similar among both sets of mutants, we have suggested here that charge substitutions involving the back of the C-terminal lobe allow association between the back of the C-terminal lobe of one molecule and the front of a neighboring molecule, concomitant with hydrophobic association with the surface. This may explain in part a higher value of x_2 estimated for charge mutants as opposed to stability mutants. Excellent agreement was obtained with regard to both x_1 and x_2 for each set of mutants at hydrophilic surfaces, consistent with the thought that the process of orienting the regions of most mobile positive charge to the interface plays a large role in the adsorption process, while intermolecular associations are of minor significance for either set of mutants. In other words, data for the wild type protein, the stability mutants Ile 3 → Trp and Ile 3 → Cys (S-S), and the charge mutants 135 and 135-147 allow each to be considered as adopting one of two states upon adsorption, where state 1 molecules are completely elutable by DTAB while about 58% of state 2 molecules resist elution; i.e., only the populations of molecules adsorbed in each of the two states differed among those five proteins, according to the model.

Acknowledgments

We are indebted to Professor Brian Matthews and co-workers, particularly Sheila Snow and Joan Wozniak, of the Institute of Molecular Biology, University of Oregon, for providing the bacterium strains in addition to the technical support needed for successful lysozyme production. We are grateful to Nilobon Podhipleux and Brijesh Singla of the Department of Bioresource Engineering, Oregon State University, for their role in isolation of the lysozymes selected for these tests. This material is based in part upon work supported by the National Science Foundation under Grant CTS-9216798, The Whitaker Foundation, and the Swedish Research Council for Engineering Sciences (TFR). Support from the Swedish Natural Science Research Council (NFR) is also gratefully acknowledged.

Literature Cited

1. Arai, T.; Norde, W. *Colloid Surf.* **1990**, *51*, 1.
2. Arai, T.; Norde, W. *Colloid Surf.* **1990**, *51*, 17.
3. Shirahama, H.; Lyklema, J.; Norde, W. *J. Colloid Interface Sci.* **1990**, *139*, 177.
4. Wei, A.P.; Herron, J.N.; Andrade, J.D. In *From Clone to Clinic*; Crommelin, D.J.A.; Schellekens, H., Eds.; Kluwer Academic Publishers: Amsterdam, 1991; pp. 305-313.
5. Elbaum, D.; Harrington, J.; Roth, E.F.Jr.; Nagel, R.L. *Biochim. Biophys. Acta.* **1976**, *427*, 57.
6. Horsley, D.; Herron, J.; Hlady, V.; Andrade, J.D. In *Proteins at Interfaces: Physicochemical and Biochemical Studies*; Brash, J.L.; Horbett, T.A., Eds.; Symposium Series No. 343; American Chemical Society: Washington, D.C., 1987; pp. 290-305.
7. Xu, S.; Damodaran, S. *J. Colloid Interface Sci.* **1993**, *159*, 124.
8. Kato, A.; Yutani, K. *Protein Engineering.* **1988**, *2*, 153.
9. Asakura, T.; Ohnishi, T.; Friedman, S.; Schwartz, E. *Proc. Nat. Acad. Sci. USA*. **1974**, *71*, 1594.
10. Adachi, K.; Asakura, T. *Biochemistry.* **1974**, *13*, 4976.
11. Bohnert, J.L.; Horbett, T.A. *J. Colloid Interface Sci.* **1986**, *111*, 363.
12. Rapoza, R.J.; Horbett, T.A. *J. Biomater. Sci. Polym. Ed.* **1989**, *1*, 99.
13. Rapoza, R.J.; Horbett, T.A. *J. Colloid Interface Sci.* **1990**, *136*, 480.
14. Rapoza, R.J.; Horbett, T.A. *J. Biomed. Mater. Res.* **1990**, *24*, 1263.
15. Ertel, S.I.; Ratner, B.D.; Horbett, T.A. *J. Colloid Interface Sci.* **1991**, *147*, 433.
16. Wahlgren, M.C.; Arnebrant, T. *J. Colloid Interface Sci.* **1991**, *142*, 503.
17. Wahlgren, M.C.; Arnebrant, T. *J. Colloid Interface Sci.* **1992**, *148*, 201.
18. Wahlgren, M.C.; Paulsson, M.A.; Arnebrant, T. *Colloids Surf. A: Physicochem. Eng. Aspects.* **1993**, *70*, 139.
19. Wahlgren, M.C.; Arnebrant, T.; Askendal, A.; Welin-Klintström, S. *Colloids Surf. A: Physicochem. Eng. Aspects.* **1993**, *70*, 151.
20. Alber, T.; Matthews, B.W. *Meth. Enzymol.* **1987**, *154*, 511.
21. Weaver, L.H.; Matthews, B.W. *J. Mol. Biol.* **1987**, *193*, 189.
22. Grütter, M.G.; Matthews, B.W. *J. Mol. Biol.* **1982**, *154*, 525.
23. Matthews, B.W.; Dahlquist, F.W.; Maynard, A.Y. *J. Mol. Biol.* **1973**, *78*, 575.
24. Matsumura, M.; Matthews, B.W. *Science.* **1989**, *243*, 792.
25. Matsumura, M.; Becktel, W.J.; Matthews, B.W. *Nature.* **1988**, *334*, 406.
26. Dao-pin, S.; Söderlind, E.; Baase, W.A.; Wozniak, J.A.; Sauer, U.; Matthews, B.W. *J. Mol. Biol.* **1991**, *221*, 873.
27. Alber, T.; Dao-pin, S.; Nye, J.A.; Muchmore, D.C.; Matthews, B.W. *Biochemistry.* **1987**, *26*, 3754.
28. Lundström, I.; Elwing, H. *J. Colloid Interface Sci.* **1990**, *136*, 68.
29. Krisdhasima, V.; Vinaraphong, P.; McGuire, J. *J. Colloid Interface Sci.* **1993**, *161*, 325.
30. Krisdhasima, V.; McGuire, J.; Sproull, R. *J. Colloid Interface Sci.* **1992**, *154*, 337.

RECEIVED December 22, 1994

Chapter 5

Structure and Adsorption Properties of Fibrinogen

Li Feng and Joseph D. Andrade

Department of Bioengineering, University of Utah, Salt Lake City, UT 84112

This paper correlates molecular structure of fibrinogen to its adsorption properties at the solid-water interface. These properties play an integral role in helping fibrinogen fulfill its biological functions. We first introduce some unique surface (interfacial) attributes of fibrinogen by comparing it to other plasma proteins. We consider its concentration on solid surfaces, effects of solid substrates, adsorption kinetics and isotherms, competitive adsorption, structural adaptivity, and platelet-binding ability. We then examine its structural organization by utilizing available data from the primary structure, immunochemical properties, structural stability, solubility, microscopic images, X-ray diffraction data, and information on proteolytic fragments. By analyzing the amino acid sequences, we estimate the flexibility and hydropathy along individual polypeptide chains. We inspect the net charge and hydrophobicity of individual structural domains. We pay special attention to the less addressed Aa chain, especially to its contribution to fibrinogen adsorptivity. While discussing its characteristics, we try to correlate the structure of fibrinogen to its properties.

Fibrinogen plays an indispensable role in hemostasis of vertebrate animals: blood coagulation (conversion of fibrinogen to fibrin through proteolysis of thrombin and polymerization of fibrin monomers into fibrin clots) and platelet aggregation (binding to platelets and linking them together or immobilizing them on a surface) (1-4). These two processes usually happen simultaneously, resulting in a platelet-fibrin plug (hemostatic thrombus) (5).

Besides interacting with solution proteins (thrombin, plasmin), fibrinogen has other interfacial interactions: binding to cells (platelets, endothelium (6) and leukocyte (7)) and associating with non-biological surfaces. Because of its cell adhering propensity, fibrinogen is categorized as a cell adhesive protein, together with fibronectin, vitronectin, and von Willebrand Factor (vWF) (6,8-10), all containing the Arg-Gly-Asp (RGD) sequence (11-13). Fibrinogen is a "sticky" protein, having a strong tendency to adsorb onto various surfaces. These properties earn fibrinogen the reputation of causing thrombogenesis on biomaterials.

0097-6156/95/0602-0066$12.00/0

The origin of the surface activity of fibrinogen has not been thoroughly studied. Since it is important for our understanding of blood compatibility, we wish to explore these areas: How is the surface activity of fibrinogen related to its molecular structure? How does its structure help its platelet binding function? Is fibrinogen unique or is it similar to other surface active proteins?

Recently a large volume of data concerning structure and properties of proteins have become available. We can address the molecular structure of fibrinogen and its relation to the adsorption properties although we still lack the details of its secondary and tertiary structures. In this paper we focus on its interactions with solid surfaces, antibodies and platelets and correlate its primary and domain structures to its adsorptivity at the solid-water interface and to platelet binding induced by surface adsorption. Our analysis of amino acid sequences is based on human fibrinogen.

Adsorption Properties

Although most proteins are amphipathic and surface active, the degree of their surface activity is different. Fibrinogen shows high surface activity at solid-water interfaces, manifested from its high surface concentration, high adsorption competitivity, and persistence on most surfaces. Strong association with a solid surface seems to be a property of many of the proteins participating in blood coagulation, including Hageman factor, high molecular weight kininogen (HMWK), and plasma prekallikrein *(14)*.

High Surface Concentration. More fibrinogen is usually adsorbed than most other proteins *(15-21)*. Part of this is likely due to its high molecular weight. However, as will be discussed later, other contributing factors include its strong lateral interactions, producing close packing of adsorbed fibrinogen films and its surface activity, resulting in multilayer adsorption *(17,22,23,24)*.

Low Substrate Influence. The adsorption of fibrinogen is less affected by surface nature of materials *(21,25-28)*. Like most proteins, fibrinogen is usually adsorbed more on hydrophobic surfaces *(27)*, but the difference is smaller than in the case of most other proteins. An increase in hydrophilicity of a surface may not substantially reduce fibrinogen adsorption *(29)*. For example, fibrinogen can adsorb on hydrophilic as well as on hydrophobic surfaces in similar amounts *(21,30)*. It adsorbs more on sulfonated polyurethane, which is more hydrophilic than polyurethane *(16,31)*. Fibrinogen can bind to heparin, a negative but hydrophilic substance, as do fibronectin, vitronectin, and vWF *(32)*. It can also adhere to some protein "repelling" surfaces *(29,31,33,34)*.

Transient Adsorption Kinetics. The adsorption kinetics of fibrinogen from dilute plasma onto some hydrophilic surfaces often shows a maximum at some point, followed by decreasing surface concentration with time. Called the Vroman effect *(20,27,35)*, this phenomenon is believed to reflect a process where the adsorbed fibrinogen is gradually displaced by other even more surface active proteins on these surfaces, such as HMWK, Hageman factor, or high density lipoprotein *(36,37)*.

Maximum in Isotherms. When adsorbed from dilute plasma, the isotherm of fibrinogen sometimes has a maximum *(20,35)*. The plasma concentration at which the maximum occurs seems to be higher on more hydrophobic surfaces *(36)*. This pattern is also considered to be related to displacement of fibrinogen.

Competitive Adsorptivity. Perhaps the best way to describe the high surface activity of fibrinogen is to compare the ratio of its surface concentration to solution concentration to those of other proteins in a multi-protein adsorption process *(38-42)*. Some proteins, such as albumin, can show considerable adsorption from single protein solutions, but its surface concentration may be drastically reduced in the presence of other proteins, even trace proteins *(17,40)*. Albumin has a low affinity for many surfaces and is easily displaced by many plasma proteins, including fibrinogen *(9,43-45)*. Acid glycoprotein is another representative protein with low surface activity *(46)*. On the other hand, the adsorption of fibrinogen is much less affected by the presence of other proteins*(17)*. Fibrinogen often displaces other already adsorbed proteins *(45)*.

Structurally Labile Protein. Some parts of fibrinogen are rather compliant, as deduced from ease in surface induced denaturation *(27,39,47,48)* and low thermal denaturation temperature *(49,50)*. Conformational change is more easily revealed by means of proteolysis or antigen-antibody bonding. After local cleavage by an enzyme or bonding to an antibody, fibrinogen often undergoes a global conformational change, expressing many neo-binding sites.

Platelet Binding. Fibrinogen is very active in binding to platelets or other cells. The binding not only immobilizes platelets onto a surface or brings them together to form a gel network, it can also activate the bound cells *(51,52)*.

Structural Characteristics And Their Consequences

A Massive Molecule with Many Molecular Domains. Figure 1 presents a molecular model of human fibrinogen, adapted from several sources and based on electron microscopic images *(53-56)*, X-ray diffraction analyses of modified fibrinogens *(55,57)*, and calorimetric measurements *(49)*. A dimer of molecular weight (Mw) of 340,000 daltons, fibrinogen consists of three pairs of non-identical polypeptide chains and two pairs of oligosaccharides. The Mw of the Aα chain is 66,066 (610 residues), Bβ chain 52,779 (461 residues), γ chain 46,406 (411 residues) *(4)*, and each oligosaccharide 2,404 *(58)*. The negatively charged and hydrophilic carbohydrate moiety may contribute a repulsive force between fibrinogen or fibrin molecules, enhancing the solubility of fibrinogen and structurally setting fibrin to assemble into a normal clot *(58,59)*. They do not have a large contribution to fibrinogen clottability *(60)*. There are no x-ray diffraction data on the secondary and tertiary structures of fibrinogen. Measurements by CD, Raman spectroscopy, and FTIR indicate that the native molecule contains about 35% α-helix, 10-30% β-sheet, and 14% β-turn *(61,62)*.

The structurally distinguishable regions of fibrinogen are: a lone central E domain, two distal D domains, two α helical coiled coils, two αC domains, and a pair of junctions between them. The E domain contains all the N terminal ends of the six polypeptide chains linked by disulfide bonds. The D domain consists of the C terminal ends of the Bβ and the γ chains, and a small portion of the Aα chain. The coiled coil, made of three α-helices, connects the D domain to the E domain. It is interrupted in the middle by a small non-helical, plasmin sensitive region *(2,63)*. In the context of this paper, the αC domain contains only the C terminal third of the Aα chain (391-610). Although the middle third of the Aα domain (200-390) has no definite domainal structure *(63-66)*, we call it the αM domain (M for middle). The masses are 32,600 for the E domain, 67,200 for the D domain, 42,300 for the 2/3 αC chain (the αC and αM domains), and 39,100 for the coiled coil *(1)*.

Figure 2 shows the occurrence of the three chains in the different domains and sub-domains for a half molecule *(65,67)*, and the sites of carbohydrate attachment *(2)*. This figure helps us deduce domain properties from analysis of the polypeptide chains. There are four proteolytically splittable sub-domains in the D domain, two of which are formed by the Bβ chain and two by the γ chain *(50,57,61)*. Here the defined E or C domains are molecular domains, not equivalent to fragment E (Mw 45,000) or D (Mw 100,000), degradation products from plasmin digestion *(62)*. Fibrinogen is susceptible to plasmin cleavage, producing different sized fragments depending upon the digestion conditions *(50,68,69)*. The plasmin cleavage sites along individual chains have been illustrated *(70)*.

Structure and Properties of Individual Chains and Domains. Assuming that the adsorption properties of fibrinogen are largely determined by its somewhat independently acting domains, we utilize a domain approach to facilitate our analysis of this complex protein *(71)*.

Net charge. Figure 3 shows the net charges of individual domains at pH 7.4. The data are calculated according to the amino acid composition of each domain and the charge of the carbohydrate *(72)*. The values shown are an approximation because of the uncertainty of the charge of His and of the exact positions of charged residues *(71)*. The result of pH titration of fibrinogen gives a net charge of -7 at the pH 7 *(22)*, which is close to our calculated value: -10. Figure 3 assumes that the charge of Glu and Asp is -1, Lys and Arg +1, His +0.33 (one out of three His residues bears a unit positive charge).

The E and D domains are negative while the αC domain is positive. The D sub-domains have different net charges (Figure 3). In cases where electrostatic interactions play a major role, fibrinogen may orient its appropriate domains to interact with the surface. For example, the D (or less probably E) domains may be favored to adsorb on positively charged surfaces. Likewise the positively charged αC domain may be more favorably attracted to negative surfaces. Even though the global fibrinogen molecule has a net negative charge (-10), a negative surface may adsorb via the αC domains. HMWK and Hageman factor can effectively displace adsorbed fibrinogen from negatively charged glass *(73)*, perhaps because both have domains with much higher positive charge density. Figure 4 clearly shows the high density positively charged region along the peptide chains of HMWK and Hageman factor, compared with the Aα chain of fibrinogen. Low molecular weight kininogen does not have such a sequence and does not displace fibrinogen. Considering its size, fibrinogen has a low charge density.

Structural stability. Figure 5 shows the thermal denaturation temperatures of fibrinogen at pH 8.5 *(49)*. The E domain and coiled coils are relatively thermostable whereas the D and αC domains are thermolabile, suggesting that the D and αC domains have lower structural stability *(74)*. They are thus "soft" domains, readily changing their conformations *(75,76)*. Such softness may provide the domains with high adaptivity to maximize their contact area with a surface during adsorption *(41)*. Calculated according to Ragone scale *(77)*, Figure 6 shows that the αC domain is flexible, especially in the αM domain. The disappearance of the αC domain under electron microscopy may be due to its structural collapse in the presence of substrate. Structural adaptivity of proteins is an important parameter contributing to the adsorptivity. The D and αC domains should be more surface active than the other domains *(74,78)*.

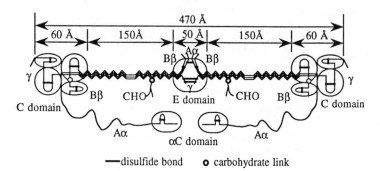

Figure 1. Molecular model of fibrinogen and its individual domains. (Fibrinogen actually exists in an antiparallel arrangement. The shown parallel structure is just for convenience).

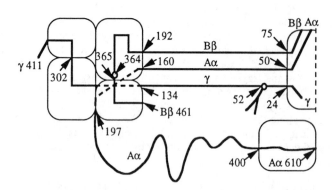

Figure 2. Polypeptide segments in individual domains of human fibrinogen.

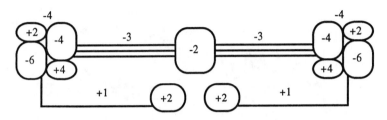

Figure 3. Net charges on individual domains and sub-domains.

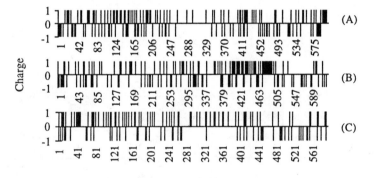

Amino acid sequence

Figure 4. Charge distributions along three polypeptides of human fibrinogen Aα chain (A), human HMWK(B), and human Hageman factor (C). The charge of His is shown as +1. HMWK and Hageman factor have very dense positive charge between 400 to 510 and 290 to 400, respectively.

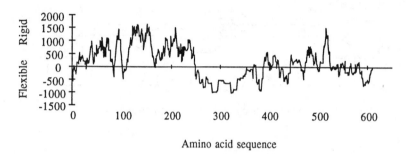

Figure 5. Thermal denaturation temperatures of fibrinogen after Privalov *(49)*.

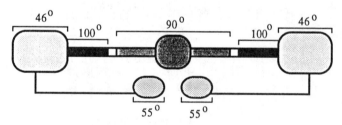

Figure 6. Flexibility of human fibrinogen Aα chain, computed using Ragone*(77)* scale.

Hydrophobicity. Figure 7 plots the hydrophathy of the 3 polypeptide chains according to the Ponnuswamy scale *(79)*. The hydrophobicity order is γ chain > Bβ chain > Aα chain. The E and D domains apparently have comparable hydrophobicity, the αC domain is slightly more hydrophilic, and the αM domain has the highest hydrophilicity. The hydrophobic region of the αC domain may provide interactions between the αC domains or with the E domain *(64)*. The hydrophobic, soft D domain should be preferentially adsorbed onto hydrophobic surfaces via hydrophobic interactions.

Immunogeneity and Surface Topography. Fibrinogen can bind to a large number of antibodies *(5)*. It is estimated that a single fibrinogen molecule can accommodate about 35 antibodies simultaneously *(80)*. It has two kinds of epitopes: exposed ones on the native protein and latent ones that do not bind to their antibodies unless the native molecule is conformationally changed *(5,51)*. The Aα chain is the most surface exposed and readily cleaved by a wide variety of proteases *(52,65,70,81-83)*. The γ chain is the least exposed *(65,84)*, hidden in the E and D domains, and in the coiled coil *(85)*. The accessibility of the three chains in native fibrinogen can be produced by analyzing their binding activity to a variety of antibodies *(5,52,81,84,85)*. The exposures of domains can be ranked as highest for the αM domain, followed by the αC domain, then the D domain, with the E domain last. The coiled coil native fibrinogen is inactive in binding to antibodies. Once fibrinogen is perturbed this region seems to express more neo-epitopes than any other domain whereas the αM domain has probably the fewest changes.

Immunochemical analysis is a sensitive tool to interrogate allosteric effects among the domains. Delicately balanced in structure, the global fibrinogen molecule has an amazing ability to respond to very local structural alterations. For example, when fibrinogen is converted to fibrin, some latent epitopes of the γ chain in the C domain become expressed. That suggests that information on conformational changes induced at the N termini of Aα and Bα chains in the E domain is transmitted to the γ chain in the C domain *(82)*. Reversibly, removal of the C terminal end of the Aα chain in the αC domain will conformationally rearrange the central domain *(86)*. These phenomena suggest that there is good communication among the domains *(65,81)*. They function cooperatively, and are not totally independent in action *(49)*. The coiled coil is likely a conduit for the transmission of allosteric changes from one domain to another *(65)*.

Platelet Binding Sites. Native fibrinogen can bind to stimulated platelets *(8)*. Conformationally perturbed fibrinogen can also bind to unstimulated platelets and activates the bound platelets, causing them to aggregate and secrete *(51,52,87)*. Fibrinogen possesses six platelet binding sites, three different pairs located at the D domains (γ 400-411), the αC domains (RGDS of Aα 572-575), and the coiled coils (RGDF of Aα 95-98), respectively *(12)*. While the RGD sequence is a common cell recognition region *(10)*, some cells, like endothelium, do not recognize γ 400-411 *(6)*. Although not available to unstimulated platelets *(12)*, the RGDF sequence of native fibrinogen can bind to ADP-stimulated platelets *(11)*. The C terminus of γ 400-411 and the RGDS sequence of immobilized fibrinogen actively participates in binding to unstimulated platelets *(12)*. Both regions also directly interact with ADP-stimulated platelets *(11)*. However, the Aα chain is only 20-25% as effective as the γ chain, and the RGD sequences of fibrinogen are thought not to be essential to platelet aggregation *(88)*. In fact, the Aα chain of fibrinogens of some species do not have the RGD sequence at all *(14)*. It has been speculated that these cell-binding sequences in a fibrinogen molecule cooperatively enhance the affinity of fibrinogen for platelets *(89)*.

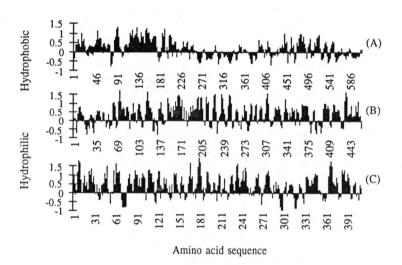

Figure 7. Hydropathy of three polypeptide chains of human fibrinogen, computed using Ponnuswamy scale *(19)*. (A) Aα, (B) Bβ and (C) γ.

We suggest the such broad distribution of these sites can ensure that fibrinogen maintain its cell binding ability for different situations. Evidence shows that the native or near native tertiary structure is essential for fibrinogen to bind to platelets *(28,51,90,91)*, and to express its polymerization sites *(92)*. The multi-binding ability allows adsorbed fibrinogen in various adsorption states to interact with platelets. Some of the platelet binding sites are likely protected from alteration by surfaces. Usually only certain domains are favored by a particular surface. Some domains are adsorbed and probably denatured, anchoring fibrinogen on the surface. Others not directly interacting with the surface have their platelets binding sites preserved so fibrinogen can interact with platelets <u>while</u> adhering to a surface. Depending on the type of surface, immobilization can be provided by a particular domain and binding by another. For instance, on glass it is likely that the αC domain adheres to the surface and the site on the D domain bind to platelets. On a hydrophobic surface the D domain more likely adsorbs to the surface, leaving the αC domain available to interact with platelets.

Platelet binding is likely faster than surface-induced denaturation. Immobilized fibrinogen may bind to and activate platelets prior to its full scale denaturation. Surfaces with microphase separation *(87)*, with balanced polar and non-polar characters *(73,93)*, or with high protein affinity *(94)* show better blood compatibility, presumably because they can interact with all the domains of fibrinogen, destroy the binding sites, and thus disabling its ability to bind to platelets.

Importance of the αC Chain (αC and αM domains). The αC domain is often near or on the E domain *(95)*, forming nodules observed by electron microscopy*(51,96)*. Studies indicate that the αC domains show a considerable homology among species *(97)*. Although the αM domains vary, the nature of their tandem repeats are conserved *(14,66)*. This overall similarity suggests that the αC chain plays an important role. Unfortunately, its biological function has not been clearly defined *(98)*. The αC chain is not considered as a structural component for the formation of a fibrin clot, but it may be a factor promoting *(66,99)*, branching *(63)*, and stabilizing *(100)* the normal assembly of fibrin. Fibrinogen with the αC chain portions deleted show retarded polymerization and lower clottability *(64)*.

The conformation of the αC chain has not been completely solved. Although the αC chain seems to have neither α-helix nor β-sheet *(2)*, it does show a compact structure *(55)*, with a defined thermal denaturation temperature *(49,64)*. The αM domain has high hydrophilicity and low intra-molecular cohesion (Figs. 7 and 6), providing high solubility, flexibility and mobility. The αC domain contains a hydrophobic (Aα 470-520) and hydrophilic (Aα 550-610) region (Figure 7).

Although less attention has been given to it *(66,72)*, more evidence is showing that the αC domain may be partly responsible for the surface activity of fibrinogen. The αC domain seems to have a strong tendency to interact with another αC domain *(72,95)*, enhancing lateral interactions between fibrinogen molecules (39,40). Electron microscopy reveals that a free 40,000 αC fragment cleaved by plasmin can bind to the αC domains of fibrin *(95)*. It is reported that the αC domains have a complementary site so that they can be strongly tied together intra-molecularly *(49,66,64,101)*, forming a super αC domain, often seen as a fourth domain by electron microscopy (57). They also bind to the corresponding regions inter-molecularly, bringing individual fibrinogen or fibrin molecules together, guiding fibrin monomers to polymerize, and regulating the structure of fibrin clots *(63)*. Covalent bonding between two αC chains from two fibrin molecules by Factor XIIIa further stabilizes fibrin polymers *(63,102)*. The αC domain can attach to the E domain, producing an apparently larger central domain (56) and blocking the antibody accessibility of the γ chain in the E domain *(65)* . Because of its uncertain position related to the bulk molecule, the αC chain prevents fibrinogen from crystallization*(55)*.

Inter-molecular association of the αC domains may contribute to fibrinogen's susceptibility to precipitation and thus its low solubility *(103)*. The interaction between the αC chains can also enhance lateral interactions among adsorbed fibrinogen molecules *(104)*. Although the αC chain is more hydrophilic than the other chains, fibrinogen actually becomes more soluble in aqueous solutions without it. The cleavage starts at the C terminus of the αC chain and gradually extends towards the N-terminus, producing modified fibrinogen with the αC chain of different lengths. The longer the αC segment chopped off, the higher the solubility of the modified fibrinogen *(105,106)*.

We suggest that the αC chain acts as a "pioneer" for adsorption. The αM domain has high flexibility and mobility, permitting high collision frequencies of the αC chain with a surface. The αC domain easily changes its conformation and adheres to a surface by either the hydrophobic or hydrophilic regions, or is attracted to negative surfaces. The cooperation of the two domains results in a highly mobile "hand". The αM domain can also locally associate with themselves or with the αM domain of another molecule. This segment possesses an accordion structure composed of a series of tight turns, a series of imperfect 13-residue repeats. Most the β-turns contain a Trp residue at the fourth position, which acts as a "meager toe-hold" for inter-chain interaction. It is proposed that the stickiness of the αC chain is due to the exposure of the "caged" Trp residues on the Aα chain *(66,72)*. Fibrinogen is seen to attach to the solid surface by one end from a solution droplet and extend in the direction of the receding edge as the droplet continues to dry, suggesting that the αC domain is the first to adhere to the surface *(96)*. Sometimes presumed dimers on hydrophilic silica may actually be single fibrinogen molecules with the αC chain stretched from the D domain *(30)*. In addition to electrostatic forces, the interaction between a hydrophilic surface and the hydrophilic αC chain may be due to the formation of hydrogen bonds between their mutual hydrophilic groups.

Based on the above discussion, we are able to make a number of predictions. The interaction between the αC domain and a surface should not be strong since the chain is neither very hydrophobic nor highly positive. Fibrinogen adsorbed on a surface via its αC domain alone can be readily displaced by other surface active proteins *(27)*, or eluted by sodium dodecyl sulfate (SDS) *(90,97)*. Only when other domains later begin to adhere to the surface does fibrinogen adsorption become increasingly strong, reducing its displacement and elutability. The presence of albumin in solution can inhibit this process and thus slows down the decrease in SDS elutability *(90,97)*. At low plasma concentration, fibrinogen has sufficient time to use all its domains interacting with surfaces. Thus the surface concentration increases with plasma concentration. When the plasma concentration is high enough, adsorption of other proteins prevents fibrinogen from contacting the surface through domains other than the αC domain. Such adsorbed fibrinogen can be easily displaced, resulting in a decrease in surface fibrinogen with increasing plasma concentration. This can explain the maximum in adsorption isotherms of fibrinogen vs. plasma concentration. Surfaces strongly interacting with the αC domain may improve the blood compatibility. For example, despite more fibrinogen adsorbed sulfonated polyurethane has a longer thrombin time *(16,107)*. In addition to the suspected heparin like behavior of the sulfonated polyurethanes, the interaction may cause a change in adsorbed fibrinogen and increase its resistance to thrombin attack.

Very large proteins tend to show high surface activity. However, mass is not necessarily the major driving force. For example, adsorbed fibrinogen (Mw 340,000) can be displaced by Hageman factor (Mw 78,000) from glass *(108)*, or by hemoglobin (Mw 64,650) from several surfaces *(36)*. What is really important is that high Mw proteins usually contain highly diverse structures and multi-domain organization. These characteristics usually offer them the necessary ingredients for high surface activity, such as structural flexibility, high charge density, hydrophobicity, etc. As we have seen, higher heterogeneity of structure can result in higher surface activity, as it provides the opportunity for proteins to optimize their interactions with different surfaces.

Postscript

Fibrinogen is a well built molecule. Its primary and domain structures dictate its physicochemical properties. It contains domains ensuring rapid, strong interactions with non-biological surfaces. Although fibrinogen is not very hydrophobic, it is surface active enough to accumulate at interfaces. Fibrinogen can cooperatively change its conformation, providing a sensitive structure which "senses" foreign surfaces and permits quick reactions. It has the ability to activate platelets through adsorption and is often only slowly inactivated by surfaces.

Acknowledgments

We thank the Center for Biopolymers at Interfaces, University of Utah, for supporting this work.

Literature Cited

1.   Doolittle, R.F. *Ann. Rev. Biochem.* **1984**, *53*, 195.
2.   Doolittle, R.F. In *The molecular basis of blood diseases, 2nd Ed.*; Stamatoyannopoulos, G., Saunders, Philadelphia, PA, 1994; pp. 701.
3.   Smith, G.F. In *The thrombin;* Machovich,R., CRC-Press, Boca Raton, FL, 1984, *Vol.1.*, Chapter 4.
4.   Shafer, J.A.; Higgins, D.L. *CRC-Crit. Rev. Clin. Lab. Sci.* **1988**, *26*, 1.
5.   Plow, E.F.; Edgington, T.S. *Seminar Thromob. Hemostasis* **1982**, *8*, 36.
6.   Cheresh, D.A.; Berliner, S.A.; Vicente, V.; Ruggeri, Z.M. *Cell* **1989**, *58*, 945.
7.   Altieri, D.C.; Plescia, J.; Plow, E.F. *J. Biol. Chem.* **1993**, *268*, 1847.
8.   Phillips, D.R.; Charo, I.F.; Parise, L.V.; Fitzgerald, L.A. *Blood* **1988**, *71*,831.
9.   Fabrizius-Homan, D.J.; Cooper, S.L. *J. Biomater. Sci., Polym. Ed.* **1991**, *3*, 27.
10.   Berliner, S.A. *Pept. Res.* **1988**, *1*, 60.
11.   Hawiger, J. *Biochemistry* **1989**, *28*, 2909.
12.   Savage, B.; Ruggeri, Z.M. *J. Biol. Chem.* **1991**, *266*, 11227.
13.   Lambrecht, L.K. *Thromb. Res.* **1986**, *41*, 99.
14.   *Blood compatibility;* Williams, D.F., Ed.; CRC, Boca Raton, FL, 1987, Vol.1., Chapter 2.
15.   Absolom, D.R.; Zingg, W.; Neumann, A.W. *J. Biomed. Mater. Res.* **1987**, *21*, 161.
16.   Santerre, J.P. *J. Biomed. Mater. Res.* **1992**, *26*, 39.
17.   Feng, L.; Andrade, J.D. *Biomaterials* **1994**, *15*, 323.
18.   Baszkin, A.; Lyman, D.J. In *Biomaterials 1980*. Winter, G.D. Eds.; Wiley, Chichester, 1982, pp. 393-397.
19.   Young, B.R.; Pitt, W.G.; Cooper, S.L. *J. Colloid Interface Sci.* **1988**, *124*, 28.
20.   Wojciechowsky, P.; ten Hove, P.; Brash, J.L. *J. Colloid Interface Sci.* **1986**, *111*, 455.
21.   Weathersby, P.K.; Horbett, T.A.; Hoffman, A.S. *J. Bioengineering* **1977**, *1*, 393.
22.   Valerio, F.; Balducci, D.; Lazzarotto, A. *Environ. Res.* **1987**, *44*, 312.

23. Lahav, J. *J. Colloid Interface Sci.* **1987**, *119*, 262.
24. Pankowsky, D.A. *J. Vasc. Surg.* **1990**, *11*, 599.
25. Schmitt, A. *J. Colloid Interface Sci.* **1983**, *92*, 145.
26. Kibing, W.; Reiner, R.H. *Chromatographica* **1978**, *11*, 83.
27. Slack, S.M.; Horbett, T.A. *J. Biomed. Mater. Res.***1992**, *26*, 1633.
28. Rapoza, R.J.; Horbett, T.A. *J. Biomed. Mater. Res.* **1990**, *24*, 1263.
29. Maechling-Strasser, C. *J. Biomed. Mater. Res.* **1989**, *24*, 1385.
30. Nygren, H.; Stenberg, M.; Karlsson, C. *J. Biomed. Mater. Res.* **1992**, *26*, 77.
31. Okkema, A.Z.; Visser, S.A.; Cooper, S.L. *J. Biomed. Mater. Res.* **1991**, *25*, 1371.
32. Mohri, H.; Ohkubo, T. *Arch. Biochem. Biophys.*, **1993**, *303*, 27.
33. Bergstrom, K. *J. Biomater. Sci., Polym. Ed.* **1992**, *3*, 375.
34. Amiji, M.; Park, K. *Biomaterials* **1992**, *13*, 682.
35. Slack, S.M.; Horbett, T.A. *J. Colloid Interface Sci.* **1933**, *124*, 535.
36. Poot, A. *J. Biomed. Mater. Res.* **1990**, *24*, 1021.
37. Brash, J.L. *Blood* **1988**, *71*, 932.
38. Kim, S.W.; Lee, R.G. In *Applied chemistry at protein interfaces*, Baier, R.E.., Ed.; ACS, Washington, DC, 1975, pp. 218-229.
39. Bagnall, R.D. *J. Biomed. Mater. Res.* **1978**, *12*, 203..
40. Lensen, H.G.W. *J. Colloid Interface Sci.* **1984**, *99*, 1.
41. Norde, W. *Adv. Colloid Interface Sci.* **1986**, *25*, 267.
42. Park, K.; Mao, F.W.; Park, H. *J. Biomed. Mater. Res.* **1991**, *25*, 407.
43. Fabrizius-Homan, D.J.; Cooper, S.L. *J. Biomed. Mater. Res.* **1991**, *25*, 953.
44. Sunny, M.C.; Sharma, C.P. *J. Biomater. Appl.* **1991**, *6*, 89.
45. Voegel, J.C.; Pefferkorn, E.; Schmitt, A. *J. Chromatogr.* **1988**, *428*, 17.
46. Feng, L.; Andrade, J.D. *Colloids and Surfaces*, **1994**, in press.
47. Katona, E.; Neumann, A.W.; Moscarello, M.A. *Biochim. Biophys. Acta* **1978**, *534*, 275.
48. Feng, L.; Andrade, J.D. *J. Biomed. Mater. Res.*, **1994**, in press.
49. Privalov, P.L. *J. Mol. Biol.* **1982**, *159*, 665.
50. Medved, L.V.; Litvinovich, S.V.; Privalov, P.L. *FEBS* **1986**, *202*, 298.
51. Shiba, E. *Am J. Physiol.* **1991**, *260*, C965.
52. Ugarova, T.P. *J. Biol. Chem..* **1993**, *268*, 21080.
53. Mosesson, M. *J. Mol. Biol.* **1981**, *153*, 695.
54. Gollwitzer, R.; Bode, W.; Karges, H.E. *Thromb. Res.* **1983**, *suppl 5*, 41.
55. Weisel, J.W. *Science* **1985**, *230*, 1388.
56. Beijbom, L. *J. Ultrastruct. Mol. Strucut. Res.* **1988**, *98*, 312.
57. Rao, S.P.S. *J. Mol. Biol.* **1991**, *222*, 89.
58. Langer, B.G. In *Fibrinogen 3: biochemistry, biological functions, gene regulation and expression*; Mosesson, M.W. , Eds., Excerpta Medica, Amsterdam, 1988, pp. 63-68.
59. Townsend, R.R. *J. Biol. Chem.* **1982**, *257*, 9704.
60. Nishibe, H.; Takahashi, N. *Biochim. Biophys. Acta* **1981**, *661*, 274.
61. Doolittle, R.F. *Protein Sic.* **1992**, *1*, 1563.
62. Azpiazu, I.; Chapman, D. *Biochim. Biophys. Acta* **1992**, *1119*, 268.
63. Weisel, J.W.; Papsun, D.M. *Thromb. Res.* **1987**, *47*, 155.
64. Medved, L.V.; Gorkun, O.V.; Privalov, P.L. *FEBS Lett.* **1983**, *160*, 291.
65. Plow, E.F.; Edgington, T.S.. Cierniewski, C.S. *Ann. N.Y. Acad. Sci.* **1983**, *408*, 44.
66. Doolittle, R.F. In *Fibrinogen, thrombosis, coagulation, and fibrinolysis*; Liu, C.Y; Chien, S., Eds.; Plenum, New York, NY, 1990, pp. 25-37.
67. Litvinovich, S.V.; Medved, L.V. *Mol. Biol.* **1988**, *22*, 744.
68. Marder, V.J.; Budzynski, A.Z. *Thrombos. Diathes. Haemorrh.* **1975**, *33*,199.

69. Nieuwenhuizen, W. In *Fibrinogen and its derivatives*. Muller-Berghaus, G. , Eds.; Excerpta Medica, Amsterdam, 1986, pp. 245-256.
70. Henschen, A. *Thromb. Res.* **1983**, *Suppl. 5*, 26.
71. Andrade, J.D. et al. *Croatica Chem. Acta* **1990**, *63*, 527.
72. Doolittle, R.F. et al. *Nature* **1979**, *280*, 464.
73. Tengvall, P. *Biomaterials* **1992**, *13*, 367.
74. Wahlgren, M.C.; Paulsson, M.A.; Arnebrant, T. *Colloid and Surfaces A: Physcochem. Eng. Aspects* **1993**, *10*, 139.
75. Arai, T.; Norde, W. *Colloids and Surfaces* **1990**, *51*, 1.
76. Wei, A.P.; Herron, J.N.; Andrade, J.D. In *From Clone to Clinic*, Crommelin, D.J.A.; Schellekens, H., Eds.; ,Kluwer, Amsterdam, 1990, pp.305-313.
77. Ragone, R. *Protein Engineering* **1989**, *2*, 479.
78. Norde, W.; Favier, J.P. *Colloids and Surfaces* **1992**, *64*, 87.
79. Ponnuswamy, P.K. *Prog. Biophys. Mol. Biol.* **1993**, *59*, 57.
80. Telford, J.N. *Proc. Natl. Acad. Sci. USA* **1980**, *77*, 2372.
81. Fair, D.S.; Edgington, T.S.; Plow, E.F. *J. Biol. Chem.* **1987**, *256*, 8018.
82. Zamarron, C.; Ginsberg, M.H.; Plow, E.F. *Thromb. Haemostats* **1990**, *64*, 41.
83. Schielen, W.J. *Blood* **1991**, *15*, 2169.
84. Tanswell, P. *Thrombos. Haemostas.* **1979**, *41*, 702.
85. Cierniewski, C.S.; Budzynski, A.Z. *J. Biol. Chem.*, **1987**, *262* 13896.
86. Cierniewski, C.S. *Thromb. Res.* **1979**, *14*, 747.
87. Elam, J.H.; Nygren, H. *Biomaterials* **1992**, *13*, 3.
88. Farrell, D.H. *Proc. Natl. Acad. Sci., USA* **1992**, *89*, 10792.
89. Mohri, H.; Ohkubo, T. *Peptides* **1993**, *14*, 353.
90. Chinn, J.A.; Ratner, B.D.; Horbett, T.A. *Biomaterials* **1992**, *13*, 322.
91. Lindon, J. *Blood* **1986**, *68*, 355.
92. Cierniewski, C.S.; Kloczewiak, M.; Budzynski, A.Z. *J. Biol. Chem.* **1986**, *261*, 9116.
93. Mori, A *Biomaterials* **1986**, *7*, 386.
94. Feng, L.; Andrade, J.D. *J. Biomat. Sic., Polym. Ed.*, **1994**, in press.
95. Veklich, Y.I. *J. Biol. Chem.* **1993**, *268*, 13577.
96. Rudee, M.L.; Price,.T.M. *Ultramicroscopy* **1981**, *7*, 193.
97. Chinn, J.A. *J. Biomed. Mater. Res.* **1991**, *25*, 535.
98. Hirschbaum, N.; Budzynski, A.Z. In *Fibrinogen 3: biochemistry, biological functions, gene regulation and expression*; Mosesson, M.W. Eds.; Excerpta Medica, Amsterdam, 1988, pp. 297-300.
99. Cierniewski, C.S.; Budzynski, A.Z. *Biochemistry* **1992**, *31*, 4248.
100. Koopman, J. *J. Clin. Invest.* **1993**, *91*, 1637.
101. Siebenlist, K.R. *Blood Coagul. Fibrinolysis* **1993**, *4*, 61.
102. Sobel, J.H. *Biochemistry* **1990**, *29*, 8907.
103. Young, B.R.; Pitt, W.G.; Cooper, S.L. *J. Colloid Interface Sci.* **1988**, *125*, 246.
104. Murthy, K.D. *Scanning. Microsc.* **1987**, *1*, 765.
105. Mosesson, M. *Biochemistry* **1966**, *5*, 2829.
106. Mosesson, M. *Ann. N.Y. Acad. Sci.* **1983**, 97.
107. Silver, J.H. *Trans. Soc. Biomat.* **1990**, *13*, 139.
108. Tengvall, P. *Biomaterials* **1992**, *13*, 367.

RECEIVED December 22, 1994

Chapter 6

Macroscopic and Microscopic Interactions Between Albumin and Hydrophilic Surfaces

C. J. van Oss[1,2], W. Wu[1,3], and R. F. Giese[3]

Departments of [1]Microbiology, [2]Chemical Engineering, and [3]Geology, State University of New York at Buffalo, Buffalo, NY 14214-3078

On a macroscopic level, proteins such as serum albumin (HSA), dissolved in water, should in theory be sufficiently repelled by clean glass or silica particles not to adsorb to them at pH values significantly higher or lower than the PI of HSA of 4.85. Yet at pH < 8, HSA does adsorb to a moderate extent to such hydrophilic surfaces. An explanation is that on a microscopic level, discrete heterogeneous sites in the hydrophilic surface that are positively charged and electron acceptors (e.g., Ca^{++}), locally attract negatively charged and electron donor moieties of the HSA surface. It can be shown via extended DLVO analysis (which includes Lewis acid-base interactions) that such moieties must be situated on HSA sites with a small radius of curvature. The adsorption of HSA to hydrophilic particles could be decreased by the admixture of Na_2EDTA. In the presence of this complexing agent, the HSA adsorption was 60% less with montmorillonite and 51% less with silica particles at pH 7.2. There was no free Ca^{++} in the particles or in the HSA solution.

The net macroscopic interaction between hydrophilic proteins and hydrophilic surfaces, such as glass or montmorillonite clay particles, immersed in aqueous media, at neutral pH, is strongly repulsive. Thus, under conditions where the macroscopic-scale rules of Lifshtz-van der Waals (LW), Lewis acid-base (AB), and electrical double layer (EL) interactions are applicable, adsorption of hydrophilic proteins onto hydrophilic mineral surfaces should not occur. However, hydrophilic proteins, dissolved in water, do adsorb onto glass [albeit more sparsely than the adsorption of hydrophilic proteins onto hydrophobic surfaces (1)], as well as onto montmorillonite surfaces (2).

The adsorption of proteins onto hydrophilic, high-energy surfaces, appears to occur through interactions between microscopic attractor sites on the high-energy surface, and the protein surface. For the microscopic attraction between such attractor sites to prevail, it must be relatively long-range in nature as well as strong enough to surmount the overall macroscopic repulsion. The most likely macroscopic attractors are plurivalent cations present in glass and montmorillonite surfaces, acting with

0097–6156/95/0602–0080$12.00/0

anionic sites on the protein. Concomitantly with the attraction caused by differences in electrostatic signs of charge, there also is a Lewis acid-base attraction between the accessible cationic sites which also function as Lewis acids, and the Lewis base sites that dominate the surface of dissolved protein.

To determine the conditions that have to be fulfilled to enable the interaction energies between discrete microscopic attractor sites to overcome the overall macroscopic repulsion field, an analysis must be made of the interplay of all the free energy elements involved in these interactions, as a function of the distance between the interacting sites and surfaces. To that effect energy versus distance diagrams must be elaborated, taking into account LW and El (3), as well as AB forces, as a function of distance, in the guise of extended DLVO diagrams (4, 5, 6)

To test the putative influence on protein adsorption by plurivalent cationic sites imbedded in the surfaces of high-energy hydrophilic mineral materials, attempts were made to block the action of such cationic sites by means of complexing agents; see below.

Theory

On a macroscopic level, at all distances ℓ between the outer surface of protein (1), and the adsorbing surface (2), immersed in water (w):

$$\Delta G_{1w2}^{TOT} = \Delta G_{1w2}^{LW} + \Delta G_{1w2}^{AB} + \Delta G_{1w2}^{EL} \tag{1}$$

However as the LW, AB and EL free energies decay as a function of distance ℓ according to different regimes, $\Delta G_{1w2}^{LW}(\ell)$, $\Delta G_{1w2}^{AB}(\ell)$ and $\Delta G_{1w2}^{EL}(\ell)$ must each be determined separately before being combined into $\Delta G_{1w2}^{TOT}(\ell)$.

The expressions for the different types of $\Delta G_{1w2}(\ell)$ are, for a sphere with a radius R and a flat surface:

$$\Delta G_{1w2}^{LW}(\ell) = -\frac{A_{1w2}R}{6\ell} \tag{2}$$

$$\Delta G_{1w2}^{AB}(\ell) = 2\pi R\lambda \Delta G_{1w2}^{AB''}(\ell_0)\exp[\frac{(\ell_0-\ell)}{\lambda}] \tag{3}$$

and

$$\Delta G_{1w2}^{EL}(\ell) = R\epsilon \psi_0^2 \ln[1+\exp(-\kappa\ell)] \tag{4}$$

where: the van der Waals, or Hamaker constant,

$$A_{1w2} = -12\pi\ell_0^2\Delta G_{1w2}^{LW''} \tag{5}$$

and the apolar (LW) energy of interaction between two flat parallel bodies, $\Delta G_{1w2}''(\ell_0)$ at the minimum equilibrium distance, $\ell_0=0.157$ nm, is:

$$\Delta G_{1w2}^{LW''} = \gamma_{12}^{LW} - \gamma_{1w}^{LW} - \gamma_{2w}^{LW} \tag{6}$$

and the apolar interfacial tension components:

$$\gamma_{ij}^{LW} = (\sqrt{\gamma_i^{LW}} - \sqrt{\gamma_j^{LW}})^2 \qquad (7)$$

The free energy of polar (AB) interaction between two flat parallel bodies, $\Delta G_{1w2}^{AB''}$, at $\ell_0 = 0.157$ nm is:

$$\Delta G_{1w2}^{AB''}(\ell_0) = -2[\gamma_w^{\oplus}(\gamma_1^{\ominus} + \gamma_2^{\ominus} - \gamma_w^{\ominus}) + \gamma_w^{\ominus}(\gamma_1^{\oplus} + \gamma_2^{\oplus} - \gamma_w^{\oplus}) - \sqrt{\gamma_1^{\oplus}\gamma_2^{\ominus}} - \sqrt{\gamma_1^{\ominus}\gamma_2^{\oplus}}] \qquad (8)$$

$\gamma_i^{\oplus}$ and $\gamma_i^{\ominus}$ respectively are the electron-acceptor and the electron-donor parameters of the surface tension of material i. γ_i^{LW}, $\gamma_i^{\oplus}$ and $\gamma_i^{\ominus}$ can be determined together, by means of contact angle measurements with a number of apolar and polar liquids (7, 8, 9); λ is the decay length of water, here taken to be 0.6 nm (6, 7); ϵ is the dielectric constant of the liquid medium (for water at 20°C, $\epsilon=80$), ψ_0 is the potential of a molecule or particle at the potential-determining surface and κ is the reciprocal thickness of the diffuse ionic double layer such that:

$$\kappa = [\frac{4\pi e^2 \sum v_i^2 n_i}{\epsilon kT}]^{\frac{1}{2}} \qquad (9)$$

where e is the charge of the electron (e=4.8 x 10^{-10} esu), v_i is the valency of each ionic species, n_i is the number of counterions per cm^3 of bulk liquid, k is Boltzmann's constant (k=1.38 x 10^{-20} J per degree Kelvin) and T is the absolute temperature in degrees Kelvin.

For the interaction between two equal spheres the right-hand sides of equations 2-4 should be halved. For cylinders of radius R and length L, interacting with a flat plate, with the cylinders' axes parallel to the plate, equations 2-4 are replaced by:

$$\Delta G_{1w2}^{LW}(\ell) = -\frac{A_{1w2}L\sqrt{R}}{12\sqrt{2}\ell^{\frac{3}{2}}} \qquad (10)$$

(11), and as a first approximation:

$$\Delta G_{1w2}^{AB}(\ell) = 2\pi\lambda\sqrt{2LR} \; \Delta G_{1w2}^{AB''}(\ell_0)\exp[\frac{(\ell_0-\ell)}{\lambda}] \qquad (11)$$

and

$$\Delta G_{1w2}^{EL}(\ell) = \epsilon\sqrt{2LR} \; \psi_0^2\ln[1+\exp(-\kappa\ell)] \qquad (12)$$

The net macroscopic interaction between water-soluble proteins and strongly electron-donating surfaces (such as glass) is repulsive (9, 10).

For the microscopic interaction between discrete sites the following data are utilized. The EL free energy of one cationic site (with one available valency) interacting with one corresponding anionic site, at close range, is about -12.5 kT (12). The AB free energy resulting from the interaction between the electron-donating protein and the glass surface (8, 9, 13), is of the order of magnitude of the energy of one hydrogen bond (e.g., in water, at 20°C), which is maximally about - 7.6 kT [c.f. the free energy of cohesive H-bonding in water of -102 mJ/M^2, which per surface area of ≈0.3 nm^2 (*14*), corresponds to -7.6 kT]. Microscopic AB interactions are taken to decay as a function of ℓ as $\exp[(\ell_0-\ell)]$ (cf. Equations 3 and 11) and EL interactions as $\ln[1+\exp(-\kappa\ell)]$ (cf. Equations. 4 and 12). The LW attraction remains unchanged.

Materials and Methods

Clean glass microscopic slides (VWR Scientific, Media, PA) were used for the estimation of the interactions between human serum albumin (HSA) and glass. Its γ_i^{LW}, $\gamma_i^{\oplus}$ and $\gamma_i^{\ominus}$ values were determined by contact angle measurements, and the ψ_0 potential was determined via electrophoretic mobility measurements on ground-up glass slides, by microelectrophoresis (*15*). The resulting surface properties of the glass thus were: $\gamma_i^{LW} = 33.7$, $\gamma_i^{\oplus} = 1.3$ and $\gamma_i^{\ominus} = 62.2$ mJ/M^2; $\psi_0 = - 59.3$ mV. HSA (Cutter, Berkeley, CA) has the following surface properties: $\gamma_i^{LW} = 26.8$, $\gamma_i^{\oplus} = 6.0$ and $\gamma_i^{\ominus} = 51.5$ mJ/M^2 [HSA with two layers of water of hydration, (*13, 16*)]; $\psi_0 = - 31.8$ mV (*17*). HSA was exhaustively dialyzed against a diluted phosphate saline buffer, at pH 7.2, ionic strength $\Gamma/2 = 0.015$.

The surface properties of montmorillonite clay particles (SWy-1, Crook County, Wyoming, from the Source Clay Minerals Repository of the Clay Minerals Society, Dept. of Geology, University of Missouri, Columbia, MO) were measured in the same manner as used for glass. For contact angle measurements clay particles were deposited in a thin flat layer on porous (0.45 μm pore-size) silver membranes (Hytrex Filter Division, Osmonics, Minnetonka, MN). Microelectrophoresis of SWy-1 particles was done by the same methods as used for the glass particles. The surface properties are: $\gamma_i^{LW} = 44.1$, $\gamma_i^{\oplus} = 0.2$ and $\gamma_i^{\ominus} = 34.1$ mJ/M^2; $\psi_0 = - 72.2$ mV.

Energy-distance diagrams, taking LW, AB and EL forces into account, were elaborated for hydrated HSA interacting with glass microscope slides, at neutral pH and $\Gamma/2 = 0.015$ (PBS/10). The dimensions and shape of the HSA molecule were taken from the recent X-ray diffraction results obtained by He and Carter (*18*). The cylindrical sections of HSA are taken to be 4.0 nm long, with R=1.3 nm. The "elbows" of the HSA molecule are taken to be (half) spherical, with R=1.3 nm. The five "elbows" of HSA are similar in shape; for the repulsive interactions the macroscopically measured properties of HSA may be reasonably taken to be the same everywhere.

The SWy-1 particles were purified by dispersal in distilled water and settling for four hours, followed by removal of the sedimented particles and storage in 0.075 M NaCl at a particle concentration of 1% (w/v). 10 ml aliqots of the stock SWy-1 suspension were diluted and washed three times with PBS/10 by centrifuging at 1,200 G for 2 minutes, and then suspended in 20 ml PBS/10 or in 20 ml PBS/10 containing 2.5, 5.0, 10.0, 20.0, 30.0, 40.0 mg Na$_2$EDTA in test tubes, respectively, with a few drops 0.3 M NaOH added to maintain the pH at 7.2 ±0.05. Adsorption measurements of HSA were done onto the above prepared SWy-1 particles, in PBS/10. By addition of 5 ml 0.1% HSA in PBS/10 to the above SWy-1 suspensions, HSA concentrations of 0.02% (w/v) were obtained in a volume of 25 ml, also containing 100 mg washed SWy-1 particles, of a size less than 2 μm, with total surface area of 3.0 M^2 (*19*). After equilibration for 45 minutes, the suspension of the SWy-1 and HSA was centrifuged at 2,600 G for 5 minutes to remove the SWy-1 particles. The remaining liquid part was then transferred to another tube and further centrifuged at 2,600 G for 15 minutes. The UV absorption of the liquid was then measured at 280 nm with a UV spectrophotometer (Gilford Instruments, Oberlin, OH).

Silica particles, silica TK 800 (SCI-IUPAC-NPL, code No. 6A/31/22) with a sepecific surface area of 166 M^2/g, were used in this study. 0.050 g aliquots of the silica particles were dispersed in 20 ml PBS/10 or in 20 ml PBS/10 containing 20.0 mg Na$_2$EDTA, at pH 7.2. The silica suspensions were then mixed with 5 ml 0.1%

HSA in PBS/10 and equilibrated for an hour. The procedures followed with the silica suspensions were the same as those used for SWy-1 particles.

Results

Macroscopic Scale Interactions. Both with a cylinder segment parallel to the flat glass plate (Figure 1) and with a spherical ("elbow") moiety interacting with the glass plate (Figure 2), a strong macroscopic repulsion prevails which is at close-range (up to $\ell \approx 1.5$ nm) mainly due to AB forces and at intermediate range ($\ell \approx 4$ nm) due to slightly longer-range EL forces (at the relatively low ionic strength of 0.015). The range of action of this repulsion, i.e., the distance at which the net repulsion energy is still of the order of +1 kT, is about 4 nm for the sphere-flat plate (Figure 2) and about 6 nm for the cylinder-parallel flat plate configuration (Figure 1).

In both configurations the overall strong short-range and long-range repulsion and the fact that in the first case (Figure 1) the short-range repulsion is more than 100 kT and in the second case (Figure 2) more than 40 kT, would rule out any possible adsorption of HSA onto clean glass, if macroscopic interactions were the only operative factor.

Microscopic and Macroscopic Scale Interactions Combined. Even though the microscopic-scale attraction to one attractor site on the glass surface is energetically much feebler (ΔG_{1w2}^{TOT} = -20 kT at $\ell = \ell_0$) than the close-range macroscopic repulsion in either of the configurations shown in Figure 1 and 2, in the case of the sphere-flat plate configuration, the microscopic discrete-site attraction will prevail. Figure 3 illustrates that in the case of the cylinder segment-parallel to the glass plate, the macroscopic repulsion still dominates at all distances ℓ; however in Figure 4 it can be seen that for the sphere-flat plate configuration, at close range, the microscopic-scale single site attraction is stronger. This is because at $\ell < R$, i.e., at $\ell < 1.3$ nm, the overall macroscopic repulsion is locally superseded by the microscopic site's attraction. However, for the parallel cylinder-flat plat configuration, the attraction by one microscopic attraction site per cylinder segment of 4.0 nm length does not suffice to overcome the overall macroscopic repulsion (Figure 3). Even two such microscopic attraction sites per cylinder-segment would be insufficient to surmount the macroscopic repulsion field. Only when three microscopic attraction sites are grouped within a rectangle of about 4.0 x 1.3 nm, would they be able to bind the protein via such a cylinder-segment. Nonetheless, given a calcium content of about 7% by weight (which is typical for this type of glass), such a configuration is rather rare, so that HSA adsorption to glass via albumin "elbow" interactions with the glass surface will be the prevalent mode by far. In addition, the greater ease with which the "elbow" can penetrate the macroscopic repulsion field, would cause the parallel cylinder segments to dip toward the flat plate as a consequence of the incipient attraction of one or the other of its "elbows" to one of the attractor sites on the glass, thus causing the configuration to revert to the "elbow"-flat plate situation.

Influence of EDTA on HSA Adsorption to Mineral Particles. From the top row of Table 1 it can be seen that 78.1% of 5 mg=3.90 mg HSA adsorbed onto 100 mg, or 3.0 M^2 of SWy-1 montorillonite clay particles, or 1.3 mg HSA per M^2 SWy-1. The silica suspension (Table 2) of 8.3 M^2 adsorbed 2.275 mg HSA, or 0.274 mg HSA per M^2 (cf. MacRitchie, 1972, who found a similar degree of adsorption onto silica under comparable conditions).

The presence of Na_2EDTA, in both cases, caused a significant decrease in HSA adsorption (Table 1 and 2). From Table 1 it appears that there is a minimum of HSA adsorption at the intermediate Na_2EDTA concentration of 10 mg EDTA per 25 ml (or 0.04%). At higher EDTA concentration its adsorption-suppressive effects seem to diminish (Table 1), but this is most probably a consequence of the propensity of

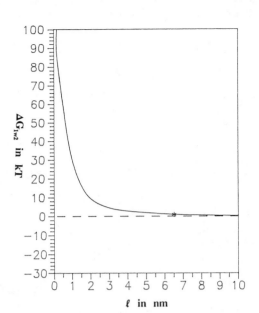

Figure 1. Energy vs. Distance Balance I. Macroscopic interaction between a parallel cylindrical HSA domain and the glass surface. The asterisk indicates the average closest approach, at +1kT. A net short-range repulsion prevails.

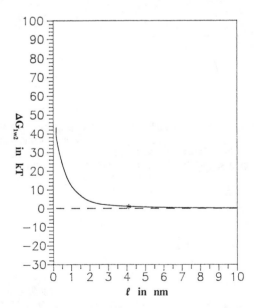

Figure 2. Energy vs. Distance Balance II. Macroscopic interaction between an HSA "elbow" and the glass surface. The asterisk indicates the average closest approach, at +1kT. A net short-range repulsion prevails.

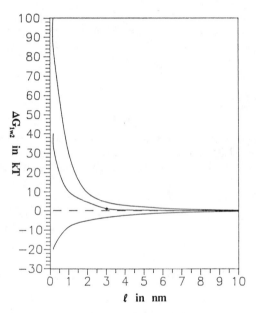

Figure 3. Energy vs. Distance Balance III. Microscopic attraction vs. macroscopic repulsion between a parallel cylindrical HSA domain and the glass surface. The asterisk indicates the average closest approach, at +1kT. A net short-range repulsion still prevails. The middle curve represents the composite interaction.

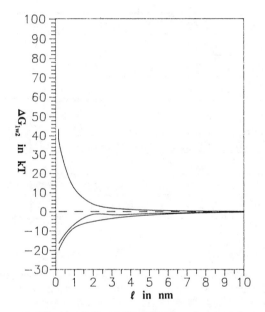

Figure 4. Energy <u>vs</u>. Distance Balance IV. Microscopic attraction vs. macroscopic repulsion between an HSA "elbow" and the glass surface. <u>A net short-range attraction now obtains between microscopic attractor sites on the "elbow" and the glass surface.</u>
The middle curve represents the composite interaction.

EDTA to render SWy-1 montmorillonite particles more hydrophobic, in proportion to the EDTA-concentration, see Table 3. It is well established that albumin adsorbs much more avidly to low than to high-energy surface (1).

Table 1. Adsorption of human serum albumin (0.02%) in 25 ml of a 0.4% Swy-1 suspension in PBS/10 at pH 7.2 in the presence of varous amounts of Na_2EDTA

	pH±0.05	Γ/2	Absorbed (%)
no EDTA	7.2	0.015	78.1
2.5 mg EDTA	7.2	0.015	69.7
5.0 mg EDTA	7.2	0.017	44.7
10.0 mg EDTA	7.2	0.018	31.1
20.0 mg EDTA	7.2	0.019	48.2
30.0 mg EDTA	7.2	0.020	61.8
40.0 mg EDTA	7.2	0.023	71.9

Table 2. Adsorption of human serum albumin (0.02%) in 25 ml of a 0.2% Silica suspension in PBS/10 at pH 7.2 in the presence of of Na_2EDTA

	pH±0.05	Γ/2	Absorbed (%)
no EDTA	7.2	0.015	45.5
20.0 mg EDTA	7.2	0.019	22.2

Discussion

Competition between Macroscopic and Microscopic Interactions. Taking the known three-dimensional properties of serum albumin into account, as well as the properties of a typical glass surface, it could be shown by means of energy (LW+AB+EL) versus distance diagrams (Figures 1-4), that an albumin molecule would only be prone to surmount the net macroscopic (AB+EL) repulsion field, through attraction by a discrete-site attractor in the glass surface, when approaching such an attractor via one of its "elbows", with a small radius of curvature (Figure 5). Such an attractor on a glass (or clay) surface would be, e.g., a calcium, or other plurivalent cation. [The presence of calcium and/or other plurivalent cations in all types of glass is well-documented (21)]. To test this hypothesis we attempted to inhibit the action of such a specific attractor by means of a complexing agent, see below.

Effect of Complexing Agents on HSA Adsorption to Mineral Particles. The adsorption (at Γ/2 = 0.015 and PH = 7.2) of HSA onto montmorillonite particles could be decreased 60% (Table 1) and the adsorption onto silica particles could be decreased 51% by the admixture of Na_2EDTA (Table 2). However, the addition of large amounts of Na_2EDTA appeared counterproductive, most likely as a consequence of an increasingly hydrophobizing influence on montmorillonite of higher EDTA concentrations (Table 3). EDTA itself also appears to adsorb onto SWy-1, so that the decrease in HSA adsorption at the lower EDTA concentration may also be due, partly

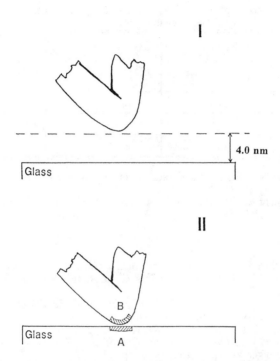

Figure 5. Schematic presentation of the mechanism of adsorption of human serum albumin (HSA) onto glass. In I only the averaged macroscopic repulsion between one "elbow" of the HSA molecule and glass surface is operative: even a moiety with a small radius of curvature cannot approach the glass surface more closely than to $\ell \approx 4$ nm. In II it is shown that the attraction between a microscopic attractor site (A) imbedded in the glass surface and an HSA site on a point of maximum curvature (B) or "elbow" can surmount the macroscopic repulsion, and lead to a local point of attachment where B meets A. From C.J. van Oss, Interfacial Forces in Aqueous Media, Marcel Dekker, New York, 1994, Fig. XXI-1, with permission from the Publisher.

or entirely, to competition for adsorption sites between EDTA and HSA. However, the attachment sites on SWy-1 for EDTA are most likely the same plurivalent cation sites which are also the most probably sites for HSA adsorption. It thus would appear that a tentative conclusion is warranted that at neutral pH, albumin mainly adsorbs onto hydrophilic mineral surfaces, immersed in aqueous media, via plurivalent cation (e.g., calcium) sites for electron-donating (and often significantly negatively charged) hydrophilic proteins. The cause of this attraction lies not only in the (local) excess

Table 3. The surface properties of SWy-1 particles, as is, and in the presence of different amounts of Na$_2$EDTA, at neutral pH

	Natural SWy-1	SWy-1 + 0.04% (10 mg) EDTA	SWy-1 + 0.16%(40mg) EDTA
γ_S^{LW}(mJ/M^2)	44.1	39.9	39.1
$\gamma_S^{\oplus}$ (mJ/M^2)	0.2	1.3	1.3
$\gamma_S^{\ominus}$ (mJ/M^2)	34.1	30.0	23.9

of positive charges of the imbedded plurivalent cations, but also in their concomitant electron-acceptity (*9, 19, 20*).

Very recent results show that HSA can be completely eluted with EDTA from glass and similar surfaces (*22*).

Literature Cited

1.   MacRitchie, F. *J. Colloid Interface Sci.*, **1972**, *vol. 38*, p. 484.
2.   Costanzo, P.M.; Giese, R.F.; van Oss, C.J. *J. Adhesion Sci. Technol.*, **1990**, *vol. 4*, pp. 267-275.
3.   Verwey, E.J.W.; Overbeek, J.Th.G. *Theory of the Stability of Lyophobic Colloids*, Elsevier, Amsterdam, 1948.
4.   van Oss, C.J. *Cell Biophys.*, **1989**, *vol. 14*, pp. 1-16.
5.   van Oss, C.J. in *Biophysics of the Cell Surface*; Glaser, R. and Gingell, D. ED.; Springer-Verlag, Berlin, 1990, pp. 131-152.
6.   van Oss, C.J., Giese, R.F. and Costanzo, P.M. *Clays and Clay Minerals*, **1990**, *vol. 2*, pp. 151-159.
7.   van Oss, C.J. *Colloids and Surfaces A*; **1993**, *vol. 78*, pp. 1-49.
8.   van Oss, C.J., Chaudhury, M.K. and Good, R.J. *Adv. Colloid and Interface Sci.*, **1987**, *vol. 8*, pp. 35-64.
9.   van Oss, C.J., Chaudhury, M.K. and Good, R.J. *Chem. Rev.*, **1988**, *vol. 88*, pp. 927-941.
10.   van Oss, C.J., in *Protein Interaction*, Visser, H., ED., VCH, Weinheim, Germany, 1992, pp. 25-35.
11.   Israelachvili, J.N. *Intermolecular and Surface Forces*, 2nd, 1992, p. 177.
12.   Gabler, R. *Electrical Interactions in Molecular Biophysics*, Academic Press, New York, 1978, p. 245.
13.   van Oss, C.J. in *Structure of Antigens*, Van Regenmortel, M.H.V., ED., CRC Press, Boca Raton, FL, 1992, pp. 99-125.
14.   Eisenberg, D. and Kauzmann, W. *The Structure and Properties of Water*. Clarendon Press, Oxford, 1969, pp. 145; 169.
15.   van Oss, C.J., Fike, R.M., Good, R.J.; Reinig, J.M. *Analyt. Biochem.*, 1974, vol. 60, pp. 242-251.
16.   van Oss, C.J.; Good, C.J. *J. Protein Chem.* **1988**, *vol. 7*, pp. 179-183.

17. Abramson, H.A.; Moyer, S.L.; Gorin, H.M. *Electrophoresis of Proteins*, Hafner, New York, 1964, pp. 173-194.
18. He, X.M.; Carter, D.C. *Nature*, **1992**, *vol. 358*, pp. 209-215; Carter, D.C.; Ho, J.X., *Advan. Protein Chem.*, **1994**, *vol. 45*, pp. 155-203.
19. van Oss, C.J., Wu, W and Giese, R.F. (**1994**) Macroscopic and Microscopic Interactions between Albumin and Hydrophilic Surfaces, (Abstract), 207th ACS National Meeting, San Diego.
20. van Oss, C.J., Chaudhury, M.K. and Good, R.J. in *Membrane Fusion*, Ohki, S.; Doyle, D.; Flanagan, T.D.; Hui, S.W.; Mayhew, E. ED., Plenum, New York, 1988, pp. 113-122.
21. Budd, S.M. in *Materials and Technology*, vol. 2, Codd, L.W.; Dijkhoff, K.; Fearon, J.H.; van Oss, C.J.; Roeberson, H.G.; Stanford, E.G.; van Thoor, T.J.W., EDS., Longman. London, 1971, pp. 331-412.
22. van Oss, C.J.; Wu, W.; Giese, R.F., *Colloids Surfaces B* (accepted).

RECEIVED January 31, 1995

Chapter 7

A New Hydrophobicity Scale and Its Relevance to Protein Folding and Interactions at Interfaces

Dan W. Urry[1] and Chi-Hao Luan[1,2]

[1]Laboratory of Molecular Biophysics, School of Medicine, University of Alabama, 1670 University Boulevard, VH525, Birmingham, AL 35294–0019
[2]Bioelastics Research, Ltd., 1075 13th Street South, Birmingham, AL 35205

A hydrophobicity scale, based on the temperature, T_t, at which an inverse temperature transition of hydrophobic folding and assembly occurs in a model protein, has been used for hydrophobicity plots of protein primary structures of fibronectin, fibrinogen and two apolipoproteins. The periodicity of recurrence of hydrophobic residues could be used to identify amphiphilic β-sheets and α-helices. Evaluation of relative stability of hydrophobic domains can use the calculated hydrophobicity of the domain, i.e., $\sum_d T_t/N = \langle T_t \rangle_d$. Domains with the lowest values of $\langle T_t \rangle_d$ would be the most stable and those with values closer to 37°C would be the least stable corresponding respectively to the rigid and soft nomenclature of Norde for protein adsorption at interfaces. Non-uniformity within a given hydrophobic domain can be analyzed to assess a path of hydrophobic unfolding in the process of binding at a hydrophobic interface.

Over the past several decades there has been an increasing interest in the interaction of proteins at interfaces *(1-13)*. This short report will neither attempt to review those previous efforts nor to emulate any of the previous approaches. Rather, this effort will involve an analysis of protein interactions at interfaces based on a new hydrophobicity scale developed directly on the hydrophobic folding and association transition of relevance to proteins.

In a purely aqueous environment, it is difficult to build a strong case for charge-charge interactions as being a primary effect for protein folding and interaction. For example, charge-charge interactions in protein-based polymers, where proximity can be forced by primary structure, result in limited pKa shifts *(14-16)*. This is not to say that charge-charge interactions cannot be significant in proteins and in protein interactions but rather in particular well-defined examples that they can be shown to derive their significance due to hydrophobic environments imposed by polymer or surface structures *(17-19)* which cause large repulsive free energies of interaction *(17)*. The large apolar-polar repulsive free energies of interaction can be relieved in part by ion-pairing or even by ion-ion cloud interactions.

Proteins in serum or in the extracellular matrix, unless activated in some way or otherwise structurally disturbed, are generally well contained structures

0097–6156/95/0602–0092$12.00/0

resulting from having arrived at their properly folded and associated minimal free energy states *(20)*. Increasingly, over the past several decades hydrophobic folding and assembly of proteins have been considered more prominently in arriving at the structural global minimum in free energy *(21-28)* and in achieving function *(17)*.

In a particularly instructive series of protein folding experiments on hen lysozyme, Dobson and colleagues *(29,30)* have provided strong evidence for the prominent role of hydrophobic folding. In stopped-flow far ultraviolet circular dichroism experiments, they observed that the native circular dichroism pattern is obtained after four milliseconds of refolding, a result generally interpreted to mean that the α-helices and β-sheet secondary structures are already well formed within this short period. Using pulsed hydrogen exchange labeling studies, however, no stable hydrogen bonding, that is, no slowing of peptide NH exchange, was found as deduced from the observation that all backbone NH moieties were still rapidly exchanging after the first several milliseconds of refolding. The third experiment followed the time development of fluorescent intensity of intrinsic tryptophan residues where it was found that the tryptophans were already buried in a hydrophobic domain within the first few milliseconds. The conclusion is that hydrophobic collapse can be an initial event *(29,30)* followed by the sorting out or stabilizing of the most favorable secondary structures that would support the hydrophobic domain.

In related theoretical arguments, Dill and colleagues *(31)* have shown that hydrophobic folding in the sense of a "hydrophobic zipper" can be a cooperative process, and work of this Laboratory going back for more than a decade has described a cooperative hydrophobic folding and assembly transition with the formation of β-spiral structures as arising entirely from a helical hydrophobic zipper in elastic model proteins *(17)*.

It is the dependence on amino acid composition of the temperature, T_t, at which the hydrophobic folding and assembly transition occurs in elastic model proteins that is the basis for a new hydrophobicity scale *(32)*. And it is the new hydrophobicity scale that provides for hydrophobicity plots of protein primary structure to analyze for the occurrence of structural features such as β-sheets and α-helices, that can be used to scale proteins from rigid to soft in the Norde classification *(12)* and that can be used to deduce the hydrophobic unfolding of soft proteins (or protein domains) in the process of hydrophobic association with interfaces.

The T_t-based Hydrophobicity Scale

The Model Protein System. The T_t-based hydrophobicity scale has been developed employing an elastic repeating sequence of mammalian elastic fibers, $(Val^1\text{-}Pro^2\text{-}Gly^3\text{-}Val^4\text{-}Gly^5)_n$ where n = 11 in bovine elastin without a single substitution, but using chemically synthesized polymers with values of n greater than 120. Written for simplicity as poly(VPGVG) or equivalently as poly(GVGVP), this sequential elastomeric polypeptide, or elastic protein-based polymer, cannot be substituted at the Pro^2, Gly^3 and Gly^5 positions without loss of elastic property or without causing other significant higher order structural perturbations. Some substitution is possible at the Val^1 position. Fortunately, however, all of the naturally occurring amino acids, and many chemical modifications thereof, can be substituted at the Val^4 position. The general formula for the family of model proteins can be written as $poly[f_v(VPGVG),f_x(VPGXG)]$ where f_v and f_x are mole fractions and $f_v + f_x = 1$.

When these elastic protein-based polymers are dissolved in water at sufficiently low temperatures, they are miscible with water in all proportions with the formation of clear solutions. On raising the temperature, there is an onset of turbidity which defines the onset of a hydrophobic folding and assembly transition. The temperature at which this onset of folding and assembly occurs is concentration dependent, occurring at lower and lower temperatures as the concentration is increased, but only up to a high concentration limit of about 40mg/ml. The temperature at which half intensity in the turbidity versus temperature curve is achieved for the 40mg/ml concentration is designated as T_t, the temperature of an inverse temperature transition resulting in increased order as hydrophobic folding and assembly occurs *(17,32)*.

Dependence of T_t on Composition. Hundreds of elastic protein-based polymers were synthesized using different amino acids in position four and doing so at different values of f_x, i.e., at different extents of substitution. Plots of f_x versus T_t were found to be essentially linear up to $f_x = 0.5$ for the various substitutions as shown in Figure 1 *(17,32)*. Importantly, it is seen that addition of more hydrophobic residues lowers the value of T_t, whereas addition of more polar residues increases the value of T_t. As a standardized point of reference, all plots of f_x vs. T_t are extrapolated to $f_x = 1$, i.e., to poly(VPGXG), and this value of T_t is taken as an index of the relative hydrophobicity as shown in Table I *(17)*.

The first hydrophobicity scale for amino acid residues was due to Nozaki and Tanford *(33)*; the most commonly used is that of Kyte and Doolittle *(34)*, and there are now more than forty such scales *(35,36)*. So there is little need for yet another hydrophobicity scale. The T_t-based hydrophobicity scale, however, is the only scale based directly on the hydrophobic folding and assembly of interest; it explicitly delineates between charged and uncharged states of functional side chains; but most importantly, it is an integral part of understanding a mechanism whereby free energy transduction can be achieved involving all of the intensive variables of mechanical force, temperature, pressure, chemical potential, electrochemical potential and light. All of these energy conversions are achieved by controlling the value of T_t. It is easy to recognize how lowering T_t would drive hydrophobic folding and result in the performance of mechanical work, but just as clearly lowering T_t has been shown to change the pKa of Glu, Asp and Lys residues *(14-19)* in a manner that can result in the performance of chemical work. Other hydrophobicity scales have not been demonstrated to be integral to such functional roles. Those scales which are based on distribution of residues buried versus on the surface of a globular protein appear to be most similar to the T_t-based hydrophobicity scale.

T_t-based Single Residue Hydrophobicity Plots for Protein Primary Structures

Human Fibronectin. The T_t-based hydrophobicity plot for human fibronectin sequence *(37)* is given in Figure 2 where 37°C is taken as the reference temperature. All of those residues which are hydrophobic and could contribute to hydrophobic folding are positive deflections as plotted from the 37°C line and all of those more-polar residues which would disrupt or not significantly contribute to hydrophobic folding, i.e., those with T_t values greater than 37°C, are plotted as negative deflections from 37°C. At the right-hand side of the figure are the one-letter codes for the amino acid residues corresponding to their T_t values, and superscripts are used where appropriate to indicate the charged state of the functional side chain.

Even though there are more than 2000 residues in the fibronectin sequence *(37)*, it is possible to see simply by glancing at Figure 2 that there are repeating motifs which are most apparent due to the periodicity of the most hydrophobic tryptophan(W) residue. It is also seen that the repeats differ in hydrophobicity by noting the number of tyrosine(Y), phenylalanine(F) and leucine(L) residues that occur between the tryptophans. In order to use the distribution of hydrophobic

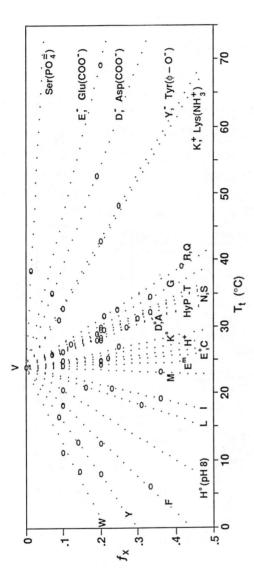

Figure 1. Plots of f_X versus T_t (the temperature of a hydrophobic folding transition) for each of the amino acid residues and some chemical modifications thereof for poly[f_V(VPGVG),f_X(VPGXG)] where f_V and f_X are mole fractions of occurrence of the respective pentamers in the polymers with $f_V + f_X = 1$. More hydrophobic residues lower the value of T_t and more polar or less hydrophobic residues than Val increase the value of T_t in a regular manner which allows for an evaluation of the relative hydrophobicities. Reproduced with permission from reference 17.

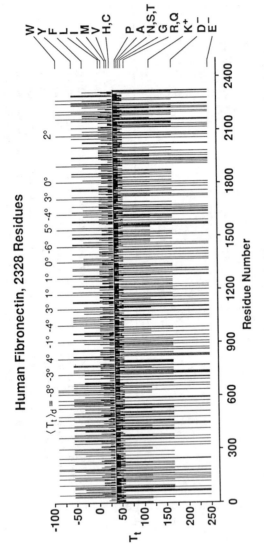

Figure 2. T_t-based hydrophobicity plot for human fibronectin showing the recurrence of different types of sequences. With the Trp residue as residue 22 of each Type III domain, the numbers above each Trp line for the Type III domains are the sum of the T_t values of 25°C or less divided by the number of residues in the sum. See text for a more complete discussion.

residues to aid in analysis of the conformations of the repeats, it will be helpful to expand the residue number scale and to look at a single repeat. This will be done below for the specific repeat sequence (the tenth Type III domain) in fibronectin that contains the RGD cell attachment sequence.

Human Fibrinogen. The hydrophobicity plots of the three chains of human fibrinogen, the α-, β- and γ-chains are given in Figure 3. Again, the hydrophobicity plots provide much information even at a glance. In all three chains, the hydrophobic leader sequences at the amino ends are apparent. These are utilized and cleaved during transport of the chains out of the cell into the serum. In the α-chain there is a striking 80-residue sequence near the center of the sequence in which there is only one charged residue, a lysine(K) residue. There are also significant runs where a dominant motif is the recurrence of a hydrophobic residue every third or fourth residue which is often referred to as a heptad repeat *(38)*. This is the structure required for an amphiphilic α-helix with resulting supercoiled α-helices. The example of an amphiphilic α-helix is considered in more detail below for a particularly interesting apolipoprotein.

Analysis of Single Residue Hydrophobicity Plots in Terms of Conformational Features

If it is correct that a dominant primary structural feature for determining conformation is the distribution of hydrophobic residues with which to form a hydrophobic domain, as suggested above in the Introduction, it should be possible to analyze for the periodicity of hydrophobic residues and thereby to deduce preferred conformational features.

β-turns and Cross-β-structures. The β-turn is a conformational feature (actually a secondary structural feature) whereby a chain can fold back on itself in the formation of a cross-β-structure as shown in Figure 4 *(39)*. Such a pair of β-chains defines a plane with a distinct sidedness in which alternate residues in the primary structure occur on the same side. Thus, if one side were to be hydrophobic, this would be apparent in the primary structure by the occurrence of alternating hydrophobic residues. Also, if hydrophobic association of side chains was the initial association in protein folding and assembly, then the aligning of pairs of hydrophobic side chains in the formation of hydrophobic microdomains would precede the stabilizing of interchain hydrogen bonding. Furthermore, the formation of the first hydrophobic microdomain as seen in Figure 4 would facilitate the formation of the second and third pairs and so on in a cooperative folding transition as in the hydrophobic zipper of Dill *(31)* and as early appreciated by Matheson and Scheraga *(40)*. This would be the case to the extent that hydrophobic residues occur in ready juxtaposition.

Antiparallel and Parallel β-sheets and β-barrels. Whether the chains run antiparallel as in the cross-β-structure of Figure 4 or parallel to form β-sheet structures, there occurs a sidedness such that an entire side of a β-sheet could be hydrophobic. Of course the β-chains could form a β-barrel with the hydrophobic residues on the inside in the formation of a water soluble globular structure, or the hydrophobic residues could be directed outward as might occur in the folding of a transmembrane protein with a pore structure.

The Tenth Type III Domain of Fibronectin (The RGD-containing Sequence). The β-barrel topology of the fibronectin Type III domain, an approximately 90-amino acid residue sequence, is found in many proteins: in

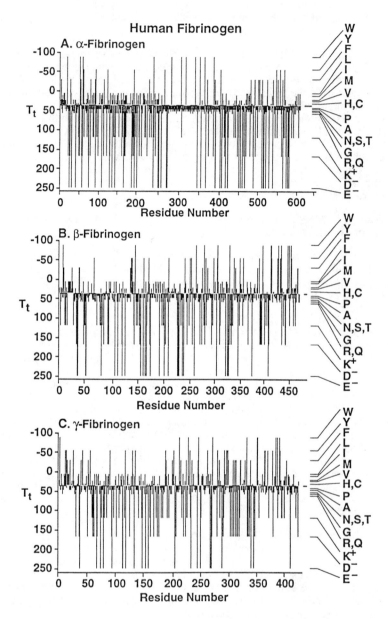

Figure 3. T_t-based hydrophobicity plots for the α-, β- and γ-chains of human fibrinogen.

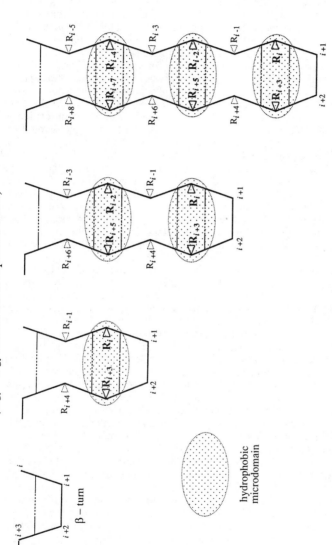

Hydrophobic Folding of Cross - β - Structures
(Cα – Cα virtual bond representation)

Figure 4. Hydrophobic folding of cross-β-structures wherein hydrophobic microdomains form on pairing of hydrophobic residues. When alternating hydrophobic residues occur in pairs of chains, there develops a hydrophobic sidedness which can result in an amphiphilic β-sheet. Reproduced with permission from reference 39.

immunoglobulins, in tenascin, in titin (the elastic third filament of muscle), in cytokine receptors, etc. *(41)*. The β-barrel-type structure is comprised of seven β-chains with a hydrophobic interior. It is perhaps better described as four antiparallel β-chains forming one sheet and three antiparallel β-chains forming a second sheet and with hydrophobic residues at the interface between the two sheets whose chains are at an angle of about 30°. Accordingly, not all of each β-chain contributes to the hydrophobic interface between the sheets. The structure of the third Type III fibronectin-like domain of tenascin which is analogous to the RGD-containing tenth Type III domain of fibronectin is given in Figure 5 *(41)*.

The T_t-based hydrophobicity scale for the tenth Type III domain of fibronectin, residues 1420 through 1513, is given in Figure 6A. If, as anticipated, for the β-chain that hydrophobic residues should appear at alternate positions to form a hydrophobic side to a β-sheet, the plot of Figure 6A can be usefully sequence edited. The odd residue plot is given in Figure 6B, and the even residue plot is shown in Figure 6C. Indeed in Figure 6B are observed six of the seven β-chains. The two most hydrophobic sequences are β-chains B and C, followed by G and F, with A and C′ contributing less to the hydrophobic fold and with E contributing least with a single Ile residue on the even residue plot. The β-chains and the hydrophobic residues, Trp, Tyr, Phe, Leu, Ile and Val of the analogous domain of tenascin, can be followed in Figure 5. What is of particular interest is that the β-chains with hydrophobic sidedness are readily observed from the analysis of the primary structure. Furthermore, when interested in the adsorption at a hydrophobic interface, those chains with the weakest hydrophobic interactions would be expected to open (separate) most readily in the required unfolding step. This is discussed below.

Amphiphilic α-helices. The classical α-helix is characterized by having 3.5 residues per turn such that one side of the helix would be defined by every third or fourth residue and most effectively by each seventh residue as in the heptad repeats *(38)*. When hydrophobic collapse is an initial, important step in protein folding, then sequences which become α-helices should be characterized with a periodicity of hydrophobic residues resulting in amphiphilic α-helices having one side hydrophobic. There are amphiphilic α-helices in human fibrinogen *(38)*, and it would be of interest to see how readily the hydrophobicity plots of Figure 3 provide argument for such structures and for the association of particular sequences. The situation is, however, more complex than the interaction of hydrophobic residues between turns of the helix that would give a helical zippering effect; there is the more general hydrophobic domain in which the side of a single α-helix is but one component. In the case of fibrinogen, this translates into the additional problem of which amphiphilic α-helices associate with which other amphiphilic α-helices.

For purposes of this introductory discussion of using the T_t-based hydrophobicity scale to analyze for protein hydrophobic folding and interaction at interfaces, two apolipoproteins provide an interesting case study. These are single protein sequences which form a hydrophobically clustered bundle of five α-helical segments. The T_t-based hydrophobicity plots for the proteins, apolipoprotein III from *Locusta migratoria* and *Manduca sexta* *(42,43)*, are given in Figure 7. In both plots and perhaps even more dramatically in the *Manduca sexta* sequence, there is a surprisingly regular recurrence of a strongly hydrophobic (or non-polar) residue at every third or fourth position. The regularity is so extensive that the challenge is to identify the non-helical sequences, that is, to determine where given helical segments would begin and end to allow the helices to fold back and forth to form a common hydrophobic domain.

The crystal structure for apolipoprotein III from *Locusta migratoria* *(44)* is given in Figure 8 using the α-carbon to α-carbon representation. In the structure, as presented, the side chains are explicitly included for the more hydrophobic Trp, Phe,

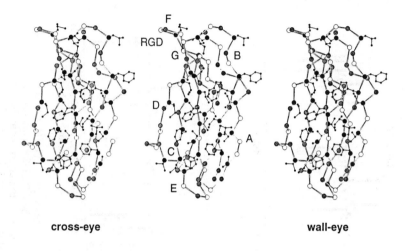

cross-eye   wall-eye

● W, Y, F, L, I, M, V
○ all other residues
○ D, E, K, R

Using the protein data bank coordinates of
Leahy, Hendrickson, Aukhil and Erickson
Science, 1992

Figure. 5.   Stereoplots of the molecular structure of the third fibronectin Type III domain of human tenascin using the protein data bank coordinates due to Leahy et al. *(41)* for the structure which is analogous to the tenth Type III domain of fibronectin. It is an α-carbon representation of a seven-chain β-barrel-like structure with inclusion of the hydrophobic side chains of Trp, Tyr, Phe, Leu, Ile and Val showing the hydrophobic interaction between a three-chain β-sheet and four-chain β-sheet. Also indicated is the β-turn containing the RGD cell attachment sequence. The left-hand pair of plots are for cross-eye viewing, and the right-hand pair of plots are for wall-eye viewing.

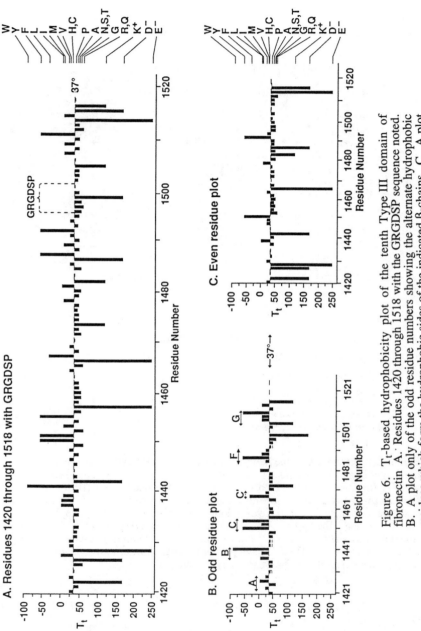

Figure 6. T_t-based hydrophobicity plot of the tenth Type III domain of fibronectin A. Residues 1420 through 1518 with the GRGDSP sequence noted. B. A plot only of the odd residue numbers showing the alternate hydrophobic residues which form the hydrophobic sides of the indicated β-chains. C. A plot of the even residues only.

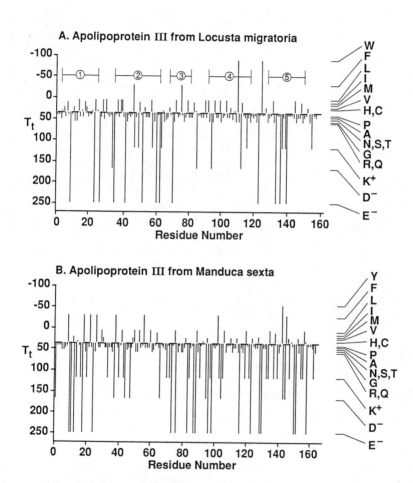

Figure 7. T_t-based hydrophobicity plots for two primary structures. A. Apolipoprotein III from Locusta migratoria where the five amphiphilic α-helical sequences are indicated as known from crystal structure. B. Apolipoprotein III from Manduca sexta. In both plots the preponderance of hydrophobic (or non-polar) residues at every third or fourth position in the sequence is noted. See text for discussion of the folding to form a five helix bundle resulting from the optimization of a hydrophobic domain.

Leu, Ile and Val residues. In Figure 7A it is indicated where the five α-helical segments occur in the primary structure based on the crystal structure *(44)*.

Perhaps the fundamental consideration in analyzing the T_t-based hydrophobicity plot to arrive at the three-dimensional structure is to address the means whereby the best hydrophobic domain might result. With the packing volume required for hydrophobic side chains, a five helical bundle appears to work well; six would be too many as they would leave a central void. If five helical segments are the most that could form without a void, then the issue becomes one of choosing the five amphiphilic helical segments which would form the best hydrophobic domain given a 161-residue protein. With interconnecting segments between amphiphilic helices, there should be no more than about 30 residues per helical segment. To have one amphiphilic α-helical segment too long would leave some of its hydrophobic side exposed to water. Accordingly, one might naturally look in the region of residues 30, 60, 90 and 120 for the break regions between α-helical segments.

Next, there are two specific ways that one might consider for disrupting an amphiphilic α-helix. One would be to place one or two very polar residues, e.g., charged residues, on the hydrophobic side. A second way would be to have a periodicity of hydrophobic residues which directed for β-chain formation, i.e., to have alternating hydrophobic residues, rather than a periodicity which directed for an amphiphilic α-helix by every third or fourth residue being hydrophobic. Looking near residue 30 in Figure 7A, there is seen an alternating pair of hydrophobic residues followed by a charged residue in the fourth position after the second hydrophobic residue of the pair. Accordingly, the break between the first and second helix seems to be quite well defined. For the break between the second and third helical segment, there are four charged residues at the end of the second helix followed by a weak hydrophobic directing force which would contribute little to a hydrophobic domain thereby defining the second break sequence. For the third break which would occur in the vicinity of residue 90, the most obvious issue is the alternate pair of hydrophobic residues. Near residue 120, with the important exception of a single tryptophan residue, there is little hydrophobicity such that if the Trp side chain could be directed into the end of the hydrophobic core, a break could readily occur here. As seen in Figure 8, this folding is observed. Accordingly, it seems that the bundle of five amphiphilic α-helixes can be understood by noting the periodicity of hydrophobic residues and by maximizing the hydrophobic domain intensity. As a test of this understanding, one might predict that the breaks between amphiphilic α-helixes for the apolipoprotein III from Manduca sexta would involve residues 30 to 35, 65 to 75, 95 to 105 and 130-135.

Binding of Serum Proteins at Interfaces

When designing materials for medical devices with preferred properties, a particular concern is the adsorption of serum or extracellular matrix proteins onto the surface of the device which would thereby impart their own properties to the surface. In this regard, the study of a series of bioelastic matrices of different compositions has been instructive. The study suggests that significant serum protein adsorption is proportional to the matrix hydrophobicity, and the study demonstrates that hydrophobicity dependent adsorption can alter the properties of the surface of the matrix in undesirable ways.

The intended purpose for certain bioelastic material compositions is to achieve a matrix or surface with non-interactive, non-adhesive properties *(45-47)*. Once this is achieved for certain applications, then for other applications the matrix could be selectively modified with the addition of specific sequences, such as the RGD cell attachment sequence of fibronectin *(48,49)*, to confer on the matrix desired specific cell attachment properties.

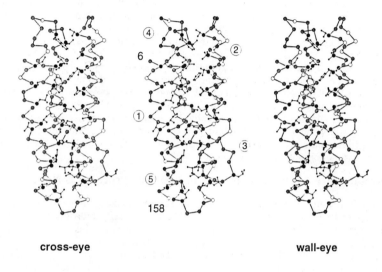

cross-eye   **wall-eye**

- ● W, Y, F, L, I, M, V
- ◉ all other residues
- ○ D, E, K, R

Using the protein data bank coordinates of
Breiter, Kanost, Benning, Wesenberg, Law, Wells, Rayment and Holden
Biochemistry, 1991

Figure 8.   Molecular structure of Apolipoprotein III from Locusta migratoria residues 6 through 158 using the protein data bank crystal structure data of Breiter et al. *(44)*. It is an α-carbon representation with the side chains of the hydrophobic residues included to show the hydrophobic domain.   It is a stereoplot in which the left-hand pair of structures are for cross-eye viewing and the right-hand pair are for wall-eye viewing. The circled numbers are to identify the helices which correspond to the sequences indicated in Figure 7A.

The elastic protein-based polymer, poly(VPGVG) which was used to develop the hydrophobicity scale, when γ-irradiation cross-linked by 20 Mrads to form an elastic matrix, i.e., X^{20}-poly(VPGVG), is remarkably biocompatible, eliciting neither fibrous capsule formation when implanted *(45-47)* nor cell attachment when in the appropriate culture medium with bovine serum albumin added *(50-52)*. When fetal bovine serum (20%) is added to the cell culture medium, limited cell attachment does occur without significant spreading and growth. When the GRGDSP fibronectin cell attachment sequence is included in the primary structure, as in X^{20}-poly[40(GVGVP),(GRGDSP)], cells then attach firmly, spread and grow to confluence even in the absence of serum *(50-52)*.

In addressing the effect of serum on inducing limited cell attachment, another several elastic matrices were formed, X^{20}-poly(GGIP), X^{20}-poly(GGVP) and X^{20}-poly(GGAP). In the absence of serum, like X^{20}-poly(GVGVP) these matrices did not support cell adhesion. In the presence of serum, limited cell attachment occurred on X^{20}-poly(GGIP), a poorer attachment occurred on X^{20}-poly(GGVP) but no attachment whatever occurred even in serum on X^{20}-poly(GGAP) *(53)*. Noting the hydrophobicity scale of Figure 1 and Table I, it seems apparent that cell adhesion in serum is proportional to hydrophobicity. The interpretation is that serum proteins, such as fibronectin, are adsorbed by hydrophobic association and then the adsorbed proteins promote limited cell attachment.

Relevance of a Mean T_t to the Rigid and Soft Proteins of Norde

Poly(GVGIP) and poly(GVGVP) can be dissolved in an aqueous solution at 5°C. On raising the temperature from 10 to 15°C poly(GVGIP) hydrophobically folds, self-assembles and settles out in the formation of a viscoelastic phase leaving poly(GVGVP) alone in solution. As the temperature is raised above 25°C, poly(GVGVP) begins to hydrophobically fold and assemble. At 28°C poly(GVGIP) remains well folded and assembled, but poly(GVGVP) is actively folding and unfolding. At 28°C poly(GVGIP) with a $\langle T_t \rangle$ of 10°C could be considered a less soft protein in the Norde classification *(12)*, whereas poly(GVGVP) with a $\langle T_t \rangle$ at 25°C would be considered a very soft protein as it can readily unfold and associate with an interface. A low mean value of T_t for the hydrophobic domain (d) of a protein, i.e., a minimal $\langle T_t \rangle_d$, would correspond to a rigid protein and the closer that the value of $\langle T_t \rangle_d$ is to the working temperature, the softer the protein would be. Thus $\langle T_t \rangle_d$ can be used to provide a continuous series of values to the Norde classification with increasing adhesiveness to a hydrophobic surface as $\langle T_t \rangle_d$ became lower until the hydrophobic domain would no longer unfold.

Given the human fibronectin hydrophobicity plot of Figure 2, it is apparent that there are different $\langle T_t \rangle_d$ values for the Type III domains because of the differing number of Tyr, Phe and Leu residues. It should therefore be possible to calculate the value of $\langle T_t \rangle_d$ for each domain and thereby to evaluate the relative contribution of each domain to interaction at a hydrophobic interface.

In the comparison of Type III domains, it is necessary, of course, to decide upon an appropriate means in which to calculate $\langle T_t \rangle_d$ such that the relative magnitudes will provide a scaling from the more rigid to the softer domains. It would be best to analyze, for example, the structure of Figure 5 to determine which residues can contribute side chains to the hydrophobic domain, to sum the value of T_t for each residue, $\sum_d T_t$, and then to divide by the number of residues(N) to get $\sum_d T_t/N = \langle T_t \rangle_d$. For a quick look at relative stabilities, it is sufficient to simply sum the hydrophobic residues W, Y, F, L and I and divide each sum by a common N or by one.

A more detailed quick look is possible by taking the single Trp residue of each Type III domain of fibronectin as residue 22 and summing over those residues

Table I. T_t-Based Hydrophobicity Scale for Proteins

T_t =Temperature of Inverse Temperature Transition for poly[f_v(VPGVG),f_x(VPGXG)]

Residue X		T_t, linearly extrapolated to $f_x = 1$	Correlation Coefficient
Lys(NMeN, reduced) [a]		−130°C	1.000
Trp	(W)	−90°C	0.993
Tyr	(Y)	−55°C	0.999
Phe	(F)	−30°C	0.999
His (imidazole)	(H°)	−10°C	1.000
Pro	(P) [b]	(−8°C)	calculated
Leu	(L)	5°C	0.999
Ile	(I)	10°C	0.999
Met	(M)	20°C	0.996
Val	(V)	24°C	reference
Glu(COOCH$_3$)	(E^m)	25°C	1.000
Glu(COOH)	(E°)	30°C	1.000
Cys	(C)	30°C	1.000
His (imidazolium)	(H$^+$)	30°C	1.000
Lys(NH$_2$)	(K°)	35°C	0.936
Pro	(P) [c]	40°C	0.950
Asp(COOH)	(D°)	45°C	0.994
Ala	(A)	45°C	0.997
HyP		50°C	0.998
Asn	(N)	50°C	0.997
Ser	(S)	50°C	0.997
Thr	(T)	50°C	0.999
Gly	(G)	55°C	0.999
Arg	(R)	60°C	1.000
Gln	(Q)	60°C	0.999
Lys(NH$_3^+$)	(K$^+$)	120°C	0.999
Tyr(ϕ–O$^-$)	(Y$^-$)	120°C	0.996
Lys(NMeN, oxidized) [a]		120°C	1.000
Asp(COO$^-$)	(D$^-$)	170°C	0.999
Glu(COO$^-$)	(E$^-$)	250°C	1.000
Ser(PO$_4^=$)		1000°C	1.000

NMeN is for N-methyl nicotinamide pendant on a lysyl side chain, i.e., N-Methyl nicotinate
attached by amide linkage to the ε-NH$_2$ of Lys and the reduced state is N-methyl-1,6-dihydronicotinamide.
The calculated T_t value for Pro comes from poly(VPGVG) when the experimental values of Val and Gly
are used. This hydrophobicity value of -8°C is unique to the β-spiral structure where there is
hydrophobic contact between the Val$_i^1$γCH$_3$ and the adjacent Pro$_i^2$δCH$_2$ and the interturn Pro$_{i+3}^2$ βCH$_2$
moieties.
The experimental value determined from poly[f_V(VPGVG),f_P(PPGVG)].

Adapted with permission from reference [17]

with T_t values of 25°C or less to give the $\langle T_t \rangle_d$ values for the domains indicated above the Trp of each domain in Figure 2. The most rigid domains would be the first domain (–8°C) and the tenth domain (–6°C) which contains the RGD cell attachment sequence. The softer domains, those most likely in this analysis to open up and adhere to a hydrophobic interface, would be the eleventh domain (+5°C) followed by the third (+4°C), sixth (+3°C) and thirteenth (+3°C) domains.

Furthermore, a given hydrophobic domain as in Figure 5 for a fibronectin-like Type III domain and in Figure 8 for the apolipoprotein III of Locusta migratoria can be analyzed to determine the softer region of the domain which would be expected to be the section where unfolding would begin in the process of hydrophobic association with an interface. By the location of the two Trp and two Phe residues of Figure 8, it would be concluded that the upper part of the hydrophobic domain as presented would have the lower $\langle T_t \rangle_{d'}$ where d' signifies a more hydrophobic region of the domain. Furthermore, the less hydrophobic, lower part of the domain as printed has a Leu residue that is directed outward, not into the hydrophobic domain, and it has other readily accessible hydrophobic residues. Accordingly, it is expected that unfolding would begin from this end with an unfolding of helices ③ and ④. Interestingly, in the analysis of the lipid transport role of this apolipoprotein III, Brieter et al *(44)* suggested a similar unfolding for extramolecular hydrophobic interaction.

The Type III fibronectin domains are not individual globular proteins and therefore there may be hydrophobic association between domains as suggested by the Leu and Tyr residues at the upper right-hand side of the structure in Figure 5. This does complicate consideration of where unfolding might begin but the softer end would be the RGD end with the F and G β-chains separating from the A and B β-chains. Presumably, this would occur with the RGD sequence yet available to the serum side for some cell attachment which is less than optimal.

Conclusions

A hydrophobicity scale, based directly on the hydrophobic folding process of interest and characterized by the temperature T_t at which the hydrophobic folding and assembly transition occurs, can be utilized in analyzing for different periodicities of hydrophobic residues which direct for different folded structures such as α-helices and β-sheets with hydrophobic sidedness. Protein interactions at hydrophobic interfaces can be viewed as resulting from hydrophobic unfolding initially at less hydrophobic regions of a given hydrophobic domain. The rigid and soft classification of Norde can be given a continuous scaling by means of a calculated mean hydrophobicity, $\langle T_t \rangle_d$, for the hydrophobic domain that unfolds in order to interact at the interface.

Acknowledgments

This work was supported in part by Contract Nos. N00014-89-J-1970 from the Department of the Navy, Office of Naval Research and F33615-93-C-5378 from Wright-Patterson Air Force Base. The authors wish to thank Johanna Kahalley and Mark Logan of the McCormick/Cannon Biopolymer Research Group at The University of Southern Mississippi for calling our attention to the insect apolipoprotein structures.

Literature Cited

1. Vroman, L. *Biomat. Med. Dev., Art. Org.* **1984-85**, *12*, 307-323.
2. Vroman, L.; Adams, A. L. *J. Biomed. Mater. Res.* **1969**, *3*, 46-47.
3. Vroman, L. *Fed. Proc.* **1971**, *30*, 1703-1704.

4. Vroman, L.; Adams, A. L.; Klings, M. *Fed. Proc.* **1971**, *30*(5), 1492-1502.
5. Brash, J. L.; Lyman, D. J. In *The Chemistry of Biosurfaces;* Hair, M. L., Ed.; Marcel Dekker: New York, NY, 1971; p. 177.
6. MacRitchie, F. *Advan. Protein Chem.* **1978**, *32*, 283.
7. Ivarsson, B.; Lundström, I. *CRC Critical Reviews in Biocompatibility* **1985**, *2*, 1.
8. Norde, W. *Advan. Colloid and Interface Sci.* **1986**, *25*, 267.
9. Dong, D. E.; Andrade, J. D.; Coleman, D. L. *J. Biomed. Mater. Res.* **1987**, *21*(6), 683-700.
10. Ho, C-H.; Hlady, V.; Nyquist, G.; Andrade, J. D.; Caldwell, K. D. *J. Biomed. Mater. Res.*, **1991**, *25*(4), 423-441.
11. Li, J.; Caldwell, K. D.; Huang, S-C.; Yan, G. *Calorimetric Studies of Protein Conformations at Solid/Liquid Interfaces* **1994**. This symposium.
12. Norde, W.; *Clinical Materials* **1991**, *11*, 85-91.
13. Pettit, D. K.; Horbett, T. A.; Hoffman, A. S. *J. Biomed. Mater. Res* **1992**, *26*, 1259-1275.
14. Urry, D. W.; Peng, S. Q.; Parker, T. M. *J. Am. Chem. Soc.* **1993**, *115*, 7509-7510.
15. Urry, D. W.; Peng, S. Q.; Parker, T. M.; Gowda, D. C.; Harris, R. D. *Angew. Chem. (German)* **1993**, *105*, 1523-1525; *Angew. Chem. Int. Ed. Engl.* **1993**, *32*, 1440-1442.
16. Urry, D. W.; S. Q. Peng; Gowda, D. C.; Parker, T. M.; Harris, R. D. *Chem. Phys. Ltrs.* **1994** *225* 97-103.
17. Urry, D. W. *Angew. Chem. (German)* **1993**, *105*, 859-883; *Angew. Chem. Int. Ed. Engl.* **1993**, *32*, 819-841.
18. Urry, D. W.; Gowda, D. C.; Peng, S. Q.; Parker, T. M.; Harris, R. D. *J. Am. Chem. Soc.* **1992**, *114*, 8716-8717.
19. Urry, D. W.; Gowda, D. C.; Peng, S. Q., Parker, T. M.; Jing, N.; Harris, R. D. *Biopolymers*, **1994**, *34* , 889-896.
20. Anfinsen, C. B. *Science*, **1973**, *181*, 223-230.
21. Edsall, J. T. *J. Am. Chem. Soc.* **1935**, *57*, 1506-1507.
22. Frank, H. S.; Evans, M. W. *J. Chem. Phys.*, **1945**, *13*, 507-532.
23. Kauzmann, W. *Adv. Protein Chem.* **1959**, *14*, 1-63.
24. Némethy, G.; Scheraga, H. A. *J. Phys. Chem.* **1962**, *66*, 1773-1789.
25. Némethy, G.; Scheraga, H. A. *J. Chem. Phys.* **1962**, *36*, 3382-3417.
26. Tanford, C. *The Hydrophobic Effect: Formation of Micelles and Biological Membranes*; John Wiley & Sons: New York, NY, 1980.
27. Ben-Naim, A. *Hydrophobic Interactions*; Plenum Press: New York, NY, 1980.
28. Edsall, J. T.; McKenzie, H. A. *Adv. Biophys.* **1983**, *16*, 53-183.
29. Evans, P. A.; Radford, S. E. *Curr. Op. Struct. Biol.* **1994**, *4*, 100-106
30. Dobson, C. M.; Evans, P. A.; Radford, S. E. *Trends in Biochem. Sci* . **1994**, *19*, 31-37.
31. Dill, K. A.; Fiebig, F. M.; Chan, H. *Proc. Natl. Acad. Sci. USA* **1993**, *90*, 1942-1946.
32. Urry, D. W.; Gowda, D. C.; Parker, T. M.; Luan, C-H.; Reid, M. C.; Harris, C. M.; Pattanaik, A.; Harris, R. D. *Biopolymers* **1992**, *32*, 1243-1250.
33. Nozaki, Y.; Tanford, C. *J. Biol. Chem.* **1971**, *246*, 2211-2217.
34. Kyte, J.; Doolittle, R. F. *J. Mol. Biol.* **1982**, *157*, 105-132.
35. Cornette, J. L.; Cease, K. B.; Margalit, H.; Spouge, T. L.; Berzofsky, T. A.; DeLisi, C. *J. Mol. Biol.* **1987**, *195*, 659-685.

36. Esposti, M. D.; Crimi, M.; Venturoli, G. *Eur. J. Biochem.* **1990**, *190*, 207-219.
37. Kornblihtt, A. R.; Umezawa, K.; Vibe-Pedersen, K.; Baralle, F. E. *Embo J.* **1985**, *7*, 1755-1759.
38. Conway, J. F.; Parry, D. A. D. *Int. J. Biol. Macromol.* **1991**, *13*, 14-16.
39. Urry, D. W.; *Int. J. Quant. Chem.: Quant. Biol. Symp.* **1994** (in press).
40. Matheson, Jr., R. R.; Scheraga, H. A. *Macromolecules* **1978**, *11*, 819-829.
41. Leahy, D. J.; Hendrickson, W. A.; Aukhil, I.; Erickson, H. P. *Science* **1992**, *258*, 987-991.
42. Kanost, M. R.; Boguski, M. S.; Freeman, M.; Gordon, J. I.; Wyatt, G. R.; Wells, M. A. *J. Biol. Chem.* **1988**, *263*, 10568-10573.
43. Cole, K. D.; Fernando-Warnakulasuriya, G. J. P.; Boguski, M. S.; Freeman, M.; Gordon, J. I.; Clark, W. A.; Law, J. H.; Wells, M. A *J. Biol. Chem.* **1987**, *262*, 11794-11800.
44. Breiter, D. R.; Kanost, M. R.; Benning, M. M.; Wesenberg, G.; Law, J. H.; Wells, M. A.; Rayment, I.; Holden, H. M. *Biochemistry*, **1991**, *30*, 603-608.
45. Urry, D. W.; Gowda, D. C.; Cox, B. A; Hoban, L. D.; McKee; A.; Williams, T. *Mat. Res. Soc. Symp. Proc.* **1993**, *292*, 253-264.
46. Hoban, L.; Pierce, M.; Quance, J.; Hayward, I.; McKee, A.; Gowda, D. C.; Urry, D. W.; Williams, T. *J. Surgical Res.* **1994**, *56*, 179-183.
47. Elsas, F. J.; Gowda, D. C.; Urry, D. W. *J. Pediatr. Ophthalmol. Strabismus* **1992**, *29*, 284-286.
48. Ruoslahti, E *J. Clin. Invest* **1991**, *87*, 1-5.
49. Pierschbacher, M. D.; Ruoslahti, E. *Nature* **1984**, *309*, 30-33.
50. Nicol, A.; Gowda, D. C.; Urry, D. W. *J. Biomed. Mater. Res.* **1992**, *26*, 393-413.
51. Nicol, A.; Gowda, D. C.; Parker, T. M.; Urry, D. W. In *Biotechnol. Bioactive Poly.*; Gebelein, C. G.; Carraher, Jr., C. E., Eds.; Plenum Press: New York, NY, 1994; pp 95-113.
52. Urry, D. W.; Nicol, A.; Gowda, D. C.; Hoban, L. D.; McKee, A.; Williams, T.; Olsen, D. B.; Cox, B. A. In *Biotechnological Polymers: Medical, Pharmaceutical and Industrial Applications*; Gebelein, C. G., Ed.; Technomic Publishing Co., Inc.: Atlanta, GA, 1993; pp 82-103.
53. Nicol, A.; Gowda, D. C.; Parker, T. M.; Urry, D. W. *J. Biomed. Mater. Res.* **1993**, *27*, 801-810.

RECEIVED August 1, 1995

COMPETITIVE ADSORPTION OF PROTEINS

Chapter 8

The Vroman Effect

A Critical Review

Steven M. Slack[1] and Thomas A. Horbett[2]

[1]Department of Biomedical Engineering, University of Memphis,
Memphis, TN 38152
[2]Department of Chemical Engineering, University of Washington,
Seattle, WA 98195–1750

Fibrinogen adsorption from plasma to solid surfaces passes through
a maximum when studied as a function of adsorption time, plasma
dilution, or column height in narrow spaces.   Adsorption from
plasma increases rapidly and then passes through a maximum after
short contact times, and is much lower at steady state than the
maximum seen in the transient phase. If adsorption time is held
constant but the plasma is diluted, maximal fibrinogen adsorption
occurs at intermediate plasma dilutions. Finally, if both time and
plasma concentration are held constant but adsorption occurs in the
narrow space under a lens placed on a surface, the fibrinogen
maximum occurs as a ring corresponding to a particular column
height of liquid and distance from the center.   These time,
concentration, and spatially dependent maxima are all related to the
displacement of fibrinogen known as the Vroman effect. Although
this phenomenon was initially believed to be unique to the
adsorption behavior of fibrinogen, subsequent studies have
demonstrated that the Vroman effect is a general phenomenon
reflecting competitive adsorption of proteins for a finite number of
surface sites. In this review, the experimental factors affecting
fibrinogen displacement, the underlying mechanisms responsible
for the Vroman effect, and the significance of the Vroman effect
with respect to blood-material interactions will be examined.

The adsorption of plasma proteins from blood onto the surface of artificial
materials occurs very rapidly and is considered to be an important event affecting
the subsequent adhesion of blood cells, especially platelets, that are involved in
thrombogenesis on foreign surfaces (1-4). Because blood plasma contains well
over one hundred distinct proteins, only the adsorption behavior of the most
prevalent species has been studied extensively. Of particular interest has been the
adsorption behavior of fibrinogen, an abundant plasma protein that plays a central
role in blood coagulation (5) and that also supports platelet adhesion and
aggregation through its binding to the glycoprotein (GP) IIb-IIIa integrin receptor
(6,7).

0097–6156/95/0602–0112$12.00/0

Almost a quarter of a century ago Leo Vroman and his colleagues made the observation that fibrinogen adsorbed from plasma to several materials, including glass, anodized tantalum, and oxidized silicon, appeared to undergo a transformation that rendered it immunologically undetectable within seconds or minutes of surface contact (*8-10*). This loss in reactivity with antifibrinogen sera was subsequently termed "conversion" and was the first described example of the fibrinogen displacement phenomenon now referred to as the Vroman effect in recognition of Vroman's initial studies (*11,12*). At the time, the ellipsometric methods used by Vroman and his co-workers could not discriminate between different potential causes of this "conversion". The loss in immunological reactivity could have resulted from loss of epitope reactivity due to structural or orientational changes induced by immobilization of fibrinogen on a solid substrate, replacement of adsorbed fibrinogen by other plasma proteins, proteolytic digestion of adsorbed fibrinogen by an enzyme such as plasmin, or masking of fibrinogen by adsorbing proteins. The latter two events might result in marked changes in the thickness of the adsorbed protein film, but because no significant variations in the film thickness were noted, it appeared as though the likely causes of the "conversion" involved either structural alterations in adsorbed fibrinogen resulting in diminished reactivity with antisera or replacement of fibrinogen by one or more different plasma proteins (*8*).

Since those first observations were made, numerous investigators have shown, primarily with the use of ^{125}I-labeled fibrinogen but also with the Enzyme-Linked Immunoassay (ELISA) technique, that initially adsorbed fibrinogen is displaced from the surface of many synthetic materials, presumably by plasma proteins with a higher affinity for the surface (*11-24*). Maxima in fibrinogen adsorption have been observed as a function of adsorption time, plasma dilution, and column height in narrow spaces, and the term "Vroman effect" is now used to describe all these situations. Although the Vroman effect has been observed on many materials, the magnitude of fibrinogen displacement varies greatly. Moreover, some substrates, e.g., certain sulfonated polyurethanes (*25*) and copolymers containing large amounts of polyhydroxyethylmethacrylate (HEMA) (*23*) exhibit no peak in fibrinogen adsorption as a function of plasma dilution.

The identity of the protein(s) displacing adsorbed fibrinogen is not clear, at least on some surfaces. Surprisingly, certain clotting factors participating in the intrinsic coagulation pathway that are present in plasma at very low concentrations appear to play a role in the process under certain conditions, as will be discussed in a later section (*20,26-29*). In addition, the significance of the Vroman effect with respect to its influence on the blood compatibility of artificial materials has yet to be fully clarified. And finally, although a few conceptual and mathematical models have been constructed to explain the existence of the Vroman effect and its variation among materials of differing surface chemistries (*30-33*), the complexity of protein interactions with other proteins and with solid surfaces has hindered the development of a completely satisfactory description of this phenomenon.

The aims of this chapter are to review the progress that has been achieved over the last decade in understanding various aspects of the Vroman effect, beginning with a summary of factors influencing fibrinogen adsorption from blood plasma, an examination of the proposed role of contact activation clotting factors in the displacement of adsorbed fibrinogen, a brief review of the transitions in adsorbed fibrinogen that probably play a role in the Vroman effect, and finally a discussion of the mathematical models developed to describe the Vroman effect.

Factors Affecting Fibrinogen Adsorption From Plasma

The amount of fibrinogen adsorbed from plasma to artificial materials depends on several factors including adsorption time, plasma dilution, material surface properties, buffer composition, and temperature. The effects of these factors on fibrinogen adsorption from plasma will be reviewed here because it is not possible to understand the results from separate laboratories unless it is appreciated that the different experimental conditions used in each lab strongly influence the results.

Since Vroman's early studies demonstrating that fibrinogen appeared to become immunologically unreactive with increasing adsorption time (8,9), several investigators have demonstrated that this "conversion" reflects the displacement of initially adsorbed fibrinogen by other plasma proteins (17,18,22,34-36). The time course of fibrinogen adsorption from plasma to two materials, glass and poly(ethylmethacrylate), is shown in Figure 1 and clearly illustrates the rapid displacement of initially adsorbed fibrinogen, especially for the latter polymer. On a few materials, fibrinogen adsorption from plasma increases slightly with time (18). Occasionally, adsorption and displacement are so rapid, especially using undiluted plasma, that the amount of adsorbed fibrinogen appears not to change over time (18). By diluting the plasma prior to exposure to surfaces, the rates of adsorption and subsequent displacement of adsorbed fibrinogen can be reduced. From highly diluted plasma, fibrinogen displacement is not observed and the time course of fibrinogen adsorption is similar to that observed from pure fibrinogen solutions although the amount of adsorption is substantially less.

As alluded to above, the processes of fibrinogen adsorption and displacement can be studied by diluting the plasma with buffer prior to exposure to the material. In this way fibrinogen adsorption isotherms from plasma can be measured and this method has been exploited extensively in studies of the Vroman effect. Plots of the amount of adsorbed fibrinogen versus plasma dilution yield curves that exhibit a maximum at intermediate plasma dilutions, and representative results measured on several different materials are shown in Figures 2 and 3. Because of the nature of the fibrinogen displacement process, the plasma dilution at which peak adsorption occurs as well as the amount of adsorbed fibrinogen differ considerably depending on the adsorption time; lengthier exposure times give rise to smaller amounts of adsorbed fibrinogen as well as adsorption maxima that occur at greater plasma dilutions (24).

The adsorption of fibrinogen from plasma has been studied on a variety of materials, including glass, numerous polyurethanes such as Biomer and Tecoflex, copolymers of hydroxyethylmethacrylate (HEMA) and ethylmethacrylate (EMA), silicone rubber, polystyrene, polyvinylchloride, and polyethylene. Adsorption is typically greater on more hydrophobic materials such as polystyrene or poly(ethylmethacrylate) and maximal adsorption usually, but not always, occurs at

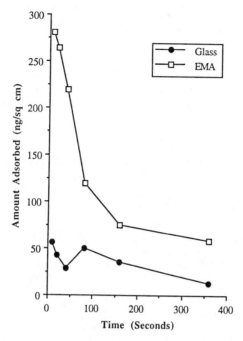

Figure 1. The time course of fibrinogen adsorption to glass and poly(ethylmethacrylate) (EMA) from undiluted blood plasma.

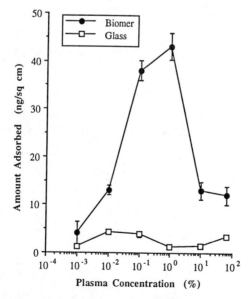

Figure 2. Fibrinogen adsorption to Biomer and glass from various concentrations of blood plasma.

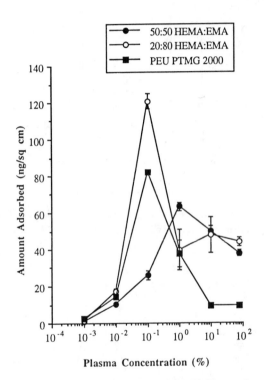

Figure 3. Fibrinogen adsorption to two HEMA:EMA copolymers and a segmented polyetherurethane (PEU) from various concentrations of blood plasma.

greater plasma concentrations (smaller plasma dilutions) on such materials. Widely different amounts of fibrinogen adsorption have been reported by different investigators studying apparently identical materials (*11,12,17,31,37*), a result caused by the use of differing experimental conditions such as adsorption time and temperature. Attempts to correlate the amount of fibrinogen adsorbed from plasma with surface properties such as wettability or contact angle have been unsuccessful (*17*).

The nature of fibrinogen adsorption to and displacement from artificial materials is further complicated on the basis of results demonstrating a strong effect of the choice of plasma diluent on fibrinogen adsorption. For instance, fibrinogen adsorption to glass from plasma diluted with Tris-buffered saline was approximately twice as great as that obtained with plasma diluted with citrate-phosphate buffered saline (*24*).

Temperature has also been shown to influence fibrinogen adsorption to several surfaces. Vroman and Adams (*9*) noted that the "conversion" of fibrinogen on oxidized silicon was significantly greater and occurred more rapidly at 37 °C compared to either 10 °C or 25 °C. Temperature also strongly affects fibrinogen adsorption to polyethylene and glass, from both pure solutions (*38*) and plasma (*19,24*). Specifically, fibrinogen adsorption from plasma to glass was markedly greater at 25 °C compared to that at 37 °C, an effect proposed to result from an inhibition of fibrinogen displacement at the lower temperature (*24*). Le et al. (*39*) report similar observations regarding the effects of temperature on fibrinogen adsorption in a separate chapter within this volume. An alternative explanation, that proteolysis of adsorbed fibrinogen by an enzyme such as plasmin was inhibited at 25 °C, thereby resulting in increased amounts of adsorbed fibrinogen, seems unlikely based on studies indicating that fibrinogen displacement from glass was similar from normal and plasminogen-deficient plasma (*28*).

Based on this short summary, then, it is clear that a number of factors influence fibrinogen adsorption to synthetic materials. Thus, a comparison of fibrinogen adsorption measurements (from both pure solutions and blood plasma) reported in the literature can only be made after careful consideration of the varied experimental conditions used in the different laboratories.

Role of the Intrinsic Coagulation Proteins

Hematologists have known for many years that negatively charged materials, such as glass and kaolin, can initiate blood coagulation through the so-called intrinsic (or contact-phase) pathway. The mechanism appears to involve the autoactivation of adsorbed Factor XII (Hageman factor), which in turn proteolytically activates two other clotting factors, prekallikrein (PK) and Factor XI, to their enzymatically active forms, kallikrein and Factor XIa, respectively. Another protein, high molecular weight kininogen (HK), circulates as a complex with PK and Factor XI (*40*) and serves to facilitate the localization of these factors at the surface of the material where they can be activated by Factor XIIa. Newly formed kallikrein then exerts positive feedback on the system through its ability to generate additional Factor XIIa (*5*). More recently, it has been demonstrated that a two-chain activated form of HK (HKa), generated by proteolytic cleavage of HK by kallikrein (41) and/or Factor XIIa (42), exhibits a higher affinity for negatively charged surfaces and thus greater procoagulant activity. Also, subsequent work has shown that HKa can be proteolytically degraded into an inactive form by the action of Factor XIa (43).

At approximately the same time that many of the details of the intrinsic pathway of blood coagulation were being elucidated, Vroman and his colleagues (26) reported that the "conversion" of adsorbed fibrinogen was greatly delayed when plasma congenitally deficient in HK (Fitzgerald trait) was used. They further reported that a slight delay was present using Factor XII-deficient plasma whereas "conversion" appeared normal with plasmas deficient in either Factor XI or PK. Slack et al. (19) also showed that the Vroman peaks on glass and polyethylene were unaffected by a deficiency of PK in the plasma. In a later study, Schmaier et al. (27) showed that the abnormal "conversion" observed with HK-deficient plasma could be reversed by adding purified HK back to the plasma. Scott et al. (41) extended this work by presenting evidence that HK circulates as a procofactor which must first be proteolytically cleaved by kallikrein prior to manifesting maximal procoagulant activity. In this study, it was shown that proteolytically activated HK (HKa) possessed a higher affinity for kaolin than the uncleaved form, and also that fibrinogen (3 mg/ml) competitively inhibited the adsorption of HK but not HKa to kaolin.

Additional evidence for a role of HK in fibrinogen displacement came from Vroman et al. (15,16) who measured fibrinogen deposition in a convex lens on slide system from normal and HK-deficient plasma. In this system, which is illustrated in a separate contribution in this volume by Elwing et al. (44), plasma is injected into the narrow space formed by placing a convex lens atop a glass slide, allowed to reside for a period of time, rinsed away, and finally probed for the presence of fibrinogen using antisera against fibrinogen. In the most narrow regions, they noted the presence of adsorbed fibrinogen; at distances farther away from the center, little or no fibrinogen was detected. They interpreted these data to mean that in the very narrow regions, sufficient fibrinogen was available to coat the glass but insufficient HK was present to displace the fibrinogen owing to HK's extremely small plasma concentration. In the larger gaps, enough of both proteins was available and HK displaced the initially adsorbed fibrinogen. Using HK-deficient plasma, the distance from the center where fibrinogen could not be detected was increased compared to that observed for normal plasma. As the authors noted, the experiments were qualitative since "the relationship between surface concentration of fibrinogen, and density of antibody deposited on it, may not be linear...", but they do provide evidence that HK displaces fibrinogen adsorbed to glass.

A more recent study by Poot et al. (45), again using antibodies and a two-step ELISA assay to detect adsorbed proteins, also showed that HK played a role in displacing fibrinogen adsorbed from plasma to glass. On polyethylene, however, no effect of HK was observed but their data did suggest that high density lipoprotein (HDL) might be responsible for the low adsorption of fibrinogen. Additional observations by the same group are presented in a separate chapter in this volume (46).

Taken together, these observations suggested that HKa played a central role in fibrinogen displacement from negatively charged surfaces such as glass and kaolin. Thus, an intriguing relationship between fibrinogen displacement, or the Vroman effect, and the initiation of contact activation began to emerge. This apparent role for HK in the Vroman effect presented an interesting dilemma to investigators in the biomaterials field. On the one hand, a material exhibiting diminished fibrinogen adsorption at normal plasma concentrations because it had evoked a strong Vroman effect implied that it also initiated a potent response by the intrinsic pathway of blood coagulation. Conversely, reduced amounts of adsorbed fibrinogen also would suggest that the material would adhere fewer platelets and therefore might be less thrombogenic. Numerous studies had previously demonstrated that surfaces pre-adsorbed with fibrinogen could readily support platelet adhesion (47-50). Thus, an assessment of a material's potential blood compatibility could differ significantly depending on whether initiation of blood coagulation or platelet adhesion was considered the most important predictive factor.

Despite the extensive evidence implicating a role for contact phase clotting factors in the displacement of adsorbed fibrinogen, subsequent studies by Brash et al. (28) demonstrated that even with plasma congenitally deficient in HK, a concentration domain Vroman effect on glass was clearly discernible, although less fibrinogen was displaced than that from normal plasma. These results indicated to the authors "that other components of plasma are also active in displacing fibrinogen." Indeed, there are several lines of evidence suggesting that contact activation factors play little, if any, role in the Vroman effect observed on other artificial materials. First, Slack et al. (19) demonstrated that fibrinogen adsorption to two other materials, silicone rubber and polyethylene, did not differ from normal or HK deficient plasma. Because these polymers do not activate the intrinsic pathway to any significant degree, little HKa generation is expected, and yet both exhibit distinct Vroman effects, with maximal fibrinogen adsorption occurring at a plasma concentration of ~ 1%. Furthermore, because many other hydrophobic materials that would be expected to be poor activators of the intrinsic coagulation pathway also exhibit Vroman peaks, it seems unlikely that contact activation is necessary for a material to exhibit a Vroman effect. For example, a recent study (51) has shown that certain titanium alloys, which are used extensively in orthopedic applications, exhibit a Vroman effect but minimally activate the intrinsic pathway compared to the positive glass control. Thus, the relationship between the intrinsic coagulation cascade and fibrinogen displacement became blurred.

Further evidence has been provided by Elwing and colleagues that suggests HK does not play a role in the displacement of fibrinogen from hydrophobic surfaces (20,52). In their study, silicon wafers were treated in such a way as to generate a surface wettability gradient and these materials were then exposed to plasma for 1 minute or 1 hour. Using ellipsometry and antisera to fibrinogen and HK, they observed decreased amounts of anti-fibrinogen binding at the hydrophobic end of the wafer after 1 hour compared to 1 minute but the decrease was not accompanied by an increase in anti-HK binding. Conversely, at the hydrophilic end of the material, the decreased amount of anti-fibrinogen binding observed after 1 hour was accompanied by a concomitant increase in anti-HK binding. These studies indicate that HK may play a role in effecting fibrinogen displacement from hydrophilic surfaces but that other plasma proteins are responsible for displacing fibrinogen from hydrophobic surfaces.

The previous studies by Vroman and others, demonstrating that proteolytically activated HKa displaced adsorbed fibrinogen, were carried out at room temperature. Because enzymatic activity is strongly influenced by temperature, Slack et al. (*19*) compared the adsorption of fibrinogen from normal and HK-deficient plasma to glass at room temperature (~ 25 °C) and 37 °C. At 25 °C significantly more fibrinogen adsorbed to glass from HK deficient plasma than from normal plasma, in agreement with earlier studies, whereas at 37 °C no differences were observed. Inasmuch as the influence of HKa on fibrinogen adsorption from plasma to glass is nonexistent at physiological temperature, the importance of the contact activation pathway with respect to the Vroman effect is unclear at 37 °C.

Perhaps the clearest evidence indicating that contact activation factors are not necessarily important with regard to the Vroman effect stem from studies of fibrinogen adsorption from binary and ternary protein solutions (*19,21,53*). In one study, mixtures of fibrinogen and either oxyhemoglobin or bovine albumin were prepared and fibrinogen adsorption to polyethylene from serial dilutions of the binary mixtures was measured. Adsorption passed through maxima from both solutions, with the peak shifting to smaller mixture dilutions as the ratio of competing protein to fibrinogen increased. A peak in fibrinogen adsorption from these mixtures was also observed to occur as a function of adsorption time. Interestingly, less oxyhemoglobin than albumin was required for a given amount of inhibition, a result in agreement with a previous study demonstrating that oxyhemoglobin was a better competitor than albumin for adsorption to glass (*11*). In the other study, the adsorption of albumin, IgG, and fibrinogen from various dilutions of a ternary mixture of the three components to polystyrene (PS) and polyvinylchloride (PVC) was measured using immunological techniques (*54*). Both albumin and IgG, but not fibrinogen, oddly, exhibited adsorption maxima at intermediate mixture dilutions. The authors also measured the adsorption of the same proteins from various plasma dilutions to PS and PVC and demonstrated a Vroman effect for each protein.

It should be noted that detection of adsorbed fibrinogen using immunological techniques, i.e., antibodies, presents subtle difficulties with respect to interpretation of the data. For instance, the process of immobilizing a protein on a solid surface may result in changes or even losses in the reactivity of its epitope with the antibody. Also, antibodies may be inhibited from binding their antigen due to steric considerations. On the other hand, at least one study has shown that measurements of fibrinogen adsorbed to several HEMA/EMA copolymers with both ^{125}I-fibrinogen and an ELISA technique (using a polyclonal antifibrinogen antibody) were well-correlated (*23*). In general, results of immunological studies of the Vroman effect must be interpreted cautiously because of the complexities associated with antibody-antigen interactions.

The results of binary and ternary competition studies are consistent with the hypothesis that the Vroman effect reflects the competition between numerous proteins of differing surface affinities for a limited number of adsorption sites. This does not mean that contact activation proteins do not or cannot displace adsorbed fibrinogen; indeed, because of their role in initiating coagulation on surfaces, it is expected that they would exhibit high surface affinities, particularly for negatively charged materials. However, their relative role in displacing fibrinogen, at least at physiological temperature, appears to be a minor one, with other plasma proteins being primarily responsible for the displacement, especially on hydrophobic materials.

In this view of the Vroman effect, fibrinogen displacement occurs as a result of the action of numerous plasma proteins, each with a unique concentration and surface affinity. Hence, proteins other than fibrinogen would be expected to exhibit maximal adsorption from complex protein mixtures such as plasma. This has been demonstrated with several other proteins, including albumin (*34,54*) and IgG (*12,54*). Moreover, Grinnell and others have shown that fibronectin adsorption to tissue culture polystyrene (*55*) and glass (*56*) passes through a maximum from intermediate dilutions of serum. Thus, the Vroman effect appears to be a somewhat general phenomenon reflecting competitive adsorption of proteins for a finite number of surface sites rather than an effect unique to fibrinogen or initiation of the contact activation pathway.

Transitions in Adsorbed Fibrinogen and Other Proteins

During the last ten to fifteen years, evidence has been collected in various laboratories that protein molecules undergo rearrangements following their adsorption to solid surfaces (*2,57*). The transitions seem to be important in the Vroman effect because they affect the tightness with which the proteins are bound, i.e., their displaceability is related to the transitions. In addition, the transitions modulate the biological activity of the adsorbed adhesion proteins. The increases in tightness of binding and the modulation of biological activity have been shown to be affected by the post-adsorptive residence time, surface chemistry, and co-adsorbed proteins.

Rearrangements in adsorbed proteins are indicated by losses in enzymatic activity (*58,59*), residence time dependent decreases in protein elutability by various detergents (*36,60*), reduction in the ability of plasma components to displace fibrinogen (*31,61*), and expression of neoantigenic epitopes recognized by monoclonal antibodies (*62*). Representative results of fibrinogen displacement by plasma as a function of residence time are shown in Figure 4 for polyethylene and the more hydrophilic 50:50 HEMA:EMA copolymer. As seen in the figure, fibrinogen adsorbed to PE rapidly becomes resistant to displacement by plasma with increasing residence time whereas for the copolymer the process occurs at a much slower rate. These differences most likely reflect the nature of the interaction of fibrinogen with surfaces of differing hydrophobicity, as discussed below. Fibrinogen also undergoes a time-dependent transition following adsorption to certain surfaces, from both pure solutions (*60*) and plasma (*36*), that results in a reduction in its elutability by the anionic surfactant sodium dodecyl sulfate (SDS) (*60*). In addition, changes in the amide II frequency of adsorbed fibrinogen with increasing residence time have been observed (*63*). Modulation of

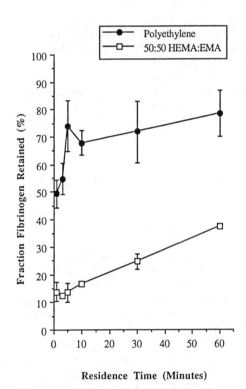

Figure 4. The fraction of initially adsorbed fibrinogen retained by polyethylene and a 50:50 HEMA:EMA copolymer as a function of the residence time of fibrinogen on the surface of the materials.

the biological activity of adsorbed fibrinogen is indicated by residence time dependent reductions in platelet and antibody binding to the adsorbed fibrinogen (*64,65*). Modulation of the biologic activity of adsorbed fibronectin has also been shown in several studies in which the ability of fibronectin adsorbed to various surfaces to support cell attachment or spreading was found to differ. Thus, several groups have shown that substrata of varying chemical composition to which fibronectin was preadsorbed varied considerably in their ability to induce cell spreading and other morphological features (*66*). Lewandowska et al concluded that the "experiments reveal the considerable flexibility that FNs utilize in binding to inert surfaces in order to effect various responses from cells. Overall, they reveal that cell responses are altered, not because of limiting amounts of pFN adsorbed to any particular derivatized surface, but because of the varied conformation of pFN molecules as they interact with the specific chemical endgroups of the self-assembled monolayer." (*66*).

The co-adsorption of albumin has been shown to affect transitions in adsorbed fibrinogen and fibronectin. Thus, the addition of BSA to the solution used to store the fibrinogen adsorbed surfaces during the residence time prevents the loss of platelet adhesion, an effect we have called albumin "trapping". Lewandowska et al. showed that on hydrophobic surfaces preadsorbed with fibronectin, there was poor cell spreading unless albumin was coadsorbed, an effect called albumin "rescuing" (*66*). Grinnell's group showed that fibronectin adsorbed to tissue culture grade polystyrene was able to support cell attachment and spreading, whereas fibronectin adsorbed to ordinary polystyrene did not support spreading unless some albumin was added to the fibronectin solution (*55,67*).

The concept of surface-induced transitions in adsorbed fibrinogen that evolved from these studies is relevant to the Vroman effect for several reasons. These transitions often resulted in a tighter attachment to the surface that endow fibrinogen with the capability to resist displacement. An intrinsic feature of this idea is that such rearrangements may occur at different rates on the various materials, which suggested that the variation of the Vroman effect among materials with diverse surface chemistries might be related to these transformations. Finally, the effect of coadsorbed albumin on molecular spreading of fibrinogen that would presumably accompany these transitions suggested a role for the availability of vacant surface sites.

Mathematical Models of the Vroman Effect

Protein adsorption to solid surfaces has traditionally been described by a simple Langmuir isotherm, i.e.,

$$\Gamma = \Gamma_{max} \frac{KC_p}{1 + KC_p} \qquad (1)$$

where Γ = adsorbed amount (mass or moles per unit area), Γ_{max} = maximum adsorption (typically estimated as the amount present in a monolayer), K is the adsorption equilibrium constant, and C_p is the solution protein concentration (mass or moles per unit volume). In some cases, the Freundlich adsorption isotherm has been found to more accurately describe protein adsorption data. Neither model is especially realistic, however, since assumptions regarding reversibility and the absence of interactions between co-adsorbed molecules are probably not valid in protein systems.

Approaches for the modeling of the Vroman effect pose additional problems in that fibrinogen displacement, and its dependency on plasma dilution and substrate surface chemistry, must be considered. Because the amount of adsorbed fibrinogen as well as the plasma dilution at which peak adsorption occurs vary widely among synthetic materials, any model of the Vroman effect should incorporate information regarding the interactions of fibrinogen with solid substrates. Thus, the transitions in adsorbed proteins, perceived as a process of molecular spreading in which fibrinogen (and presumably other proteins) maximized its contact with the surface resulting in tighter attachment, were conceived to be fundamental in explaining the displacement process. An empirical model was therefore developed that incorporated the processes of adsorption, transition, and displacement of weakly-bound fibrinogen (*31*). A simplification was made that the remainder of the plasma proteins, many of which probably contributed to fibrinogen displacement, could be treated as one "hypothetical" protein with a surface affinity equivalent to the cumulative affinity of the plasma proteins. Transition rate constants were determined for four materials from mathematical fitting of the experimental data to the relevant equations and the model was found to qualitatively explain the observed differences in the Vroman effect. Large differences in the magnitudes of the transition rate constants on the different substrates were observed, helping explain the variations in the Vroman effect on these materials.

An alternative mathematical model has been proposed (*30*) to explain the Vroman effect in which adsorption and desorption rate constants were assumed to depend on the total surface protein concentration. Additionally, an interaction term was included to describe the effects of repulsive and/or attractive forces between co-adsorbed protein molecules. Computer simulations indicated that maxima in protein adsorption could be obtained and, depending on the magnitude of the interaction term, the influence of plasma dilution on the amount of adsorbed fibrinogen could be qualitatively explained. Subsequent Monte Carlo simulations of protein adsorption by the same group (*32*) indicated that the decline in adsorption rates with increasing surface coverage could differ considerably among various proteins and that the Vroman effect could be explained by accounting for surface exclusion effects and the lateral mobility of adsorbed proteins.

More recently, Chittur and his colleagues (*33*) have developed a mathematical model to describe protein adsorption at the solid/liquid interface that incorporates the diffusion and reversible adsorption of the protein followed by its transformation to an irreversibly bound form. Computer simulations based on a ternary mixture of albumin, IgG, and fibrinogen (present at concentrations similar to those expected in 0.1% plasma) demonstrated that, depending on the relative magnitude of the rate constants assumed for the three proteins, adsorption maxima as a function of contact time could be predicted and the amounts of adsorbed protein were comparable to those observed experimentally.

Future mathematical treatments that successfully describe fibrinogen adsorption from plasma and the resulting Vroman effect will likely require elements from each of the models presented above. In particular, experimental evidence for the transitions undergone by adsorbed fibrinogen and quantitative measurements of the variation in the rates of those transitions on different synthetic materials, such as those presented by Slack and Horbett (*31*), will need to be combined with data regarding interactions between co-adsorbed proteins and related information before the Vroman effect can be adequately modeled.

Summary and Conclusions

The adsorption of fibrinogen to artificial materials and its subsequent displacement (the Vroman effect) continues to inspire research efforts in the area of blood-material interactions. The interest can be traced to the fact that fibrinogen plays a central role in the coagulation pathway and in the processes of platelet adhesion and aggregation, essential events in normal hemostasis but undesirable with respect to the performance of synthetic materials used in blood-contacting devices. Hence, an understanding of the factors governing fibrinogen adsorption to and displacement from artificial materials should help to provide a rational framework for the development of new materials exhibiting improved blood compatibility. Although progress in this area has been achieved over the last ten years, much remains to be elucidated. In particular, future studies will need to address the issues of the heterogeneity of adsorbed fibrinogen, i.e., weakly versus tightly bound molecules, and the surface characteristics contributing to the transition of adsorbed fibrinogen. In addition, the significance of conformational changes in fibrinogen, induced by immobilization on solid substrates, with respect to platelet reactivity needs to be considered. It is not at all clear that materials adsorbing less fibrinogen will be less thrombogenic than those adsorbing large amounts. It is conceivable, for instance, that on the latter material the fibrinogen is oriented in such a way (or becomes oriented over time) that the regions recognized by platelets are unavailable with the opposite being true for the former material. In such a case, a judgment regarding the potential blood compatibility of a new material based solely on the amount of adsorbed fibrinogen would be questionable. Finally, it is anticipated that an understanding of the Vroman effect at a molecular level, combined with a more sophisticated mathematical treatment of this phenomenon, will contribute not only to studies in the field of blood-material interactions but also to related research areas in which proteins and other macromolecules interact with solid surfaces.

Acknowledgments

The authors acknowledge the financial assistance of the National Heart, Lung, and Blood institute through grants HL19419 and HL48244.

Literature Cited

1. Horbett, T. A. In *Biomaterials: Interfacial Phenomena and Applications*; S. L. Cooper and N. A. Peppas, Eds.; American Chemical Society: Washington, D.C., 1982; pp 233–244.
2. Horbett, T. A.; Brash, J. L. In *Proteins at Interfaces: Physicochemical and Biochemical Studies*; J. L. Brash and T. A. Horbett, Eds.; American Chemical Society: Washington, D.C., 1987; Vol. 343; pp 1–33.
3. Packham, M. A. *Proc. Soc. Exp. Biol. Med.* **1988**, *189*, 261–274.
4. Brash, J. L. In *Blood Compatible Materials and Devices*; C. P. Sharma and M. Szycher, Eds.; Technomic: Lancaster, 1991; pp 3–24.
5. Colman, R. W.; Marder, V. J.; Salzman, E. W.; Hirsh, J. In *Hemostasis and Thrombosis: Basic Principles and Clinical Practice*; 2nd ed.; R. W. Colman, J. Hirsh, V. J. Marder and E. W. Salzman, Eds.; J.B. Lippincott: Philadelphia, 1987; pp 3–17.
6. Ruoslahti, E. *J. Clin. Invest.* **1991**, *87*, 1–5.
7. Phillips, D. R.; Charo, I. F.; Scarborough, R. M. *Cell* **1991**, *65*, 359–362.
8. Vroman, L.; Adams, A. L. *Surf. Sci.* **1969**, *16*, 438–446.
9. Vroman, L.; Adams, A. L. *J. Biomed. Mater. Res.* **1969**, *3*, 43–67.
10. Vroman, L.; Adams, A. L.; Klings, M. *Fed. Proc.* **1971**, *30*, 1494–1502.
11. Horbett, T. A. *Thromb. Haemostas* **1984**, *51*, 174–181.
12. Brash, J. L.; Hove, P. t. *Thromb. Haemostas.* **1984**, *51*, 326–330.
13. Weathersby, P. K.; Horbett, T. A.; Hoffman, A. S. *J. Bioeng.* **1977**, *1*, 395–410.
14. Ihlenfeld, J. V.; Cooper, S. L. *J. Biomed. Mater. Res.* **1979**, *13*, 577–591.
15. Vroman, L.; Adams, A. L.; Fischer, G. C. *Adv. Chem.* **1982**, *199*, 265–276.
16. Adams, A. L.; Fischer, G. C.; Munoz, P. C.; Vroman, L. *J. Biomed. Mater. Res.* **1984**, *18*, 643–654.
17. Wojciechowski, P.; Hove, P. t.; Brash, J. L. *J. Colloid Interface Sci.* **1986**, *111*, 455–465.
18. Horbett, T. A.; Cheng, C. M.; Ratner, B. D.; Hanson, S. R.; Hoffman, A. S. *J. Biomed. Mater. Res.* **1986**, *20*, 739–772.
19. Slack, S. M.; Bohnert, J. L.; Horbett, T. A. *Ann. N.Y. Acad. Sci.* **1987**, *516*, 223–243.
20. Elwing, H.; Askendal, A.; Lundstrom, I. *J. Biomed. Mater. Res.* **1987**, *21*, 1023–1028.
21. Slack, S. M.; Horbett, T. A. *J. Colloid Interface Sci.* **1988**, *124*, 535–551.
22. Brash, J. L.; Scott, C. F.; Hove, P. t.; Wojciechowski, P.; Colman, R. W. *Blood* **1988**, *71*, 932–939.
23. Slack, S. M.; Posso, S. E.; Horbett, T. A. *J. Biomat. Sci. Polym. Edn.* **1991**, *3*, 49–67.
24. Slack, S. M.; Horbett, T. A. *J. Biomater. Sci. Polymer Edn.* **1991**, *2*, 227–237.
25. Santerre, J. P.; Hove, P. t.; VanderKamp, N. H.; Brash, J. L. *J. Biomed. Mater. Res.* **1992**, *26*, 39–57.
26. Vroman, L.; Adams, A. L.; Fischer, G. L.; Munoz, P. C. *Blood* **1980**, *55*, 156–159.

27. Schmaier, A. H.; Silver, L.; Adams, A. L.; Fischer, G. C.; Munoz, P. C.; Vroman, L.; Colman, R. W. *Thromb. Res.* **1983**, *33*, 51–67.

28. Brash, J. L.; Scott, C. F.; Hove, P. t.; Colman, R. W. *Trans. Soc. Biomater.* **1985**, *11*, 105.

29. Gustafson, E. J.; Lukasiewicz, H.; Wachtfogel, Y. T.; Norton, K. J.; Schmaier, A. H.; Niewiarowski, S.; Colman, R. W. *J. Cell Biol.* **1989**, *109*, 377–387.

30. Cuypers, P. A.; Willems, G. M.; Hemker, H. C.; Hermens, W. T. *Ann. N.Y. Acad. Sci.* **1987**, *516*, 244–252.

31. Slack, S. M.; Horbett, T. A. *J. Colloid Interface Sci.* **1989**, *133*, 148–165.

32. Willems, G. M.; Hermens, W. T.; Hemker, H. C. *J. Biomater. Sci. Polym. Edn.* **1991**, *1*, 217–226.

33. Lu, C. F.; Nadarajah, A.; Chittur, K. K. *J. Coll. Interf. Sci.* **In Press,**

34. Horbett, T. A. *ACS Org. Coat. Plast. Chem. Prepr.* **1979**, *40*, 642–646.

35. Horbett, T. A. In *Adhesion and Adsorption of Polymers*; L. Lee, Ed.; Plenum: New York, 1980; pp 677–682.

36. Rapoza, R. J.; Horbett, T. A. *J. Biomat. Sci. Polymer Edn.* **1989**, *1*, 99–110.

37. Vroman, L.; Adams, A. L. *Thromb. Diath. Haemorrh.* **1967**, *18*, 510–.

38. Brynda, E.; Houska, M.; Kalal, J. *Ann. Biomed. Engr.* **1980**, *8*, 245–252.

39. Le, M. T.; Mulvihill, J. N.; Cazenave, J. P.; Dejardin, P. In *ACS Symposium Series Proteins at Interfaces*; T. A. Horbett and J. L. Brash, Eds.; American Chemical Society: Washington, DC, 1995.

40. Wiggins, R. C.; Bouma, B. N.; Cochrane, C. G.; Griffin, J. H. *Proc. Natl. Acad. Sci. USA* **1977**, *74*, 4636–4640.

41. Scott, C. F.; Silver, L. D.; Schapira, M.; Colman, R. W. *J. Clin. Invest.* **1984**, *73*, 954–962.

42. Wiggins, R. C. *J. Biol. Chem.* **1983**, *258*, 8963.

43. Scott, C. F.; Silver, L. D.; Purdon, A. D.; Colman, R. W. *J. Biol. Chem.* **1985**, *260*, 10856.

44. Elwing, H. B.; Li, L.; Askendal, A. R.; Nimeri, G. S.; Brash, J. L. In *ACS Symposium Series Proteins at Interfaces*; T. A. Horbett and J. L. Brash, Eds.; American Chemical Society: Washington, DC, 1995.

45. Poot, A.; Beugeling, T.; Aken, W. G. v.; Bantjes, A. *J. Biomed. Mater. Res.* **1990,**

46. Turbill, P.; Beugeling, T.; Poot, A. A. In *ACS Symposium Series Proteins at Interfaces*; T. A. Horbett and J. L. Brash, Eds.; American Chemical Society: Washington, DC, 1995.

47. Packham, M. A.; Evans, G.; Glynn, M. F.; Mustard, J. F. *J. Lab. Clin. Med.* **1969**, *73*, 686–697.

48. Zucker, M. B.; Vroman, L. *Proc. Soc. Exp. Med.* **1969**, *131*, 318–320.

49. Mason, R. G.; Shermer, R. W.; Zucker, W. H. *Amer. J. Pathol.* **1973**, *73*, 183–200.

50. Barber, T. A.; Lambrecht, L. K.; Mosher, D. L.; Cooper, S. L. *Scan. Electron Microsc.* **1979**, *III*, 881–890.

51. Yun, Y. H.; Slack, S. M.; Turitto, V. T.; Daigle, K. P.; Davidson, J. A. **Submitted,**

52. Elwing, H.; Welin, S.; Askendal, A.; Nilsson, U.; Lundstrom, I. *J. Coll. Interf. Sci.* **1987**, *119*, 203–210.

53. Lensen, H. G. W.; Breemhaar, W.; Smolders, C. A.; Feijen, J. *J. Chromatog.* **1986**, *376*, 191–198.

54. Breemhaar, W.; Brinkman, E.; Ellens, D. J.; Beugeling, T.; Bantjes, A. *Biomaterials* **1984**, *5*, 269–274.

55. Grinnell, F.; Feld, M. K. *J. Biol. Chem.* **1982**, *257*, 4888–4893.

56. Horbett, T. A.; Schway, M. B. *J. Biomed. Mater. Res.* **1988**, *22*, 763–793.

57. Horbett, T. A. *Coll. Surf. B: Biointerfaces* **1994**, *2*, 225–240.

58. Sandwick, R. K.; Schray, K. J. *J. Colloid Interface Sci.* **1987**, *115*, 130–138.

59. Sandwick, R. K.; Schray, K. J. *J. Colloid Interface Sci.* **1988**, *121*, 1–12.

60. Bohnert, J. L.; Horbett, T. A. *J. Coll. Interface Sci.* **1986**, *111*, 363–377.

61. Slack, S. M.; Horbett, T. A. *J. Biomed. Mater. Res.* **1992**, *26*, 1633–1649.

62. Soria, J.; Soria, C.; Mirshahi, M.; Boucheix, C.; Aurengo, A.; Perrot, J. Y.; Bernadou, A.; Samama, M.; Rosenfeld, C. *J. Coll. Interf. Sci.* **1985**, *107*, 204–208.

63. Lenk, T. J.; Horbett, T. A.; Ratner, B. D.; Chittur, K. K. *Langmuir* **1991**, *7*, 1755.

64. Chinn, J. A.; Posso, S. E.; Horbett, T. A.; Ratner, B. D. *J. Biomed. Mater. Res.* **1991**, *25*, 535–555.

65. Chinn, J. A.; Posso, S. E.; Horbett, T. A.; Ratner, B. D. *J. Biomed. Mater. Res.* **1992**, *26*, 757–778.

66. Lewandowska, K.; Balachander, N.; Sukenik, C. N.; Culp, L. A. *J. Cell. Physiol.* **1989**, *141*, 334.

67. Grinnell, F.; Feld, M. K. *J. Biomed. Mater. Res.* **1981**, *15*, 363–381.

RECEIVED October 20, 1994

Chapter 9

Transient Adsorption of Fibrinogen from Plasma Solutions Flowing in Silica Capillaries

M. T. Le[1], J. N. Mulvihill[2], J.-P. Cazenave[2], and P. Déjardin[1,3]

[1]Institut Charles Sadron, 6 rue Boussingault, 67083 Strasbourg, France
[2]Institut National de la Santé et de la Recherche Médicale U311, 10 rue Spielmann, 67085 Strasbourg, France

The adsorption of fibrinogen on silica capillaries from diluted human plasma was studied under laminar flow conditions using radiolabeled ^{125}I-fibrinogen. Exchange of fibrinogen at the interface with displacing species in plasma (Vroman effect) was followed by continuous recording of radioactivity during flow and different behavior was observed for a plasma pool as compared to single donor plasma, in particular with regard to the rate of the exchange process. In neither case was an extremum in interfacial fibrinogen concentration detected with increasing dilution above d = 10^{-3}. The influence of wall shear rate on the exchange reaction demonstrated the significant role of transport under flow conditions, while temperature was also found to be an important parameter as previously reported for baboon plasma.

Some years ago L. Vroman demonstrated the existence of changes in interfacial populations when human plasma contacted a surface, in particular the transient presence of fibrinogen (1). This resulted from exchange processes at the interface, fibrinogen being detected at short contact times but not at longer times. Several further studies have indicated that high molecular weight kininogen (HK) is responsible for the displacement of fibrinogen (2,3). However, it may not be the only displacer (4). Recently a model of C. Scott (5) proposed a complex of activated kininogen with prekallikrein or factor XI as the main species contributing to the replacement of fibrinogen on contact activating surfaces. Apart from numerous isolated articles, in 1991 two issues of *J. Biomater. Sci.* were dedicated to the Vroman effect (6).

Let us consider fibrinogen adsorption and displacement. It is recognized that hydrophilic surfaces facilitate exchange reactions, an effect which could be related to

[3]Corresponding author

0097–6156/95/0602–0129$12.00/0

a minimization of protein denaturation and to lower interaction energies than at hydrophobic surfaces. Examination of adsorbed fibrinogen *in situ* or after desorption by circular dichroism allows quantitation of alterations in the native structure. Whereas no denaturation of adsorbed fibrinogen could be detected on silica beads *in situ* (7), a significant loss of α-helix content was observed in fibrinogen eluted from glass surfaces (8). The mean residence time on the surface before displacement is also an important parameter (9). Although many different static systems have been employed to study the Vroman effect, quantitative description is not always easy especially at low concentrations where exchange is readily observed, as the depletion due to adsorption can be high and vary with protein type, leading to changes in relative bulk concentrations. Moreover, to start and end an experiment under static conditions, there is an obligation to fill and empty the system. Convection then occurs and hence the interpretation of results from such static models requires corrections (10). Using the "lens-on-slide" method (11,12), similar problems arise with the additional effects of undesirable lateral diffusion under the lens and possible local convection due to small thermal gradients leading to minimize the contact time in experiments (12).

Therefore, a study under flow conditions would seem easier to interpret as there is always an arrival of fresh solution in a well defined velocity field. Transport may control the initial interfacial events, while with increasing surface coverage, we would expect a gradual predominance of processes controlled by surface phase reactions. The present work describes application of the CRAFS technique (Continuous Recording of Adsorbance in Flowing Systems) to the adsorption of radiolabeled fibrinogen (13,14) from flowing plasma on hydrophilic silica fibers.

Materials and Methods

Silica capillaries. High quality fused silica capillaries of diameter 530 μm and length 100 m (SGE, Australia) were purchased from Perichrom (France). An average length of 10m was treated with diluted sulfochromic acid (1/10) at 50°C followed by a mixture of 30% (w/w) aqueous H_2O_2 and 25% (w/w) aqueous NH_3 with water (respective volume ratios 25 /5 / 70) at 80 °C under flow conditions for one hour. This cleaning procedure was completed by thorough rinsing with deionised water (SuperQ, Millipore) at 20°C for 2 hours with a low flow of 10^{-2}M Tris buffer overnight. This procedure makes the surface very hydrophilic with a high density of SiO^- groups. The capillary was then cut into 22 cm long sections and the streaming potential ΔE_s was measured under varying pressure drops ΔP to deduce the ζ potential of the interface from the slope dE_s/dP. Maximal values of about -80 mV (Tris 10^{-2} M; pH 7.4) were stable only over one to two days and the subsequent decrease in ζ potential with time was not accurately reproducible.

A chosen number of fibers were assembled in a polystyrene pipet of internal diameter 3 mm by injecting an epoxy type glue at its extremities, which then were cut cleanly before determining the streaming potential of the fiber bundle. In fact the ζ potential was often smaller (-45 to - 60 mV) than for individual fibers. Flow rate was measured to determine whether Poiseuille's law could be derived from the pressure drop data and to verify that no fiber was plugged.

Fibrinogen and plasma. Purified human fibrinogen was provided by the *Centre Régional de Transfusion Sanguine de Strasbourg.* A first series of experiments was carried out using a human plasma pool prepared from citrated anticoagulated blood collected from 20 healthy donors, while a second series was performed with plasma from a single donor. Characteristics of the pool and single donor plasma are given in Table I and both preparations were stored at -80°C until use. Plasma dilution was defined by a factor d $(0 < d < 1)$, high dilution corresponding to a small d value, as for instance d $= 10^{-3}$ for a dilution of 0.1%. Fibrinogen was labeled with ^{125}I by the iodogen technique (15). Radioactive fibrinogen was added to plasma in proportions giving an increase in total fibrinogen of 8 %, except in one experiment (d $= 10^{-3}$, single donor) where the increase was 11 %.

Table I: Concentrations of high molecular weight kininogen (HK, ChromoTimeSystem Behring), fibrinogen and prekallikrein in undiluted plasma

	Plasma pool	Single donor
Fibrinogen (g/l)	2.8	1.8
HK (mg/l)	20	76
Prekallikrein (mg/l)	43	32
Factor XI (U/ml)	-	0.96
Factor XII (U/ml)	-	1.0

Adsorption experiments. Plasma solutions at different dilutions were passed by aspiration with a syringe pump (Harvard Apparatus 22, Mass. USA) at a flow rate corresponding to a wall shear rate of 200 s^{-1} over a period of one hour. Passage of buffer before and after the solutions allowed to estimation of the background and desorption kinetics respectively. Radioactivity is detected over a 4 cm length whose middle is positioned at 7.5 cm from the capillary entrance.

Results and Discussion

Influence of plasma dilution. Figure 1 shows the kinetics of fibrinogen deposition on silica at varying plasma dilutions. It is clearly seen that almost all fibrinogen is removed from the surface at 2% dilution, while a significant amount remains adsorbed at the highest dilutions. In accordance with Brash (16,17), the maximum of interfacial concentration appears at a time which is a decreasing function of d: at about 15 minutes for d$=10^{-3}$, at about 2 minutes for d$= 5 \times 10^{-3}$ and below one minute for d $= 0.01$ and 0.02. Whereas at high dilution there is a slow decline in interfacial fibrinogen concentration, probably due to low concentrations of the displacing species and to a longer residence time of the initially adsorbed molecules, at d=5 x 10^{-3} and above the interfacial concentration decreases more steeply over 30 minutes until complete removal of fibrinogen from the surface. In general, low bulk concentrations tend to favor transport controlled processes. To verify this assumption, another experiment at $d = 10^{-3}$ was performed at a wall shear rate of 500 s^{-1} (Fig. 2) and under these higher shearing conditions stable adsorption could not be attained. The initial experimental adsorption constant was an increasing function of shear rate: 3.8 x 10^{-5} at 200 s^{-1} and 5.0 x 10^{-5} at 500 s-1. A plot of ln Γ *vs* time, where Γ is the interfacial concentration, to estimate the exchange rate constant - displacer concentration included - gave slopes of 1.9 and 2.1 x 10^{-4} s^{-1}, values sufficiently close to assume that transport did not play a crucial role in the exchange process over this range of shear rates and for this plasma pool. Comparison with the theoretical Lévêque model for initial adsorption (6.7 and 9.1 x 10^{-5} cm s^{-1}) would suggest the process to be only partially transport controlled and complementary experiments would be required to draw more precise conclusions with regard to the dependence of exchange on wall shear rate.

It is of interest here to consider the differences between static and flow conditions. Assuming under static conditions complete adsorption of the displacing species, present in solution at concentration C_{dis}, in a capillary of radius R, the interfacial concentration would be given by 0.5 R C_{dis}. In the case of kininogen, this value ($\approx$ 1 ng cm^{-2} at d = 10^{-3}) would be too low to allow observation of significant exchange. Complete surface coverage by fibrinogen would be likewise impossible below a critical dilution factor d^{*}. Approching this value from above, the spatial boundary conditions far from the interface do not remain constant as depletion is not negligible, while since this phenomenon is not identical for all solutes their relative solution concentrations vary with time and the dilution factor d loses its initial significance. Under flow conditions, relative bulk concentrations and spatial boundary conditions are maintained constant even at small d values throughout the adsorption process.

Similar results were obtained in a second series of experiments performed at the same shear rate (200 s^{-1}) using single donor plasma (Fig.3) although the exchange of fibrinogen was slower. In this series, sensitivity was improved with respect to the preceding experiments by increasing the number of capillaries in a bundle from five to ten, while dispersion was reduced by increasing the count time from 60s to 99s and data are presented as average values over 3 to 5 minutes. A maximal interfacial concentration was observed at 45 minutes for d = 10^{-3} followed by a gradual decrease over two hours without return to a plateau level. Adsorbance of fibrinogen was once again an increasing function of shear rate as an experiment at 360 s^{-1} showed an extremum about 40% higher than at 200 s^{-1}. Since from this plasma the difference between the two curves Γ(t) was larger than in the first series due to slower exchange, it was possible to obtain experimental points before and after the extremum even at d = 10^{-2}. Extrema occurred at about 15 minutes for d = 2 x 10^{-3} and at 5 minutes or below for d larger than 10^{-2}, an increase in the concentration of the displacing species leading to sharper extrema with final values at 80, 30 and 15 minutes for d = 0.002, 0.004 and 0.010 respectively. Determination of the bulk kininogen (HK) concentration (Table I) unexpectedly showed higher levels in the single donor plasma than in the first plasma pool. These results emphasize the

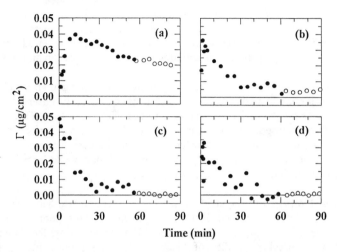

Figure 1. Kinetics of adsorption of fibrinogen on silica capillaries from plasma (pool) at varying dilutions: $d = 10^{-3}$ (a); 5×10^{-3} (b); 10^{-2} (c); 2×10^{-2} (d). Closed symbols refer to passage of plasma, open symbols to rinsing with buffer. $T = 37°C$. $\gamma = 200$ s^{-1}.

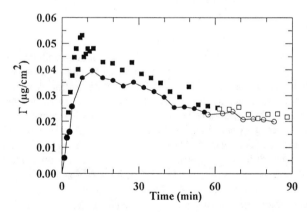

Figure 2: Kinetics of adsorption of fibrinogen from diluted plasma (pool, $d = 10^{-3}$) at wall shear rates of 200 s^{-1} (lower curve, o) and 500 s^{-1} (upper curve, □). Closed symbols refer to passage of plasma, open symbols to rinsing with buffer. $T = 37°C$.

complexity of the process and could be related to different bulk concentrations of other factors, in particular those taking part in contact phase activation. Initial adsorption constants lay slightly below the Lévêque limit and tended to decrease with increasing dilution factor d. This could indicate transition from a transport to an interfacial controlled reaction with increasing concentration, but a greater number of points at early adsorption times would be required to determine an accurate initial adsorption rate at high plasma concentrations where surface occupation takes place more rapidly.

Figure 4 shows the interfacial concentrations of fibrinogen after one hour of contact with plasma in both series of experiments. Contrary to observations under static conditions on glass (17) extrema do not occur above $d = 10^{-3}$. This effect may be due to faster transport processes in presence of convection. Surface nature and preparation are also a relevant parameter. As the kinetic curves do not show a true plateau after one hour, still lower interfacial concentrations would be expected with increasing contact times.

Influence of temperature. Figure 5 presents an adsorption "isotherm" at one hour for two temperatures 23°C and 37°C using the first plasma pool. No maximum appears above $d = 10^{-3}$ at either temperature. The temperature effect is equally as important as in previous studies of Slack et al. using baboon plasma (19) on borosilicate glass coverslips and Pyrex rectangles. At this stage, let us note that in those static experiments the ratio volume / area is about 1 cm while with a tube of diameter 0.25 cm (17) it is 0.06 cm. Kinetic analysis at $d = 5 \times 10^{-3}$ (Fig. 6) suggests the initial fibrinogen adsorption to be only slightly dependent on temperature, possibly due to transport control. The exchange process is nevertheless slower at 23°C than at 37°C as likewise shown by the maximal interfacial concentrations.

Conclusions

Study of the adsorption of fibrinogen on silica capillaries under flow conditions demonstrates the important influence of transport phenomena on the transient state of fibrinogen at the silica / solution interface. Plasma composition is probably another significant parameter, especially in the case of phase contact activating materials. The relevance of transport has already been mentioned in earlier work (20), where the time-window concept was introduced and the importance of irregularities in the shape of conducts and cavities was emphasized as a source of molecular exchange essentially involving diffusion. A transient state is more readily observed under static conditions at low dilutions, due to the limiting law $\Gamma \sim t^{1/2}$ as compared to $\Gamma \sim t$ in a coupled convection - diffusion model. However, convection ensures constant bulk solution concentrations of proteins and should facilitate the interpretation of results. The interfacial fibrinogen concentration showed no extremum above $d = 10^{-3}$, while temperature appeared to influence the final surface concentration through a facilitated exchange at 37°C relative to 23°C. A more accurate quantitative description of the exchange process should be possible using the addition of labeled fibrinogen to an afibrinogenic plasma.

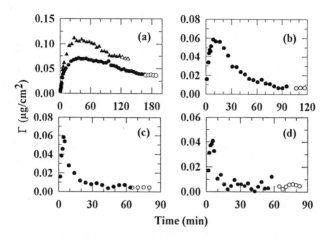

Figure 3: Kinetics of adsorption of fibrinogen on silica capillaries from plasma (single donor) at $\gamma = 200$ s^{-1} and varying dilutions: d = 10^{-3} (a) (upper curve $\gamma = 360$ s^{-1}); d = 2×10^{-3} (b); d = 4×10^{-3} (c); d = 10^{-2} (d). Closed symbols refer to passage of plasma, open symbols to rinsing with buffer. T = 37°C.

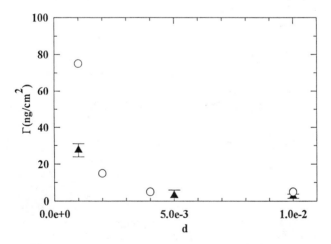

Figure 4: Adsorption "isotherm" after one hour of contact with plasma at 37°C. Plasma pool (▲). Single donor (o). Standard deviation with n=2 (d = 10^{-3}) and n=3 (d = $5 \ 10^{-3}$ and d = 10^{-2})

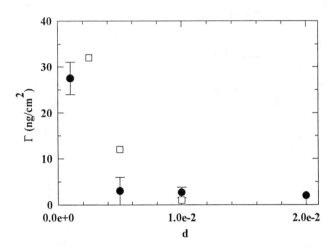

Figure 5: Adsorption "isotherm" after one hour of contact with plasma (pool) at 23°C (□) and 37°C (•). Standard deviation with n=2 (d = 10^{-3}) and n=3 (d = 5 10^{-3} and d = 10^{-2})

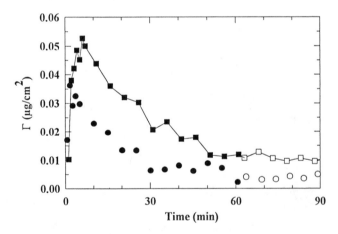

Figure 6: Kinetics of adsorption of fibrinogen from diluted plasma (pool, d = 5 x 10^{-3}) at 23°C (□) and 37°C (o). Closed symbols refer to passage of plasma, open symbols to rinsing with buffer.

Literature Cited

1. Vroman, L.; Adams, A. L. *Surf. Sci.* **1969**, *16*, 438
2. Schmaier, A. H.; Silver, L.; Adams, A. L.; Fischer, G. C.; Munoz, P. C.; Vroman, L.; Colman, R. W. *Thromb. Res.* **1983**, *33*, 51-67
3. Poot, A.; Beugeling, T.; Van Aken, W. G.; Bantjes, A. *J. Biomed. Mater. Res.* **1990**, *24*, 1021-1036
4. Brash, J. L. *Annals New-York Acad. Sci.*, **1987**, *516*, 206-222
5. Scott, C. F. *J. Biomater. Sci.: Polymer Edn.*, **1991**, *2*, 173-181
6. *J. Biomater. Sci.: Polymer Edn.* **1991**, *2*, 161-237 ; ibid. **1991**, *3*, 1-114
7. Mc Millan, C. R.; Walton, A. C. *J. Colloid Interface Sci.* **1980**, 48, 345-349
8. Chan, B. M. C.; Brash, J. L. *J. Colloid Interface Sci.* **1981**, 84, 263-265
9. Slack, S. M.; Horbett, T. A. *J. Colloid Interface Sci.* **1989**, 133, 148-165
10. Wojciechowski, P.; Brash, J. L. *J. Biomater. Sci.: Polymer Edn.* **1991**, *2*, 203-216
11. Vroman, L.; Adams, A. L. *J. Colloid Interface Sci.* **1986**, *111*, 391-402
12. Elwing, H.; Tengvall, P.; Askendal, A.; Lundstrom, I. *J. Biomater. Sci.: Polymer Edn.* **1991**,*3*, 7-15
13. Yan, F.; Déjardin, P. *Langmuir* **1991**, *7*, 2230-2235
14. Boumaza, F.;Déjardin, P.; Yan, F.; Bauduin, F.; Holl, Y. *Biophys. Chem.* **1992**, *42*, 87-92
15. Regoeczi, E. *Iodine-labeled Plasma Proteins*, CRC Press, Boca Raton, Florida, 1984 pp. 49-56
16. Brash, J. L.; ten Hove, P. *Thromb. Haemostas.* **1984**, *51*, 326-330
17. Wojciechowski, P.; ten Hove, P.; Brash, J. L. *J. Colloid Interface Sci.* **1986**, 111, 455-465
18. Wojciechowski, P. W.; Brash, J. L. *Colloids Surfaces B: Biointerfaces* **1993**, *1*, 107-117
19. Slack, S. M.; Horbett, T. A. *J. Biomater. Sci.: Polymer Edn.* **1991**, *2*, 227-237
20. Leonard, E. F.; Vroman, L. *J. Biomater. Sci.: Polymer Edn.* **1991**, *3*, 95-117

RECEIVED March 1, 1995

Chapter 10

Protein Displacement Phenomena in Blood Plasma and Serum Studied by the Wettability Gradient Method and the Lens-on-Surface Method

Hans B. Elwing[1,3], Liu Li[1], Agneta R. Askendal[1], Ghada S. Nimeri[1], and John L. Brash[2]

[1]Interface Biology Group, Department of Physics and Measurement Technology, Linköping Institute of Technology, S–581 83 Linköping, Sweden
[2]Department of Chemical Engineering, McMaster University, Hamilton, Ontario L8S 4L7, Canada

Protein exchange or displacement phenomena on silicon-based surfaces in heparinized blood plasma or serum was studied using the lens-on-surface method and the wettability gradient method. Ellipsometry was used for the quantification of adsorbed proteins. Selected proteins were detected with the use of antibodies. Using the lens-on-surface method it was found that the lack of FG (fibrinogen) in serum did not facilitate binding of anti-HSA (albumin), anti-IgG (immunoglobulin G) or anti-HMWK (high molecular weight kininogen) following incubation of serum with silicon oxide surfaces. Instead, the binding of these antibodies was virtually the same in serum as in plasma. Thus it seems that the behavior of fibrinogen in the "Vroman sequence" is essentially independent of these other proteins. With the wettability gradient method it was found that binding of anti-HSA increases with increasing incubation time at the hydrophobic end of the gradient when either plasma or serum was used. Anti-FG was found to bind to plasma-incubated gradient surfaces at both the hydrophobic and hydrophilic ends of the gradient at short incubation time. With prolonged incubation, binding of anti-FG decreased at the hydrophilic end of the gradient. Binding of anti-HMWK occurred mostly at the hydrophilic end of both plasma and serum incubated gradient surfaces with a binding optimum at about 10 min. With prolonged incubation the binding of anti-HMWK decreased. Binding of anti-IgG was relatively low on serum-incubated as well as plasma-incubated gradient surfaces.

[3]Current address: Laboratory of Interface Biophysics, University of Göteborg, Lundberg Building, Medicinaregatan 9c, S–413 90 Göteborg, Sweden

Solid surfaces are spontaneously covered with a layer of protein after seconds of contact with blood (*1,2*). Because this accumulation of protein occurs so rapidly, it precedes the arrival of cells on implant surfaces. Cells are therefore likely to interact with the protein coated biomaterial rather than directly with the material. Therefore, much research has been done to understand the process of protein adsorption at solid surfaces. The physicochemical aspects of protein adsorption at solid surfaces have been investigated extensively with wide implications in different areas of research (*3,4*). Of particular interest in biomaterials oriented research is that many proteins change biological activity and conformation upon adsorption (*5-9*), and form an increasingly irreversible attachment to the surface with time (*10,11*).

The subject of this investigation is the phenomenon of surface exchange or displacement of adsorbed proteins, an important effect in complex protein "solutions" like blood. This phenomenon has created much interest since Vroman's original observation of the "transitory" adsorption of fibrinogen from blood plasma on glass surface (*12-18*). According to Vroman, blood protein adsorption at glass surface has the characteristics of a sequence starting with albumin, followed by IgG, fibrinogen and finally high molecular weight kininogen (19). Additional proteins were included in the sequence in Vroman's original publication but these have been excluded in this investigation for reasons of simplicity. The "lens-on-surface" method was used by Vroman and was found to be very sensitive for these types of study (*19*). With this method we have been able to reproduce some of Vroman's original observations and to quantify objectively the displacement reaction with a supplementary ellipsometric method (*20,21*).

The blood plasma concentration and the time of incubation have an influence on the speed of the displacement reaction (*14,15,22*). The degree of surface hydrophobicity is also an important factor since more hydrophobic surfaces bind proteins less reversibly (*23,24*). Therefore we have also investigated the "Vroman effect" in human plasma on the so-called wettability gradient surface (*25*), formed by diffusion of methylsilane over a silicon oxide surface. In this system we have demonstrated relations between both time of incubation and surface hydrophobicity and the exchange between fibrinogen and high molecular weight kininogen on gradient surfaces incubated in human plasma (*26*).

Fibrinogen (FG) is an important protein in biomaterial interactions due to its many biological effects. It is therefore important to understand the adsorption of this protein in more detail. To this end we have studied the Vroman effect in fibrinogen-poor serum using the wettability gradient and lens-on-surface methods. By comparing the results of the serum and plasma experiments, it was hoped to get more information, both qualitative and quantitative, on the relative importance of albumin, IgG and HMWK in the "Vroman sequence".

This work is based partly on original observations that HMWK could be detected at hydrophilic silicon surfaces incubated in serum. This observation was first reported by J. Warkentin at the San Diego meeting in 1984. See also Warkentin, *et al.*, this volume.

Materials and Methods

Plasma, serum, and antibodies. Human heparinized plasma and serum were obtained from healthy donors and stored at -70°C until use. The plasma and serum were diluted to desired concentration with phosphate buffered saline (PBS). Rabbit anti-human albumin (a-HSA), rabbit anti-human IgG (a-IgG), rabbit anti-human fibrinogen (a-FG), and rabbit anti-goat immunoglobulin were obtained from DAKO Immunoglobulin a/s, Glostrup,

Denmark. Goat anti-human high molecular weight kininogen (a-HMWK) was from Nordic Immunochemical Laboratories, the Netherlands. The antibodies were diluted in PBS buffer to a working concentration of 1/50. The concentration of HMWK in plasma and serum was determined using a double diffusion immunoassay. No difference in precipitation ability between plasma and serum was found, indicating that HMWK in serum had at least 50% of its concentration in plasma.

Preparation of wettability gradient surfaces. Polished silicon wafers, 0.3 mm thick (Okmetic OY, Finland) were used as solid substrate. The wafers were cut into rectangle (10 x 25 mm) which were washed as previously described (25). Wettability gradient surfaces were prepared on silicon substrates arranged vertically in xylene (Merck p.a.) Dimethyldichlorosilane (0.05%, v/v, Sigma, USA) in 20 ml trichloroethylene (Merck p.a. was added at the bottom and allowed to diffuse into the upper xylene phase for 90 minutes. The wettability gradient formed was approximately 7 mm long (25). The wettability properties can be determined by measuring the adsorption of fibrinogen (27) or by measuring the advancing water contact angle. The contact angle was found to vary from 90° at the hydrophobic end to 10° at hydrophilic end (Figure 1b).

Protein adsorption experiments on wettability gradient surfaces. The experiments were performed at room temperature (23°C). The surfaces were immersed in serum or plasma diluted 1/10 with PBS for 2.5 min, 10 min, 40 min, and 160 min, respectively After rinsing with PBS, some of the surfaces were further incubated in the appropriate antibody solution for 30 minutes. The surfaces exposed to a-HMWK were then incubated in rabbit anti-goat immunoglobulin solution for 15 min in order to amplify the response of a-HMWK. Surfaces were finally rinsed in distilled water and dried in nitrogen.

The lens-on-surface method. The details of this method have been described previously (20,28). Briefly, a glass lens with a focal length of about 200 mm was placed with the convex side down on the hydrophilic silicon surface. Serum or plasma (0.1 ml) at dilutions of 1/10 and 1/100 was injected into the space between the lens and the test surface. After 10 min incubation at room temperature, the plasma and the lens were removed. The test surface was then incubated in the appropriate antibody for 15 min after rinsing in PBS. The surfaces treated with a-HMWK were further incubated in rabbit anti-goat immunoglobulin for 15 min. All surfaces were finally rinsed with distilled water and dried in flowing nitrogen.

Ellipsometric quantification of adsorbed proteins on surface. In ellipsometry the change in polarization of light reflected from a surface is measured, from which the thickness of the adsorbed film is calculated as described elsewhere (25). The dry gradient surfaces were monitored in an automatic ellipsometer (Auto Ell 3, Rudolph Research, NY USA) equipped with a device for stepwise lateral scanning. The resolution of the lateral measurements was 0.635 mm and scanning was performed stepwise at 0.635 mm interval (Figure 1c). The experiments presented in Figures 2-4 were done at least three times. No major qualitative differences between replicate experiments were found.

Some properties of the silicon oxide and methylated silicon oxide surfaces. The silicon of electronic device quality used in this investigation has a layer of spontaneously grown, hydrophilic, silicon oxide. The top layer of the surface contains a large number of silanol groups which are amphoteric, being both proton acceptors and donors. The zeta

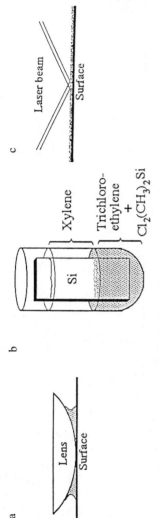

Figure 1. Methods used in this investigation. a) The Lens-on-surface method. A convex glass lens is placed on a silicon oxide surface and blood plasma or serum is pipetted under the lens. After incubation the glass is removed from the surface. Detection of specific adsorbed proteins is made by subsequent incubation of the surface in antiserum. b) The wettability gradient surface is a silicon surface with a thin layer (5Å) of silicon dioxide. Controlled hydrophobic gradients are made on these surfaces by diffusion of dichlorodimethylsilane. The gradient surfaces are incubated in plasma or serum. Detection of specific adsorbed proteins is made by subsequent incubation in antiserum. c) Quantification of protein adsorption and antibody adsorption is made by "scanning ellipsometry" across interesting sections of the surface.

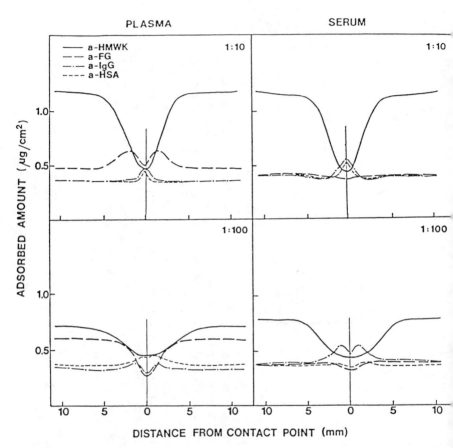

Figure 2. Adsorbed amount of organic material on silicon oxide surface as obtained by the lens on surface method. Plasma or serum diluted 1/10 or 1/100 was incubated under the lens for 10 min. After removal of the lens, the surfaces were incubated with specific antibodies for 15 min. Scanning ellipsometry determinations of the adsorbed amounts of organic material were performed across the contact point of the lens. The very high amounts of organic material deposited, about 0.8 µg/cm², in the "a-HMWK" group may be explained by the fact that polyclonal anti-immunoglobulin was used to amplify the reaction.

otential of these surfaces has been found to be zero at pH 3.0 to 3.2 (*29*). The silicon xide surface is made hydrophobic by means of reaction with dichlorodimethylsilane DDS). DDS reacts with the silanol groups of the surface and the resulting -Si(CH$_3$)$_2$ roups are bound covalently to the silicon oxide forming a densely packed array of methyl roups (*30*). Silicon of electronic device quality has a surface that is flat at the nanometer cale. The extreme flatness makes it highly reflective and thereby suitable as a substrate r ellipsometry. In a recent investigation of plasma protein adsorption on solid surfaces f different origin it was found that both silicon oxide and methylated silicon oxide enerally adsorb less protein per unit area than some polymer surfaces, for example olystyrene (*31*). It is possible that the extreme flatness (as measured by scanning force icroscopy) of the silicon surface is responsible for these differences (*30*).

Results

The lens-on surface method and the "Vroman effect" in serum and plasma. The rinciple of the lens method is that a gradient in fluid height or volume above the surface, created from the contact point of the lens at the surface and outwards (Figure 1a). √hen serum or plasma is incubated under the lens, the protein solution close to the ontact point will be depleted of protein due to adsorption at the surface. Such depletion ill have an effect which is similar (but not identical) to dilution of the protein solution ear the contact point of the lens (*21*) .

Serum or blood plasma diluted 1/10 and 1/100 were used in these experiments ith the hydrophilic silicon surface. After washing off unbound protein and removal of ie lens, the surfaces were incubated in various antisera followed by ellipsometric ieasurement of the amount of adsorbed protein. Representative data are shown in Figure . At a dilution of 1/10 binding of anti-HMWK occurred at distances far from the ontact point of the lens. There was no apparent difference between plasma and serum in ie binding pattern of anti-HMWK. Binding of anti-FG on surfaces incubated in plasma iowed maxima at some distance from the contact point (2-3 mm) as previously observed *0*). On surfaces incubated in serum, little binding of anti-FG was observed as expected. inding of anti-HSA and anti-IgG was low on both plasma- and serum-incubated surfaces.

In experiments with plasma and serum diluted 1/100 the binding of anti-HMWK as very similar. The total bound anti-HMWK was however lower than in the 1/10 lution experiments. In addition, there were no binding maxima of anti-FG on plasma-cubated surfaces. The binding of anti-IgG or anti-HSA was insignificant in the 1/100 lution experiments. The double peaks representing binding of anti-IgG on serum-cubated surfaces were frequently, but not always, observed.

eposition of plasma and serum proteins on wettability gradient surfaces. The ns-on-surface method can be used only on hydrophilic surfaces; it is precluded on ydrophobic surfaces due to capillary effects. Investigation and assessment of protein xchange phenomena using the gradient method were done at different times of cubation.

Gradient surfaces were incubated with plasma or serum, diluted 1/10, for 2.5 min,) min, 40 min and 160 min at room temperature. After rinsing and drying the surfaces, e adsorbed amounts of organic material were determined with the ellipsometer. The sults of a representative experiment are shown in Figure 3. Approximately 0.25 µg/cm^2 as deposited at the surface independent of the time of adsorption. However, at the

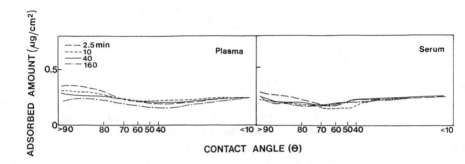

Figure 3. Adsorbed amounts of organic material on gradient surfaces incubated in plasma or serum. Scanning ellipsometry was used to quantify the adsorbed amounts along the gradient surface. The wettability distribution, expressed as advancing contact angle with water, is given on the x-axis. Human plasma or serum was incubated for different times as given in the Figure.

hydrophobic end of the plasma-incubated gradient the total amount of material deposited decreased with increasing incubation time. This phenomenon may be associated with the so called "total protein" Vroman effect involving maxima in total adsorbed protein as function of plasma dilution or time (Brash, J.L., unpublished observation).

Antibody binding to wettability gradient surfaces incubated in plasma or serum. In Figure 4 the amounts of antibodies bound to gradients incubated in plasma or serum for different times are shown. The adsorbed amount of organic material in the first layer (adsorbed from plasma or serum) has been subtracted.

The binding of anti-HSA was low at the hydrophilic end of the gradient at all incubation times. At the hydrophobic end, binding of anti-HSA gradually increased with increasing incubation time. At short incubation time a reproducible peak of antibody binding around 60° contact angle was observed. The binding of anti-HMWK reached maximum at 10 min, then gradually decreased to a relatively low level at 140 min. Binding of anti-HMWK also occurred on the hydrophobic part of the gradient, especially at short incubation times. However it must be kept in mind that the sensitivity of anti-HMWK detection was increased by incubation with anti-immunoglobulin. Binding of anti-IgG was low on all parts of the gradient and at all incubation times. As expected binding of anti-FG was low at all incubation times on gradient surfaces incubated in serum, since this medium contains very little fibrinogen. The binding of anti-FG on gradient surfaces incubated in plasma was significant at the hydrophobic end of the gradient at short incubation time but gradually decreased with increasing incubation time. At the hydrophilic end of the gradient, the binding of anti-FG was significant at short incubation times but gradually decreased with increasing incubation time. A substantial amount of anti-FG was, however, still bound after 140 min incubation.

In some gradient experiments (data not shown) binding of anti-FG was found to decrease also at the far hydrophobic end of the gradient as reported previously (26). This phenomenon seems to be related to structural change in the adsorbed fibrinogen (32).

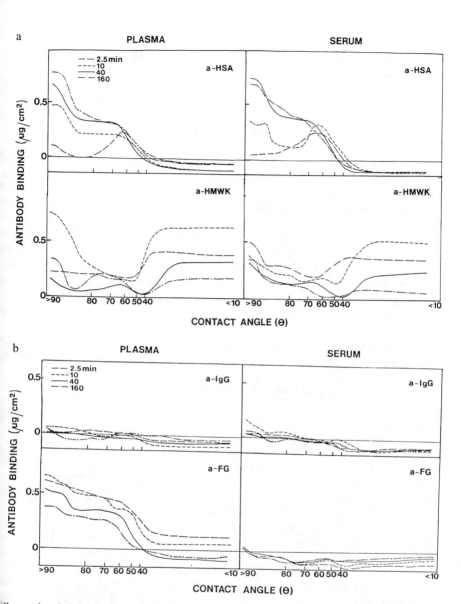

Figure 4. a) Adsorbed amounts of antibodies, a-HSA and a-HMWK (y-axis), on gradient surfaces incubated in serum or plasma for different times. The wettability distribution, expressed as advancing contact angle with water, is given on the x-axis. Human plasma or serum was incubated for different times as given in the Figure. Antibody incubation was performed for 15 min. b) Adsorbed amounts of antibodies, a-IgG and a-FG (y-axis), on gradient surfaces incubated in serum or plasma for different times.

Discussion

There has been much discussion of a possible molecular explanation for the transitory adsorption phenomena of proteins in multicomponent solutions such as plasma. It has been suggested that albumin, the protein in plasma with the highest concentration, will initially dominate the adsorbed layer but will eventually be displaced by IgG which has lower concentration but higher surface affinity than albumin. A sequence of adsorption and displacement involving all the proteins will occur based on these considerations. In the latter stages in this overlapping "sequence"   HMWK will dominate the adsorption (19).

Thus one might have expected that binding of anti-IgG or anti-HSA would increase on serum- compared to plasma-incubated surfaces, since in serum IgG and albumin would potentially fulfil the role of fibrinogen in the Vroman sequence. This is not an unrealistic expectation since IgG and HSA constitute a large part of the total protein content of plasma. However there are few indications that this was occurring in the lens-on-surface or the wettability gradient experiments. It is obvious that the presence of fibrinogen at the surface is not a prerequisite for the subsequent adsorption of HMWK. However it has been shown that plasma devoid of HMWK deposits fibrinogen after long exposure time (23,24). This observation along with the present results suggests that the HMWK in normal plasma is involved in the removal of adsorbed fibrinogen, but a "specific" exchange such that adsorbed fibrinogen is especially susceptible to displacement by HMWK is not indicated.

It is important to point out that in the present investigation the adsorbed proteins were detected by antibody binding. A binding reaction, judged to be positive using proper controls, is normally a reliable measure of the presence of a protein. However, the absence of antibody binding does not exclude the presence of an adsorbed protein. An example is the decreased antibody binding properties of adsorbed fibrinogen that has been incubated in buffer (11). On the other hand, it has been concluded that antibody binding to adsorbed fibrinogen is a better criterion of biological activity (eg platelet adhesion) than the actual mount of adsorbed fibrinogen (33,8). Another weakness of the antibody method is that, in principle, the antiserum itself may be active in exchanging with protein on the test surface. However, we found in a previous study that antibody binding to adsorbed protein antigens seems to prevent the antigen from being desorbed, possibly through cross linking of the protein antigen (28).

Some possible biological consequences of the results using the gradient method are also of interest. Albumin adsorption on a biomaterial is an important effect and albumin is often found in substantial quantities on explanted biomaterials (34,35). seems to have a passivating effect on thrombogenic response (36). In addition, adsorbed albumin on implanted polymer surfaces has been shown to down-regulate the acute inflammatory reaction in mice (37). The present investigation shows that anti-albumin binding increases with time at the hydrophobic end of the gradient when exposed to either plasma or serum. The hydrophobic end of the gradient surface consists of -Si-CH$_3$ groups and possibly also moieties such as -Si-CH$_3$-Si-. Such groups are also present on polymer surfaces modified by radio frequency gas plasma polymerization of hexadimethylsiloxane (HMDSO). In addition, such HMDSO surfaces have a high preference for albumin adsorption compared to other plasma proteins in mixtures as indicated by a recent ELISA investigation (*31*). This may reflect the specific binding of albumin to siloxane based materials.

A point of interest also is the "peak" of anti-albumin binding in the middle part of the gradients. This highly reproducible phenomenon has also been observed by Hlady et al (*38*) using wettability gradient surfaces made by diffusion of dimethyldisiloxane as in the present work. The method for detection of adsorbed albumin was total internal reflection fluorescence spectroscopy (TIRF). This is a direct detection method for fluorescently labeled albumin and thus demonstrates that the "peak" in the middle of the gradient is not an indirect antibody effect but represents a true increase in adsorption of albumin in the middle of the gradient.

The adsorption of fibrinogen is a very important effect in biomaterial interactions due to its many biological consequences, for example adhesion and activation of platelets (*39,40*), triggering of the acute inflammatory reaction in experimental animals (*37*), and provision of binding sites for bacteria (*21*). As observed in this investigation the fibrinogen adsorption pattern on wettability gradient surfaces was different at the hydrophilic and hydrophobic ends of the gradient. It may therefore be of interest to perform implantations of gradient surfaces and measure various tissue response parameters along the gradient surface on the explanted material. Such investigations are currently under way.

Very little binding of anti-IgG to surfaces incubated in serum or plasma was observed in this investigation. Adsorbed IgG is known to cause complement activation and to be involved in the attachment of neutrophils to biomaterials. Preadsorption of IgG on hydrophobic surfaces, eg the hydrophobic end of a wettability gradient surface, is known to cause complement activation. However the wettability gradients do not activate complement unless they are preadsorbed with IgG (*28*). Complement activation does occur on some surfaces, eg cuprophane dialysis membranes, but seems to proceed predominantly through the alternative pathway without involvement of adsorbed IgG. In addition Tang and Eaton have concluded that IgG is not a major cause of inflammation in their experimental in vivo model (*41*). Thus it seems that adsorption of IgG out of blood or plasma has little significance for biocompatibility.

It is obvious from the present work that HMWK has a very high affinity for silicon oxide or glass-like surfaces in general. HMWK contains a positively charged histidine rich domain which may be related to its high affinity for the negatively charged silicon oxide surface (*42*). HMWK is involved in both the contact activation system of blood coagulation and the kinin generation system. Its adsorption seems to provide a good indicator of intrinsic pathway activation based on previous studies using the wettability gradient method (*43,44*). It has also been reported recently that HMWK has pronounced anti-adhesive properties for both bacteria (*21*) and eukaryotic cells (*45*).

Conclusions

Some important protein exchange phenomena occurring during plasma or serum contact were studied on flat silicon surfaces using gradient methods such as the lens-on-surface and the wettability gradient. With the use of these rather simple methods we were able to measure the effect of continuous gradients in "space" and surface chemistry. Pronounce protein displacement effects were found on hydrophilic silicon surfaces both in serum ar plasma. The fact that serum lacks fibrinogen did not affect the binding of anti-albumin ʌ the hydrophobic end of the wettability gradient or anti-HMWK at the hydrophilic end.

Acknowledgments

The financial support of this work by the following agencies is acknowledged. Swedis Research Council for Engineering Sciences; Swedish National Board for Industrial ar Technical Development; Swedish Biomaterials Consortium; Medical Research Council ɪ Canada; Natural Sciences and Engineering Research Council Of Canada.   Heart ar Stroke Foundation of Ontario.

Literature Cited

1. Baier, R. E.; Dutton, R. C. J Biomed Mater Res 1969, 3, 191-206.
2. Sevastianov V I   CRC Critical Review in Biocompatibility 1988, 4, 109-154.
3. Norde, W.; Lyklema, J. J Biomater Sci Polym Ed 1991, 2, 183-202.
4. Norde, W. J Disper Sci Tech 1992, 13, 363-377.
5. Nilsson, U. R.; Storm, K. E.; Elwing, H.; Nilsson, B. Mol. Immunol. 1993, 30, 211-19.
6. Elwing, H.; Askendal, A.; Lundstrom, I. J Colloid Interface Sci 1989, 128, 296-300.
7. Elwing, H.; Nilsson, B. O., Svensson, K. E.; Askendahl, A.; Nilsson, U. R.; Lundstrom, I. J Colloid Interface Sci 1988, 125, 139-45.
8. Shiba, E.; Lindon, J. N.; Kushner, L.; Matsueda, G. R.; Hawiger, J.; Kloczewiak, M.; Kudryk, B.; Salzman, E. W. Am J Physiol 1991, 260, C965-74.
9. Kondo, A.; Oku, S.; Higashitani, J. J Colloid Interface Sci 1991, 143, 214-221.
10. Rapoza, R. J.; Horbett, T. A. J Biomed Mater Res 1990, 24, 1263-87.
11. Chinn, J. A.; Posso, S. E.; Horbett, T. A.; Ratner, B. D. J Biomed Mater Res 1992, 26, 757-78.
12. Vroman, L.; Adams, A.L. 1969, 16, 438-446.
13. Andrade, J. D.; Hlady, V. J Biomat Sci Polym Edn 1991, 2, 161-72.
14. Brash, J.L.; Scott, C.F.; ten-Hove, P.; Wojciechowski, P.; Colman, R.W. Blood 1988, 71, 932-9.
15. Brash, J. L.; ten-Hove, P. J Biomed Mater Res 1989, 23, 157-69.
16. Elwing, H.; Askendal, A.; Lundstrom, I. Progr Colloid Polymer Sci 1987, 74, 103-107.
17. Leonard, E. F.; Vroman, L. J Biomat Sci Polym Edn 1991, 3, 95-107.
18. Lundstrom, I.; Elwing, H. J Colloid Interface Sci 1990, 136, 68-84.
19. Vroman, L.; Adams, A. L. J. Colloid Interface Sci. 1986, 111, 391-402.

20.Elwing, H.; Tengvall, P.; Askendal, A.; Lundstrom, I. J Biomat Sci Polymer Edn 1991, **3**, 7-15.

21.Elwing, H.; Askendal, A. J Biomed Mat Res 1994, in press.

22.Slack, S. M.; Bohnert, J. L.; Horbett, T. H. Ann NY Acad Sci 1987, **516**, 223-243.

23.Vroman, L.; Adams, A. L.; Fischer, G.C.; Munoz, P.C. Blood 1980, **55**, 156-159.

24.Brash, J.L. Ann NY Acad Sci 1987, **516**, 206-222.

25.Elwing, H.; Welin, S.; Askendal, A.; Nilsson, U.; Lundstrom, I. J Colloid Interface Sci 1987, **119**, 203 - 210.

26.Elwing, H.; Askendal, A.; Lundstrom, I. J Biomed Mater Res 1987, **21**, 1023-1029.

27.Elwing, H.; Welin, S.; Askendal, A.; Lundstrom, I. J Colloid Interface Sci 1988, **123**, 306-308.

28.Li, L.; Elwing, H. J Biomed Mater Res 1994, in press.

29.Bousse, L.; Mostarshed, S.; van der Shoot, B.; de Rooij, N. F.; Gimmel, P.; Gopel, W. J Colloid Interface Sci 1991, **147**, 22-32.

30.Welin-Klintstrom, S. Ph.D. Thesis, University of Linkoping, 1992.

31.Nimeri, G.; Lassen, B.; G_lander, C.G.; Nilsson, U.; Elwing, H. J Biomat Sci Polymer Edn 1994, in press.

32.Elwing, H.; Golander, C. G. Adv Coll and Interface Sci 1990, **32**, 317-339.

33.Salzman, E.W.; Mcmanama, G.; Ware, A.J. Ann NY Acad Sci 1987, **516**, 184-195.

34.Pankowsky, D.A.N.; Ziats, N.P.; Topham, N.S.; Ratnoff, O.D.; Anderson, J. M. J Vasc Surg 1990, **11**, 599-606.

35.Nojiri, C.; Okano, T.; Koyanagi, H.; Nakahama, S.; Park, K. D.; Kim, S.W. J Biomat Sci Polymer Edn 1992, **4**, 75-88.

36.Guidoin, R.; Snyder, R.; Marin, L.; Botzko, K.; Marois, M.; Awad, J.; King, M.; Bedros, M.; Gosselin, C. Ann Thorac Surg 1984, **37**, 457-462.

37.Tang, L.; Eaton, J.W. J Exp Med 1993, **178**, 2147-2156.

38.Hlady, V.; Golander, C.G.; Colloids Surfaces 1987, **25**, 185-190.

39.Feuerstein, I.A.; McClung, W.G.; Horbett, T.A. J Biomed Mater Res 1992, **26**, 221-237.

40.Gaebel, K.; Feuerstein, I. A. Biomaterials 1991, **12**, 597-602.

41.Tang, L.; Lucas, A.; Eaton, J. W. J Lab Clin Med 1993, **122**, 292-300.

42.Chang, J.; Scott, J. F.; Colman, R. W. Blood 1986, **67**, 805-810.

43.Tengvall, P.; Askendal, A.; Lundstrom, I.; Elwing, H. Biomaterials 1992, **13**, 367-74.

44.Tengvall, P.; Olsson, L.; Walivaara, B.; Askendal, A.; Lundstrom, I.; Elwing, H. In Biomaterial-Tissue Interfaces; P. Doherty, R. Williams, D. Williams and A. Lee, Eds.; Elsevier: 1992; Vol. 10; pp 511 - 519.

45.Ziats, N.P.; Jablonski-Bernasconi, M.; Anderson, J.M. *Trans Soc Biomat* 1994, **17**, 74.

RECEIVED February 10, 1995

Chapter 11

Competitive Adsorption of Proteins During Exposure of Human Blood Plasma to Glass and Polyethylene

P. Turbill, T. Beugeling, and A. A. Poot

Department of Chemical Technology, University of Twente, P.O. Box 217, 7500 AE Enschede, Netherlands

Fibrinogen adsorption to glass from various human blood plasmas has been measured as a function of time. The plasmas were 11 single donor plasmas, pooled plasma, a single donor HWMK-deficient plasma and HMWK-deficient plasma, which had been reconstituted with HWMK. For adsorption times between 1 min and 1 h more fibrinogen adsorbed from HWMK-deficient plasma compared to the amounts of fibrinogen which adsorbed from the other plasmas. This result supports the conclusion of several authors that HWMK is involved in the displacement of fibrinogen, initially adsorbed from normal human plasma to glass.

Glass surfaces, preexposed to solutions of plasma and subsequently exposed to 1:1 diluted plasma, give rise to a relatively high adsorption of HMWK which is independent of the plasma concentration of the precoating solution. This result demonstrates that HMWK displaces preadsorbed plasma proteins from the glass surface. The experiments with polyethylene as a substrate reveal that HDL displaces preadsorbed plasma proteins from the polyethylene surface. Moreover, evidence is presented that substantial amounts of albumin and fibrinogen, adsorbed from 1:1,000 diluted plasma to glass and polyethylene, are displaced from the material surfaces by proteins different from HMWK and HDL, when these surfaces are subsequently exposed to 1:1 diluted plasma.

Several studies strongly suggest that fibrinogen, initially adsorbed from blood plasma to glass or a glass-like surface, is subsequently displaced from the surface by HMWK (high molecular weight kininogen) (1-6). These studies include experiments in which the time dependent adsorption of fibrinogen from single donor HMWK-deficient plasma had been compared with the adsorption of fibrinogen from normal single donor plasma or from pooled plasma. The amount of adsorbed fibrinogen may, however, depend on the plasma composition. In order to get more insight into the real differences between normal donor plasmas and HMWK-deficient plasma with respect to fibrinogen adsorption, we determined the adsorption of fibrinogen to glass from 11 donor plasmas and pooled plasma as a function of time. The 11 donors did not belong to the group of 15 donors, who donated blood for the preparation of pooled plasma. In addition, we measured fibrinogen adsorption from single donor HMWK-deficient plasma and the same plasma, which had been reconstituted with HMWK.

The displacement of adsorbed plasma proteins by HMWK was also studied in another way. Glass surfaces were first preexposed to solutions with different concentrations of plasma, resulting in protein layers having different protein compositions. The precoated glass surfaces were subsequently exposed to 1:1 diluted plasma. Thereafter the amounts of HMWK in the newly formed protein layers were determined and compared with the amounts of this protein adsorbed to the preexposed glass surfaces. Measurements of albumin and fibrinogen adsorption were also included in this study. Similar experiments were carried out with polyethylene instead of glass, but in this study the adsorption of HDL (high density lipoprotein) was investigated because it has been reported that HDL preferentially adsorbs to hydrophobic polymers such as polyethylene (*6,7*).

Materials and Methods

Test device. Protein adsorption from plasma solutions to the material surfaces was studied by means of a two step enzyme-immunoassay (*6,8*). In this enzyme-immunoassay (EIA) a special test device is used (Fig.1). The device consists of a stainless steel bottom plate (13 x 9.5 x 0.2 cm) provided with nine screw pins, and a Teflon upper part (13 x 9.5 x 1 cm) containing 24 cylindrical holes with a diameter of 10 mm as well as holes for the screw pins. At the bottom side, each of the 24 holes has a stepped recess (15.5 mm ID, depth 2.0 mm) in which a silicone sealing ring (10.77 ID x 2.62 mm; Eriks, Alkmaar, The Netherlands) is placed. Either a polymer sheet or two glass plates (in order to prevent breakage) can be placed between the bottom plate and the sealing rings in the Teflon upper part. After the components have been pressed together by means of wing nuts, a 24 wells test device is formed which allows the adsorption as well as the detection of proteins adsorbed to the surface of a polymer or glass. The test surface area and the maximum content of each   well are 0.9 cm^2 and 800 μL respectively.

Material surfaces. Glass plates were obtained from Corning, New York, USA (hard glass, type 7059, thickness 2mm). Polyethylene sheet (low density polyethylene, thickness 0.05 mm) was obtained from TALAS, Zwolle, The Netherlands. Glass plates and polyethylene sheets were cleaned as described by Poot et al (*6*).

Plasmas and HMW kininogen. Pooled normal human plasma was obtained from 15 healthy male donors. From each donor 100 mL of venous blood was collected via a 1.5 mm needle and 'Silastic' Medical-Grade tubing (length 15 cm, 3/16 in ID) into two polypropylene centrifuge tubes (50 ml each), containing

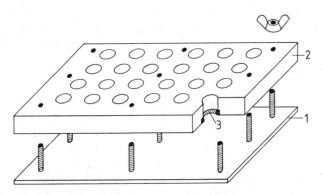

Figure 1. Test device for studying protein adsorption with the aid of an enzyme-immunoassay.
(1) Stainless steel bottom plate, (2) Teflon upper part, (3) silicone rubber sealing ring.

anticoagulant. The anticoagulant was 130 mM trisodium citrate and the anticoagulant to blood ratio was 1:9 (v/v). The tubes were centrifuged for 15 min at 1,570 g and the remaining plasmas were centrifuged for 15 min at 3,000 g. Thereafter the plasmas were pooled in a polypropylene beaker of 1,000 mL and the pooled plasma was transferred into polypropylene vessels of 2.2. mL. Vessels with plasma were kept at -30°C. Just before use, plasma was thawed in a water bath of +37°C.

Serial plasma dilutions were made with phosphate buffered saline, pH 7.4 (PBS) (NPBI, Emmer-Compascuum, The Netherlands), and put into polypropylene vessels of 2.2 mL before (diluted) plasma was transferred into these vessels.

The 11 single donor plasmas were prepared from buffycoats which were obtained from the Blood Bank Twente-Achterhoek (Enschede, The Netherlands). These buffycoats had been prepared from citrated/dextran A blood collected in PVC blood bags. The buffycoats were centrifuged in polypropylene tubes for 15 min at 1,570 g, and the remaining plasmas were centrifuged for 15 min at 3,000 g. The single donor plasmas were put into polypropylene vessels of 2.2 mL and kept at -30°C. Just before use, the plasma was thawed in a water bath of +37°C.

Congenitally HMWK-deficient plasma was obtained from George King Biomaterials, Overland Park, KS, USA, and kept at -30°C. Just before use, the plasma was thawed in a water bath of +37°C.

Purified native (single chain) HMWK (0.6 mg.mL^{-1}) was kindly provided by Dr B.N. Bouma (Department of Hematology, University Hospital, Utrecht, The Netherlands). The purified HMWK has been extensively characterized (9). This protein was also kept at -30°C until use. HMWK-deficient plasma, reconstituted with HMWK, had a HMWK concentration of 70 μg.mL^{-1}.

Protein adsorption and EIA. A description of protein adsorption experiments and the subsequent enzyme-immunoassay (EIA) of adsorbed proteins with the aid of the test device has been given by Poot et al (6) and Van Damme et al (8).

The adsorption experiments as well as the EIA's were carried out at 20°C ± 1.5°C. In order to prevent an air-plasma-solid interface which might induce protein denaturation, 200 μL of PBS was pipetted into the wells of a test device. An adsorption experiment was started by adding 200 μL of undiluted or diluted plasma, depending on the experiment, into three wells of a test device. During pipetting of the plasma (solutions), the end of the pipettor tip was kept under the liquid surface but did not touch the test surface. The liquid in the well was gently mixed using the pipettor tip.

200 μL PBS (instead of plasma) was added into two or three wells of each test device. EIA's performed with the contents of these wells served as blanks.

Adsorption experiments and the subsequent EIA's in which the adsorption of fibrinogen from a single donor plasma (or HMWK-reconstituted plasma) was compared with fibrinogen adsorption from pooled plasma, were simultaneously carried out in threefold. Seven couples of experiments in which fibrinogen adsorption from single donor plasmas (Fig. 3) and fibrinogen adsorption from pooled plasma were compared, were carried out in seven days. Four other couples of experiments were performed in two days. The shortest adsorption time in these experiments was 30 seconds.

In experiments in which glass or polyethylene (mounted in a test device) were preexposed to plasma solutions and subsequently exposed to 1 : 1 diluted plasma, the preexposed surfaces were rinsed 4 times with 800 μl PBS containing 0.005% (v/v) Tween-20. After rinsing, 200 μL of PBS was pipetted into the wells. Thereafter 200 μL of undiluted plasma was pipetted into these wells. The same precautions for the pipetting of plasma solutions into the wells of a test device were taken as described above.

Antibodies and buffers used in the EIA. Rabbit serum directed against human fibrinogen (first antibody) was obtained from the Central Laboratory of the Netherlands Red Cross Blood Transfusion Service (CLB), Amsterdam, The Netherlands. This serum was diluted 100 fold with first antibody buffer. The serum, as well as the other sera and the enzyme-labelled antibodies, were kept at -30°C and thawed in a water bath of +37°C just before use.

The first antibody buffer consists of 8.7 g/L NaCl, 6.1 g/L Tris (Merck, Darmstadt, Germany), 0.02% (v/v) Tween-20, 0.20% (w/v) gelatin (Merck) and 0.5% (w/v) bovine serum albumin (BSA, obtained from Sigma) with pH adjusted to 7.5.

The serum directed against human HDL, i.e. against apolipoprotein A-1 of HDL, was purchased from Behringwerke AG (Marburg, FRG) and diluted 10 fold with first antibody buffer.

Purified goat antibody (1.8 mg/mL) directed against the light chain of human HMWK was kindly provided by Dr F. van Iwaarden (Department of Hematology, University Hospital, Utrecht, The Netherlands). This solution was diluted 2,000 fold with first antibody buffer.

Sheep anti-rabbit IgG and rabbit anti-goat IgG, both conjugated to horse-radish peroxidase (United States Biochemical Co., Cleveland, USA) were the enzyme-labelled antibodies; these conjugates were diluted 1:200,000 and 1:6,000, respectively, in conjugate buffer. When polyethylene was used as a substrate for the adsorption of proteins, the conjugate buffer had the same composition as the first antibody buffer except for a tenfold higher concentration of BSA (5%). When glass is used as a substrate, non-specific adsorption of the enzyme-labelled second antibody may occur. In order to prevent this, BSA in the conjugate buffer was replaced for 10% (v/v) of normal sheep serum (CLB).

Results and Discussion

Protein adsorption data are expressed as absorbances (A450) of the generated dye in the EIA. These data are not directly related to amounts of adsorbed protein/cm^2. Protein molecules generally undergo conformational changes during and after adsorption. As a consequence several antigenic determinants of a protein molecule may lose their specific structure and are not able to react with binding sites of the applied antibody (8). Moreover, antigenic determinants of protein molecules may be masked by other protein molecules.

In order to prevent an air-plasma-solid interface which might induce protein denaturation, buffer was brought into the wells of a test device before plasma was added (see Materials and Methods). For this reason the time dependent adsorption of various proteins from 1:1 diluted plasma instead of undiluted plasma was determined.

The adsorption of fibrinogen to glass from a single donor plasma (donor 6) and from pooled normal human plasma as a function of time are shown in Fig. 2. A two way analysis of variance (two plasmas and different adsorption times) revealed that the curve representing fibrinogen adsorption from this single donor plasma as a function of time does not significantly differ from the curve obtained for pooled plasma. Both fibrinogen adsorption curves show a strong decrease during the first 10 min of adsorption and reach a plateau level after about 10 minutes. A similar phenomenon has also been found by Brash et al (5) and Poot et al (6). In the study of Poot et al 1:1 diluted plasma was used, while Brash et al used plasma diluted to 2.5 and 0.5%. In the last case, however, the amount of fibrinogen adsorbed to glass still decreased after 3 hr, probably because an equilibrium between fibrinogen in solution and adsorbed fibrinogen had not yet been reached.

Fibrinogen adsorption curves determined for 7 of the 11 donor plasmas showed small but significant differences compared with the corresponding curves for pooled plasma. This is not surprising because it is well known that the protein composition of human plasmas varies (10) and as a result the adsorbed amounts of a particular protein may be different. Larger differences were observed between the fibrinogen adsorption curves of the 11 donor plasmas (Fig. 3). The real differences between these single donor curves are most probably smaller than the differences represented in Fig. 3, because the results of EIA experiments show a day-to-day fluctuation (8) and the majority of fibrinogen adsorption curves were determined at different days.

The two mean fibrinogen adsorption curves for the single donor plasmas and pooled plasma, presented in Fig. 4, show a small difference during the first 10 min of adsorption. No significant difference between the plateau values of the curves is found. The reason for the difference between the decreasing parts of the adsorption curves is not known.

The A 450 value of 0.12, corresponding with the amount of HWMK adsorbed from 1:1 diluted HMWK-deficient plasma to glass (not shown), is small compared with the A450 value of about 0.80 for a normal plasma. Therefore it may be concluded that the deficient plasma contained only a small amount of HMWK.

The fibrinogen adsorption curve for HMWK-deficient plasma (Fig. 5) is located at a much higher level than the curve for pooled plasma. For adsorption times larger than (about) 1 min, the curve for the deficient plasma is also located at a significantly higher level than the curves obtained with the 11 single donor plasmas. This finding strongly suggests that fibrinogen adsorption curves obtained with normal donor plasmas will always be located at a lower level than the adsorption curve for a single donor HMWK-deficient plasma. The fibrinogen

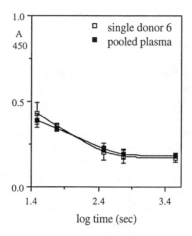

Figure 2. Adsorption of fibrinogen to glass from 1:1 diluted human plasmas as a function of time. Adsorption times were 30 s, 1 min, 5 min, 10 min and 1 hour. □ single donor (donor 6) plasma, ■ pooled plasma. Adsorption values (A450) with regard to the two adsorption curves were simultaneously determined in threefold (n=3 ± SD).

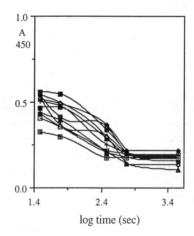

Figure 3. Adsorption of fibrinogen (A450) to glass from 11 donor plasmas (1:1 diluted) as a function of time. Most of the single donor curves were determined at different days (see Materials and Methods). For adsorption times, see caption of figure 1.

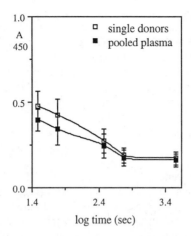

Figure 4. Mean adsorption curves for the adsorption of fibrinogen to glass from 11 single donor plasmas (1:1 diluted) and pooled plasma (1:1 diluted) as a function of time (n=33 ± SD). For adsorption times, see caption of figure 1.

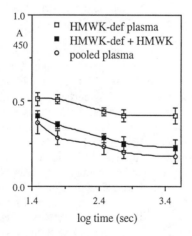

Figure 5. Adsorption of fibrinogen (A450) to glass from 1:1 diluted human plasmas as a function of time. □ single donor HMWK-deficient plasma, ■ HMWK-deficient plasma, reconstituted with HMWK. O pooled plasma (n=3 ± SD). For adsorption times, see caption of figure 1.

adsorption curve obtained with the HMWK-deficient plasma, reconstituted with HMWK to a physiological level of 70 μg. mL^{-1}, is located at a somewhat higher level than the curve for pooled plasma. Some of the plateau levels of the curves obtained with single donor plasmas are, however, not significantly different compared to the plateau level of the curve for the reconstituted plasma. This means that addition of HMWK to HMWK-deficient plasma is mainly responsible for the fact that fibrinogen adsorption from the reconstituted plasma to glass is about the same as fibrinogen adsorption from a normal plasma. This result supports the conclusion of several authors (2-6), including ourselves, that HMWK is involved in the displacement of fibrinogen, initially adsorbed from normal plasma to glass (the Vroman effect). The decrease of fibrinogen adsorption from HMWK-deficient plasma during the first 10 min of exposure to glass indicates, however, that one or more other proteins are also involved in the displacement of adsorbed fibrinogen.

At low plasma concentrations the amount of HMWK, adsorbed to glass, increases with increasing plasma concentration (Fig. 6). Above a plasma concentration of about 0.1% a plateau value for the amount of adsorbed HMWK is reached. These results are in agreement with those of earlier experiments (6). In very diluted normal plasma the concentration of HMWK is apparently decreased below such a low value that other proteins, which are rapidly adsorbed, cannot be replaced anymore by HMWK (11). At relatively high plasma concentrations relatively small amounts of other plasma proteins, like albumin, immunoglobulin G, fibrinogen, fibronectin and Factor VIII adsorb to glass (6). The high HMWK adsorption plateau, which is reached when glass is exposed to a solution with a plasma concentration higher than about 0.1%, indicates that HMWK preferentially adsorbs onto glass and may displace initially adsorbed plasma proteins from the glass surface.

The displacement of adsorbed plasma proteins by HMWK was also studied in another way. Glass surfaces were first preexposed (1 h) to solutions with different concentrations of plasma, resulting in protein layers having different protein compositions. The precoated glass surfaces were subsequently exposed to 1:1 diluted plasma for 1 hour. Thereafter the amounts of HMWK in the newly formed protein layers were compared with the amounts of this protein present on the preexposed glass surfaces. Earlier experiments (6) demonstrated that glass surfaces, which were exposed to 1:100,000 to 1:1,000 diluted plasma, i.e. precoated surfaces to which smaller amounts of HMWK had been adsorbed compared to the amount of HMWK represented by the above mentioned adsorption plateau, had adsorbed relatively large amounts of proteins like albumin, immunoglobulin G, fibrinogen, fibronectin, Factor VIII, and most probably many other proteins. In the present experiments, the preexposed glass surfaces, which were subsequently exposed to 1:1 diluted plasma, give rise to the same amount of adsorbed HMWK. This amount is equal to the amount of HMWK represented by the HMWK adsorption plateau, mentioned earlier. This indicates that HMWK, adsorbing from the 1:1 diluted plasma, is involved in the displacement of proteins from glass surfaces which had been preexposed to solutions with a plasma concentration below 0.1%.

From Fig. 7 it may be concluded that about 50% of the amounts of albumin and fibrinogen, adsorbed to glass from 1:1,000 diluted plasma, are displaced from the glass surface by one or more other plasma proteins after a subsequent exposure of the surface to 1:1 diluted plasma. In this case HMWK is not responsible for the displacement of albumin and fibrinogen, because the amount of adsorbed HMWK did not significantly change after the subsequent exposure to 1:1 diluted plasma. This is in agreement with Fig. 6, which shows that the amount of HMWK, adsorbed to glass from 1:1,000 diluted plasma, already reaches the HMWK plateau value.

In order to investigate the displacement of proteins, adsorbed to polyethylene from plasma solutions, similar experiments were carried out as discussed above, but in those experiments the adsorbed amounts of HDL instead of HMWK were determined. The amount of HDL adsorbed to polyethylene increases as a function of the plasma concentration and reaches a plateau value at a plasma concentration of about 1% (Fig. 8). This result agrees well with the result of experiments reported earlier (*12*). A similar result was found for the adsorption of HDL from plasma solutions to PVC (*7*).

The relatively low adsorption plateau level of HDL may be due to the fact that only parts of the HDL molecule consists of apolipoprotein A-1, against which the first antibody is directed. This results in a relatively low surface concentration of antibody molecules, which in turn leads to a low A450 value (*8*).

Earlier experiments (*6*) revealed that polyethylene which had been exposed to 1:100,000 to 1:100 diluted plasma, i.e. precoated polyethylene surfaces to which smaller amounts of HDL had been adsorbed compared to the amount corresponding with the HDL adsorption plateau, had adsorbed relatively large amounts of other proteins like albumin, immunoglobulin G, fibrinogen, fibronectin and HMWK. The polyethylene surfaces which were exposed to solutions with different plasma concentrations and subsequently exposed to 1:1 diluted plasma show the same amount of adsorbed HDL. This amount is equal or practically equal to the amount represented by the HDL adsorption plateau mentioned above. A similar result was obtained in earlier experiments (*12*). These results reveal that HDL, adsorbing from the 1:1 diluted plasma, is involved in the displacement of proteins from polyethylene surfaces, which had been preexposed to solutions with a plasma concentration below 1%.

From Fig. 9 it may be concluded that the major part of albumin and fibrinogen, adsorbed to polyethylene from 1:1,000 diluted plasma, is displaced from the polyethylene surface by one or more other plasma proteins after a subsequent exposure of the surface to 1:1 diluted plasma. Fig. 9 also shows that there is only a small increase of the amount of adsorbed HDL after exposure of the precoated polyethylene surface to 1:1 diluted plasma. This indicates that one or more proteins, which differ from HDL, is (are) involved in the displacement of adsorbed albumin and fibrinogen from the preexposed polyethylene surface.

In the foregoing text the displacement of preadsorbed proteins by proteins from 1:1 diluted plasma was discussed. However, similar displacement phenomena will occur when a material surface is directly exposed to plasma or a plasma solution, because it probably makes little difference whether a certain surface concentration of a particular protein has been obtained by 'precoating' or

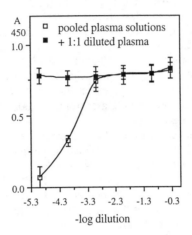

Figure 6. Adsorption of HMWK (A450) to glass as a function of the plasma concentration (□). Adsorption time 1 hour. Pooled human plasma was used to make plasma dilutions of 1:1 to 1:100,000. In a simultaneous experiment glass surfaces were first preexposed for 1 h to solutions with the same plasma dilutions. The preexposed glass surfaces were subsequently exposed to 1:1 diluted pooled plasma for 1 h and the amounts of adsorbed HMWK (■) were determined (n=3 ± SD).

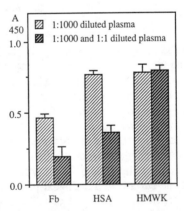

Figure 7. Adsorption of fibrinogen (Fb), albumin (HSA) and HMWK to glass from 1:1,000 diluted pooled plasma (▨). Adsorption time 1 hour.
In a simultaneous experiment glass surfaces were first preexposed to 1:1,000 diluted pooled plasma for 1 h and subsequently exposed to 1:1 diluted pooled plasma for 1 hour. Thereafter the adsorbed amounts of fibrinogen, albumin and HMWK were determined (▨) (n=3 ± SD).

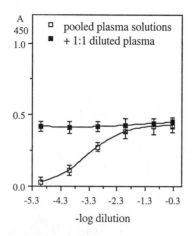

Figure 8. Adsorption of HDL (A450) to polyethylene as a function of the plasma concentration (□). Adsorption time 1 hour. Pooled human plasma was used to make plasma dilutions of 1:1 to 1:100,000. In a simultaneous experiment polyethylene surfaces were first preexposed for 1 h to solutions with the same plasma dilutions. The preexposed polyethylene surfaces were subsequently exposed to 1:1 diluted pooled plasma for 1 h and the amounts of adsorbed HDL (■) were determined (n=3 ± SD).

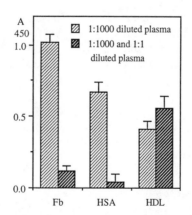

Figure 9. Adsorption of fibrinogen (Fb), albumin (HSA) and HDL to polyethylene from 1:1,000 diluted pooled plasma (▨). Adsorption time 1 hour. In a simultaneous experiment polyethylene surfaces were first preexposed to 1:1,000 diluted pooled plasma for 1 h and subsequently exposed to 1:1 diluted pooled plasma for 1 hour. Thereafter the adsorbed amounts of fibrinogen, albumin and HDL were determined (▨) (n=3 ± SD).

by a direct exposure to plasma during a definite time. The results obtained thus far indicate that many proteins are involved in the Vroman effect during exposure of a material surface to plasma.

References

1. Vroman, L. and Adams, A.L., *J. Biomed. Mater. Res.*, **1969**, *3*, 43-67.
2. Vroman, L., Adams, A.L., Fisher, G.C. and Munoz, P.C., *Blood*, **1980**, *55*, 156-159.
3. Schmaier, A.H., Silver, L., Adams, A.L., Fischer, G.C., Munoz, P.C., Vroman, L. and Colman, R.W., *Thromb. Res.*, **1983**, *33*, 51-67.
4. Elwing, H., Askendal, A. and Lundström, I., *J. Biomed. Mater. Res.*, **1987**, *21*, 1023-1028.
5. Brash, J.L., Scott, C.F., Ten Hove, P., Wojciechowski, P. and Colman, R.W., *Blood*, **1988**, *71*, 932-939.
6. Poot, A., Beugeling, T., Van Aken, W.G. and Bantjes, A., *J. Biomed. Mater. Res.*, **1990**, *24*, 1021-1036.
7. Breemhaar, W., Brinkman, E., Ellens, D.J., Beugeling, T. and Bantjes, A., *Biomaterials*, **1984**, *5*, 269-274.
8. Van Damme, H.S., Beugeling, T., Ratering, M.T. and Feijen J., *J. Biomater. Sci. Polymer Edn.*, **1991**, *3*, 69-84.
9. Van Iwaarden, F., De Groot, Ph.G., Sixma, J.J., Berrettini, M. and Bouma, B.N., *Blood*, **1988**, *71*, 1268.
10. Ritchi, R.F. in F.W. Putman (ed), *The Plasma Proteins*, Academic Press, New York, **1975**, p. 418.
11. Brash, J.L. and Ten Hove, P., *Thromb. Haemostas.*, **1984**, *51*, 326-330.
12. Beugeling T. in Y.F. Missirlis and W. Lemm (eds), *Modern Aspects of Protein Adsorption on Biomaterials*, Kluwer Academic Publishers, Dordrecht, The Netherlands, **1991**, p. 211-212.

RECEIVED February 10, 1995

Chapter 12

Protein–Protein Interactions Affecting Proteins at Surfaces

Peter H. Warkentin, Ingemar Lundström, and Pentti Tengvall

Biomaterials Consortium, Laboratory of Applied Physics, Linköping University, S–581 83 Linköping, Sweden

Protein adsorption to methylated silicon gradient and hydrophilic silicon surfaces as single, binary, and tertiary purified components as well as in association with normal plasma were studied. Antibody and albumin incubations of protein and plasma treated surfaces were noted to affect ellipsometric thickness and antibody binding patterns. BSA blocking procedures on surfaces treated with pure fibrinogen were able to mimic, to a degree, the same ellipsometric patterns as plasma treated gradients when probed by specific antibodies. Indications are that serum albumin influences and may even participate in the removal of IgG and fibrinogen from certain surfaces, even as a subsequent, non-competitive step. HMWK adsorbs differently to surfaces dependent on whether it is a serum component or a single protein. Specific antibody incubations of protein and plasma treated surfaces have been observed to cause a reduction in the ellipsometrically observable surface bound protein layer. These relationships may be due to components other than those indicated by the Vroman sequence.

Protein adsorption at the blood-material interface is a complex interaction that has yet to be thoroughly defined. It is known to be a function of the surface with which proteins interact but it is also true that protein-surface associations are not the same in mixed protein situations as in pure protein solutions. Increasingly we have had to address the fact that proteins respond not only to the surfaces that they have been introduced to but also each other. In the following studies we see how proteins compete in limited systems and how this behaviour differs if the competitive edge is removed.

In a previous paper (1), we have presented studies showing that protein-protein interactions affect protein interactions towards the surface in limited protein systems. These surface interactions vary according to the order and combination of the other proteins involved in the experimental model. Previous findings by Deyme et al (2) and Baszkin and Boissonade (3) indicate similar interactive behaviour of proteins where the presence of collagen resulted in increased albumin adsorption to solution-air and solution-polyethylene interfaces. These researchers support the approach of the study of limited defined protein systems stating that "detailed studies of binary solutions must be carried out with consideration to both components in such a system". It is through studies of simple systems such as these that the behaviour of more complex protein mixtures such as blood serum and blood plasma can be better understood.

0097–6156/95/0602–0163$12.00/0

This paper offers further evidence of protein-protein interactions towards surfaces focusing largely, but not exclusively, on albumin affects. Plasma and serum system experiments illustrate that while these complex mixtures do not give results identical to those observed in defined protein systems, there are effects produced by the introduction of albumin that result in similar protein detection patterns.

While there have been many models presented to explain protein adsorption onto surfaces they are often dependent on proteins behaving strictly as "particles" bearing average physical characteristics that have been verified. Some of these particle properties are however assigned for the convenience of the model (4). Other researchers have offered models of more complex proteins by partitioning physical characteristics thereby allowing differential behaviour by the particle model (5). There is also a tendency to regard the model particle only with respect to the material interface. While these assumptions aid the model, they tend to ignore subtle interactions and differences in real proteins that may limit the models applicability especially in real world multi-component systems.

In order to study variable surface characteristics we have often employed a methylation gradient over a negatively charged hydrophilic silicon surface (6) such that the degree of surface energy can be modified in a controlled manner. This results in a surface that is extremely hydrophobic at one end (water contact angle >85°) to very hydrophilic (water contact angle <10°) at the other. This gives a useful gradient for mimicking the hydrophobic/hydrophilic glass surface although it should be noted that the properties expressed are unique to this system. By looking at the protein detectability over a continuous gradient rather than on discrete surfaces, the resulting pattern can offer a protein behaviour "fingerprint" where localized changes in binding and antigenicity can be more easily identified. This is because these differential chemical and surface energy properties affect interacting blood proteins differently, particularly for single proteins. In more complex mixtures however the behaviour of a protein may not demonstrate the same behaviour based upon the interactions or influences of proteins that surround it. Proteins may even be affected by other proteins that precede them to the interface or come after it. The effects may be the result of displacement, competition, or cooperation and are further studied on discrete hydrophilic silicon surfaces in this paper.

The surface deposition and desorption of fibrinogen in dilute protein mixtures has been termed the Vroman effect (7) and a sequence of plasma protein surface adsorption has been proposed (8). What we have noted in our recent studies (1) is that displacement is not always as indicated in the Vroman sequence. In particular, bovine serum albumin and human serum albumin have been used to affect the surface interactions of IgG and fibrinogen after adsorption. This effect has been most notable at the hydrophilic end of the gradient.

Experimental

Phosphate buffered saline, (PBS) was prepared as 8.0 gm NaCl, 0.2 gm KH_2PO_4, 2.9 gm $Na_2HPO_4 \cdot 12H_2O$, 0.2 gm KCl, and 0.2 gm NaN_3. This was brought to 1 liter, pH 7.4, without adjusting. Tris buffered saline, (TBS) was prepared as 20 mM Tris(hydroxymethyl)-aminomethane, 0.5 M NaCl, adjusted to pH 7.5 with HCl.
Bovine serum albumin (BSA), a product of the Sigma Chemical Company, U.S.A., was of a quality sufficient for use in ELISA with a fatty acid composition of 0.009%. Human serum albumin (HSA) was obtained commercially as Fraction V, 96-99% albumin from Sigma Chemical Company, U.S.A. A purified form of fibrinogen was obtained from Calbiochem with a coagulability of >95% of the protein.

The human IgG used was a commercial gammaglobulin fraction from KabiVitrum AB, Sweden of a quality suitable for human injection. The source of purified high molecular weight kininogen (HMWK) used in gradient studies was Calbiochem,

U.S.A. while discrete surface energy studies were performed using a purified single chain high molecular weight kininogen obtained from Enzyme Research Laboratories, U.S.A.

Citrated human plasma was pooled from five apparently healthy donors. HMWK deficient plasma was obtained from Sigma as lyophilized powder which was reconstituted immediately before use with deionized water.

Rabbit anti-human IgG, rabbit anti-human albumin, rabbit anti-human fibrinogen, and alkaline phosphatase conjugated swine anti-rabbit IgG were all polyclonal IgG fractions from Dakopatts, Denmark. Rabbit anti-human HMWK was a polyclonal IgG fraction from Calbiochem while the goat anti-human HMWK antiserum was purchased from Nordic, Finland and an alternate goat anti-human HMWK antiserum was obtained as a gift from Enzyme Research Laboratories, U.S.A.

Commercially produced Type P(100) boron doped silicon wafers with a diameter of 76.2 ± 0.5 mm were obtained from Okmetic, Finland, cut to required sizes (1 cm x 1 cm) and used in hydrophilic and hydrophobic flat surface experiments. Silicon gradient surfaces, 10 mm x 37 mm, were also cut from the same silicon wafers.

Ellipsometry. The ellipsometric one-zone measurements were made on dried protein surfaces, and using a Rudolph Research Auto El III ellipsometer. The protein concentration was then estimated assuming a constant film refractive index $n_{ox} = n_f = 1.465$ according to the method of Stenberg and Nygren (9),

$$\text{Surface concentration (ng/mm}^2) \approx K \times \text{thickness (nm)}$$

where K is the density of the protein ≈ 1.35 g/l. Since the refractive index of proteins, $n_p \approx 1.55$, the thickness of the protein film is decreased by a factor:

$$[\, (n^2_{ox}-1)/n^2_{ox} \,] \, / \, [\, (n^2_p-1)/n^2_p \,] \qquad \text{thus giving } K \approx 1.2 \ (9)$$

As the refractive indices, n_f, may vary slightly for different proteins and on different substrates, the calculated values reported in this paper are effective values on silicon and DDS-methylated silicon, respectively.The error bars presented in the graphs indicate the statistical errors of the measured values at the surface. They do not include systematic discrepancies due to the approximations introduced by the refractive indices or mathematical modelling techniques. A previous comparison of *in situ* ellipsometric and radiotracer experimental values of adsorbed amounts of proteins onto silicon surfaces demonstrate reasonable agreement between the two techniques (10).

Silicon Surface Protein Treatments. Wettability gradients were produced on silicon surfaces using the method of Elwing et al (6). Silicon wafers were cut into sizes of 10 x 37 mm and made hydrophilic (water contact angle < 5°) by heating to 80°C in a 1:1:5 (volume) solution of $NH_3:H_2O_2:H_2O$ (deionized). They were washed 3 times in deionized water heated to 80°C in a 1:1:6 solution (volume) of $HCl:H_2O_2:H_2O$ (deionized) and finally washed 3 times in deionized water and stored until use in acidified H_2O. These were placed vertically in a container filled with xylene and a 0.05% solution of DDS in trichloroethylene which was carefully layered below the xylene phase and permitted the DDS to diffuse into the upper phase for 90 minutes. The contents were then drained from the bottom and the surfaces washed sequentially with ethanol, trichloroethylene, and ethanol and finally dried under nitrogen stream prior to use.

Depending on the particular experiment, proteins were either used at human physiological concentrations or brought to 1/10 physiological concentrations in PBS as indicated in the review of Andrade and Hlady (11) or as indicated by Müller-Esterl (12) with respect to human HMWK. Some researchers have, in the past, chosen to use Tris buffered saline to overcome the problem of phosphate interfering with absorption of some proteins, particularly fibrinogen (13). On the other hand, Tris buffers are less physiological and, when used in adsorption steps, interfered with HMWK detectability (14). Since it has never been established how the bias introduced by the use of various buffer differs from the absolute physiological adsorption conditions, we use PBS as a standard buffer. Physiological concentrations were uniformly defined as 12.5 mg/ml for IgG, 2.5 mg/ml for fibrinogen and 40 mg/ml for albumin. HMWK was diluted to final physiological concentration of 0.074 mg/ml. Plasma was diluted 1:9 in PBS for comparisons with the single protein solutions where indicated. Incubation times are either 10 minutes or 1 hour depending on the experiment. In BSA blocked plasma experiments, citrated plasma diluted to 10% physiological in PBS was incubated for 1 hour followed by 1 hour in 1% (10 mg/ml) BSA in PBS.

In experiments using BSA, HMWK, and plasma treated gradients, duplicate surfaces were used and representative surfaces are shown in the figures. In all other experiments triplicates were used and results are represented statistically. Experimental results were corrected for background DDS methylation and oxide layer by measuring untreated surfaces and subtracting these values.

Gradients were washed with PBS and incubated in a 1.8 ml solution per gradient in either the plasma or protein solution for 1 hour. In experiments using discrete surfaces, 3 surfaces of 1 cm^2 were incubated in 1.8 ml. Surfaces were washed in PBS followed by deionized water, dried under nitrogen and the initial layer read by ellipsometry. In blocking experiments surfaces were treated as above except that they were blocked for 1 hour in 1% BSA after exposure to the initial protein treatment and rinsed in PBS and deionized water prior to the first ellipsometric reading. In controls and other experiments where surfaces were checked for amplified secondary antibody binding or non-specific antibody adsorption they were incubated in swine anti-rabbit IgG-alkaline phosphatase, 1/50 dilution in a PBS solution for 1 hour and washed in TBS followed by water, dried under nitrogen and read again. The conjugated antibody was used to keep the reagent consistent with previous ELISA and surface imaging work (14). Where there was primary antibody binding the surfaces were checked for further specific ellipsometric increases while in the case of no specific primary antibody incubation, they indicate only non-specific background binding.

BSA Treatment of Protein Adsorbed Gradients. Gradients were incubated with IgG, 1.25 mg/ml or fibrinogen, 0.25 mg/ml in PBS for 1 hour. Half the number of surfaces were blocked for 1 hour in 1% BSA in PBS and read by ellipsometry. All surfaces, blocked and non blocked, were incubated in duplicate for 1 hour at room temperature in 1.8 ml/surface of the following antibody solutions diluted 1/50 in PBS: rabbit anti-human IgG, rabbit anti-human fibrinogen, and rabbit anti-human HMWK. Surfaces were then washed in PBS followed by deionized water, dried under nitrogen and read by ellipsometry. These represented surfaces after primary antibody incubation. The surfaces were then washed in PBS and incubated for 1 hour at room temperature in a 1/50 PBS solution of swine anti-rabbit IgG-alkaline phosphatase as secondary antibody.

After the second antibody treatment, surfaces were washed 3 times in TBS and rinsed in deionized water. The surface thickness was read after every incubation step by ellipsometry.

Results and Discussion

Plasma Treated Gradients. Plasma treated gradient surfaces displayed a fair degree of variation in surface thicknesses as measured by ellipsometry (± 1.9 ng/mm^2, not shown). It is not certain whether this is due to variability in gradient preparation or changes in plasma protein adsorption. The differences in plasma treated surfaces alone and surfaces plasma treated followed by BSA blocking, however, were not statistically conclusive (not shown). When the same surfaces are incubated with the amplifying antibodies alone there was no alteration in thickness indicating that the secondary antibodies in the absence of specific primary antibodies did not add to ellipsometric thickness.

In surfaces which were both treated with plasma alone (not shown) and those followed by BSA blocking (Fig. 1), there was no statistically conclusive indication that rabbit anti-IgG altered in its ability to bind the surface adsorbed IgG, although there appeared to be a reduction from 8.4 ng/mm² without BSA to 7.2 ng/mm² with BSA on the hydrophobic end. In surfaces which were both treated with plasma alone and those which were blocked, the anti-fibrinogen demonstrated a maximal ellipsometric thickness subterminally with respect to the hydrophobic end. When BSA incubations are introduced prior to primary antibody treatment there is a significant reduction in thickness of about 4.8 ng/mm² from peak maxima. This reduction is not as great at the hydrophilic end of plasma treated surfaces but is still noticeable.

Single Protein Treated Gradients. The single protein adsorption profiles are in agreement with previous observations (6, 1). IgG treated surfaces gave a smooth decrease in thickness along the gradient from an average of 5.0 ng/mm² at the hydrophobic side to 1.9 ng/mm² at the hydrophilic side (Fig. 2a). On BSA blocked surfaces this pattern is much the same, 4.8 ng/mm² to 2.5 ng/mm² (Fig. 2b). Primary antibody incubation of surfaces treated with IgG in the absence of BSA blocking (Fig. 2a) gives an average increase in thickness to 12.1 ng/mm² at the hydrophobic end to 10.1 ng/mm² at the hydrophilic end. This indicates confluent, albeit differential, IgG binding across the entire gradient. Surfaces that were exposed to BSA after IgG incubations display a remarkably different pattern however upon primary antibody incubation. The thickness at the hydrophobic end is unchanged (12.1 ng/mm²) but very much lower (2.5 ng/mm²) at the hydrophilic end (Fig. 2b). This drop-off is most pronounced from 5 to 11 mm from the hydrophobic end. The BSA in this case has definitely affected the preexisting IgG layer, either by removing and/or replacing it or by compromising its antigenicity. The degree of the BSA influence is a function of the surface characteristics, as well as the protein on the surface, otherwise IgG across the entire gradient would have been affected to the same extent. Human serum albumin in plasma may act in the same manner. Although it is in lower concentration than is these blocking experiments, it may be more competitive since it is in solution at the same time as IgG as opposed to the post-incubation exposure of BSA in this experimental model.

The pure fibrinogen story on gradients resembles IgG in may ways. As a pure protein on the gradient (Fig. 3a) it resides as two distinct ellipsometric thicknesses, one at the hydrophobic end of 4.9 ng/mm² and one at the hydrophilic end of 3.5 ng/mm². From 8 to 12 mm from the hydrophobic end there is a steeper transition in thickness than at the two extremes. The BSA blocked profile (Fig. 3b) is essentially the same displaying no apparent protein thickness changes. Upon application of the rabbit anti-fibrinogen there is a confluent shift upward of the entire fibrinogen profile to 14.5 ng/mm² at the hydrophobic end and 12.5 ng/mm² at the hydrophilic end (Fig. 3a). In the BSA blocked surfaces (Fig. 3b) the thickness is submaximal at the hydrophobic end and increases to a maximum again at a position 3 mm from the hydrophobic terminus. This low terminal antigenicity is similar to the BSA blocked or unblocked plasma and not noted in anti-fibrinogen activity where albumin is not a component. It may be a direct effect of the presence of albumin. The anti-fibrinogen remains maximal and begins to drop at a distance 8-9 mm along the gradient which reflects the transition point of the fibrinogen alone. The drop in antibody binding thickness at this point is similar to the effect on IgG-anti-IgG treated surfaces but in this case the drop in thickness is not as great. The pattern is similar to the anti-fibrinogen binding to BSA

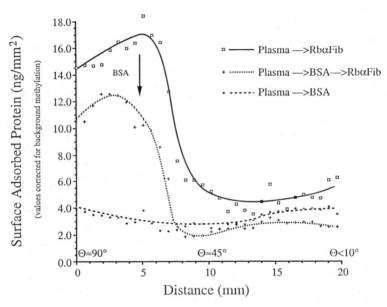

Figure 1. Ellipsometric readings of plasma adsorption and fibrinogen antigenicity changes due to BSA blocking of plasma treated methyl-silica gradient surfaces.

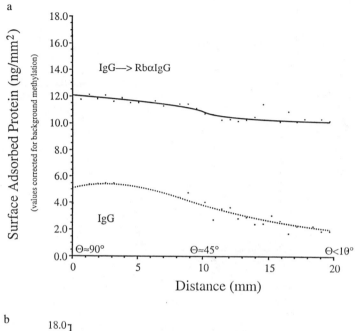

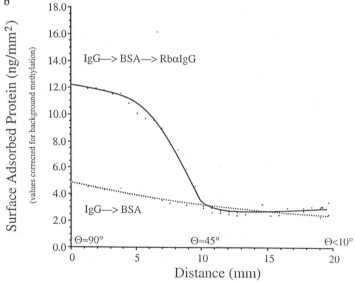

Figure 2. Ellipsometric readings of IgG adsorption and antigenicity on methylation gradients. a) Lower curve: IgG, 1.25 mg/ml. Upper curve: IgG followed by rabbit anti-IgG b) Lower curve: IgG, 1.25 mg/ml followed by 1% BSA. Upper curve: IgG, 1.25 mg/ml followed by 1% BSA followed by rabbit anti-IgG.

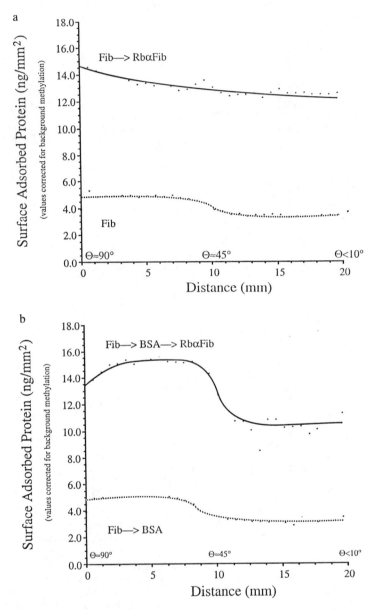

Figure 3. Ellipsometric readings of fibrinogen adsorption and antigenicity on methylation gradients. a) Lower curve: Fibrinogen, 0.25 mg/ml. Upper curve: Fibrinogen followed by rabbit anti-fibrinogen. b) Lower curve: Fibrinogen, 0.25 mg/ml followed by 1% BSA. Upper curve: Fibrinogen, 0.25 mg/ml followed by 1% BSA and then rabbit anti-fibrinogen.

blocked and unblocked plasma treated surfaces but the hydrophilic end of the pure protein still does not display as low a thickness (10.1 ng/mm^2, Fig. 3b) as the BSA non-blocked plasma treated surface (6.1 ng/mm^2) or the BSA blocked plasma treated surface (2.5 ng/mm^2, Fig. 1). This may reflect a greater tenacity of the pure fibrinogen for the surface in a non-competitive situation. Nonetheless albumin must be playing a role in displacing or masking the antigenicity of both IgG and fibrinogen.

Human Serum Albumin Effects. While it is not possible to present all work done involving human serum albumin, the results thus far indicate that there is no significant difference compared to those previously observed with BSA. Gradients (in triplicate) treated with 1/10th physiological fibrinogen, 1 hour followed by 1/10th physiological HSA, 1 hour, and then anti-fibrinogen for 30 minutes resulted in a profile that is essentially the same as in Figure 3b but having lower overall thicknesses (13.2 ng/mm^2, hydrophobic end, 7.8 ng/mm^2, intermediate region, 9.0 ng/mm^2, hydrophilic end, not shown).

The albumin effects on fibrinogen are once again demonstrated in work done using discrete hydrophilic surfaces (Fig. 4). In this case, rather than having albumin at high concentrations as a blocking component, it is presented to the experimental system in concentrations physiologically proportional to that of fibrinogen. In both normal and 1/10th physiological concentrations for 1 hour (triplicate surfaces, 5 readings per surface) the effects are essentially the same. If HSA is introduced to the fibrinogen treated surfaces prior to anti-fibrinogen, there is significantly less protein thickness. One might conclude that masking is involved but as shown in the physiological concentration experiment this decrease in thickness is statistically lower than the initial layer of fibrinogen alone. This would imply then that HSA may somehow be involved not simply in altering fibrinogen antigenicity, but in the removal of loosely bound fibrinogen itself.

HMWK as Single Protein vs Serum and Plasma Component. Previous studies (15) have indicated that while the most abundant plasma proteins are antigenically detectable predominantly at the hydrophobic end of the gradient, goat anti-HMWK (Nordic) binds more to the hydrophilic portion. Figure 5 indicates that the same is true for serum. In this case it appears as a major detectable protein.

Figure 6 shows that as a pure protein however, HMWK at the same 1/10th physiological concentration is predominantly indicated at the hydrophobic end, similar to other proteins when introduced to surfaces as a single protein component. Its behaviour in the mixed protein systems is a function of the presence of the other proteins as well as the surface. This could be the result of a displacement by HMWK similar to that proposed by Vroman and Adams (16) but could also be the consequence of other factors or proteins not previously considered. In serum, fibrinogen is not present and its deposition is not a prerequisite to HMWK adsorption. In Figure 5 there appears to be very little antigenic indication that IgG is present over the gradient and even anti-albumin does not bind in substantial amounts compared to plasma. If HMWK adsorbed to the surface through sequential displacement then it bypassed fibrinogen involvement. This would mean that other serum components provided HMWK better access to the hydrophilic surfaces and limited deposition on hydrophobic side. Fibronectin may be a candidate as indicated by Vroman and Adams (16) but until it is firmly established what factors these are and what role they play, application of the Vroman displacement concept to proteins other than fibrinogen is open to challenge by other cooperative protein-surface association models which may or may not involve displacement. In another experiment (Fig. 7a), a physiological three protein mixture of HSA, fibrinogen, and HMWK were incubated for an hour on hydrophilic silicon and then probed with specific antibodies. Results indicated that no anti-fibrinogen above the initial protein thickness was detectable by ellipsometric methods, 0.7 ng/mm^2 as a

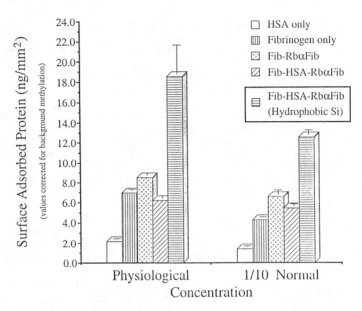

Figure 4. HSA effects on fibrinogen antigenicity on discrete hydrophilic silicon. Left block: proteins at physiological concentration. Right block: proteins at 10% physiological concentration. Rightmost column for each: fibrinogen followed by HSA and rabbit anti-fibrinogen on hydrophobic silicon for reference.

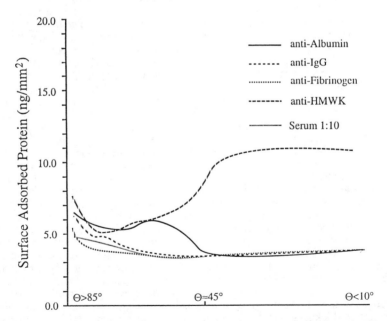

Figure 5. Ellipsometric readings of antibody treatment of methylation gradients incubated for 10 minutes in human serum diluted to 10% physiological in PBS.

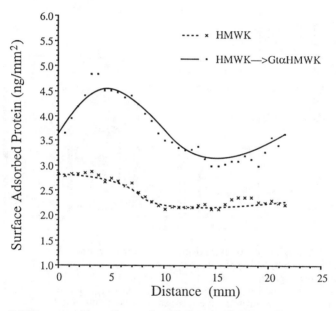

Figure 6. Ellipsometric readings of methylated silicon gradients treated with 0.0075 mg/ml high molecular weight kininogen for 1 hour.

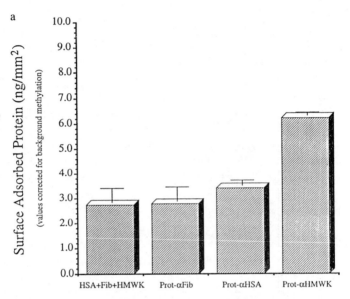

Figure 7. Antigenicity of mixed protein adsorption onto discrete hydrophilic silicon surfaces. a) Ellipsometric readings of: i) adsorbed proteins from a mixture of HSA, 40 mg/ml, Fibrinogen, 2.5 mg/ml, and HMWK, 0.075 mg/ml. ii) Protein layer probed with rabbit anti-fibrinogen. iii) Protein layer probed with rabbit anti-HSA. iv) Protein layer probed with goat anti-HMWK. b) i) HSA, 40 mg/ml. ii) HMWK, 0.075 mg/ml. iii) a mixture of HSA and HMWK. iv) HSA/HMWK mixture probed with rabbit anti-HSA. v) HSA/HMWK mixture probed with goat anti-HMWK. vi) HMWK, 0.075 mg/ml probed with goat anti-HMWK. c) i) a mixture of Fibrinogen and HMWK. ii) Fibrinogen/HMWK mixture probed with rabbit anti-fibrinogen. iii) Fibrinogen/HMWK mixture probed with goat anti-HMWK.

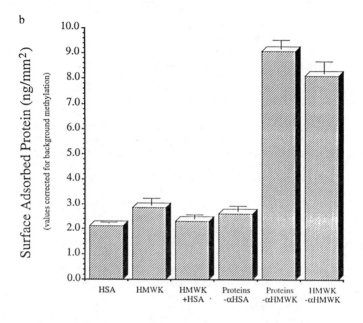

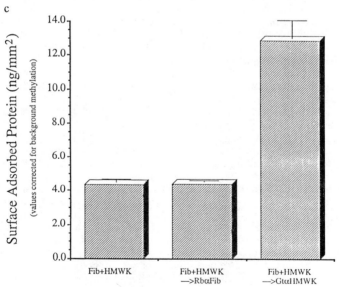

Figure 7. Continued.

result of anti-albumin, and 3.4 ng/mm^2 from anti-HMWK incubation. HMWK may have displaced the fibrinogen as it is very surface active and efficient and without necessarily removing all the albumin. Another possibility is that albumin is accommodating, and may even be displaced by, HMWK without hydrophilic surface involvement of fibrinogen. Indeed in similar experiment using physiological HSA and HMWK only (Fig. 7b), it was found that while there was no statistically significant increase in thickness from anti-HSA incubations, anti-HMWK added a further 6.7 ng/mm^2 over the protein deposition thickness of 2.3 ng/mm^2. HMWK alone resulted in 2.9 ng/mm^2 followed by anti-HMWK giving a further increase of 5.2 ng/mm^2. Consequently HMWK does interact with hydrophilic surfaces without involving fibrinogen. This is not to say that fibrinogen does not demonstrate the Vroman effect. When fibrinogen and HMWK are the only two proteins in the system (Fig. 7c) once again fibrinogen (4.3 ng/mm^2) is not indicated above the protein layer (4.3 ng/mm^2) using ellipsometric antibody adsorption methods. The anti-HMWK level now adds a further 8,5 ng/mm^2 in the presence of fibrinogen which is a 65% greater increase in anti-HMWK thickness than when HMWK is the only protein in the system. This means that both albumin and fibrinogen are not simply exchanged with HMWK, they apparently make the surface more accessible to HMWK or anti-HMWK than would otherwise be the case and that HMWK has a preference for interacting with fibrinogen. If one were to assume that HMWK behaves similarly to the other plasma proteins considered by Andrade and Hlady (11) and estimate its diffusion coefficient as 4.8 x 10^{-7} cm^2/s, then its CD$^{1/2}$ can be approximated as ≈ 0.2 compared to 11 for fibrinogen and 1500 for albumin. This implies a very large surface activity for HMWK in order for it to virtually monopolize the hydrophilic silicon surface. These experiments reflect this property of the molecule which is believed to be largely a function of its hydrophilic histidine region interacting with negatively charged surfaces.

Schmaier et al (17) had previously indicated that using total kininogen deficient plasma to which HMWK had been added back, anti-fibrinogen was present at 10 seconds but not 10 minutes while anti-HMWK was indicated at both times. In studies performed in this laboratory, when 10% HMWK deficient plasma was incubated for 10 minutes on hydrophilic silicon and analysed by the use of antibodies (not shown), neither HMWK nor fibrinogen were detectable. After adding back HMWK and probing again, anti-HMWK and trace anti-HSA were indicated and only a low, statistically insignificant thickness from anti-fibrinogen. It is not entirely clear why fibrinogen was not detectable prior to HMWK addition in our studies. Since fibrinogen was a component of HMWK deficient plasma and still was not indicated on the surface then factors other than HMWK must be responsible for its absence or poor antigenicity. Schmaier et al (17) had also suggested a second mechanism for fibrinogen removal but that involved exposure times greater than 15 minutes. Although factor XII has been suggested to play a role (8) this has not been established and requires further study.

Antibody-Antigen Induced Desorption. The experiment shown in Figure 8 demonstrates the range of ellipsometric effects possible when using antibodies. Triplicate gradients were run using physiological concentrations of fibrinogen followed by a PBS wash and then HSA and employing 1 hour incubation times for each protein. After washing once more ellipsometric readings were made of the two protein layer. Triplicates of the same procedure followed by incubating with rabbit anti-human serum albumin for 30 minutes was also performed. The pre-antibody results indicate that the protein curve resembles the fibrinogen profile although it is greater in thickness than for fibrinogen alone. Upon antibody incubation however the profiles change significantly. The anti-albumin indicates the midrange albumin peak observed in previous experiments (1) while the hydrophilic region indicates no change. This may lead one to conclude that no albumin is present at the hydrophilic end and while this may be the case, other experiments (not shown) indicate that some albumin remains on this

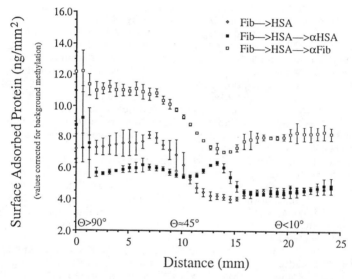

Figure 8. Antigenicity of sequential fibrinogen and albumin treated gradients as analysed by ellipsometry: i) (open diamond) 1 hour fibrinogen, 2.5 mg/ml, followed by 1 hour HSA, 40 mg/ml. ii) (closed square) (i) treated with rabbit anti-HSA. iii) (open square) (i) treated with rabbit anti-fibrinogen.

surface. The observation from this experiment that argues in favour of caution when interpreting results is in regard to the hydrophobic anti-albumin region. The values obtained here, give total protein thicknesses lower than before antibody incubation. This could in part be due to reorganization and surface packing leading to a change in refractive index but this could not account for a 21% reduction in observed thickness. We have interpreted these results as to mean a true desorption of surface associated protein. Since the specific albumin-anti-albumin interaction is involved, it is likely that albumin is the predominant protein removed but specific experiments have not been performed to confirm or refute the possibility that albumin may also affect fibrinogen-surface interactions in some manner. The anti-fibrinogen curve shown in this experiment does however appear to be lower than in Figure 3a or 3b but the results cannot be directly compared since the fibrinogen and albumin concentrations in this experiment is higher. Other observations in our laboratory (unpublished) also indicate that fibrinogen-surface antigenicity is being affected by albumin. It is likely that the specific anti-albumin at the hydrophobic region is desorbing albumin that has been prevented from forming strong surface interactions by fibrinogen.

The reduction of observed organic layer thickness is probably more common in adsorption experiments where target molecules are not involved in covalent or strong electrostatic surface interactions. These conditions differ from most standardised immunological test methods (such as ELISA) where the antibody or target antigen is firmly bound to the surface prior to assay. The type of antisera used in ellipsometric measurements is therefore important. It is generally found that precipitating antisera give ellipsometrically measurable increases in measured organic layer thicknesses when incubated with a surface containing its antigen. More purified antisera with reduced crossreactivity and increased specificity do not always give measurable increases and may even result in a thickness decrease.

Antibody induced decreases in surface thickness have been also been observed by other researchers but often goes unnoticed since many researchers amplify the thickness once more when they assay with secondary antibodies or when the net accumulation of antibody exceeds the tendency of the antibody-antigen to dissociate from the surface bound layer (Baszkin, A., Université Paris-Sud personal communication, 1994). This antibody-antigen interaction is another example of protein-protein interactions affecting protein-surface interactions.

In the work presented here, only the goat anti-HMWK serum from Nordic was seen to give specific ellipsometrically measurable increases in the organic layer thickness when probing for HMWK. When other sources were incubated in the same way they caused a decrease in thickness on both hydrophobic and hydrophilic surface (not shown) although they do exhibit high specificity and sensitivity when used in ELISA and direct imaging methods (14). The reasons for such differences are not clear but the different behaviour dependent on antibody sources (eg. antiserum versus monoclonal) have been indicated by other workers (16). If, after antibody incubation the newly formed antigen-antibody arrangement possesses an increase in solubility the the ellipsometric thickness would be reduced, otherwise such a complex would add t the overall ellipsometric thickness.

"Vroman" Interactions vs Other Protein Interactions. The Vroman effect was first proposed by Brash and ten Hove (18) and Horbett, (19) as cited by Scott (20) who summarizes their definition as the "phenomenon of the absorption and subsequent 'conversion' of fibrinogen from plasma that occurred on artificial, hydrophilic surfaces" and certainly substantial research exists to support this effect as defined. This has often been reinterpreted inappropriately by others however to apply to any surface situation in which one protein interacts to replace another regardless of the surface properties. In addition, the Vroman effect is often confused with a proposed sequence of displacement as indicated by Vroman and Adams (8) based upon experiment performed in narrow spaces (16). It must be noted however that the only proteins indicated in their model were those being probed for and observations only indicate the sequential dominance in antigenicity of those components. This may not accurately reflect their true presence and other serum or plasma proteins may be also be involved in surface interactions.

Results involving albumin, fibrinogen, and IgG shown here and in the previous paper (1) show some similarities to those demonstrated by other researchers (21) although, in light of results presented here, we would interpret their observations differently. The Vroman effect is recognized to take place with respect to fibrinogen but it cannot account for all protein surface-interactive protein behaviour. We agree with the assessment by Wojciechowski and Brash (7) that "other components must be involved in the competition for surface sites and that the concept of a sequential and specific displacement is not realistic". The albumin effects on other proteins noted here and in other studies are a prime example of protein-protein interactions that take place outside of the present protein-surface models. The role of this protein we feel has been underestimated and may in fact be heavily involved in mediating both biological and non-biological surface interactions. There also exist specific enzymatic and antibody-antigen activities that affect how proteins behave with respect to interfaces that must be considered.

Conclusions

Protein-protein interactions affect surface-protein interactions and therefore can alter observed results from the pure protein/surface situation. They are involved both in mixed protein studies and as a result of attempts to analyse the original surface adsorbed status. The albumin and specific antibody interactions noted in this work can be responsible for undesirable experimental bias and therefore such results should not be considered a definitive basis upon which to base conclusions. This does not negate their use as investigative tools but it should be noted that these proteins are the same types of biological molecules as those being probed for. The antigenicity indicated in these assays can reflect changes in amount of protein under investigation or can be the result of conformational alterations that may increase as well as decrease protein detectability in disproportion to its actual presence. The interactions imply that one mechanism involved may be a detergent-like behaviour as a result of protein-protein associations. Displacement and removal of certain protein species are not necessarily the only activities involved. Competition may also play a role and, in the case of HMWK as a pure protein versus a hydrophilic serum component, albumin may act in a cooperative surface association capacity. These aspects deserve further investigation using simple protein systems and such knowledge will ultimately contribute to the understanding of complex protein behaviour.

Acknowledgements

The authors would like to thank Agneta Askendal for her technical assistance. The research presented in this paper is supported by grants from the Swedish Biomaterials Consortium founded by the Swedish National Board for Industrial and Technical Development (NUTEK) and the Swedish Natural Science Research Council (NFR)

Literature Cited.

1. Warkentin, P.H., Wälivaara, B., Lundström, I., and Tengvall, P., *Biomaterials* **1994**, *15*, pp 786-795

2. Deyme, M., Baszkin, A., Proust, J.E., Perez, E., Albrecht, G., and
    Boissonnade, M.M., *J. Biomed. Mater. Res.* **1987**, *21*, pp 321-328
3. Baszkin, A. and Boissonade, M.M., *J. Biomed. Mater. Res.* **1993**, *27*, pp 145-
    152
4. van Straaten, J. and Peppas, N.A., *J. Biomater. Sci. Polymer Edn.* **1991**, *2*, pp
    91-111
5. Andrade, J.D., Hlady, V., and Wei, A.P., *Pure and Appl. Chem.* **1992**, *64*,
    pp 1777-1781
6. Elwing, H., Welin. S., Askendal, A., Nilsson, U., and Lundström, I., *J. Colloid
    Interface Sci.* **1987**, *119*, pp 203-210,
7. Wojciechowski, P. and Brash, J.L., *J. Biomater. Sci. Polymer Edn.* **1991**, *2*,
    pp 203-216
8. Vroman, L. and Adams, A.L., In *Proteins at Interfaces*, Brash, J.L. and Horbett,
    T.A., Eds., American Chemical Society Symposium Series, 192nd Meeting,
    American Chemical Society, Washington, DC, 1987, pp 154-164
9. Stenberg M., and Nygren H., *Jour. de Physique* **1983**, *44*, pp 83-86
10. Jönsson U., Malmqvist M., and Rönnberg I, *J. Colloid Interface Sci.*, **1985**,*103*
    pp 360-372
11. Andrade, J.D. and Hlady, V., In *Blood in Contact with Natural and Artificial
    Surfaces*; Leonard, E.F., Turitto, V.T., and Vroman, L., Eds.; Annals of the New
    York Academy of Sciences; The New York Academy of Sciences: New York,
    NY, 1987, Vol. 516; pp 158-173
12. Müller-Esterl, W., In *Meth. Enzymatic Analysis* ; Bergmeyer, H.U., Editor-in-
    Chief, Bergmeyer, J. and Graßl, M., Eds.; VCH Publishers: Deerfield Beach, FL
    1986, Vol. 9; pp 304-316
13. Giaever, I. and Keese, C. R., In *Proteins at Interfaces*, Brash, J.L. and Horbett,
    T.A., Eds., American Chemical Society Symposium Series, 192nd Meeting,
    American Chemical Society, Washington, DC, 1987, pp 582-602
14. Warkentin, P.H., Lundström, I., and Tengvall, P., *J. of Mater. Sci.: Mater. in
    Med.* **1993**, *4*, pp 318-326
15. Tengvall, P., Askendahl, A., Lundström, I., and Elwing, H., *Biomaterials*
    1992, *13*, pp 367-374
16. Vroman, L. and Adams, A.L., *J. Colloid Interface Sci.* **1986**, *111*, pp 391-401
17. Schmaier, A.H., Silver, L., Adams, A.L., Fischer, G.C., Munoz, P.C., Vroman
    L., and Colman, R.W., *Thromb. Res.* **1983**,*33*, pp 51-67
18. Brash, J.L. and ten Hove, P., *Thromb. Haemostas.* **1984**, *51*, pp 391-402
19. Horbett, T.A., *Thromb. Haemostas.* **1984**, *51*, pp 174-181
20. Scott, C.F., *J. Biomater. Sci. Polymer Edn.* **1991**, *2*, pp 173-181
21. Gölander, C-G., Lin, Y-S., Hlady, V., and Andrade, J.D., *Colloids and Surface*
    1990, *49*, pp 289-302

RECEIVED February 10, 1995

Chapter 13

Modeling the Dynamics of Protein Adsorption to Surfaces

A. Nadarajah, C. F. Lu, and K. K. Chittur

Department of Chemical and Materials Engineering, University of Alabama, Huntsville, AL 35899

The adsorption and desorption of proteins from surfaces is explained by consideration of relevant molecular interactions between protein molecules in solution and the adsorbing surface. A macroscopic rate equation providing a realistic description of the process was incorporated into a dynamic model of protein adsorption, which fully accounts for the unequal mass transport rates from the bulk solution to the surface of various proteins, and the competitive adsorption on this surface. The resulting equations were solved numerically for a simulated plasma solution and the simulations clearly showed the process of adsorbed protein turnover. Simulations of the adsorption of fibrinogen from this mixture at various dilutions resembled experimental adsorption measurements of fibrinogen from blood plasma at various dilutions. These results suggest the validity of the dynamic macroscopic model of protein adsorption presented.

The adsorption of proteins to surfaces is now recognized to be a critical event in a number of fields including biomaterials, biological separations and biosensors. This process occurs whenever an aqueous protein solution comes into contact with a phase interface. The proteins diffuse to the surface and, given sufficient time, will adsorb and form strong bonds with sites on the surface. However, this seemingly simple process can in reality be extremely complicated and, in spite of numerous investigations, is still not completely understood.

One example of this complexity is the observed turnover of the adsorbed protein during adsorption from protein mixtures. This process was first noticed by Vroman and co-workers when investigating the response of surfaces to blood plasma (1,2). Fibrinogen was found to adsorb onto surfaces, only to be displaced soon after, and this phenomenon has become known as the "Vroman effect" (3,4). Early explanations for this were sought from biochemical mechanisms, such as the binding of high molecular weight kininogen to the surface displacing the fibrinogen (2). Subsequent studies have shown that such mechanisms alone are not adequate to explain the process (5), and that the process of turnover was a general one occurring for adsorption from any protein or protein-detergent mixture (6,7). These studies are discussed in greater detail in the article by Slack in this volume. More recent studies have attempted to understand the adsorption process by investigating the molecular

0097–6156/95/0602–0181$12.00/0

contributions to the adsorption process. Although the process as a whole is still not completely understood, these contributions to protein adsorption can now be detailed with some confidence. The primary driving force for protein adsorption seems to be the hydrophobic interactions. These interactions are also the dominant force driving protein folding and their importance in protein adsorption was first indicated by studies showing that adsorption increased with the hydrophobicity of the surface (4,8,9). Other evidence include an investigation in which a surface with a hydrophobicity gradient was prepared by varying the silane density on it, and the adsorption was shown to be the greatest on the most hydrophobic end (10). In a recent study, the hydrophobicity of the solvent was increased by the addition of alcohol and this was shown to lead to a decrease in protein adsorption (11).

Although hydrophobicity has been established as the primary factor driving protein adsorption, it is by no means the only one. Electrostatic interactions play an important role as well, particularly for the more hydrophilic surfaces. The experiments of Norde and co-workers (12,13) and others (14) have shown that protein adsorption is decreased when the surface and the protein carry the same charge. Increasing the pH of the solution towards the isoelectric point of the protein results in increased adsorption (12-14). Electrostatic interactions can also be more subtle, such as in the formation of hydrogen bonds, salting in of proteins and the binding of counterions (15,16).

Surface geometry and steric considerations are likely to play a role as well. This is especially true for complex solid surfaces such as block co-polymers and surfaces with flexible polyethylene oxide molecules immobilized on them. The topography of the surface is also related to another factor in protein adsorption, namely the stability of the adsorbing protein molecule itself. Proteins that denature easily due to the presence of the surface are likely to have a stronger surface affinity than the more structurally stable ones. These proteins will become more tightly bound to the surface by undergoing conformational changes influenced by the topography and the hydrophobicity of the surface (17,18). This is likely to occur for most proteins on hydrophobic surfaces, but the less conformationally stable ones will undergo such changes more quickly and are more likely to be retained. This process is believed to be primarily responsible for making protein adsorption, which is reversible initially, to become irreversible over longer times (19).

Finally, the temperature will play a role in the process by varying the thermal energy or entropy of the protein molecules as well as their structures. Higher temperatures should decrease protein adsorption by increasing the entropy of the molecules and keeping them in motion. However, such temperatures are also likely to cause the proteins to be more irreversibly bound by increasing their tendency to denature (9). There are other contributions to protein adsorption, such as those due to van der Waals forces, but they are known to be much smaller in magnitude (20,21).

A complete understanding of the adsorption process will require that all the relevant contributions be quantified in a model. There have been only limited efforts to quantify some of these factors. The first attempts involved determining the electrostatic interactions. The protein was treated as a rigid body and detailed 3-D molecular models of the protein near a simple surface, *in the absence of the solvent*, were developed (22,23). As expected these models showed that electrostatic interactions alone were inadequate to describe protein adsorption. Subsequently, Park and co-workers (24) improved their model for lysozyme adsorption by including hydrophobic interactions and were able to show that increasing the hydrophobicity of the surface would make protein adsorption energetically more favorable. However, in this calculation the hydrophobicity was determined from the solvation energy for an unfolded lysozyme molecule which would overestimate this interaction and the effect of counterions in the solvent were neglected as well.

In a different attempt at estimating the strength of various forces involved in protein adsorption on a surface coated with polyethylene oxide, Andrade and co-workers relied on simple protein geometries such as plates and spheres (*21,25*). Electrochemical interactions were neglected in the calculations and the attractive hydrophobic force between the protein molecule and the surface was determined from an empirical expression based on measurements of other investigators (*20*). Andrade and co-workers also estimated the effect of steric hindrance of the polyethylene oxide strands by calculating their repulsive force. Based on the magnitude of these forces, they concluded that increasing the density of packing and the length of the polyethylene oxide chains would lead to a decrease in protein adsorption (*21,25*).

Clearly these models provide some insight into the fundamental mechanisms of protein adsorption. However, it is obvious that not all the factors involved in protein adsorption have been considered in each model. This difficulty in estimating all the contributions to protein adsorption has led to empirical approaches to the problem. Two approaches, one based primarily on the characteristics of the surface (*26*) and the other primarily on the protein are being pursued (*9,27*) In both cases, a number of experimentally measurable quantitities are statistically correlated for various surfaces or for various proteins (*9,26,27*). For proteins one such correlation has been christened as a "Tatra plot" (*9,27*). Thus, if these quantities are known for any protein or surface, a measure of its adsorption properties can be obtained from these correlations. General predictions of protein adsorption for any protein-surface combination are difficult, since this is an empirical approach with no direct physical basis.

A more fundamental problem with all the above models is that they are based on static concepts, while protein adsorption is a dynamic process. It is still not known, for example, why adsorbed protein molecules readily exchange with dissolved ones, but are less likely to simply desorb. Explaining these observations will require a model that realistically accounts for all the weak molecular forces involved and describes the dynamics of the process by their incorporation in a kinetic equation. The lack of such a fundamental kinetic equation to describe protein adsorption dynamics has resulted in the use of many empirical equations for this purpose (*4,28-31*). However, very few if any studies have been done to determine the validity of the postulates or to accurately evaluate the rate constants in the equations. Nevertheless, these equations are the only ones currently available for dynamic descriptions of the protein adsorption process.

Early models of protein adsorption dynamics relied exclusively on such equations (*4,28-31*). However, it was soon recognized that protein mass transport from the bulk solution to the surface played a critical role in the process. It was shown that for single protein adsorption, initially the adsorption tended to be mass transfer limited with adsorption kinetics becoming the limiting step later on (*3,5,32,33*). This realization led to studies of the mass transfer aspects of protein adsorption, focusing on the initial stages of the process. Several experimental and theoretical investigations employing flow cells, along with mathematical models of the system assuming concentration boundary layers and instantaneous adsorption, were carried out (*34-36*). Such assumptions effectively restricted these studies to single protein systems, but allowed the defining equations to be solved easily. These assumptions were omitted in a recent single protein study and the numerical solution obtained compared well with measured trends of the adsorbed protein concentration (*37*).

The discovery that mass transport is critical to protein adsorption suggested that the phenomenon of adsorbed protein turnover was the result of the interaction between mass transport and adsorption kinetics. Recent protein adsorption studies with stationary cells have attempted to show this by including both transport and adsorption kinetics in the models. However, given the mathematical complexity of the problem, most of these models assume that the adsorption kinetics step is near equilibrium (*38-41*). The mathematical difficulty of including adsorption kinetics in a transport model with minimal simplifying assumptions has prevented the examination of multi-protein systems.

Thus, many of the current models for protein adsorption have been unable to predict the Vroman effect, due to their many simplifications and the restriction to single protein systems. Some of the adsorption models have shown the turnover of adsorbed proteins (*4,42*), but with severe assumptions such as the neglect of mass transport which make the results somewhat less convincing. One attempt was made to simulate the turnover process by including mass transfer and adsorption kinetics from a two-protein system (*42*). Although the turnover of adsorbed protein was shown in this study, the severe assumptions used to arrive at this result, including unrealistic exponential kinetic terms and fixed protein concentration gradients in the solution, cast doubts on its validity. The current lack of a convincing explanation for adsorbed protein turnover seemed to suggest that other mechanisms should be sought. It is our position, that the premise of previous models that adsorbed protein turnover does occur due to the interaction of mass transport and adsorption kinetics is still valid. However, in order for such an explanation to be convincing, a more rigorous model may be needed.

A rigorous model of protein adsorption must consist of two parts. The first of these is the development of rate equations. The second part of the problem is the development of a model combining these rate equations with mass transport in a realistic manner. We have recently completed this part of the problem, showing conclusively that the turnover of adsorbed proteins, the effect of dilution and other experimental observations will be displayed by a model that combines mass transport and adsorption kinetics (*43*). This was possible only when numerical solutions to the model were obtained without simplifying assumptions. In this study we report a preliminary attempt at attacking the first part of the problem.

The Model

In our previous study (*43*), the rate equations used were empirical ones based on measured adsorption behavior. However, as discussed above, several studies have been done to quantitatively estimate the molecular forces acting between protein molecules in solution and the adsorbing interface (*21-25,44*). These results, along with recent experimental investigations of protein adsorption, allow us to develop more physically realistic adsorption mechanisms, valid for both stationary and flow situations. Mathematical rate equations can then be developed from these mechanisms.

The experimental evidence (*10,11*) indicates that the dominant force driving protein adsorption appears to be the hydrophobic interactions between the protein molecule and the surface. The exposure of hydrophobic groups on the protein and the surface to the solution results in an entropically unfavorable ordering of water molecules around these groups (*45*). The hydrophobic force acts to minimize the exposure of such groups to water by the adsorption of the protein molecules to the surface, effectively burying some of these groups. The hydrophobic interaction can be decreased by disrupting the hydrogen-bonded networks of the bulk water molecules. In a recent investigation protein adsorption was shown to decrease when the alcohol was added to the solution, by disrupting bulk water networks (*11*).

The principal force opposing adsorption is the entropy of the protein molecule. Thus, this entropic energy should be comparable in magnitude to the hydrophobic energy of interaction to ensure the initial reversibility of the protein adsorption process. This mechanism is illustrated in Figure 1, where for simplicity the protein and the surface are represented as being entirely hydrophobic, and completely surrounded by the unfavorably ordered water molecules. The protein molecule will move due to its thermal energy and collide with the surface (Figure 1a). This collision will result in the removal of the unfavorably ordered water molecules between the protein and the surface (Figure 1b). If the energetic benefit of this removal is greater than the thermal energy of the protein molecule, it will remain on the surface as shown in Figure 1b. Otherwise it will return to the solution as shown in Figure 1c.

There will be contributions to above mechanism of adsorption from electrostatic interactions, but quantitative estimates (*24*) and the experimental investigations of others (*12-14*) indicate these interactions have only a secondary effect. Other experiments and estimates (*20,21,25*) also indicate that contributions from van der Waals forces are even smaller and may be neglected. Thus, the rate of adsorption of a protein on a surface will be a function of the concentration of that protein in solution near the surface, the available surface area for adsorption and the adsorption rate constant. The rate constant will be a measure of the hydrophobic, electrostatic and entropic interactions between the protein molecules and the surface.

Although the driving forces for protein adsorption are reasonably well understood, less is known about the desorption process. As discussed earlier, adsorbed protein molecules may be displaced by other protein molecules. However, all adsorbed protein molecules cannot be desorbed from a surface by flushing with a buffer solution alone. Clearly, the presence of the protein molecules in solution promotes the desorption process. It is likely that the protein molecules in solution decrease the hydrophobic interaction between the adsorbed protein molecule and the surface, by disrupting the hydrogen-bonded networks of water molecules around it. In other words, protein molecules in solution act in a manner similar to alcohol molecules in protein solutions in decreasing adsorption (*11*). This decrease in the hydrophobic interaction will tilt the balance between the hydrophobic and entropic forces acting on the adsorbed protein molecule, resulting in desorption.

The above mechanism for desorption is illustrated in Figure 2 for hydrophobic proteins and surfaces. A protein molecule is adsorbed on the surface as shown in Figure 2a, because of the energetic benefit of removing the unfavorably ordered water molecules between the protein and the surface exceeds the protein molecule's thermal energy. The presence of two other protein molecules in solution in its vicinity, changes the adsorbed molecules environment as shown in Figure 2b. The hydrogen-bonded networks of bulk water molecules are excluded in its vicinity. The adsorbed molecule only faces a hydrophobic environment, made up of other protein molecules in solution and unfavorably ordered water molecules near it. Thus, the energetic benefit of remaining adsorbed on the surface is no longer present and cannot overcome the thermal energy of the adsorbed molecule. The molecule will then desorb due to its thermal energy, as shown in Figure 2c.

Although we have suggested that the primary means of desorption involve hydrophobic forces, there will also be some contributions from electrostatic interactions. Thus, the desorption rate for a protein will be a function of the surface concentration of that protein, the total concentration of all proteins in solution near the surface and the desorption rate constant. The desorption rate constant will reflect the hydrophobic, electrostatic and entropic interactions between an adsorbed protein molecule and the surface, when surrounded by other protein molecules in solution.

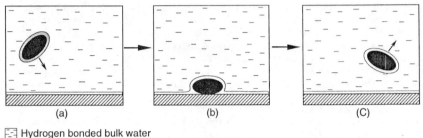

    ▭ Hydrogen bonded bulk water
    □ Ordered water surrounding hydrophobic groups
    ■ Protein molecules
    ▨ Adsorbing surface

Figure 1:   Illustration of the proposed mechanism for protein adsorption.

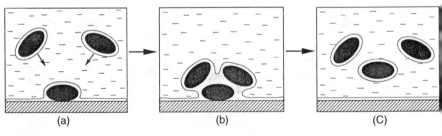

    ▭ Hydrogen bonded bulk water
    □ Ordered water surrounding hydrophobic groups
    ■ Protein molecules
    ▨ Adsorbing surface

Figure 2:   Illustration of the proposed mechanism for desorption.

Finally, it is necessary to estimate the rate at which the proteins are irreversibly bound to the surface.   As discussed earlier, this process is prompted by conformational changes of the adsorbed protein molecules, resulting in them becoming tightly bound to the surface (*17-19*). Proteins tend to denature when the surroundings become hydrophobic, with the more conformationally stable proteins less likely to undergo this process. Thus, the process of irreversible adsorption is a complex function of the topography and the hydrophobicity of the surface and the conformational stability of the protein molecule.   Although the presence of neighboring adsorbed protein molecules may contribute to this process, this is likely to be much smaller that interactions with the surface. Thus, this will mostly be a first order process with its complexity incorporated in the first order rate constant.

While the mechanism given above is a plausible explanation for the adsorption process, deriving a rate equation from it is not straightforward. Such a derivation would require (a) the rigorous determination of all the relevant molecular forces between the protein molecules and the surface, from their surface chemistry and topography, (b) incorporation of these forces in a microscopic rate equation using statistical mechanics, for the proposed adsorption/desorption mechanism, and (c) and averaging the microscopic equation to obtain the macroscopic rate equation. We are currently attempting such a derivation. In this preliminary study, we will use semi-empirical rate equations based on the above considerations. However, empirical rate equations that incorporate all the above considerations should closely resemble a rigorously derived one and should provide a realistic description of the process. The simplest of these is given by the following set of equations:

$$\frac{d\theta_i}{dt} = k_i^1 C_i (1 - \theta - \phi) - k_i^{-1} C \theta_i - k_i^2 \theta_i , \tag{1}$$

$$\frac{d\phi_i}{dt} = k_i^2 \theta_i , \tag{2}$$

$$\text{where} \quad \theta = \sum_i \theta_i \tag{3}$$

$$\text{and} \quad \phi = \sum_i \phi_i . \tag{4}$$

Here the subscript "i" refers to the ith protein in the mixture. Also C_i and C are the molar concentrations of the ith protein and total for all proteins in the mixture; θ_i and ϕ_i are the fraction of the available surface covered by reversibly and irreversibly adsorbed ith protein; the k^1, k^{-1} and k^2 are the rate constants for reversible adsorption, desorption and irreversible adsorption. This mechanism is illustrated in Figure 3, where C_i^b and C_i^{int} represent the bulk and interfacial concentrations in solution of the ith protein.

As we had shown in our previous study (*43*), the most likely mechanism for the process of adsorbed protein turnover involves the interaction between the unequal mass transport rates of proteins from the bulk solution to the surface and the competitive adsorption of these proteins on the surface. When the adsorbing surface and the protein mixture come into contact, initially all the proteins will be uniformly distributed in the solution. On contact with the surface, proteins will adsorb until all the hydrophobic groups on the surface are covered, that is up to a monolayer (*46*). Coverage of more than a monolayer implies the occurrence of strong protein-protein interactions and are likely for proteins that are normally in an aggregated state in solutions, such as insulin (*16*). The protein with the highest surface affinity (that is the highest value of k^1) should be adsorbed preferentially on this layer. However, the process will be mass transport limited initially and the first adsorbed protein will be the one with the fastest transport rate. Eventually, the next fastest diffusing protein will increase its concentration near the surface, but it will replace the first adsorbed protein only if it has a higher surface affinity. When this happens, it will produce the second protein peak. The process will continue until the protein with the highest surface affinity reaches the surface, is adsorbed and attains equilibrium.

The above description of the interaction between protein mass transport and adsorption kinetics can be represented mathematically as a system of partial differential equations, subject to initial and boundary conditions. These equations and their derivation are described in detail in our earlier study (*43*). Although the rate equations developed here resembles those used in the previous study, this must be considered fortuitous. The rate equations (1-4) are for a plausible mechanism of protein adsorption and desorption, while those in the previous study were based mostly on empirical considerations.

The effect of flow on the adsorption process described above is of interest from physiological and other considerations. Fluid flow is unlikely to effect the kinetics of the adsorption process near the surface, because of the small size of protein molecules. The particles need to be larger than a micron for flow effects to be significant for molecular events such as adsorption (*47*). Moreover, even with fluid flow in the bulk solution, there will be little motion in the fluid layers very near the adsorption surface. Thus, the adsorption process near the surface will closely resemble that in stationary solutions. However, solution flow can tremendously increase the transport rates to the surface of all proteins. This also decreases the differences in transport rates between proteins caused by their different diffusivities. A speeding up of the entire adsorbed protein turnover process will result, with all the protein peaks occurring earlier and a crowding of these peaks. Higher flow rates will increase the crowding further. In a recent study we rigorously modeled the effect of flow on adsorption from protein mixtures and showed the validity of these and other predictions (*48*).

The complete set of equations that describe protein adsorption from the above considerations, with and without flow, have to be solved numerically (*43,48*). The effect of fluid flow will not be considered here, as the focus will be on the fundamental mechanism of protein adsorption. We have shown in an earlier study that fluid flow only affects the transport rate of proteins to the surface, and introduces a variation in the adsorption rates along the surface in the flow direction (*48*). The details of the discretization and the solution scheme for the case of adsorption in stationary solutions, along with the discretized equations, are given elsewhere (*49*). The computer code of the numerical scheme was implemented on a Cray C94 of the Alabama Supercomputer Network.

Results and Discussion

Many of the protein adsorption experiments have been done with blood plasma in order to study the response of biomaterial surfaces to blood. These experiments are usually performed with highly diluted plasma solutions in order to slow down the process enough to enable measurements to be taken (*3*). Motivated by these experiments, the computer simulations in this study were performed for a solution containing albumin, immunoglobulin-γ, fibrinogen and high molecular weight kininogen (HMWK). The initial concentrations of the four proteins were chosen to be their physiological ones in human blood plasma, diluted to 1% of normal (*50*). The diffusivity data were obtained from refs. (*32,33*). The magnitudes of the rate constants of adsorption were based on measured values (*4*), but were chosen to reflect experimental observations that for several surfaces HMWK had the strongest affinity and albumin the weakest (*51*). The rate constant for irreversible adsorption was assumed to be 5×10^{-5} s^{-1} for all four proteins. The four proteins were also assumed to be oriented similarly with respect to surface coverage with the maximum surface concentration taken to be 0.18 µg/cm^2 (*3,34*). The parameters are listed in Table I.

Table I.   Parameter Values Used in the Simulations

Protein	Concentration (µgcm^{-3})	k^1 (cm^{3}µg^{-1}s^{-1})	k^{-1} (cm^{3}µg^{-1}s^{-1})	Diffusivity (cm^{2}s^{-1})
albumin	450	0.005	1.58×10^{-3}	8.5×10^{-7}
Ig-γ	150	0.03	7.93×10^{-4}	4.0×10^{-7}
fibrinogen	30	0.1	3.16×10^{-4}	2.0×10^{-7}
HMWK	0.7	2.5	1.5×10^{-6}	6.0×10^{-7}

The result of a simulation with the above parameters, for the case of adsorption from a stationary solution, is shown in Figure 4. The simulation clearly shows the process of adsorbed protein turnover. Albumin is adsorbed first, which is then displaced by immunoglobulin-γ, followed by fibrinogen and finally HMWK. The low bulk concentration of HMWK means that it will take awhile to adsorb onto the surface, displacing the other proteins, and this is shown in the inset to Figure 4. The simulation captures the complexity of the adsorption process that takes place when blood plasma comes into contact with a biomaterial surface. It also shows the generality of protein adsorption: the process of adsorbed protein turnover known as the Vroman effect is not restricted to fibrinogen alone, but occurs with other blood proteins as well. In this simulation HMWK, which had the highest surface affinity, was the final protein adsorbed on the surface. However, in blood plasma there may be proteins with still higher surface affinities which could cause the turnover process to continue further with the displacement of HMWK. Other plasma proteins with intermediate affinities may be a part of the adsorption sequence as well.

The sequence in which proteins were adsorbed in the above simulation may seem to contradict the mechanism given in the previous section. The protein that was adsorbed first, namely albumin, had the highest diffusivity and immunoglobulin-γ which had the next highest diffusivity was adsorbed next. However, HMWK did not follow this diffusivity sequence. This is because the adsorption sequence is not based on the diffusivity but on the diffusion or transport rate and the surface affinity. The transport rate is a function of the diffusivity as well as the bulk concentration of the protein. In the case of HMWK, its higher diffusivity in comparison with fibrinogen is offset by its very low physiological concentration in blood, resulting in fibrinogen adsorbing on the surface before it. If the concentration of HMWK had been much higher, not only would it have adsorbed on the surface before fibrinogen, but its higher surface affinity would have prevented fibrinogen from adsorbing at all. This simulation clearly shows that the complexity of the adsorption process precludes the adsorbed protein sequence from being predicted solely by knowing the physicochemical parameters of the blood proteins.

Figure 5 shows yet another way that the adsorbed protein sequence can be altered. This simulation was done for the protein mixture diluted tenfold, i.e., at 0.1% of normal plasma. It can be seen that the adsorption and the turnover process

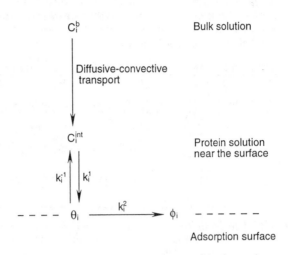

Figure 3:   Adsorption mechanism used in the study.

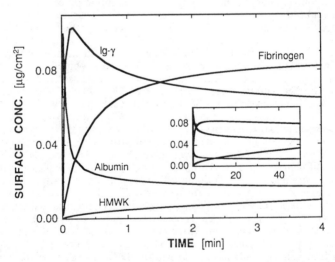

Figure 4:   Adsorbed surface concentration profiles for adsorption from a four protein solution, with concentrations corresponding to 1% plasma. The inset shows the adsorbed concentrations after long times.

have been considerably slowed down. In the previous simulation shown in Figure 4, although HMWK does adsorb and partially displace fibrinogen, it cannot completely displace it. The concentration of HMWK in solution is so depleted by adsorption, that it is insufficient to completely displace any adsorbed protein even after long times. In the simulation shown in Figure 5, the concentration of HMWK is now so low at this dilution that it hardly adsorbs at all. Moreover, even fibrinogen has some difficulty in displacing immunoglobulin-γ at this dilution, and this results in immunoglobulin-γ remaining longer on the surface and attaining a slightly higher adsorption peak. If the solution had been diluted even further, fibrinogen would have been unable to completely displace immunoglobulin-γ. Conversely, the simulation also indicates that with solutions of much higher concentrations, fibrinogen will be transported faster to the surface and directly displace albumin. Immunoglobulin-γ will hardly be adsorbed at all in this situation.

These simulations also explain a well known observation made on adsorption experiments with plasma at various dilutions. Although we have used the term Vroman effect for the process of adsorbed protein turnover in general, it more commonly refers to the adsorption and desorption of fibrinogen from blood plasma (*1-4*). In particular, it has been used to describe "adsorption isotherms" of fibrinogen from plasma at various dilutions (*3*). The results of adsorption simulations with the same four protein mixture at various dilutions is shown in Figure 6, where the adsorbed fibrinogen concentration with time is plotted, for concentrations corresponding to various plasma dilutions. At the relatively high concentration of 2.5% dilution, fibrinogen adsorbs so rapidly that there is barely any time for another protein to precede it. It is also effectively displaced by HMWK at this concentration. At the slightly lower concentration of 1.25% dilution, fibrinogen adsorbs on to the surface a little slower, but it is also only partially displaced by HMWK. At 0.5% dilution the concentration of HMWK is so low that fibrinogen is hardly displaced at all and it will be the protein that remains on the surface at equilibrium. When the solution is diluted much further, fibrinogen in solution is rapidly depleted by adsorption and now has difficulty in displacing immunoglobulin-γ and attaining its maximum possible surface concentration.

The effectiveness of the model used here to describe protein adsorption is shown by the remarkable similarity of Figure 6 to experimentally measured adsorption profiles of fibrinogen from blood plasma at various dilutions (*3*). Exact correspondence between the experimental data and the simulation cannot be expected. Only four of the plasma proteins were included in the mixture and the values of many parameters used in the simulations are not known accurately. However, the close similarity between the simulation and experimental results strongly suggests the validity of the mechanism proposed in this study. This is true for the mechanism of adsorption and desorption of proteins, as well as that for the turnover of adsorbed proteins involving the interaction of mass transport and adsorption kinetics. Thus, the model proposed here may provide the basis for future studies of protein adsorption. The focus of such future research is likely to be in relating the rate constants in equation (1) to the chemistry and physical properties of the protein and the adsorbing surface, such as their hydrophobicity, surface charge and topography.

In previous studies simulations were done examining the affect of different rate constants, fluid flow and the addition of another protein into the solution (*43,48*). Simulations were also carried out with and without irreversible adsorption. They showed that irreversible adsorption had only a small affect on the final composition of the adsorbed protein layer, and no effect on the initial adsorption and turnover of proteins. However, the adsorbed protein conformational changes that are responsible for irreversible adsorption, are likely to be critical for the subsequent cellular events following the contact of blood with the biomaterial surface of a medical implant.

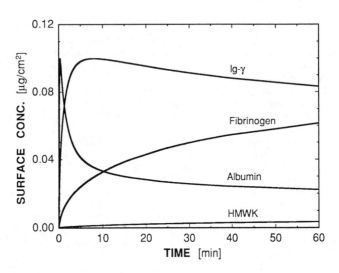

Figure 5:   Surface concentration profiles from a four protein solution, corresponding to 0.1% plasma.

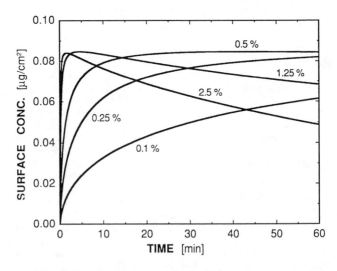

Figure 6:   Adsorbed surface concentrations of fibrinogen from a four protein solution, at concentrations corresponding to various plasma dilutions.

Conclusions

Protein adsorption from mixtures is a complex process, and requires a correspondingly complex and universal model to describe it. A complete molecular theory for protein adsorption must consider all interactions between the protein molecules and the surface. We have described the essential steps that would be required to derive a rate equation from such considerations.

The rate equation we actually used in this study is empirical and provides a realistic description of the process. The equation was used as input in a dynamic macroscopic model of protein adsorption. The resulting differential equations were solved numerically for a four protein simulated plasma solution using published values of protein diffusivities and rate constants based on measured values. The simulations clearly showed the process of adsorbed protein turnover, with albumin adsorbing to the surface first and high molecular weight kininogen last. Simulations of the adsorption of fibrinogen from the same solution at various dilutions were also done. These simulations resembled experimental adsorption measurements of fibrinogen from blood plasma at various dilutions. These results show clearly that the Vroman effect can be explained in terms of the interaction between mass transport and adsorption kinetics using a model that has no simplifying assumptions.

Literature Cited

1. Vroman, L.; Adams, G.L.; Fischer, G.C. *Adv. Chem. Ser.* **1982**, *199*, 256.
2. Schmaier, A.H.; Silver, L.; Adams, A.L.; Fischer, G.C.; Munoz, P.C.; Vroman, L.; Colman, R.W. *Thromb. Res.* **1983**, *33*, 51.
3. Wojciechowski, P.; ten Hove, P.; Brash, J.L. *J. Colloid Interf. Sci.* **1986**, *111*, 455.
4. Slack, S.M.; Horbett, T.A. *J. Colloid Interf. Sci.* **1989**, *133*, 148.
5. Slack, S.M.; Horbett, T.A. *J. Colloid Interf. Sci.* **1988**, *124*, 535.
6. Bohnert, J.L.; Horbett, T.A. *J. Colloid Interf. Sci.* **1986**, *111*, 363.
7. Horbett, T.A. In *Proteins at Interfaces*; Brash, J.L.; Horbett, T.A., Eds.; ACS Symposium Series; ACS: Washington, DC, 1987; Vol. 363.
8. *Surfaces and Interfacial Aspects of Biomedical Polymers*; Andrade, J.D., Ed.; Plenum: New York, NY, 1985; Vols. 1 & 2.
9. Andrade, J.D.; Hlady, V.; Wei, A.-P.; Ho, C.-H.; Lea, A.S.; Jeon, S.I.; Lin, Y.S.; Stroup, E. *Clinical Mater.* **1992**, *11*, 67.
10. Elwing, H.; Welin, S.; Askendal, A.; Nilsson, U.; Lundström, I. *J. Colloid Interf. Sci.* **1987**, *119*, 203.
11. Tilton, R.D.; Robertson, C.R.; Gast, A.P. *Langmuir* **1991**, *7*, 2710.
12. Koutsoukos, P.G.; Mumme-Young, C.A.; Norde, W.; Lyklema, J. *Colloids Surf.* **1982**, *5*, 93.
13. Norde, W. In *Surfaces and Interfacial Aspects of Biomedical Polymers*; Andrade, J.D., Ed.; Plenum: New York, NY, 1985; Vol. 2.
14. Kondo, A.; Higashitani, K. *J. Colloid Interf. Sci.* **1992**, *150*, 344.
15. Shibata, C.T.; Lenhoff, A.M. *J. Colloid Interf. Sci.* **1992**, *148*, 469.
16. Nylander, T.; Kékicheff, P.; Ninham, B.W. *J. Colloid Interf. Sci.* **1994**, *164*, 136.
17. Horsley, D.; Herron, J.; Hlady, V.; Andrade, J.D. *Langmuir* **1991**, *7*, 218.
18. Lu, D.R.; Park, K. *J. Colloid Interf. Sci.* **1991**, *144*, 271.
19. Lenk, T.J.; Horbett, T.A.; Ratner, B.D.; Chittur, K.K. *Langmuir* **1991**, *7*, 1755.
20. Israelachvili, J.N.; Pashley, R.M. *J. Colloid Interf. Sci.* **1984**, *98*, 500.

21. S.I. Jeon, J.H. Lee, J.D. Andrade and P.G. de Gennes, *J. Colloid Interface Sci.* **1991**, *142*, 149.
22. Lim, K.; Herron, J.N. In *Poly(Ethylene Glycol) Chemistry. Biotechnical and Biomedical Applications*; Harris, J.M., Ed.; Plenum: New York, NY, 1992.
23. Lu, D.R.; Park, K. *J. Biomater. Sci. Polymer Edn.* **1990**, *1*, 243.
24. Lu, D.R.; Lee, S.J.; Park, K. *J. Biomater. Sci. Polymer Edn.* **1991**, *3*, 127.
25. Jeon, S.I.; Andrade, J.D. *J. Colloid Interf. Sci.* **1991**, *142*, 159.
26. Pérez-Luna, V.H.; Horbett, T.A.; Ratner, B.D. *J. Biomed. Mater. Res.* **1994**, *28* , 1111.
27. Andrade, J.D.; Hlady, V.; Wei, A.P. *Pure & Appl. Chem.* **1992**, *64*, 1777.
28. Beissinger, R.L.; Leonard, E.F. *J. Colloid Interf. Sci.* **1982**, *85*, 521.
29. Aptel, J.D.; Voegel, J.C.; Schmitt, A. *Colloids Surf.* **1988**, *29*, 359.
30. Sevastianov, V.I.; Kulik, E.A.; Kalinin, I.D. *J. Colloid Interf. Sci.* **1991**, *145*, 191.
31. Lundström, I.; Elwing, H. *J. Colloid Interf. Sci.* **1990**, *136*, 68.
32. Cheng, Y.-L.; Darst, S.A.; Robertson, C.R. *J. Colloid Interf. Sci.* **1987**, *118*, 212.
33. Young, B.R.; Pitt, W.G.; Cooper, S.L. *J. Colloid Interf. Sci.* **1988**, *125*, 246.
34. Lok, B.K.; Cheng Y.-L.; Robertson, C.R. *J. Colloid Interf. Sci.* **1983**, *91*, 104.
35. Iordanski, A.L.; Polischuk, A.J.; Zaikov, G.E. *JMS Rev. Macromol. Chem. Phys.* **1983**, *C23*, 33.
36. Kim, D.; Cha, W.; Beissinger, R.L. *J. Colloid Interf. Sci.* **1991**, *159*, 1.
37. Shibata, C.T.; Lenhoff, A.M. *J. Colloid Interf. Sci.* **1992**, *148*, 485.
38. Varoqui, R.; Pefferkorn, E. *J. Colloid Interf. Sci.* **1986**, *109*, 520.
39. Schaaf, P.; Déjardin, P. *Colloids Surf.* **1987**, *24*, 239.
40. Déjardin, P. *J. Colloid Interf. Sci.* **1989**, *133*, 418.
41. Wojciechowski, P.W.; Brash, J.L. *J. Colloid Interf. Sci.* **1990**, *140*, 239.
42. Cuypers, P.A.; Willems, G.M.; Hemker, H.C.; Hermens, W.T. *Annals NY Acad. Sci.* **1987**, *516*, 244.
43. Lu, C.F.; Nadarajah, A.; Chittur, K.K. *J. Colloid Interf. Sci.* **1994**, *168*, 152.
44. Roth, C.; Lenhoff, A. *Langmuir* **1993**, *9*, 962.
45. Bergethon, P.R.; Simons, E.R. *Biophysical Chemistry*; Springer-Verlag: New York, NY, 1990.
46. Young, B.R.; Pitt, W.G.; Cooper, S.L. *J. Colloid Interf. Sci.* **1988**, *124*, 28.
47. Russel, W.B.; Saville, D.A.; Schowalter, W.R. *Colloidal Dispersions*; Cambridge Univ. Press: Cambridge, UK, 1989.
48. Nadarajah, A.; Lu, C.F.; Chittur, K.K. *J. Colloid Interf. Sci.*, submitted, 1994.
49. Lu, C.F. *Masters Thesis*; Dept. of Chemical Engineering, Univ. of Alabama in Huntsville, 1994.
50. Lehninger, A.L. *Principles of Biochemistry*; Worth: New York, NY, 1982.
51. Bale, M.D.; Mosher, D.F.; Wolfarht, L.; Sutton, R.C. *J. Colloid Interf. Sci.* **1988**, *125*, 516.

RECEIVED December 22, 1994

Chapter 14

Effect of Protein Competition on Surface Adsorption-Density Parameters of Polymer–Protein Interfaces

V. I. Sevastianov[1], Y. S. Tremsina[1], R. C. Eberhart[2,4], and S. W. Kim[3]

[1]Research Center for Blood Compatible Biomaterials, Institute of Transplantology and Artificial Organs, Shukinskaya 1, Moscow 123436, Russia
[2]Biomedical Engineering Program, University of Texas Southwestern Medical Center, 5323 Harry Hines Boulevard, Dallas, TX 75235
[3]Center for Controlled Chemical Delivery, University of Utah, 421 Wakara Way, Salt Lake City, UT 84108

We studied the adsorption of human serum albumin (HSA) and gamma globulin (HGG) onto quartz and polydimethylsiloxane (PDMS) from single and binary solutions. Total internal reflection-fluorescence (TIRF) was used to obtain continuous time-dependent protein surface concentrations from fluorescein isothiocyanate-labelled protein solutions. The results show that the character of the competitive adsorption kinetics of the binary protein mixtures, (monotonic vs. "overshoot" kinetics) of HSA-*FITC*/HGG and HGG-*FITC*/HSA depends on the molecular mass and structure of the proteins and the nature of the surface. The process of displacement of HSA by HGG is not dependent on the conformational state of adsorbed HSA on PDMS and quartz. On PDMS, it is determined by the presence of adsorption centers with affinity for HSA. In contrasts, on PDMS and quartz, the competition of HSA for HGG sites induces an HSA concentration-dependent reorientation of HGG from a side-on to an end-on configuration. The removal of HGG by HSA occurs only from an end-on adsorption state for those surfaces. The higher the surface affinity for HSA, the greater the degree of displacement of adsorbed HGG-*FITC* and the lower the threshold HSA concentration for "overshoot" kinetic curves.

Protein adsorption is one of the first and most significant stages of blood-foreign material interactions, therefore numerous studies have been done to establish relationships between the physicochemical and protein adsorptional properties of the surface and its blood compatibility characteristics [1]. Good correlations have been found between theoretical and experimental results for a number of cases, when single proteins in model solutions have been employed. However the lessons learned from

[4]Corresponding author

0097–6156/95/0602–0195$12.00/0
© 1995 American Chemical Society

these studies are limited by the complexity of the real situation in which very many proteins, not to mention water, cells and other substances [2], must be considered. At the next stage of abstraction it is necessary to take into account not only the processes of protein adsorption and desorption but also the effects of their displacement by other proteins (Vroman effect) [3,4]. The simplest model system for the study of these sequential and competitive adsorption processes is a mixture of two proteins, each with a different molar mass and affinity for the test surface.

We have studied the adsorption of fluorescein isothiocyanate (FITC)-labelled human serum albumin (HSA) and gamma-globulin (HGG) from simple solutions onto quartz and polydimethylsiloxane (PDMS) surfaces [5,6]. We employed the total internal reflectance fluorescence spectroscopy (TIRF) method [7] and applied a model of the energetic heterogeneity of protein/surface interactions to the results. The kinetics of HSA and HGG adsorption on quartz and PDMS surfaces all had a monotonicly increasing character. When the protein deposition centers derived from these kinetic curves were plotted (semilogarithmic coordinates) against the effective adsorption rate constant (k_{ads}), so-called "rectangular" and "Gaussian" distributions were obtained for quartz and PDMS, respectively. This suggested that different mechanisms might be at play for "hydrophilic" quartz and "hydrophobic" PDMS.

How would the presence of two proteins influence the protein adsorption kinetics and the surface adsorption-density parameters for these surfaces? We studied the adsorption of labelled proteins, HSA-*FITC* and HGG-*FITC*, respectively onto quartz and PDMS from single and binary solutions of unlabeled HGG and HSA. TIRF was used to obtain continuous time-dependent protein surface concentrations from simple (HSA-*FITC* or HGG-*FITC*) and binary (HSA-*FITC*/HGG or HGG-*FITC*/HSA) solutions.

Materials and Methods

Standard polydimethylsiloxane solution in hexane (Serva, Sweden) was diluted to 1% w/v, and microfiltered (pore size 20 μm) prior to casting. Quartz slides were cleaned in hot chromic acid for 30 minutes, rinsed, steeped in double distilled deionized water for 16 hours and blown dry with pure nitrogen gas. Slides were put in a frame in the horizontal position and coated dropwise with a dilute solution of the polymer. The tension between the frame and the solution permitted formation of a uniform and continuous film. Excess solvent was evaporated in air for 1 hour, then the slides were dried in vacuum at 100°C for 1 hour and at 40°C for 24 hours.

Human serum albumin (HSA, lyophilized, Fraction V, Serva) and human gamma-globulin (HGG, lyophilized, Serva) were labeled with fluorescein isothiocyanate (FITC, isomer 1, Serva) using a standard method [8]. Adsorption of these proteins on quartz and PDMS as studied by TIRF is described elsewhere [5, 6]. A spectrofluorimeter (SLM Instruments, Urbana, IL, USA) equipped with a specially designed, temperature controlled TIRF cell which permitted controlled fluid flow was used to study the kinetics of protein-FITC adsorption.

The experiments were carried out at 25°C with a calculated wall shear rate of 4800 sec^{-1} (laminar flow assumption). Competitive protein adsorption in this model system was studied from simple solution of HSA or HGG and from mixtures of these proteins. Phosphate buffered saline (PBS) (0.1 NaCl, 0.086 M KH_2PO_4, 0.041 M Na_2HPO_4, pH 7.35) was used as a standard dilution medium. The labeled proteins (HSA-*FITC* and HGG-*FITC*) were diluted to 0.1 mg/ml; the concentration of the second, unlabeled protein in the binary mixture was varied from 0.01 to 0.5 mg/ml. Calibration of TIRF was done by simultaneous adsorption of ^{131}I-labeled proteins (Medradiopreparat, Moscow, Russia).

Results

The kinetic curves of HSA-*FITC*/HGG adsorption onto PDMS (Fig. 1) and quartz (Fig. 2) both show that the greater the concentration of HGG in the mixture (C_b^{HGG}), the lesser the amount of adsorbed HSA (C_s^{HSA}). The C_s^{HSA} values decrease monotonically with C_b^{HGG} for both surfaces; but the C_s^{HSA} values obtained with PDMS are much greater than those obtained with quartz, for simple ($C_b^{HSA} = 0.1$ mg/ml) solution and all HSA-*FITC*/HGG binary solutions.

Likewise, the values of HGG-*FTIC* adsorbed onto PDMS (C_s^{HGG}, Fig. 3) are much greater than those for the quartz experiments (Fig. 4). However it must be noted, for PDMS, that in simple solution the molar adsorption of HSA is larger than is the case for HGG. The "steady-state" concentration values (C_{ss}) are $5 \cdot 10^{-3}$ and $3 \cdot 10^{-3}$ μM/cm^2 for HSA and HGG, respectively. In contrast, there are no differences for quartz between C_{ss}^{HSA} and C_{ss}^{HGG}, which both yield about $9 \cdot 10^{-4}$ μM/cm^2. The characteristics of the C_s^{HGG} - C_b^{HSA} relationships for PDMS and quartz (Figs. 3,4) are more complicated than those for C_s^{HSA} - C_b^{HGG} (Figs. 1,2). For the case of PDMS (Fig. 3), as C_b^{HSA} increases from 0.0 (curve 1) to 0.01 mg/ml (curve 2) the C_s^{HGG} values decrease. As C_b^{HSA} increases further to 0.1 mg/ml (curve 2 - curve 4) a monotonic increase of C_s^{HGG} is observed. At even higher bulk concentrations (from 0.1 to 0.3 mg/ml HSA) the kinetic curves for C_s^{HGG} (curves 5-7) take on a more complex character, with an "overshoot" and subsequent relaxation.

For quartz (Fig. 4) an analogous dependence of the C_s^{HGG} vs. t curves on C_b^{HSA} was observed, however the transitions begin at higher bulk concentrations of HSA. The decrease of C_s^{HGG} with C_b^{HSA} extended from 0.0 (curve 1) to 0.1 mg/ml (curve 4), then converted to an "increasing" character from 0.2 to 0.5 mg/ml HSA, with an "overshoot" finally appearing at $C_b^{HSA} = 0.5$ mg/ml.

Discussion

In general, the kinetic curve of protein-*FITC* adsorption from binary solution is the sum of the sorption and displacement processes. When the adsorption process becomes less intensive than the combined desorption and displacement processes, the "overshoot" appears on the kinetic curve at the corresponding concentration of the second protein [3, 9]. This was realized for HGG-*FITC*/HSA competitive adsorption onto PDMS (Curves 5-7, Fig. 3) and quartz (curve 7, Fig. 4). The broader manifestation of this phenomenon for HGG-*FITC*/HSA onto PDMS as compared with quartz may be explained by the larger number of HSA adsorption centers on PDMS, as will be shown.

The energetic heterogeneity model of protein/surface interactions [5] was used to explain the experimental data for these binary systems. The mathematical approach is based on the assumption that a "controlling band theory" [10] holds true for protein/surface interactions. We assume the following:
(i) proteins "A" (HSA) and "B" (HGG) have the noted molar mass (MM) and diffusion coefficient (D):

$$MM_{HSA} = 66,000 < MM_{HGG} = 150,000$$
$$D_{HSA} = 6.1 \cdot 10^{-7} > D_{HGG} = 4.0 \cdot 10^{-7}, \text{ cm}^2/\text{sec}$$

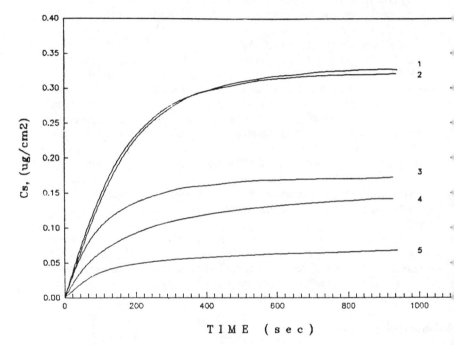

Figure 1. Kinetic curves of HSA-*FITC* adsorption onto PDMS from a simple solution of HSA (1) and from mixtures with HGG (2-5) at constant $C_b^{HSA} = 0.1$ mg/ml:

2 - $C_b^{HGG} = 0.01$ mg/ml   3 - $C_b^{HGG} = 0.02$ mg/ml
4 - $C_b^{HGG} = 0.1$ mg/ml   5 - $C_b^{HGG} = 0.2$ mg/ml

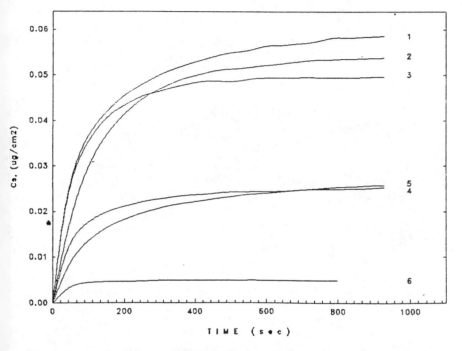

Figure 2. Kinetic curves of HSA-*FITC* adsorption onto quartz from a simple solution of HSA (1) and from mixtures with HGG (2-6) at constant $C_b^{HSA} = 0.1$ mg/ml:

2 - C_b^{HGG} = 0.01 mg/ml   3 - C_b^{HGG} = 0.02 mg/ml
4 - C_b^{HGG} = 0.1 mg/ml   5 - C_b^{HGG} = 0.2 mg/ml
6 - C_b^{HGG} = 0.3 mg/ml

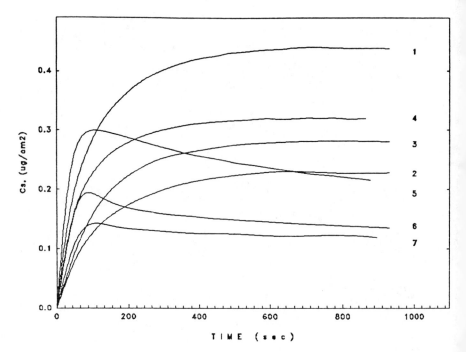

Figure 3. Kinetic curves of HGG-*FITC* adsorption onto PDMS from a simple solution of HGG (1) and from mixtures with HSA (2-7) at constant $C_b^{HGG} = 0.1$ mg/ml

2 - $C_b^{HSA} = 0.01$ mg/ml $\qquad$ 3 - $C_b^{HSA} = 0.02$ mg/ml

4 - $C_b^{HSA} = 0.03$ mg/ml $\qquad$ 5 - $C_b^{HSA} = 0.1$ mg/ml

6 - $C_b^{HSA} = 0.2$ mg/ml $\qquad$ 7 - $C_b^{HSA} = 0.3$ mg/ml

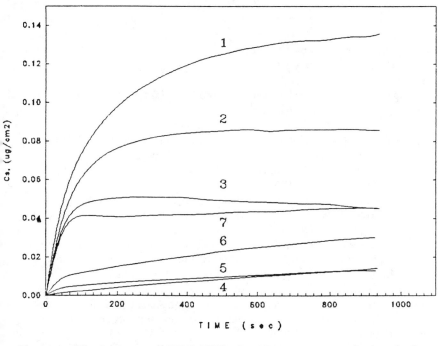

Figure 4. Kinetic curves of HGG-*FITC* adsorption onto quartz from a simple solution of HGG (1) and from mixtures with HSA (2-7) at constant $C_b^{HGG} = 0.1$ mg/ml

2 - $C_b^{HSA} = 0.01$ mg/ml   3 - $C_b^{HSA} = 0.05$ mg/ml
4 - $C_b^{HSA} = 0.1$ mg/ml   5 - $C_b^{HSA} = 0.2$ mg/ml
6 - $C_b^{HSA} = 0.3$ mg/ml   7 - $C_b^{HSA} = 0.5$ mg/ml

(ii) each protein has two adsorption states: "1" and "2" with different conformational and (or) orientational characteristics. We assume that at state "2" and (or) in a case of side-on adsorption [8], proteins occupy a larger effective area than at state "1" and (or) in the case of end-on adsorption. The ratio of the occupied centers in state "2" to those in state "1" for protein "A" (HSA) is much less in comparison with protein "B" (HGG) [5].
(iii) the protein sorption processes proceed only in state "1". The proteins can undergo conformational and (or) orientational changes (transitions from state "1" to state "2").
(iv) the rate constants for adsorption (k_{ads}), desorption (k_{des}), transition (k_t) and displacement (k_d) depend on the type of adsorption center, which is classified in terms of the activated Gibbs free energy. Protein in state "2" is not desorbed from the surface (being retained at centers of irreversible adsorption). For state "1" the probability of reversible adsorption is increased upon the decrease of k_{ads}. Thus the centers with $k_t = 0$ are centers of absolute reversible adsorption; centers with $k_d = 0$ are absolute irreversible adsorption centers for protein in state "1", which in the frame of this model is identical to state "2".
(v) displacements of adsorbed protein "A" by protein "B" from solution, and of adsorbed "B" by "A" from solution are possible.

The scheme of competitive adsorption corresponding to this model is represented in Fig. 5. According to the model, the effective density distribution of adsorption centers $\rho(k_{ads})$ can be calculated from a simple equation:

$$\rho(k^i_{ads}) = (1/RT) [dC^i_s (t)/d\ln (t)] \qquad \text{EQ. 1}$$

where $t = 1/C^i_b k^i_{ads}$ and i = HSA, HGG.

The shapes of the adsorption center density distributions for the systems HSA-*FITC*/ HGG and HGG-*FITC*/HSA, calculated from corresponding kinetic curves (Figs. 1-4), are compared in Figs. 6-9 for quartz and PDMS.

In the case of HSA-*FITC*/HGG, competitive adsorption onto quartz and PDMS, the adsorption center density for HSA-*FITC* decreases monotonically with increasing C_b^{HGG}, for both low and high values of k_{ads}, and proceeds without "overshoot" in the kinetic curves, (Figs. 6,7). Inspection of the values of MM_i and D_i for the two proteins suggests that the HSA-*FITC* molecules would be adsorbed onto PDMS and quartz at lower bulk concentrations than would be the case for the corresponding HGG experiments.

As C_b^{HGG} is increased, the relative maxima of ρRT for HSA-*FITC* onto quartz are shifted to higher k_{ads} values (Fig. 7), with a change in the type of distribution from "rectangular" (curve 1) to "Gaussian" (curves 2-6) . This suggests that the displacement, by HGG, of the HSA-*FITC* originally adsorbed onto the quartz surface proceeds more intensively for HSA in state "1" than for HSA in state "2": the two states would correspond to low k_{ads} (centers of reversible adsorption) and high k_{ads} (centers of irreversible adsorption), respectively. In contrast, for HSA-*FITC*/HGG adsorption onto PDMS, the adsorption center maximum is shifted to lower values of k_{ads} (curves 2-5, Fig. 6). This can be explained by the high affinity of HSA for PDMS centers of reversible adsorption (state "1") rather than for centers of irreversible adsorption (state "2"). This observation has been confirmed by recently obtained results concerning the effect of surface passivation with HSA, HGG or other proteins, on the kinetic adsorption of HSA onto quartz and PDMS [11]. It was noted in that publication that there is preference for irreversible adsorption centers for HSA on quartz in comparison with the types of adsorption centers on PDMS. For adsorption of HGG-*FITC*/HSA onto both PDMS and quartz, the relative maxima for the adsorption centers are shifted to higher values of k_{ads} for all values of C_b^{HSA} (Figs. 8,9). As in the case for HSA-*FITC* adsorption onto quartz,

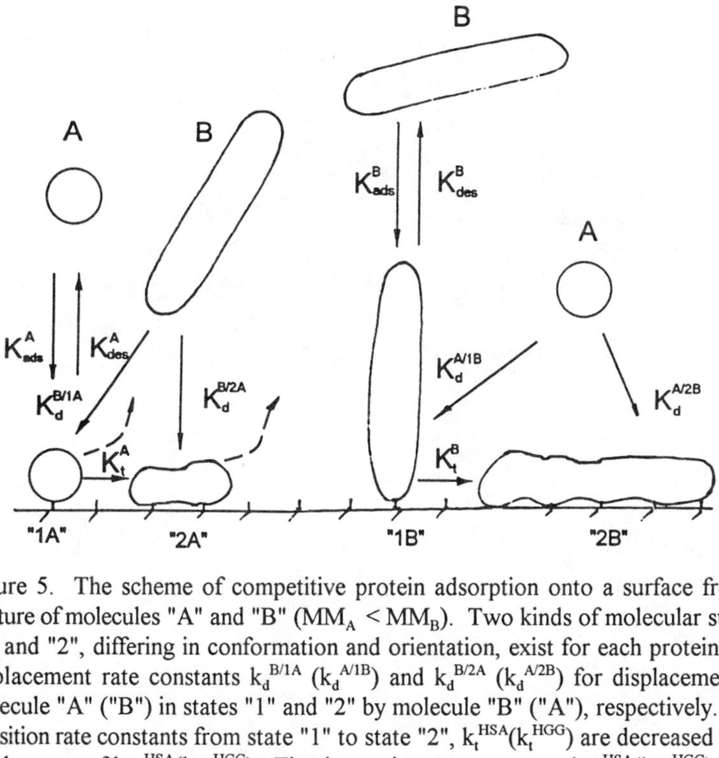

Figure 5. The scheme of competitive protein adsorption onto a surface from a mixture of molecules "A" and "B" ($MM_A < MM_B$). Two kinds of molecular states, "1" and "2", differing in conformation and orientation, exist for each protein with displacement rate constants $k_d^{B/1A}$ ($k_d^{A/1B}$) and $k_d^{B/2A}$ ($k_d^{A/2B}$) for displacement of molecule "A" ("B") in states "1" and "2" by molecule "B" ("A"), respectively. The transition rate constants from state "1" to state "2", k_t^{HSA}(k_t^{HGG}) are decreased upon the decrease of k_{ads}^{HSA}(k_{ads}^{HGG}). The desorption rate constants k_{des}^{HSA}(k_{des}^{HGG}) seem to increase upon the decrease of k_{ads}^{HSA}(k_{ads}^{HGG}).

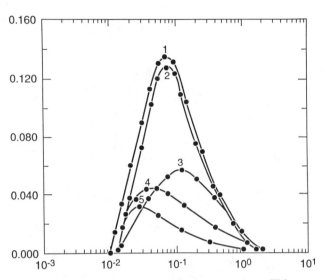

Figure 6. Normalized adsorption center distributions (ρRT vs. k_{ads}^{HSA}) for HSA-*FITC* adsorption onto PDMS from simple solution (1) and from mixtures with HGG (2-5), calculated from the corresponding values of the kinetic curves in Fig. 1.

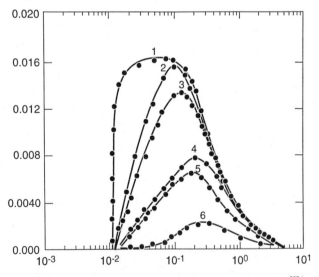

Figure 7. Normalized adsorption center distributions (ρRT vs. k_{ads}^{HSA}) for HSA-*FITC* adsorption onto quartz from simple solution (1) and from mixtures with HGG (2-6), calculated from the corresponding values of the kinetic curves in Fig. 2.

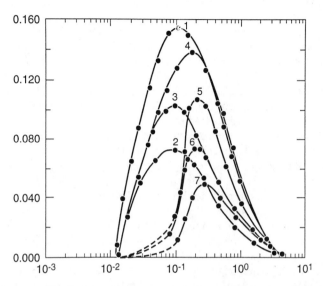

Figure 8. Normalized adsorption center distributions (ρRT vs. k_{ads}^{HGG}) for HGG-*FITC* adsorption onto PDMS from simple solution (1) and from mixtures with HSA (2-7), calculated from the corresponding values of the kinetic curves in Fig. 3.

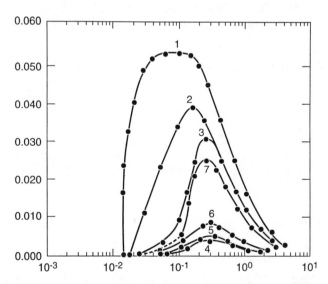

Figure 9. Normalized adsorption center distributions (ρRT vs. k_{ads}^{HGG}) for HGG-*FITC* adsorption onto quartz from simple solution (1) and from mixtures with HSA (2-7), calculated from the corresponding values of the kinetic curves in Fig. 4.

there is a change of the type of distribution from "rectangular" (curve 1) to "Gaussian" (curves 2-6, Fig. 9).

It should be noted that the mechanism underlying these shifts for HGG-*FITC* may differ for varying bulk concentrations of HSA. For C_b^{HSA} values ranging from 0.1 μM/ml (0.01 mg/ml) to 0.3 μM/ml (0.02 mg/ml), the molecules of HSA are adsorbed after HGG, but for $C_b^{HSA} > 0.3$ μM/ml, HSA is adsorbed before HGG. This shift, and the decrease of C_s^{HGG} with increasing C_b^{HSA} from 0.0 (curve 1) to 0.01 mg/ml for PDMS (curve 2, Fig. 8) and quartz (curve 2, Fig. 9), may be explained by the gradual displacement of the originally adsorbed "state 1" HGG-*FITC* (in centers of reversible adsorption) by HSA, because of the existence of a greater number of anchoring segments for HGG-*FITC* in state "2" in comparison with state "1". The subsequent decrease C_s^{HGG} with increasing C_b^{HSA} up to 0.1 mg/ml (curves 3, 4) and an increase of C_s^{HGG} with further increase of HSA to 0.5 mg/ml for quartz (curves 5-7, Fig. 9) reflects the competition of the originally adsorbed HSA with HGG-*FITC* for the adsorption centers

For $C_b^{HSA} > 0.2$ mg/ml (3.0 μM/ml), the orientation of adsorbed HGG-*FITC* induced to change from a side-on to an end-on configuration, reducing the probability of transition from state "1" to state "2". The described orientation effect seen for HGG-*FITC*/HSA on quartz was also obtained for HGG-*FITC*/HSA adsorption onto PDMS intermediate C_b^{HSA} values ranging from 0.02 to 0.03 mg/ml (curves 3, 4, Fig. 8).

A further increase of C_b^{HSA} to 0.3 mg/ml (27 μM/ml in comparison with 0 μM/ml for HGG) during adsorption of HGG-*FITC*/HSA onto PDMS leads to narrowing of the adsorption center density vs. k_{ads} profile, with a simultaneous significant decrease in the number of adsorption centers and an appreciable shift to higher values of k_{ads} (curves 5-7, Fig. 8). The narrowing of the distribution was also observed for HGG-*FITC*/HSA adsorption onto quartz at $C_b^{HSA} = 0.5$ mg/ml (curve 7, Fig.9). This phenomenon is accompanied by the appearance of the "overshoot" in the HGG-*FITC*/HSA adsorption kinetics profiles (curves 5-7, Fig. 3 and curve 7, Fig. 4). This should be explained by a sharp increase in the rate of displacement of adsorbed HGG-*FITC* by relatively high concentrations of HSA, because the main part of the adsorbed HGG-*FITC* is in state "1" (an end-on orientation) which is induced by the originally adsorbed HSA. This is depicted by the scheme (Fig. 10) of sequential, competitive adsorption between HSA (molecule "A") and HGG-*FITC* (molecule "B") with corresponding kinetics.

Unfortunately the presence of "overshoot" does not permit calculation of the adsorption center density distribution vs. k_{ads}. This case corresponds to the k_{ads} interval in which $k_{ads}^{HGG} \cdot C_b^{HGG} \cdot \rho(k_{ads}) \ll k_d^{HSA/HGG("1")} \cdot C_s^{HGG("1")} \cdot C_b^{HSA}$ (the dotted lines in Fig. 8, 9).

Conclusions

These results demonstrate that study of competitive adsorption of two proteins possible with the TIRF method. It is shown that the character of the competitive adsorption kinetics of HSA-*FITC*/HGG and HGG-*FITC*/HSA depend on the molecular mass and structure of the protein and the nature of the surface.

The process of displacement of HSA by HGG is not dependent on the conformational state of adsorbed HSA on PDMS and quartz. On PDMS, it determined by the presence of large numbers of HSA adsorption centers. The

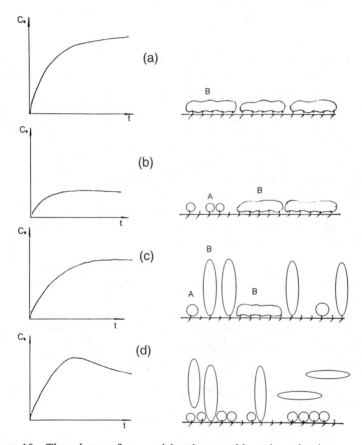

Figure 10. The scheme of sequential and competitive adsorption between HSA (molecule "A") and HGG-*FITC* (molecule "B") with corresponding kinetic curves:
a. preferential adsorption of the higher molar mass proteins "B" (HGG)
b. competition of molecule "A" with molecule "B" for the adsorption centers
c. preferential orientation of adsorbed protein "B" in state "1"
d. displacement of adsorbed protein "B" in state "1" by protein "A"

competition of HSA for HGG sites induces an HSA concentration-dependent reorientation of HGG-*FITC* from a side-on to an end-on adsorption state on both quartz and PDMS surfaces. The removal of HGG from both surfaces by HSA occurs only from an end-on adsorption state (state "1"). The greater the affinity of HSA for a surface (for example, PDMS) the higher the degree of displacement of adsorbed HGG-*FITC* and the lower the threshold HSA concentration for "overshoot" kinetic curves.

References

1.   Sevastianov, V.I. Interrelation of protein adsorption and blood compatibility of   biomaterials.   Chapter 21 in: High Performance Biomaterials: A Comprehensive Guide to Medical/Pharmaceutical Applications. Szycher, M. (ed.), Technomic Press, Lancaster, PA, 1991, pp. 313-341.

2.   Lee, R. G., Adamson, C., Kim, S.W., Competitive adsorption of plasma proteins onto polymer surfaces. Thrombosis Research 4, 485-488, 1974.

3.   Andrade, J.D., Hlady V. Vroman effects: techniques and philosophies.   J. Biomaterials Science: Polymer Edition. 2, 12-24, 1991.

4.   Shirahama, H., Lyklema, J., Norde, W.. Comparative protein adsorption in model systems. J. Colloid Interface Science. 139, 177-187, 1990.

5.   Sevastianov, V.I., Kulik, E.A., Kalinin, I.D., The model of continuous heterogeneity of protein/surface interactions for human serum albumin and immunoglobulin G adsorption onto quartz. J. Colloid Interface Science. 145, 191-206, 1991.

6.   Sevastianov, V.I., Kulik, E.A., Kim, S.W., Eberhart, R.C. Surface adsorption-center density parameters of the polymer/surface interface. Biomaterial-Living System Interactions. 1, 3-11, 1993.

7.   Cheng, Y-L., Lok, B.K., Robertson, C.R. Interactions of macromolecules with surfaces in shear fields using visible wavelength total internal reflection fluorescence. Ch. 3 in Surface and Interfacial Aspects of Biomedical Polymers. Vol 2: Protein Adsorption. Andrade, J.D. (ed.) Plenum Press, New York, NY, 1985, pp. 121-160.

8.   Goldman, M.   Fluorescent Antibody Techniques.   Academic Press, New York, NY, 1988.

9.   Morrissey, B.W. Adsorption and conformation of proteins. Ann. N.Y. Acad. Sci. 288, 50-64, 1977.

10.   Roginsky, S.Z.   Adsorption und Katalyse an Inhomogenen Oberflachen, Akad.- Ver. 1., Berlin, 1958.

11.   Sevastianov, V.I., Drushlyak, I.V., Kotov, K.N., Eberhart, R.C., Kim, S.W. Effect of protein preadsorption on albumin adsorption onto blood compatible polymers and quartz. Biomaterial-Living System Interactions. 2, 1994 (in press).

RECEIVED December 22, 1994

Chapter 15

Competitive Adsorption of Albumin and Fibrinogen at Solution–Air and Solution–Polyethylene Interfaces

In Situ Measurements

A. Baszkin and M. M. Boissonnade

Physico-Chimie des Surfaces, Unité de Recherche Associé au Centre National de la Recherche Scientifique 1218, Université Paris-Sud, 5 rue J. B. Clément, 92296 Châtenay-Malabry, France

The in situ adsorption measuring procedure based on the use of ^{14}C (β-emitting radiation) labeled proteins has been developed in our laboratory in the early 80s. The main advantage of this procedure over other methods in which (γ-emitting radiation) labeled proteins are used for adsorption studies is that no rinsing of samples is necessary to detect the surface protein concentration. We have then extended this method to sequential and competitive protein adsorption studies at the interfaces with polymers.

The adsorption of albumin and fibrinogen was studied at solution-air and solution-polyethylene interfaces. The results show that while on both studied interfaces albumin adsorption was insensitive to the pH variation, fibrinogen exhibited a strong enhancement in adsorption in the neighbourhood of its isoelectric point (IEP). The reversibility of adsorption at the polyethylene interface occured both for albumin and fibrinogen in a wide range of pHs but the effect was more pronounced for the latter.

Competitive adsorption experiments at the solution-air interface reveal that albumin was not capable of substantially reduce fibrinogen adsorption and that this tendency decreased when albumin solution concentration increased. Conversely, on polyethylene at sufficiently high solution concentrations, fibrinogen adsorption was higher than its adsorption from the single protein system.

Although a considerable amount of experimental data is available on the kinetics of irreversible adsorption of proteins, very little is known about molecular mechanisms involved in a protein ability to compete with other proteins and situate itself in a given conformational state. In general, protein molecules are assumed to unfold to a greater extent as the surface hydrophobicity increases, mainly due to the entropy driven strong hydrophobic interactions operating at these interfaces. The dynamic nature of the adsorbed layers of proteins at the interfaces leads to changes in structure of both protein and interface.

0097–6156/95/0602–0209$12.00/0

The competitive adsorption experiments from binary or multicomponent systems show that one protein can displace another adsorbed protein when its concentration is sufficiently high, the phenomenon which is representative of what has been termed the VROMAN effect (1, 2) and which is attributed to the relative affinities of proteins for surfaces. However, the information on the nature of adsorbed layers is still lacking. This is because of the time dependency of protein conformational changes and because of the insufficiency of experimental measuring techniques which are not adapted to follow adsorption of proteins in in situ conditions and to which we do not refer to in the present work.

The adsorption of albumin and fibrinogen at different interfaces has been studied abundantly over the past thirty years. McMILLAN and WALTON (3), using transmission circular dichroism (CD), showed that fibrinogen adsorbed on silica was not conformation altered, but later SODERQUIST and WALTON (4), using the same experimental technique, demonstrated that albumin, IgG and fibrinogen, after elution from synthetic polypeptide films showed a loss of secondary structure (α helix or β sheet). They have characterized adsorption and desorption of protein as a phenomenon which occurs in three stages : an initial brief period of reversible adsorption, a second stage where the adsorbed protein undergoes a slow conformational change depending on adsorption time and in which proteins are essentially irreversibly adsorbed and finally desorption of denatured material. CHAN and BRASH (5), have also used CD analytic technique and reported that human fibrinogen, first adsorbed on glass surfaces and then desorbed, showed a loss of 50% in its α-helix content provoked by the action of disrupting forces operating at the glass-adsorbed fibrinogen interface.

The conformational changes of adsorbed fibrinogen have also been studied by STUPP et al. (6) and recently by LU and PARK (7) using Fourier transform infrared spectroscopy (FTIR) coupled with attenuated total reflectance (ATR). The authors show that some α-helical structures were changed into unordered structures and that the content of β-turns was increased upon protein adsorption as the surface hydrophobicity increased. VOEGEL et al. (8) have investigated thermal desorption properties of fibrinogen adsorbed at the glass/aqueous solution interface. They have demonstrated that with the increase of the denaturation, desorption of protein occurs at different rates which are closely related to the protein structure at the interface. A thermally stabilized adsorbed fibrinogen surface was thus obtained which displayed no exchange or additional adsorption when exposed to albumin or fibrinogen solution.

LENK et al. (9), using the same experimental technique (FTIR-ATR), observed for bovine serum albumin spectral differences in the amide I, II and III regions upon adsorption of the protein at different solid surfaces. In agreement with LU and PARK, they consider that spectral changes occuring during fibrinogen adsorption result from the loss of helix and gain of β-structure. PITT and COOPER (10), using also the FTIR-ATR technique, note that albumin adsorption to C-18 alkylated polyurethanes is not affected by polymer alkylation. They attributed these findings to the desorption of the protein from its upper layer to a protein-free buffer. However, the same authors (11) showed that the desorption rate decreased in magnitude as the alkyl chain length increased from C-2 to C-18. This supports the hypothesis that the alkyl chain length influences the interaction between albumin and an alkylated polymer system.

Ellipsometry studies made by RUZGAS et al. (12) demonstrated that hydrophobization of silicon surfaces resulted in a 58-fold increase and in a 27-fold decrease in the adsorption and desorption rate constants of albumin as compared with the appropriate rate constants of a hydrophilic surface. GOLANDER et al. (13) using the total internal reflection fluorescence (TIRF) have studied adsorption of human serum albumin, IgG and fibrinogen from single and ternary mixture solutions on silica surfaces with a hydrophobicity gradient. They observe that all three protein adsorb to a larger degree on the hydrophobic than on the hydrophilic side. They also show that albumin adsorption was almost unaffected by the addition of IgG and fibrinogen to the solution. Conversely, the adsorption of IgG and fibrinogen adsorbed from ternary mixtures decreased significantly relative to their adsorption from the single proteins solutions.

The adsorption/desorption of human serum albumin and IgG has been investigated by the TIRF technique onto hydrophilic quartz surfaces by SEVASTIANOV et al. (14). The authors showed that while only 15% of adsorbed IgG were desorbed in the PBS buffer during the first 1000 seconds, about 50% of adsorbed HSA was desorbed in the same time interval. They demonstrated also that the exchange between irreversibly adsorbed HSA (protein remaining on the surface after desorption with PBS) and the protein from the solution introduced to a cell after PBS desorption leads to the additional desorption of less than 3% of HSA.

Experimentally observed differences in surface activity among different proteins clearly indicate that conformational changes in adsorbed proteins depend both on the nature of each protein-surface combination and on the manner of how these proteins compete with each other. However, the overall mechanism yielding the surface functionality is not known yet in any quantitative sense.

A great amount of confusing data reported in the scientific literature and numerous variables associated with protein adsorption seem to support the conclusions of the early work by DILLMAN and MILLER (15).These authors studied adsorption and desorption of albumin, IgG and fibrinogen at physiological conditions on a series of polymer membranes and demonstrated that protein adsorption process takes place in two separate and distinct ways simultaneously and occurs on separate surface sites. They attributed one type of adsorption as relatively hydrophilic, exothermic and easily reversible and other type as tightly bound, hydrophobic and endothermic.

The in situ adsorption measuring procedure based on the use of ^{14}C (β-emitting radiation) labeled proteins has been developed in our laboratory in the early 80s.The main advantage of this procedure over other methods in which (γ-emitting radiation) labeled proteins are used for adsorption studies is that no rinsing of samples is necessary to detect the surface protein concentration (16-18). The methodology have then been extended to sequential and competitive protein adsorption studies at the interfaces with polymers (19-21).

In this work, we report on the in situ adsorption of albumin and fibrinogen and on the competitive adsorption of fibrinogen against albumin at the solution-air and solution-polyethylene interfaces.

Materials and Methods

Proteins and protein labeling : Human Serum Albumin (HSA) was Behring ORHA 20/21 lyophilized, purified albumin for scientific laboratory use, purchased from Hoechst - Behring, France.

Fibrinogen was a lyophilized powder, prepared from human serum grade L, furnished by Kabi Diagnostica, Noisy le Grand, France. The protein was purified according to the following procedure: 90 ml of triple distilled water into a vial containing 1 g of lyophilized fibrinogen. The dissolution of the protein at 4°C, without shaking, was done during 5 hours at least and then the solution was transferred into two dialysis bags (30 cm long and 40 mm in diameter) and dialysed overnight at 4°C versus 3 liters of phosphate buffer at pH = 7.5. The content of the bags were pooled (the volume of about 100 ml) and centrifuged for 30 minutes at 3700g, the supernatants pooled again and for the concentration determination read in a 1/20 dilution concentration with the dialysis buffer, using an extinction coefficient of 1.55. Aliquots of about 10 mg/ml were stored at -40°C and then quied-thawed in a 37°C water before use.

Labeling of proteins was accomplished by reductive methylation using sodium cyanoborohydride. A slightly modified procedure to that described by JENTOFT and

DEARBORN (22) was used. To 5 ml of protein solution were added 600 μl of [^{14}C] fomaldehyde (CEA, France, specific activity 0.92 GBq/mmol) and 2 ml of NaBH$_3$CN (6 mg/ml in a phosphate buffer). The mixture was incubated at 25°C and gently stirred for 1h, and after that time 4 ml of phosphate buffer was poured into the solution. The whole mixture was dialyzed against four changes (1 liter each) of triple-distilled water for 16 h at 4°C. The concentration of albumin in solution after dialysis was 3.72 mg/ml and that of fibrinogen 1.84 mg/ml as measured spectrophotometrically at 280 nm.

The specific activity of [^{14}C]-labeled proteins was measured using as a reference [^{14}C] glucose of a known activity. A known amount of [^{14}C] glucose was deposited on a plane glass surface and evaporated. When dried, its radioactivity was measured and compared with the radioactivity of the known amount of [^{14}C]-labeled protein, deposited in the same manner on a glass surface, and counted in the same geometric conditions as used for the reference. The specific activity of [^{14}C]-labeled albumin was found to be equal to 0.94 MBq/ml and that of fibrinogen 0.23 MBq/ml.

To ascertain that [^{14}C]-labeling had no effect on the surface activity of the protein, the surface tension vs. time of labeled and unlabeled protein were measured at different concentrations with the Wilhelmy plate method. The decrease in the surface tension with the time for both labeled and unlabeled albumin and fibrinogen was almost identical (For further details see ref. 21).

The necessary and routine check was also made to ensure that labeling had no effect on the adsorption properties of [^{14}C]-labeled proteins on polyethylene surfaces. Labeled and unlabeled proteins were adsorbed in different concentration ratios, and within experimental errors not exceeding ± 5% from the corresponding means, no preferential adsorption could be detected in these experiments.

Polyethylene : The low density polyethylene film (Cryovac L film) produced by Grace, Epernon, France, was used. Its density was 0.929 g/cm^2 and its thickness was 19 μm. The samples cut out of this film were soaked in acetone for 6 h and dried under reduced pressure at room temperature overnight. The reproducible value of the contact angle obtained with water (> 95°) was in a good agreement with the litterature value of this angle for the low density polyethylene (16) and indicates that the samples were clean.

Reagents : All chemicals used, if not stated otherwise, were of analytical grade and water used was triple distilled. The NaBH$_3$CN was obtained from Aldrich and was recrystallized before use according to the following procedure: 10 g of reagent was dissolved in 25 ml of acetonitrile, the undissolved residue was removed by centrifugation, and the NaBH$_3$CN in the supernatant was precipitated by the addition of 150 ml of CH$_2$Cl$_2$. The mixture was incubated overnight at 4°C and filtered, and the precipitate was dried in a vacuum desiccator.

The aqueous substrates used in adsorption experiments were acetic buffer (0.2 NaCl - 0.1 M CH$_3$COOH adjusted to pH = 2.75 with concentrated HCl) and phosphate buffer (0.1 M KH$_2$PO$_4$ - 1M NaOH) adjusted to pH = 7.4 and to pH = 9 or pH = 11 with 1M NaOH and diluted with water to [Na] = 5.10^{-3}M.

Adsorption and Desorption Procedures : The adsorption procedures rely on a specific feature of [^{14}C]-labeled adsorbing substances emitting β-soft radiation which has a mean free path of 160 μm in aqueous solutions. This means that all radiation originating from the solution below this depth is attenuated. The gas flow counter measures this radioactivity which corresponds to the molecules adsorbed in excess at the interface and to that of a thin layer of solution (about 160 μm) adjacent to the interface. To allow for the radioactivity originating from the solution close to the interface (A_b), a separately run experiment is performed. Instead of a [^{14}C] protein, the glass container is filled with a solution of a non-adsorbing substance containing the same radioactive element ($[^{14}C]$ glucose), and its radioactivity is measured in the same geometric conditions as those in the experiments with proteins. The radioactivity A_b can be calculated from $A_b = A_{b'}$ (cp/c'p'), where c and c' are the concentrations of the protein solution and of the nonadsorbing solution, respectively; p and p' are their respective specific activities; and $A_{b'}$ is the radioactivity of the nonadsorbing solution.

Subtraction of A_b from the total measured radioactivity (A_t) gives the radioactivity of protein molecules adsorbed in excess at the interface (A_{ad}) for each of the protein concentrations studied. At protein solution concentrations lower than 0.5 mg/ml, A_b represents about 50% of the adsorbed value.

To obtain the amount of adsorbed protein on polymers, in addition to the above described corrections, the A_t value has to allow for the adsorption of radiation by the sample. The magnitude of this correction for each sample is determined with the help of a $[^{14}C]$ methyl methacrylate solid source placed above the polymer window and in the same geometric conditions as for the adsorption measurements.

The measuring devices and cells used in the adsorption measurements in static conditions have been described in detail elsewhere (16-19). Adsorption of proteins at the studied interfaces has been continuously displayed on a recorder as a function of time. The presented results are those of a single observation but three experiments have been made independently and the standard deviation of the mean never exceeded ± 0.005 mg/cm^2 and was independent of the protein solution concentration.

To measure the in situ desorption at a given time, a protein solution was pumped out from the cell and simultaneously replaced by the buffer solution. Multiple replacement cycles led to a negligible protein concentration in the cell. The loosely bound protein fraction (reversibly adsorbed protein) thus corresponds to the difference between the totally adsorbed amount and that left at the surface after rinsing. The in situ desorption experiments on polyethylene samples were carried out after 20 h of adsorption. The constant values after desorption of the loosely bound protein fraction were obtained almost instantaneously and the reported values are that recorded after 15 min.

Competitive adsorption of $[^{14}C]$-labeled albumin from binary solutions containing unlabeled fibrinogen were carried out as a function of fibrinogen in bulk and were followed by desorption with a buffer solution.

Results and Discussion

The kinetics of albumin and fibrinogen adsorption at solution-air and solution-polyethylene interfaces from their single component solutions is shown in Fig. 1 and Fig. 2, respectively. From those data it is immediately apparent that on both interfaces the amounts of adsorbed fibrinogen are higher than those of albumin. The adsorption during the first few minutes is very rapid and diffusion-controlled. Thus, while the observed difference in initial albumin and fibrinogen adsorption can be easily explained when comparing their respective diffusion coefficients, $D_{alb} = 6.1 \times 10^{-7}$ cm^2 sec^{-1} and $D_{fib} = 2.0 \times 10^{-7}$ cm^2 sec^{-1} (23), other differences in kinetics and in the amounts of adsorbed protein result from the dissimilarities in protein size, structure and surface characteristics.

Albumin (molecular weight 66.000) consists of a single peptide chain forming three small globular units. The molecule is a rotational ellipsoid negatively charged at physiological conditions and has no carbohydrate. It has a high degree of α-helicity and high disulfide cross-link content. Its isoelectric point (IEP) is 4.9 (24).

Fibrinogen (molecular weight 360.000) is built up of two equivalent parts hold together by disulfide bridges and consists of three linear, covalently bound peptide chains. The molecule carbohydrate is low and located in the coiled/coiled region near the disulfide knot and on the β chain in the terminal domain. There is no carbohydrate on the γ or more extended and probably more surface active, α chain (25). The IEP of

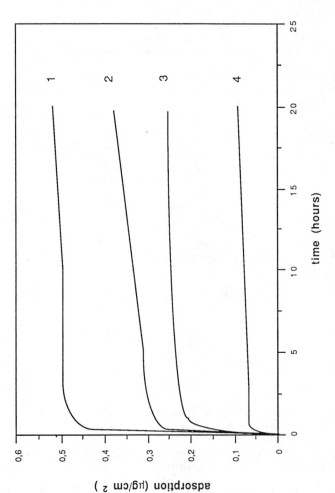

Figure 1 : Time dependence of albumin and fibrinogen adsorption at solution-air interface. Solution concentration of fibrinogen (1) 0.2 mg/ml, (2) 0.05 mg/ml. Solution concentration of albumin (3) 0.2 mg/ml, (4) 0.05 mg/ml.; pH = 7.4

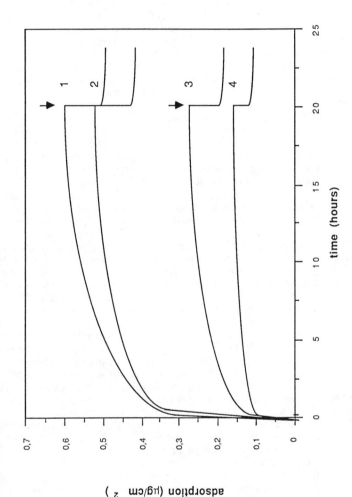

Figure 2 : Time dependence of albumin and fibrinogen adsorption at solution-polyethylene interface; pH = 7.4. Symbols as in Figure 1. The adsorption was recorded during 20 hours and then (as indicated by arrows) the cell was flushed with pure buffer.

fibrinogen is 5.5 (23). A large molecular weight and large size of fibrinogen molecule favours on its adsorption a large number of contact points with the surface. This is all the more true as fibrinogen is asymmetric, low in carbohydrate and conformationally sufficiently labile. This conformational lability of fibrinogen results in its high spreading pressure and its adaptability to the interface.

The air-water interface presents a model hydrophobic surface and one might expect to see some correlation between behavior at this interface and behavior of proteins at the polyethylene interface which is also hydrophobic. The specificity of air-water interface consists, however, in that at low protein surface coverages, the rate of adsorption is diffusion-controlled, while at higher surface coverages, the protein molecules is able to create space in the existing film, penetrate it and rearrange, and the process becomes according to GRAHAM and PHILLIPS rate determining (26). According to TER-MINASSIAN-SARAGA (27), the interfacial water may hydrate internal polar groups of globular proteins, rupture internal hydrogen and salt bonds and microunfold adsorbed globular proteins. Rapid total unfolding of a protein results in its irreversible adsorption. McRITCHIE (28) explains this irreversibility by the presence of a large activation barrier which hinders the process of desorption.

Closer examination of adsorption kinetics curves at the solution-air interface (Fig. 1) as well as of the adsorption isotherms illustrated in Fig. 3 indicates that protein adsorption at this interface is somewhat different to that at the solution-polyethylene interfaces (Fig. 4). After reaching a plateau, there is an additional increase in the amount adsorbed, something not observed during adsorption of studied proteins at polyethylene surface. This increase in adsorption is not reflected in the surface pressure isotherms (unpublished data) which across the studied concentration range exhibit constant saturation values. The formation of reversibly adsorbed layers at protein bulk concentrations larger than 0.1 mg/ml has already been observed for bovine serum albumin (26). Once the buildup of the primary layer of irreversibly adsorbed molecules is accomplished protein molecules continue to assemble in secondary and subsequent layers which stack beneath the primary monolayer. This trend seems to be more pronounced in the case of fibrinogen. Due to the voluminosity and flexibility fibrinogen molecules may interact more reversibly with air phase than albumin and thus the rate of desorption of the former may be significant, leading to a lower apparent adsorption rate.

A comparison of albumin and fibrinogen adsorption isotherms presented in Fig. 4 reveals differences in their binding strength with polyethylene surface. Although at low solution concentrations characterizing initial rise in protein adsorption isotherm both protein adsorb irreversibly (Fig. 5), the subsequently adsorbed molecules desorb more easily in the case of fibrinogen. It seems most likely that the principal protein binding mechanism on a hydrophobic surface involves irreversible, surface-induced adsorption at low solution concentration and formation of loosely-bound, reversibly adsorbed layers at high solution concentration. The existence of these layers, as has been seen with other proteins (29, 30) is clearly evidenced for albumin and fibrinogen adsorbed on polyethylene surface.

A model for adsorption that may be proposed is as follows. A protein diffuses to and collides with the surface. The interaction forces with a highly hydrophobic surface of polyethylene are strong enough to retain protein molecules at the surface. The residence time being sufficiently long, the protein starts to denature. This process is closely related to the intrinsic conformational lability of a protein. There may be a strong configurational entropy gain in going from a globular to a more extended state, particularly if the extended state can be accommodated by maintaining a degree of hydrophobic interaction comparable to that provided by the globular state. The surface denaturability of albumin and fibrinogen at polyethylene surface appears to be sufficiently high to prevent the proteins from desorbing to solution. Alike protein adsorption at the solution-air interface, the extent of this unfolding will depend on the protein surface concentration on polyethylene. As indicated by a total absence of their

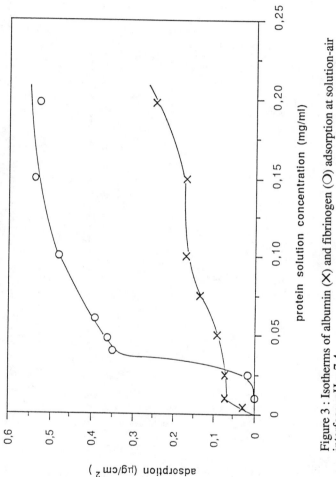

Figure 3 : Isotherms of albumin (X) and fibrinogen (O) adsorption at solution-air interface; pH = 7.4.

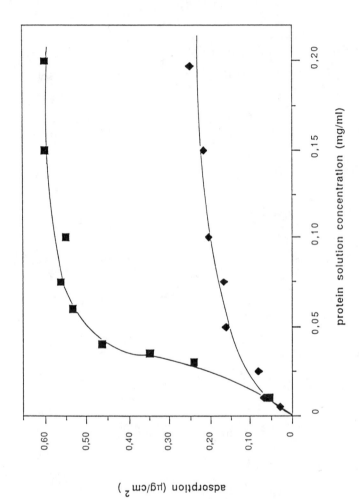

Figure 4 : Isotherms of albumin (◆) and fibrinogen (■) adsorption at solution-polyethylene interface; pH = 7.4. Time of adsorption 20 hours.

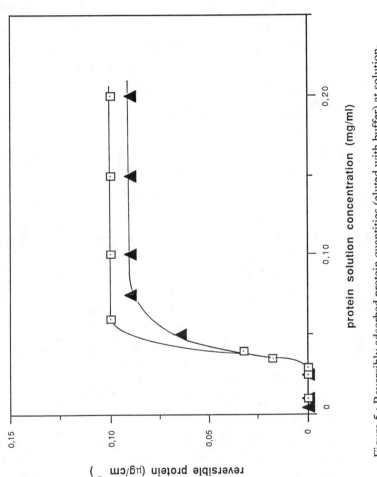

Figure 5 : Reversibly adsorbed protein quantities (eluted with buffer) at solution-polyethylene interface pH = 7.4 : albumin (▲), fibrinogen (▣). Time of adsorption 20 hours.

desorbability in the buffer (Fig. 5), albumin and fibrinogen adsorbed from solutions at low protein concentrations show stronger binding to the surface than at higher coverages when the new arriving to the interface molecules tend to adsorb on the top of the already attached molecules. This, of course, results in the decrease in their binding strength and leads to the formation of the fraction of protein population in a desorbable state. Because of its increased lability relative to albumin, fibrinogen is capable of both spreading more over the surface of polyethylene on adsorption and of forming more loosely bound reversibly adsorbed layers at higher solution concentrations. The reversibly adsorbed protein molecules which are partially denatured desorb and may rapidly renature to the equilibrium globular state (31, 32).

This mechanism of binding of albumin and fibrinogen molecules to polyethylene is further vindicated by the experiments of adsorption of these two proteins performed at different pH's (Fig. 6). While albumin adsorption at both air and polyethylene interfaces was shown to be practically pH-independent, with a very small and ill-defined maximum occuring near to its IEP, fibrinogen exhibited a clearly cut maximum at its IEP (pH = 5.5). It is also worthwhile to note that at this pH the reversibility of adsorbed fibrinogen was the highest.

The results obtained with albumin are in agreement with those previously published by BRYNDA et al. (24) and by McRITCHIE (33) who showed that the extent of albumin adsorption on polyethylene (24) and on hydrophobic silica (33) was independent of pH. They are also consistent with those reported by GRAHAM and PHILLIPS who found that albumin adsorption at the solution-air interface was insensitive to pH in the range 1-12 (26). Also, the amount of irreversibly adsorbed albumin on polyethylene at pH = 2.75 and at pH = 7.4 (0.15 mg/cm^2), coincides with the values obtained by others (24, 34).

The occurrence of a maximum in adsorption of fibrinogen at its IEP is a perfect illustration of the high flexibility of this protein. At pH's below the IEP, an important unfolding of the protein takes place due to strong hydrophobic interactions which bring about fibrinogen spreading over the polyethylene surface. These attractive hydrophobic interactions override repulsive electrostatic ones which decrease as pH of solution approaches the value of IEP. The progressive reduction in adsorbed amounts as pH is increased above the IEP is thus a consequence of repulsive coulombic forces between protein molecules which have now a net negative charge. That fibrinogen is readily denaturable and conformationally labile would also explain the highest reversibility of adsorption observed at the IEP e.g. when the molecule is electrically neutral.

The question which arises now is to know how these proteins which differ in a number of physicochemical characteristics bind to hydrophobic surfaces when adsorbed from binary mixtures. BRASH in his early works, utilising both ^{131}I and ^{125}I for two proteins, has investigated irreversible adsorption of albumin and fibrinogen from binary protein systems on polyethylene (34) and the extent and the rate of exchange (turnover) of initially adsorbed ^{125}I labeled albumin with ^{131}I albumin in solution (35). He observed that polyethylene showed a strong preference for fibrinogen such that over the composition range of these experiments (0.01 -0.14 solution mole fraction of fibrinogen) the ratio mole fraction in the surface to that in solution was always greater than one. Concerning the turnover process of adsorbed albumin on polyethylene, three characteristic features of the process were observed. Firstly, there was always a fraction of the surface layer which was exchangeable and the remainder was not; secondly both the extent and the rate of turnover increased with increased albumin concentration in solution and finally lengthy periods (up to 120 hours) were necessary to reach the steady state values.

The time dependence curves of the in situ adsorption measurements on the solution-polyethylene interface (Fig. 7) performed at constant fibrinogen concentration (0.035 mg/ml) in the presence of albumin at either low (0.05 mg/ml) or high solution concentration (0.5 mg/ml) show a considerable transient surface enrichment in

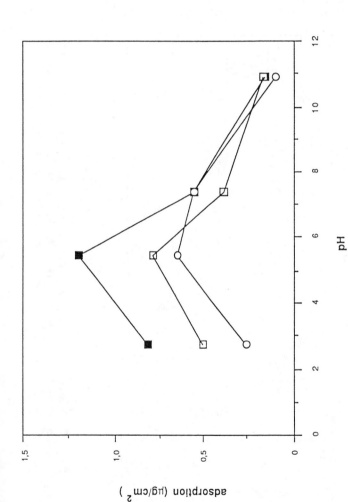

Figure 6 : pH dependence of fibrinogen adsorption at solution-air (○) and solution-polyethylene (□, ■) interfaces. Total protein adsorption (■) and irreversibly adsorbed fibrinogen (remaining at the interface after elution with buffer) (□). Fibrinogen solution concentration 0.2 mg/ml. Time of adsorption 20 hours.

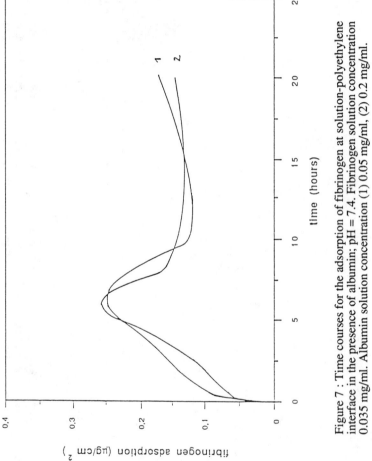

Figure 7 : Time courses for the adsorption of fibrinogen at solution-polyethylene interface in the presence of albumin; pH = 7.4. Fibrinogen solution concentration 0.035 mg/ml. Albumin solution concentration (1) 0.05 mg/ml, (2) 0.2 mg/ml.

fibrinogen after 5 hours of adsorption. Based on a simple diffusion limited mass transport view, albumin molecules which diffuse much faster than fibrinogen ones would arrive first to the surface and would largely populate it. Since the interaction between albumin and polyethylene surface is sufficiently strong, its residence time at the surface would be long and would thus explain that the observed effect occurs after a relatively long time. However, fibrinogen will gradually displace some of preadsorbed albumin. This behavior has been explained as the Vroman effect and describes adsorption in presence of two or more different proteins in solution which results in the preferential adsorption of the protein that can most optimally accommodate at the interface and displace its less optimally bound neighbours. The events are time and concentration dependent and are related to the size, collision rates, interface affinity and denaturation tendencies of competing for the surface sites proteins (36). They appear at low surface coverages when the layer is sparse i.e. at low bulk concentrations. At moderate and high surface coverages when the surface is largely populated, the process is exchange reaction controlled regardless of the transport conditions.

It is significant that in our experiments the observed effect occured only at the lower fibrinogen concentration studied (0.035 mg/ml) and at albumin solution concentrations in the 0.05 - 0.2 mg/ml range. Since no transient maximum in fibrinogen adsorption has been noticed at the solution-air interface at both studied fibrinogen concentrations (0,035mg/ml and 0,15mg/ml) it seems likely that an extensive time-dependent denaturation of albumin took place at this interface.

Other evidence of conformational lability and affinity of fibrinogen for polyethylene was obtained from the data of competitive adsorption of fibrinogen against albumin shown in Fig. 8 and Fig. 9. From the curves shown in Fig. 8 representing fibrinogen adsorption against albumin at the solution-air interface, it is immediately apparent that albumin displaced fibrinogen from adsorbing at the level of its adsorption from the single protein system. However within the range of studied albumin and fibrinogen concentrations this deplacement has never been entire. Whereas 0.05mg/ml albumin concentration appeared to be sufficient to reduce fibrinogen adsorption to approximatively 1/3 of its adsorption value from the single protein system at 0.035mg/ml, albumin concentration higher than 0.1mg/ml was necessary to decrease fibrinogen adsorption to about half of its adsorption amount from the single protein system at 0.15mg/ml. The steeper negative slope observed in Fig. 8 between abscissa 0 and 0.05mg/ml would indicate that at the lower fibrinogen concentration, albumin was more efficient to displace fibrinogen from the interface. The results correlate with a more surface active behavior of albumin relative to fibrinogen at low solution concentrations (Fig. 3) resulting in more initial contacts with the interface and increased surface denaturation. It is appropriate to note that at albumin concentration 0.2 mg/ml, the level of fibrinogen adsorption was almost the same as in the absence of albumin.

In contrast to the solution-air interface, the presence of albumin had a less straight forward influence upon fibrinogen adsorption on polyethylene (Fig. 9). At the higher fibrinogen concentration (0.15 mg/ml), the presence of albumin in solution led to a considerable enhancement of fibrinogen adsorption. Evidently, the molecule of fibrinogen exhibits a more labile behavior on polyethylene relative to the air interface and appears to dominate it completely. This effect is all the more amplified as albumin concentration in solution increases. At albumin solution concentration 0.2 mg/ml, fibrinogen adsorption was more than twofold of its adsorption value from the single protein system. All seems to indicate that albumin increases fibrinogen spreading pressure and supports a more "spread protein concept" initially proposed by WILLIAMS and BAGNALL (37) and JENNISSEN (32). However, as shown in Fig. 9, the increased amounts of adsorbed fibrinogen are, to a large degree, reversibly adsorbed quantities. This would suggest that the presence of albumin in solution did not alter the number of fibrinogen contacts with polyethylene surface.

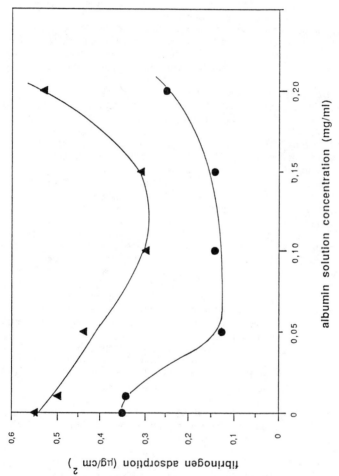

Figure 8 : Competitive adsorption of fibrinogen in the presence of albumin at solution-air interface : (●) fibrinogen concentration 0.035 mg/ml, (▲) fibrinogen concentration 0.15 mg/ml; pH = 7.4. Time of adsorption 20 hours.

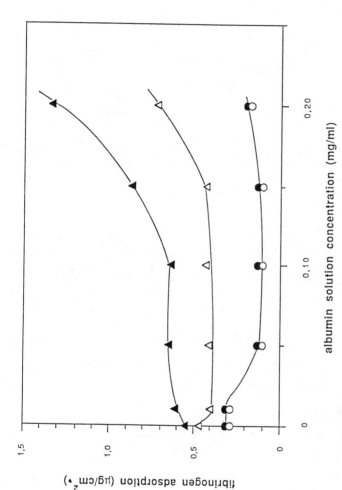

Figure 9 : Competitive adsorption of fibrinogen in the presence of albumin at solution-polyethylene interface : (●, ○) fibrinogen concentration 0.035 mg/ml, (▲, △) fibrinogen concentration 0.15 mg/ml. Open symbols refer to the irreversibly adsorbed fibrinogen (amounts remaining at the interface after elution with buffer); pH 7.4. Time of adsorption 20 hours.

The data for adsorption of fibrinogen at the lower solution concentration (0.035 mg/ml) are also consistent with the spreading pressure concept. In this case fibrinogen concentration is not enough high to ensure full spreading of molecules over the surface of polyethylene and albumin may at least partly reduce the number of contacts that it forms with the surface. Although this leads to a slightly reduced adsorption relative to fibrinogen adsorption from the single protein system, the adsorbed fibrinogen is all irreversibly adsorbed. Certainly, all adsorbed fibrinogen forms strong hydrophobic contacts with polyethylene, the contacts that can not be ruptured when the surface is flushed with the buffer solution.

In conclusion, we have shown that at both studied surfaces after the initial diffusion-controlled adsorption, further events determining the nature of this adsorption in terms of its reversibility are directly related to individual albumin and fibrinogen surface chemistry. In competitive adsorption processes, more complex events can develop with time and the amounts of protein adsorbed are related to solution concentration and the affinity and denaturation tendency of a protein to the surface. Our data clearly show that the transient maximum in adsorption of fibrinogen in competition with albumin occured only at the solution-polyethylene interface at which fibrinogen was presumably strongly labile and thus dominated the surface. Conversely, at low solution concentrations, albumin displayed increased affinity for the solution-air interface resulting in its capability to reduce fibrinogen adsorption at this interface. We discuss these phenomena in terms of the interfacial activity and spreading pressure concept of adsorption.

REFERENCES

1.   Vroman, L.; Adams, A.L. *J. Colloid and Interface Sci.* **1986**, 111, 391-402.
2.   Vroman, L.; Adams, A.L. In *Proteins at Interfaces; Physico-Chemical and Biochemical Studies* ; Brash, J.L.; Horbett, J.A.; Eds.; ACS Symposium Series 343; American Chemical Society: Washington, DC, **1987**; Ch. 19, pp.154-164.
3.   McMillan, C.R.; Walton, A.C. *J. Colloid and Interface Sci.* **1974**, 48, 345-349.
4.   Soderquist, M.E.; Walton, A.C. *J. Colloid and Interface Sci.* **1980**, 75, 386-397.
5.   Chan, B.M.C.; Brash, J.L. *J. Colloid and Interface Sci.* **1981**, 84, 263-265.
6.   Stupp, S.L.; Kauffman, J.W.; Carr, S.H. *J. Biomed. Mater. Res.* **1977**, 11, 237-246.
7.   Lu, D.R.; Park, K. *J. Colloid and Interface Sci.* **1991**, 144, 271-281.
8.   Voegel, J.C.; Dejardin, Ph.; Strasser, C.; de Baillou, N.; Schmitt, A. *J. Colloids and Surfaces* **1987**, 25, 139-144.
9.   Lenk, T.Y.; Ratner, B.D.; Gendreau, R.M.; Chittur, K.K. *J. Biomed. Mater. Res.* **1989**, 29, 549-569.
10. Pitt, W.G.; Cooper, S.L. *J. Biomed. Mater. Res.* **1988**, 22, 359-382.
11. Pitt, W.G.; Grassel, T.G.; Cooper, S.L. *Biomaterials* **1988,** 9, 36-46.
12. Ruzgas, T.A.; Razumas, V.J.; Kulys, J.J. *J. Colloid and Interface Sci.* **1992,** 151, 136-143.
13. Golander, C.G.; Lin, Y.S.; Hlady, V.; Andrade, J.D. *Colloids and Surfaces* **1990**, 49, 289-302.
14. Sevastianov, V.I.; Kulik, E.A.; Kalinin, I.D. *J. Colloid and Interface Sci.* **1991**, 145, 191-206.
15. Dillman Jr., W.J.; Miller, I.F. *J. Colloid and Interface Sci.* **1973**, 44, 221-241.
16. Proust, J.E.; Baszkin, A.; Boissonnade, M.M. *J. Colloid and Interface Sci.* **1983**, 94, 421-429.

17. Baszkin, A.; Deyme, M.; Perez, E.; Proust, J.E. In *Proteins at Interfaces, Physico-Chemical and Biochemical Studies*; Brash, J.L.; Horbett,T.A.; Eds; ACS Symposium Series 343; American Chemical Society: Washington, DC, **1987**; pp 451-467.
18. Deyme, M.; Baszkin, A.; Proust, J.E.; Perez, E.; Boissonnade, M.M. *J. Biomed. Mater. Res.* **1986**, 20, 951-962.
19. Deyme, M.; Baszkin, A.; Proust, J.E.; Perez, E.; Albrecht, G.; Boissonnade, M.M. *J. Biomed. Mater. Res.* **1987**, 21, 321-328.
20. Baszkin, A. In *Modern Aspects of Protein Adsorption on Biomaterials*; Missirlis,Y.F.; Lemm, W.; Eds; Kluwer Academic Publishers; Dordrecht/ Boston/London, **1991**, pp 19-23.
21. Baszkin, A.; Boissonnade, M.M. *J. Biomed. Mater. Sci.* **1993**, 27, 145-152.
22. Jentoft, N.; Dearborn, D.G. *J. Biol. Chem.* **1978**, 254, 4359-4365 .
23. Blomback, B.; Hanson, L.A. In *Plasma Proteins*; Eds; Wiley; New York, NY, **1979**.
24. Brynda, E.; Houska, M.; Pokorna, Z.; Cepalova, N.A.; Moiseev, Yu. V.; Kalal, J. *J. Bioengineering* **1978**, 2, 411-418.
25. Andrade, J.D.; Hlady, V. *Ann. N.Y. Acad. Sci.* **1987**, 516, 158-172.
26 Graham, D.E.; Phillips, M.C. *J. Colloid and Interface Sci.* **1979**, 70, 403-414; 415-426; 427-439.
27. Ter-Minassian-Saraga, L. *J. Colloid and Interface Sci.* **1981**, 80, 393-401.
28. McRitchie, F. *Colloids and Surfaces* **1989**, 41, 25-34.
29. Claesson, P.M.; Arnebrant, T.; Bergenstahl, B.; Nylander, T. *J. Colloid and Interface Sci.* **1989,** 130, 457-466.
30. Nylander, T.; Wahlgren, N.M. *J. Colloid and Interface Sci.* **1994**, 162, 151-162.
31. Lundstrom, I.; Ivarsson, B.; Jonsson, V.; Elwing, H. In *Polymer Surfaces and Interfaces* Feast, W.J.; Munro, H.S.; Eds; Wiley; New York, NY, **1986**, pp 201-230.
32. Jennissen, H.P. *J. Colloid and Interface Sci.* **1986**, 111, 570-589.
33. McRitchie, F. *J. Colloid and Interface Sci.* **1972**, 38, 484-488.
34. Brash, J.L.; Davidson, V.J. *Thrombosis Research* **1976**, 9, 249-256.
35. Brash, J.L. ; Saamak, Q.M. *J. Colloid and Interface Sci.* **1978**, 65, 495-504.
36. Andrade, J.D.; Hlady, V.; Wei, A.P.; Ho, C.H.; Lea, A.S.; Jeon, S.I.; Lin, Y.S.; Stroup, E. *J. Clinical Materials* **1992**, 11, 67-84.
37. Williams, D.F.; Bagnal, R.D. In *Fundamental Aspects of Biocompatibility* Williams, D.F. ; Eds; CRS Press, **1981**, Boca Raton Vol., pp 113-127.

RECEIVED February 10, 1995

Chapter 16

Ellipsometry Studies of Protein Adsorption at Hydrophobic Surfaces

Martin Malmsten and Bo Lassen

Institute for Surface Chemistry, P.O. Box 5607, S–114 86 Stockholm, Sweden

The adsorption of human serum albumin (HSA), IgG and fibrinogen, at (hydrophobic) methylated silica surfaces was investigated with *in situ* ellipsometry. By performing studies with the bare substrate at two different ambient refractive indices, and by performing 4-zone averaging in all measurement, the adsorbed amount (Γ), the adsorbed layer thickness (δ_{el}) and the mean adsorbed layer refractive index (n_f) are obtained accurately. Furthermore, the build-up of the adsorbed layers could be followed in detail. Thus, for fibrinogen both δ_{el} and n_f initially increase monotonically with the adsorbed amount, whereas at higher adsorbed amounts, a "swelling" of the adsorbed layer is observed. For IgG, on the other hand, δ_{el} is essentially independent of the adsorbed amount. Furthermore, studies of the adsorption from HSA/IgG mixtures were performed in a similar way.

The adsorption of proteins at solid surfaces is receiving increasing attention (*1,2*). This is partly motivated by fundamental issues, but also by the importance of protein adsorption in many practical and industrial applications. For example, the adsorption of proteins at biomedical surfaces, such as implants, cathethers and insulin pumps, is the first step in a complex series of biophysical/biochemical processes, which determine the biological response to the foreign material. Furthermore, there are a large number of biotechnical applications, e. g. solid-state diagnostics, which are sensitive to the state of adsorption of proteins. The immobilization of proteins at surfaces is also of large potential use in, e. g. extra-corporeal therapy and advanced bioorganic synthesis. However, despite much previous work done on protein adsorption from complex as well as simpler protein systems, much remains to be done, e. g. concerning the structure and formation of the adsorbed layer, as well as interfacial exchange processes at model surfaces. We therefore undertook investigations concerning these issues, using ellipsometry. Here we report on some of the results obtained for some serum proteins at model hydrophobic surfaces.

Experimental

Materials Human serum albumin (HSA), globulin-free, lyophilized and crystallized, was obtained from Sigma Chemical Co., USA, as was reagent grade purified immunoglobulin (IgG) and fibrinogen (Fraction I; 92% clottable protein). All proteins were used without further purification. Chemicals used for the buffer preparation were all of analytical grade, and used without further purification.

Surfaces Hydrophobized silica surfaces were prepared from polished silicon slides (Okmetic, Finland). In short, these were oxidized thermally to an oxide layer thickness of about 30 nm. The slides were then cleaned and treated with $Cl_2(CH_3)_2Si$ (Merck) as described previously (*3*). This procedure rendered the slides hydrophobic, with an advancing and receeding contact angle of 95° and 88°, respectively.

Methods The ellipsometry measurements were all performed by means of null ellipsometry. The instrument used was an automated Rudolph thin-film ellipsometer, type 436, controlled by a personal computer. A xenon lamp, filtered to 4015 Å, was used as the light source. A thorough description of the experimental setup is given in ref.4. Prior to adsorption, the ellipsometry measurements require a determination of the complex refractive index of the substrate (*5*). In the case of a layered substrate, e. g. oxidized silicon, a correct determination of the adsorbed layer thickness and mean refractive index requires an accurate determination of the silicon bulk complex refractive index ($N_2=n_2-ik_2$) as well as of the thickness (d_1) and the refractive index (n_1) of the oxide layer. This is done by measuring the ellipsometric parameters Ψ and Δ in two different media, e. g. air and buffer. From the two sets of Ψ and Δ, n_2, k_2, d_1 and n_1 can be determined separately. [The hydrophobic methyl layer is neglected, since calculations with the Bruggeman effective medium theory (*5*) show that the error in doing so is much less than 10% in δ_{el}, and even smaller in Γ.] All measurements were performed by four-zone null ellipsometry in order to reduce effects of optical component imperfections (*5*). [A thorough description of the theory of four-zone null ellipsometry experiments, as well as of adsorption studies at layered substrate surfaces, is given in refs. 4-6.] In fact, the procedure used has previously been shown to be even more accurate than the multiple angle of incidence approach (*4*). Furthermore, both the adsorbed amounts and the adsorbed layer thicknesses obtained with the present methodology agree well with results obtained with other techniques both for surfactant, polymer and protein systems (*7*). After the optical analysis of the bare substrate surface, the protein solution was added to the cuvette, and the values of Ψ and Δ recorded. The adsorption was only monitored in one zone, since the four-zone procedure is rather time-consuming and since corrections for component imperfections already had been performed. Four-zone measurements at adsorption equilibrium show that the error induced by the procedure used is less than a few percent. The maximal time-resolution between two measurements is 3-4 seconds. Stirring was performed by a magnetic stirrer at about 300 rpm.

From Ψ and Δ, the mean refractive index (n_f) and average thickness (δ_{el}) of the adsorbed layer were calculated numerically (cf Appendix) according to an optical four layer model for the proteins at hydrophobized silica (*4-6*). The refractive index and the average thickness were finally used to calculate the adsorbed amount (Γ) according to de Feijter (*8*), with dn/dc=0.188 (*9*), 0.188 (*9*), and 0.187 (*9,10*) cm^3/g for fibrinogen, IgG, and HSA, respectively. [Due to the similarity between the refractive index increments of HSA, IgG and fibrinogen, ellipsometry measurements on the binary systems will provide the total adsorbed amount.] All measurements were performed in 0.01 M phosphate buffer (0.15 M NaCl, pH 7.2) at 20°C.

NOTE: Please see Appendix on page 237.

Results and Discussion

The adsorption from HSA, IgG and fibrinogen single protein solutions at methylated silica surfaces is shown in Figure 1a. The concentrations used all correspond to (pseudo-) plateau in the respective adsorption isotherm. Although the equilibrium concentration for fibrinogen is only about 100 ppm, saturation adsorption is reached after only 1 hour. A similar finding is obtained for HSA at a 200 ppm concentration. The adsorption of IgG, on the other hand, is slower in terms of the adsorbed amunt, and 90% of saturation adsorption is reached after about 3000 s, in comparison to about 1000 s for fibrinogen. Note, however, that for all proteins, a slight but significant increase in the adsorbed amount is observed even after 4000 s adsorption, indicating slow structural rearrangements.

A parameter of large interest for protein adsorption is the adsorbed layer thickness. As can be seen from Figure 1b, the mean (optical) thickness at adsorption plateau for HSA, IgG and fibrinogen at methylated silica is 4±2 nm, 18±2 nm and 28±2 nm, respectively (7). It is interesting to compare these thicknesses with the molecular dimensions of the proteins, which for HSA, IgG and fibrinogen are approximately 4x4x14 nm, 23.5x4.5x4.5 nm and 6.0x6.0x45.0 nm, respectively. Hence, HSA adsorbs essentially side-on at hydrophobic surfaces at the present conditions, while IgG adsorbs in an essentially end-on configuration. Fibrinogen, finally, seems to adsorb in a random configuration at hydrophobic surfaces.

Apart from the adsorbed amount and the mean adsorbed layer thickness, ellipsometry provides information on the adsorbed layer mean refractive index, and hence on the average protein concentration in the adsorbed layer. For all proteins investigated here, low adsorbed layer refractive indices were obtained (7), and the average adsorbed layer protein concentrations were found to be 0.11-0.17 g/cm^3.

The adsorbed layer structure has previously been investigated for all the proteins studied here. Hence, Lee et al. (11), Norman et al. (12) and Uzgiris and Fromageot (13) obtained a hydrodynamic thickness of HSA at polystyrene (pH 7.2-7.4) of about 7, 6, and 4 nm, respectively. Furthermore, in the latter study the mean adsorbed layer refractive index was determined, and found to be comparable to that found for HSA in the present investigation. Furthermore, end-on adsorption of IgG has previously been observed (although at higher surface concentrations than in the present investigation), e. g. by Morrisey and Han (14), using polystyrene colloidal particles as substrates. Furthermore, Elwing et al. found the thickness of an adsorbed layer of IgG at methylated silica to be 17±5 nm, whereas n_f was found to be 1.39±0.02 (15). Thus, the agreement between the present and these previous results (not using the two ambient refractive indices procedure with concomitant 4-zone averaging throughout) is surprisingly good. For fibrinogen, finally, there have been several previous scanning angle reflectometry studies, although using silica as substrate (16). Although the adsorption of fibrinogen is strongly dependent on the substrate hydrophobicity, it is still interesting to note that an optical adsorbed layer thickness of 10-30 nm was obtained under otherwise similar conditions. Furthermore, it is interesting to note that the refractive index difference between the adsorbed layer and the bulk solution ($\Delta n = n_f - n_b$) is comparable ($\Delta n \approx 0.02$-0.03) to that obtained in the present investigation.

One advantage of ellipsometry as a tool for studying adsorbed protein films is the good time resolution of the method, which allows the thickness and the adsorbed layer refractive index to be studied essentially continuously as the adsorbed layer forms. As can be seen from Figure 2a, for fibrinogen the layer thickness increases linearly with the adsorbed amount up to about 4 mg/m^2. After this, there is a more pronounced growth of the adsorbed layer normal to the surface, resulting also in a slight decrease in the mean adsorbed layer refractive index (Figure 2b). Thus, as the

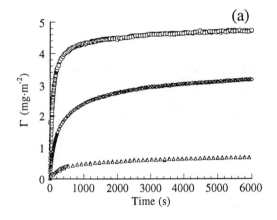

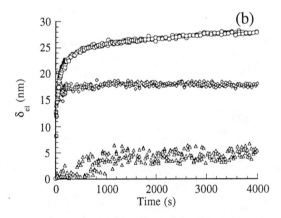

Figure 1. Adsorbed amount (a) and adsorbed layer thickness (b) of HSA (200 ppm; open triangles), IgG (100 ppm; open diamonds) and fibrinogen (100 ppm; open circles) at methylated silica from 0.15 M NaCl, pH 7.2.

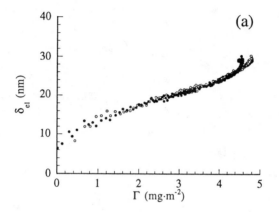

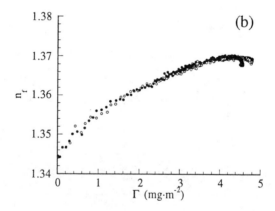

Figure 2. Adsorbed layer thickness (a) and mean adsorbed layer refractive index (b) for fibrinogen (100 ppm) versus the adsorbed amount at methylated silica from 0.15 M NaCl, pH 7.2. Two measurements are shown in order to illustrate the degree of reproducibility.

adsorbed layer builds up, this is initially achieved both by an increased average protein concentration in the adsorbed layer, and by a growth of the adsorbed layer normal to the interface. At higher surface coverages, a "swelling" of the adsorbed layer normal to the surface, most likely achieved by reorientation of the fibrinogen molecules, is the main mode for achieving an increased adsorption (7). This type of adsorbed layer formation is generally observed for high molecular weight polymers and is due to the high adsorbed layer osmotic pressure at high surface coverages.

It is interesting to note that a similar, although more pronounced, build-up of adsorbed fibrinogen layers has been observed previously with scanning angle reflectometry for silica. Thus, Schaaf et al. (16) found that as the adsorbed amount increases, either by increasing the bulk fibrinogen concentration or at increasing adsorption times, the adsorbed layer thickness initially increases only marginally, whereas the adsorbed layer refractive index increases strongly. At higher adsorbed amounts (last 10% of the adsorption) the adsorbed fibrinogen layer grows substantially normal to the surface, while the adsorbed layer refractive index decreases.

For IgG, the build-up of the adsorbed layer occurs differently. As can be seen in Figure 3a, the adsorbed layer thickness remains essentially constant with an increasing adsorbed amount above about 0.5 mg/m^2. At the same time, the adsorbed layer refractive index (average adsorbed layer protein concentration) increases essentially linearly (Figure 3b). Thus, in the case of IgG, an increasing adsorbed amount is achieved solely by packing essentially end-on adsorbed IgG molecules more densely. For HSA, finally, an increasing adsorbed amount is achieved by packing side-on adsorbed HSA molecules more densly at the interface.

In Figure 4, the total adsorbed amount at methylated silica from binary HSA/IgG protein mixtures (total concentration 200 ppm) of different compositions is shown. At IgG fractions higher than about 50%, the adsorbed amount is constant, and equal to that of IgG in the absence of HSA, which seems to indicate a complete preference for IgG adsorption at methylated surfaces at the conditions present. This is further supported by the adsorbed layer thickness being constant (within the experimental uncertainty) down to about 50% IgG, as is the adsorbed layer mean refractive index (18).

In fact, the build-up of the adsorbed layer proceeds essentially identically down to about 50% IgG in the protein mixture, as can be seen in Figure 5, illustrating the adsorbed layer thickness and mean refractive index as a function of the adsorbed amount. These results indicate adsorption of primarily IgG in a rather dilute layer with the IgG molecules adsorbed preferentially head-on. Only at an IgG fraction of less than 25%, the situation is altered, and the adsorbed layer thickness is reduced towards that of a pure HSA adsorbed layer at hydrophobic surfaces (4 nm) (18). From these considerations we conclude that at high and medium IgG fractions, this protein is strongly preferentially adsorbed at methylated silica surfaces under the conditions used. In a similar way, the adsorption from HSA/fibrinogen mixtures was studied. It was found that also fibrinogen adsorbs preferentially over HSA under these conditions, although the decrease in δ_{el} on increasing the HSA bulk fraction is more gradual than for IgG. However, the same study showed that if HSA is allowed to preadsorb at hydrophobic surfaces, it effectively reduces the IgG and fibrinogen adsorption (by about 80% and 90%, respectively), which is due to an irreversible HSA adsorption at these surfaces (18).

Acknowledgement

This work was financed by the Foundation for Surface Chemistry, Sweden, and the Swedish National Board for Industrial and Technical Development.

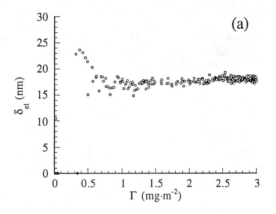

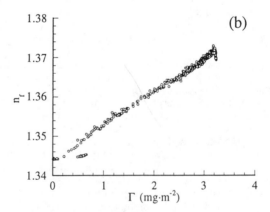

Figure 3. Adsorbed layer thickness (a) and mean adsorbed layer refractive index (b) for IgG (200 ppm) versus the adsorbed amount at methylated silica from 0.15 M NaCl, pH 7.2.

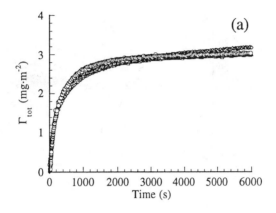

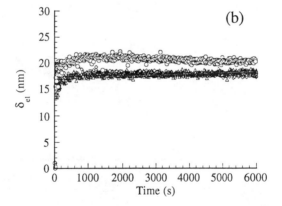

Figure 4. Total adsorbed amount (a) and adsorbed layer thickness (b) versus time at methylated silica from solutions (0.15 M NaCl, pH 7.2) with a HSA/IgG ratio of 0/100 (open diamonds), 25/75 (open circles) and 50/50 (open triangles). The total protein concentration was 200 ppm.

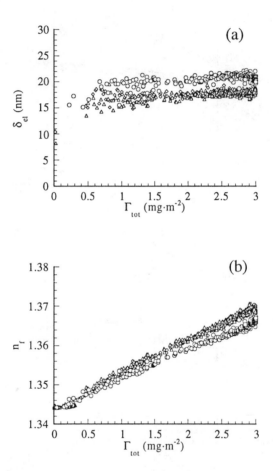

Figure 5. Adsorbed layer thickness (a) and refractive index (b) versus the total adsorbed amount at methylated silica from HSA/IgG solutions. HSA/IgG ratios of 0/100 (open diamonds), 25/75 (open circles) and 50/50 (open triangles) are shown. The total protein concentration was 200 ppm.

Appendix

In the evaluation of the ellipsometric parameters (cf ref 6), the system is modelled as consisting of four layers, i. e. bulk silicon, characterized by a complex refractive index ($N_2 = n_2 - ik_2$), an oxide layer, characterized by a refractive index (n_1) and a thickness (d_1), an adsorbed layer with refractive index n_f and a thickness d_f (referred to as δ_{el} in the paper), and finally an ambient medium of a refractive index n_0. Throughout, the oxide and the adsorbed layer, as well as the ambient medium, are assumed to be transparent, i. e. $k = 0$. The resulting reflection coefficients parallel (p) and perpendicular (s) to the plane of incidence for such a system is given by:

$$R_x = \frac{(r_{0f}^x + r_{f1}^x e^{-i2\beta_f}) + (r_{0f}^x r_{f1}^x + e^{-i2\beta_f}) r_{12}^x e^{-i2\beta_1}}{(1 + r_{0f}^x r_{f1}^x e^{-i2\beta_f}) + (r_{f1}^x + r_{0f}^x e^{-i2\beta_f}) r_{12}^x e^{-i2\beta_1}} \quad (1)$$

$x = p$ or s

where

$$\beta_1 = 2\pi \left(\frac{d_1}{\lambda}\right) n_1 \cos\phi_1 \quad (2)$$

$$\beta_f = 2\pi \left(\frac{d_f}{\lambda}\right) n_f \cos\phi_f \quad (3)$$

The (complex) reflection coefficients (r_{mn}) at the interface between layers m and n for the p and s components are given by:

$$r_{mn}^p = \frac{N_n \cos\phi_m - N_m \cos\phi_n}{N_n \cos\phi_m + N_m \cos\phi_n} \quad (4)$$

$$r_{mn}^s = \frac{N_m \cos\phi_m - N_n \cos\phi_n}{N_m \cos\phi_m + N_n \cos\phi_n} \quad (5)$$

where N_m and N_n are the complex refractive indices of layers m and n, respectively, whereas ϕ_m and ϕ_n are the angles of incidence at layers m and n, respectively. The latter are related through:

$$N_m \sin\phi_m = N_n \sin\phi_n \quad (6)$$

The measured ellipsometric parameters, Ψ and Δ, are related to R_p and R_s by:

$$\frac{R_p}{R_s} = \tan\Psi e^{i\Delta} \quad (7)$$

The bulk silicon complex refractive index (N_2), as well as the thickness (d_1) and the refractive index (n_1) of the silica layer, were all determined by measurement of Ψ and Δ in two different ambient refractive indices (in air and buffer) as discussed in detail previously (*4*). It now remains to determine n_f and d_f (δ_{el}) from measurements of Ψ and Δ after addition of protein. Although it is not possible to obtain analytical expressions for these parametres themselves, this is possible for $\exp(-i2\beta_f)$. This is done by substituting eq (1) for the p and s components into eq (7) and solving the quadratic expression for $\exp(-i2\beta_f)$ in the same way as for a homogeneous surface. The values of n_f anf d_f (δ_{el}) are found by the following iterative procedure:

(a) a value of n_f is assumed
(b) $\exp(-i2\beta_f)$ is calculated from eqs (1) and (7)
(c) d_f (δ_{el}) is calculated from $\exp(-i2\beta_f)$; generally this will be a complex number
(d) the procedure is repeated from (a) with a new value of n_f until the imaginary part of d_f (δ_{el}) is arbitrarily close to zero.

The adsorbed amount (Γ) is finally calculated from n_f and d_f (δ_{el}) according to:

$$\Gamma = \frac{\delta_{el}(n_f - n_0)}{\left(\frac{dn}{dc}\right)} \tag{8}$$

where (dn/dc) is the protein refractive index increment.

References

1.   Andrade, J.D. (Ed.) *Surface and Interfacial Aspects of Biomedical Polymers*; Vol. 2; Plenum Press: New York, 1985.
2.   Norde, W. *Adv. Colloid Interface Sci.* **1986**, *25*, 267.
3.   Jönsson, U.; Ivarsson, B.; Lundström; I.; Berghem, L. *J. Colloid Interface Sci.* **1982**, *90*, 148.
4.   Landgren, M.; Jönsson, B. *J. Phys. Chem.* **1993**, *97*, 1656.
5.   Azzam, R.M.A.; Bashara, N.M. *Ellipsometry and Polarized Light*; North-Holland: Amsterdam, 1989.
6.   Tiberg, F.; Landgren, M. *Langmuir* **1993**, *9*, 927.
7.   Malmsten, M. *J. Colloid Interface Sci.*, in press.
8.   de Feijter, J.A.; Benjamins, J.; Veer, F.A. *Biopolymers* **1978**, *17*, 1759.
9.   Gölander, C.-G. *Preparation and Properties of Functionalized Polymer Surfaces*, Thesis, The Royal Institute of Technology, Stockholm, Sweden, 1986.
10. Peters, T. *Adv. Protein Chem.* **1985**, *37*, 161.
11. Lee, J.; Martic, P.A.; Tan, J.S. *J. Colloid Interface Sci.* **1989**, *131*, 252.
12. Norman, M.E.; Williams, P.; Illum, L. *Biomaterials* **1992**, *13*, 841.
13. Uzgiris, E.E.; Fromageot, H.P.M. *Biopolymers* **1976**, *15*, 257.
14. Morrissey, B.W.; Han, C.C. *J. Colloid Interface Sci.* **1978**, *65*, 423.
15. Elwing, H.; Ivarsson, B.; Lundström, I. *Eur. J. Biochem.* **1986**, *156*, 359.
16. Schaaf, P.; Dejardin, P. *Colloids Surf.* **1988**, *31*, 89.
17. Fleer, G.J.; Cohen Stuart, M.A.; Scheutjens, J.M.H.M.; Cosgrove, T.; Vincent, B. *Polymers at Interfaces*; Chapman & Hall: London, 1993.
18. Malmsten, M.; Lassen, B. *J. Colloid Interface Sci.*, in press.

RECEIVED February 10, 1995

Chapter 17

Protein—Surfactant Interactions at Solid Surfaces

Thomas Arnebrant and Marie C. Wahlgren

Department of Food Technology, University of Lund, P.O. Box 124, 221 00 Lund, Sweden

Effects of surfactants on protein adsorption are reviewed. Differences between removal of preadsorbed proteins (elutability) and competitive adsorption are discussed and simple models are suggested. It can be concluded that surfactants may interact through solubilization or replacement mechanisms depending on surfactant- surface interactions and surfactant- protein binding. Solubilization requires complex formation between protein and surfactant, and the replacement adsorption of the surfactant to the surface. As for protein adsorption, one of the most important properties affecting the elutability appears to be the conformational stability. Differences between a competitive situation and addition of surfactant after adsorption of the protein are suggested to originate from alteration in surface activity of protein-surfactant complexes formed in solution as compared to pure protein, the difference in diffusivity of surfactants and protein, and time dependent conformational changes of the protein.

Interactions between surfactants and proteins take place in various applications involving proteins in contact with solid surfaces. Examples are found in general detergency, for example when process equipment should be cleaned from deposits of protein origin. Surfactants and proteins or peptides may be simultaneously present during protein isolation procedures and for minimizing loss of active substance during drug administration. Another area of application of these interactions is in the field of dentistry where so-called anti-plaque agents are used in the treatment of plaque related diseases (1-4). The degree of removal of adsorbed protein by surfactants has also been used as an indication of the mode of protein attachment to the surface, in particular with respect to time dependent conformational changes (5-7). The general features of protein adsorption involve what is referred to as multiple states of adsorption cf. (8). This means that adsorbed protein may exist in several adsorbed fractions with varying binding modes to the surface. These different fractions may be distinguished by their differences in binding strength as indicated by the degree of removal by rinsing with buffer, addition of surfactant or by their different susceptibility to exchange by the same or other types of protein. The importance of structural features of interfacial layers of protein is recognised in research focused on e.g. biocompatible materials, dental pellicle buildup, immuno assays and enzyme

0097—6156/95/0602—0239$12.00/0

immobilisation. A correlation between the above mentioned structural aspects and a simple measurable quantity as for example surfactant mediated elutability is therefore convenient from a practical point of view, and has proved to be valuable in the assessment of the performance of biomedical polymers (9, 10).

Surfactant Adsorption at Solid Surfaces

Surfactants, as the name implies, tend to adsorb at most interfaces and thereby strongly reduce the interfacial free energy. Surfactants are usually classified according to their head group as an-, cat- and nonionic, respectively. The surfactants discussed in this paper are presented in table I. The properties of the hydrophobic, usually hydrocarbon, part as well as the hydrophilic head group, will affect the affinity of the molecule for interfaces. Furthermore, due to the strong tendency for the hydrophobic chains to avoid contact with water, self-association will take place both in solution and at interfaces, a fact that will influence the adsorption behaviour. In bulk phase, self-association usually involves the formation of spherical micelles at low concentration and depending on surfactant structure, cylindrical micelles, lamellar structures, cubic phases and structures of the reversed type may form at higher concentrations and appropriate conditions (11). The effect is that the monomer concentration in solution will be strongly dependent on the association pattern (11) and thus have a pronounced effect on the interfacial behaviour.

Table I : A presentation of the surfactants discussed in the text

Surfactant	Abbreviation	cmc	Headgroup
Sodium dodecylsulphate	SDS	8 mM	$-SO_3^-$
Dodecyltrimethylammonium bromide	DTAB	15 mM	$-N^+-(CH_3)_3$
Tetradecyltrimethylammonium bromide	TTAB	3.6 mM	$-N^+-(CH_3)_3$
Cetyltrimethylammonium bromide	CTAB	0.9 mM	$-N^+-(CH_3)_3$
Triethylene glycol monododecyl ether	$C_{12}E_3$	-	$-(OCH_2CH_2)_3OH$
Pentaethylene glycol monododecyl ether	$C_{12}E_5$	0.065 mM	$-(OCH_2CH_2)_5OH$
Octaethylene glycol monododecyl ether	$C_{12}E_8$	0.071 mM	$-(OCH_2CH_2)_8OH$

To state briefly, a few general features of surfactant adsorption are the following:

1) At high surfactant concentrations (around the cmc) as discussed in the present work, the adsorption rate is fast, as exemplified in Fig. 1 for the adsorption of DTAB to methylated silica.

2) For water soluble surfactants, the adsorption is reversible upon dilution which also is illustrated in Fig. 1.

3) Usually a plateau in the adsorption isotherm is reached in the range of the cmc (critical micelle concentration), Fig. 1.

4) As a rule surfactants adsorb at hydrophobic surfaces. The amounts adsorbed are in the range of, or below, those corresponding to a monolayer, Fig. 2.

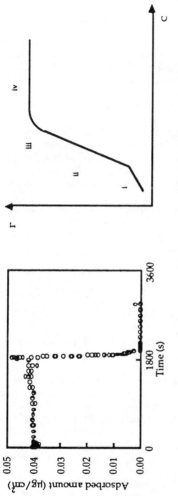

Fig. 1   The adsorption of DTAB (2*cmc in phosphate buffered saline pH 7, I=0.17) onto methylated silica. Data are from (25) (left) A schematic illustration of an isotherm for adsorption of ionic surfactants to hydrophilic surfaces, from Somasundaran (16)(right).

5) In the absence of specific chemical interactions, ionic surfactants only adsorb onto hydrophilic surfaces of opposite charge(*12-16*). At these surfaces bilayers or corresponding structures are formed (see below), Fig. 2.

6) Isotherms for surfactants adsorbing to hydrophilic surfaces usually have the features shown in Fig. 1 (*16*). The different regions will depend on the association of the surfactant molecules at the interface and in solution. There is a vast literature concerning the association of surfactants at solid/water interfaces (*12-23*). Fig. 2 is an illustration of possible association behaviour of surfactants in the different regions.

The structure of the surface aggregates at the plateau has been debated and surface micelles, finite bilayers or infinite bilayers have been suggested for hydrophilic surfaces. Indications of complete bilayers (*24*) (Fig. 2 iv a) or interpenetrating hydrocarbon chains (*16*) (Fig. 2 iv b) have been found.

Surfactant-Protein Interactions in Solution

For mixtures of proteins and surfactants there might be an interaction in bulk solution involving the formation of surfactant-protein complexes which have different properties from those of the pure protein (*26, 27*). Further, the binding of surfactant to protein will reduce the concentration of free surfactant molecules available for interaction with the protein at the interface which may show up as an apparent increase in the cmc.

Ionic surfactants are known to interact with proteins in solution, and the interaction is generally stronger for SDS than for cationic surfactants (*26-31*). Nonionic surfactants are known to generally interact poorly with soluble proteins (*26*). Three types of interactions are observed (*26*) :

i)   Binding of surfactant by electrostatic or hydrophobic interactions to specific sites in the protein, such as for β–lactoglobulin (*26, 32, 33*) and serum albumin (*26, 28, 34*).
ii) Cooperative adsorption of surfactant to the protein without gross conformational changes.
iii) Cooperative binding to the protein followed by conformational changes (*28-30, 35, 36*)

The changes i) -iii) can occur in the same system when surfactant concentration is increased. The conformational changes that occur in case (iii) involves changes in secondary structure (*28, 29, 35*). It is assumed that the surfactant molecules bind to the polypeptide chain and several models for the protein surfactant complexes have been suggested e.g., rigid rod (*37*), pearl and necklace (*38*) and flexible helix model (*39*). In the cooperative region (ii-iii), above the critical association concentration (cac), the interaction is mainly of hydrophobic character (*26, 29, 36*).

Surfactants and Proteins at Interfaces

Interaction of Surfactants with Adsorbed Proteins. The removal of preadsorbed proteins by surfactant has been extensively studied by Horbett and co-workers (*5, 6, 40, 41*) in investigations into the adsorption strength of proteins, particularly fibrinogen. They introduced the term "elutability" in order to describe the degree of removal. The degree of surfactant elutability of proteins is affected by factors that are known to influence the binding strength of a protein to a surface. Thus, surfactant elutability has been found to decrease with factors favouring conformational change

i. e. decreasing protein concentration(*5, 6, 40*), increasing temperature (*5, 6*), time of adsorption "residence time", (*5, 6, 40, 41*) and decreasing stability of the protein (*42-44*). However, surfactant elutability will not only be influenced by protein properties but also by the type of surfactant (*6, 45-47*) and surface (*7, 40, 44-46, 48, 49*) as discussed in more detail below. Of course, the self association of the surfactant plays a major role in this context and the use of non associating displacers may be more straightforward if evaluation of the binding strength is the main concern.

A discussion on the influence on protein-surfactant behaviour by protein properties, surfactant properties and surface properties, based on our own observations as well as by other workers in the field is given below.

Influence of Protein Properties. As discussed above, the interaction between surfactant and protein might involve a certain specificity, especially so at low surfactant concentrations. At higher surfactant concentrations these effects are less pronounced. The correlation of protein properties to elutability is not straightforward, as the effects of different properties overlap. However, when the DTAB induced elution of six model proteins, cytochrome c, bovine serum albumin, α–lactalbumin, β–lactoglobulin, lysozyme and ovalbumin adsorbed at a silica surface was compared, the removal of the proteins that were still adsorbed after rinsing with buffer, appear to increase with decreasing molecular weight and adiabatic compressibility (a measure of conformational stability (*50*)) and increasing thermal denaturation temperature (*44*). In the case of a methylated silica surface, the trends were weaker and differences between the proteins as regards elutability were smaller. However, increasing molecular weight and shell hydrophobicity of the protein seem to reduce elutability. It was also found that the elutability did not relate to the degree of desorption of proteins upon rinsing with buffer, indicating that the two mechanisms are different. Recent experiments on stability mutants of bacteriophage T4 lysozyme show a very convincing relation between DTAB mediated elutability and the difference in free energy of thermal unfolding of the protein in comparison with the wild type (see separate contribution within this volume (*43*)). It might thus be concluded that factors relating to the structural stability of the protein is of major importance and that an increased stability increases the degree of elution.

Influence of Surfactant Properties. It is necessary to keep in mind that the surfactants will, depending on mechanism of elution (see below), interact with the protein, the surface or usually both. Therefore knowledge of the main mechanism of removal for each combination of surfactant, protein and surface is mandatory in order to correctly interpret the effect of one component. The influence of different surfactant headgroups on the desorption of lysozyme at hydrophilic silica surfaces is presented in Fig. 3 (*25, 44, 45*).

Surfactant concentrations differ in this figure but are in all cases above the cmc, except for $C_{12}E_3$ which does not form micelles (*51*). The difference between SDS, cationic surfactants and nonionics as regards the effect on surfactant elutability of proteins is analogous to the strength of binding to protein in solution. This suggests that above the critical association concentration (cac), complex formation between surfactant and protein is involved in the removal mechanism on hydrophilic surfaces. In this connection, Blomberg and coworkers studied the removal of adsorbed lysozyme by SDSo (Sodium Dodecane Sulfonate) and SDS. They found that SDSo, which has a Krafft temperature above room temperature and hence does not form micelles, had a very minor effect on the interaction between adsorbed lysozyme layers on mica (*52*) and concluded that few surfactant molecules were bound to the adsorbed protein. SDS showed a similar low binding to lysozyme on mica at low concentrations (up to 0.5 cmc) but caused a collective desorption of the protein at the cmc of the surfactant, indicating that the cac to adsorbed lysozyme is in the range of its

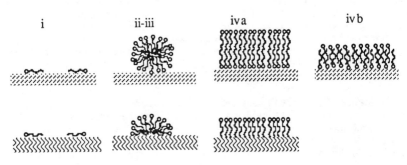

Fig. 2.   An illustration of probable arrangements of adsorbed surfactant molecules at different degrees of surface coverage. Adsorption to hydrophilic surfaces (left panels) and hydrophobic ones (right panels). The illustrations are drawn to represent structures having minimal water contact with the hydrophobic parts of the molecules. The figures should be considered as schematic and other structures, especially for ii-iii, have been suggested (*12, 16, 22, 23, 25*).

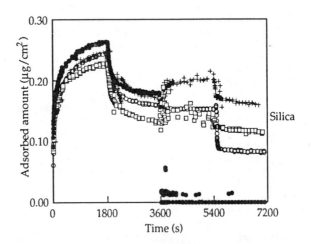

Fig. 3.   The elutability of lysozyme by different surfactants. The adsorbed amount versus time for adsorption of lysozyme to silica followed by buffer rinsing after 1800 sec., addition of surfactant after 3600 seconds and a final rinse with buffer after 5400 sec. The protein concentration is 1 mg/ml in phosphate buffered saline solution pH 7, I=0.17. The surfactants are 5 mg/ml SDS (●), 5 mg/ml DTAB (○), 0.5 g/ml $C_{12}E_3$ (□), 0.5 mg/ml $C_{12}E_5$ (+).

self-association limit in solution (cmc) (53) and further supports that the formation of surfactant- protein complexes are important in the removal process.

The non-ionic surfactants do not remove lysozyme from the hydrophilic silica surface. As mentioned above, these surfactants bind to a very low extent to protein in solution and to the protein covered surface (fig. 3).

It was found that removal of protein at methylated silica surfaces (hydrophobic) is similar for the different surfactants (25), also for non-micelle forming ones, indicating that the proteins are removed through replacement due to higher surface activity of the surfactant. The trend, that non-ionic surfactants do not affect the amount adsorbed at hydrophilic surfaces but have a considerable effect at hydrophobic surfaces has also been observed by Elwing *et al.* (54), when studying surfactant elutability of proteins adsorbed at a surface with a gradient in wettability. Rapoza and Horbett (6) found that surfactants with large headgroups such as Tween-20 gave rise to lower fibrinogen elutability levels than other surfactants at polyethylene surfaces. Welin-Klintström *et al.* (47) found that the elutability of fibrinogen adsorbed at wettability gradient surfaces decreased with the bulkiness of the hydrophobic part of the surfactant. In this connection it was also found that non-ionics showed an increased removal of fibrinogen into the more hydrophilic region of the gradient surface when the cloud point (phase separation temperature) was approached (48). Further, these general observations of removal efficiency are in line with the findings of Bäckström and co-workers (48, 55, 56) who studied the removal of fat by different surfactants and found a maximum at conditions corresponding to an optimum in the packing of surfactant molecules at a flat interface and those of Malmsten and Lindman (57) who investigated, among other variables, the effect of temperature on cleaning of hard surfaces.

The effects of chain length of alkyltrimethylammonium surfactants on the elutability of fibrinogen are presented in Fig. 4 (25). The dependence of the elutability of proteins on chain length of surfactant is small at both silica and methylated silica. The elutability of proteins by DTAB is slightly smaller than for TTAB and CTAB.

Rapoza and Horbett (6, 58) did not find any effects of chain length of sodium alkyl sulphates on the elutability for fibrinogen and albumin down to a chain length of six methyl groups. However, they found, as expected, that the chain length did influence the surfactant concentration at which the onset of protein removal started. The trend was similar to the one observed for the onset of cooperative binding events (*e.g.* micelle formation).

It may be concluded that surfactant headgroup effects are most pronounced at hydrophilic surfaces but less important at hydrophobic ones. In addition, it appears that principles for detergency in general, involving the packing efficiency of molecules at interfaces are qualitatively applicable in these systems.

Influence of Surface Properties. As described above, the adsorption and orientation of surfactant are dependent on the type of surface and it is therefore natural to expect that the way in which proteins are removed by surfactants should be influenced by the surface character as well. Elwing *et al.* (49, 54) studied the surfactant elutability of proteins adsorbed to a gradient in hydrophobicity and found large differences in the amounts removed at the hydrophilic and hydrophobic ends. In the case of a non ionic surfactant (Tween 20) the elutability was largest at the midpoint of the gradient (54), which might be attributed to enhanced conformational changes of the adsorbed protein at the hydrophobic end in combination with a lower efficiency of nonionics at hydrophilic surfaces. Horbett and co-workers (5, 40) studied the elutability of fibrinogen and albumin at different polymeric surfaces and found that the elutability and the change of elutability with time differed among the surfaces. These differences could not, however, be correlated to their critical surface tension of wetting.

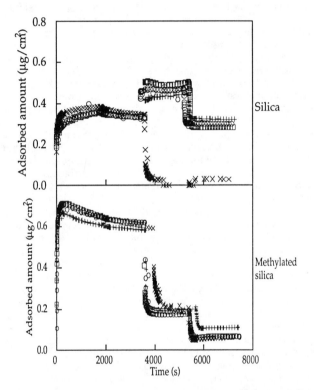

Fig. 4.   The elutability of fibrinogen by different surfactants. The protein concentration is 0.4 mg/ml in phosphate buffered saline solution pH 7. The adsorption procedure is the same as in Fig. 3 and the surfactants are SDS (x), DTAB (+), TTAB (O), CTAB (□) and surfactant concentration is twice the cmc.

Investigations into the elutability of lysozyme and β–lactoglobulin on methylated silica (hydrophobic) and oxides of silicon, chromium and nickel showed that the elutability did not follow any clear cut rules relating to the charge on the surface.

Instead, elutability of β–lactoglobulin and lysozyme by SDS decreased in the order silica>chromium oxide>nickel oxide. The order of decrease is nearly the same for CTAB, but the difference between nickel and chromium oxides was insignificant (45). The similarity between the two oppositely charged surfactants indicates that the elutability in these cases mainly reflects the binding strength of the protein to the surface.

At methylated silica surfaces the elutability was high. This should, however, not be considered as evidence for weak binding of the proteins to the surface but rather as an indication of the strong interaction between the surfactants and surface.

It might be suggested that, even though not of universal applicability, an electrostatic repulsion between surfactant and surface or a strong (hydrophobic) interaction between the hydrocarbon chain of the surfactant and the surface might favour the elution of adsorbed protein.

Simple Models and Conclusions. We have found it useful to classify the observed effects of surfactants on adsorbed proteins in four categories. A short description of these follows below (see Fig. 9 and Figs 5-8) :

i) No remaining adsorbate after addition of surfactant. At these conditions, the surfactant does not adsorb to the surface. However, it interacts with the protein and forms a complex that desorbs from the surface Fig. 5

ii) Replacement of protein by surfactant. This implies that the interaction between surfactant and surface has to be stronger than the interaction between protein or surfactant/protein complex and surface. Adsorption of the surfactant to the surface is essential in this connection, but binding of surfactant to protein may not be required Fig. 6.

iii) The surfactant might adsorb to the surface and/or to adsorbed protein but does not have any net effect on the amount of protein adsorbed Fig. 7.

iv) The adsorbate is only partly removed by the surfactant. Partial elutability of proteins has previously been suggested as an indication of the presence of multiple states of adsorbed proteins (8) Fig. 8.

It is important to note that surfactant can remove proteins without binding to the surface. This could be described as a solubilization of the proteins by the surfactant. An interesting question is whether the replacement of proteins by surfactant in category (ii) is first initiated by solubilization followed by surfactant adsorption? Non-ionic surfactants interact to a very low extent with soluble proteins (26), as for example can be concluded from the low amounts adsorbed to the adsorbed protein layer at the silica surface (Fig. 3). These surfactants still remove proteins from the methylated silica surface, and it is thus evident that in this case it occurs through replacement of adsorbed protein molecules by surfactant.

Adsorption from Mixtures of Proteins and Surfactants

The adsorption from surfactant/protein mixtures to hydrophobic solid surfaces is to some extent analogous to the adsorption at air/water or oil/water interfaces which have been the subject of frequent studies (34, 59, 60) due to the importance of stability of food emulsions and foams. Competitive adsorption between proteins and surfactant at these interfaces has recently been reviewed by Dickinson and Woskett (61). They conclude that, as expected, small surface active components above a certain critical concentration will dominate over proteins at these interfaces, as such

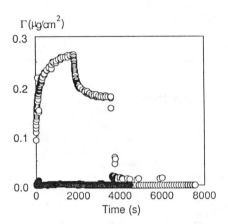

Fig. 5.   The adsorbed amount versus time for adsorption of lysozyme to silica followed by buffer rinsing after 1800 sec., addition of surfactant (SDS) after 3600 seconds and a final rinse with buffer after 5400 sec. (○). Adsorption from a mixture of the protein and surfactant for 1800 sec. followed by rinsing is also included (●). The experiments were carried out at 25° C in 0.01 M phosphate buffer, 0.15 M NaCl, pH 7.

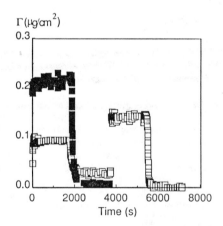

Fig. 6.   The adsorbed amount versus time for adsorption of β-lactoglobulin to silica followed by buffer rinsing after 1800 sec., addition of surfactant (CTAB) after 3600 seconds and a final rinse with buffer after 5400 sec. (□). Adsorption from a mixture of the protein and surfactant for 1800 sec. followed by rinsing is also included (■). Conditions as in Fig. 5.

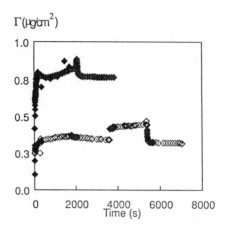

Fig. 7.   The adsorbed amount versus time for adsorption of fibrinogen to silica followed by buffer rinsing after 1800 sec., addition of surfactant (DTAB) after 3600 seconds and a final rinse with buffer after 5400 sec. (A). Adsorption from a mixture of the protein and surfactant for 1800 sec. followed by rinsing is also included (♦). Conditions as in Fig. 5.

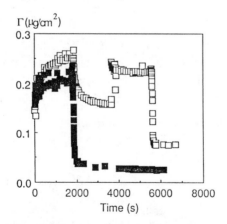

Fig. 8.   The adsorbed amount versus time for adsorption of lysozyme to silica followed by buffer rinsing after 1800 sec., addition of surfactant (CTAB) after 3600 seconds and a final rinse with buffer after 5400 sec. ((□). Adsorption from a mixture of the protein and surfactant for 1800 sec. followed by rinsing is also included (■). Conditions as in Fig. 6.

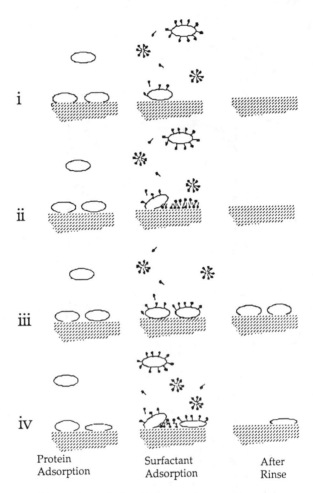

Fig. 9. An illustration of the different classes of surfactant -protein interactions at solid interfaces.

components normally have higher surface activity (superiority in lowering interfacial tension).

Experimentally it is observed that the presence of surfactants in protein solutions may influence the amount of proteins adsorbed to solid surfaces in three different ways (Figs. 5,7, 8) :

i)   Complete hindrance of protein adsorption, Fig. 5.

ii) Reduced amounts adsorbed compared to adsorption from pure protein solution, Fig. 8.

iii) Increased amounts adsorbed, Fig. 7.

In case (i), the complete lack of adsorption could be explained:

1) If a complex is formed with a surfactant that has no attraction to the surface.

2) If the surfactants adsorb, due to their higher surface activity and diffusivity, and by doing so prevent further adsorption of proteins or protein/surfactant complexes.

In (ii) and (iii), the formation of complexes in solution leads to a decrease or increase in the amounts of protein adsorbed, respectively. The presence of surfactant influences the total amount of protein by steric effects or changes in electrostatic interaction between complexes as opposed to native protein. In addition, the complex may adsorb in a different orientation than the pure protein. The surfactant molecules will desorb upon dilution, leaving protein molecules adsorbed at the surface.

Due to the different shapes of the adsorption isotherms of surfactants and proteins, the interfacial interaction is of course, as for protein mixtures, strongly dependent on the concentration of the components. The special character of the surfactant adsorption isotherms featuring the sharp increase in adsorbed amount in the range of their critical association concentration will influence these events in a very pronounced way. This fact and the reversibility upon dilution will give surfactant-protein interactions much of their uniqueness. Studies regarding these surfactant-protein "Vroman effects" have been reported, for example adsorption of fibrinogen from mixtures with Triton X-100 has been seen to go through a maximum (*63*). The adsorption from β–lactoglobulin/SDS mixtures at different degrees of dilution was studied by Wahlgren and Arnebrant (*62*) (Fig. 10). At concentrations above the cmc for the surfactant the amount adsorbed corresponded to a layer of pure surfactant and was found to increase after rinsing. At lower concentrations, the adsorbate prior to rinsing appeared to be a mixture of protein and surfactant and the total amount adsorbed passes a maximum. The composition of the adsorbate after rinsing is most likely pure β–lactoglobulin, as interactions between surface or protein and SDS are reversible.

At high degrees of dilution of the mixture, the absence of surfactant adsorption to the methylated silica, the non-reversible adsorption of β–lactoglobulin and the observed partial desorption of the adsorbate from the mixture, implies that some SDS molecules are bound to the β–lactoglobulin molecules with a higher affinity than to the surface (Fig. 10). The amount of protein adsorbed is larger, even after rinsing, than for adsorption from pure β–lactoglobulin solutions, and it can be concluded that SDS- binding facilitates the adsorption of protein.

Generally, it can be concluded that surfactants may interact through solubilization or replacement mechanisms depending on surfactant- surface interactions and surfactant- protein binding. Solubilization requires complex formation between protein and surfactant, and replacement adsorption of the surfactant to the surface. As previously concluded for protein adsorption, one of the most important protein properties affecting the elutability appear to be the conformational stability. Differences between a competitive situation and addition of surfactant after protein adsorption may be found in the alteration in surface activity of protein- surfactant complexes formed in solution as compared to pure protein, the difference in diffusivity of surfactants and proteins effecting the "race for the interface" and time dependent conformational changes resulting in "residence time" effects.

The effects at low surfactant concentration, below the cmc, involving the range of specific binding to some proteins are not fully understood. Further, the exact prerequisites for solubilization versus replacement as well as detailed information on molecular parameters such as aggregation numbers are not known. Several of the phenomena discussed in this paper resemble those observed in polymer- surfactant systems (*64*). A complete survey of the analogies is however beyond the scope of this

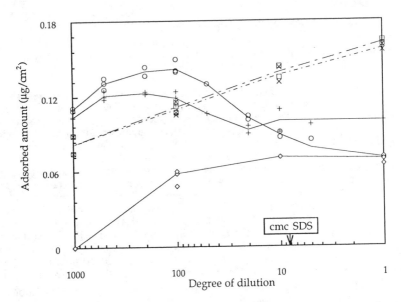

Fig. 10.   The amounts adsorbed to a methylated silica surface as a function of degree of dilution for a mixture of β-lactoglobulin and SDS (1/5 w/w), in phosphate buffered saline pH 7, I=0.17. The figure shows the adsorbed amount after 30 min of adsorption (O) and 30 min. after rinsing (+). In addition, the figure shows the adsorption of pure β-lactoglobulin, after 30 min of adsorption (□) and 30 min after rinsing (x). Finally, the adsorption isotherm of SDS is inserted (A), data from (62).

review. As pointed out in the introduction, improved knowledge concerning surfactant elution will be useful for estimation of protein attachment strength to surfaces, for optimizing detergency processes and avoiding undesired adsorption in medical applications.

Acknowledgement

The authors would like to thank Stefan Welin-Klintström and Joseph McGuire for good collaboration and Hans Elwing, Per Claesson and Eva Blomberg for valuable discussions. We also gratefully acknowledge financial support from Swedish Research Council for Engineering Sciences (TFR).

References

1.   Simonsson, T.; T. Arnebrant; L. Petersson, *Biofouling*. **1991,** *3,* 251-260
2.   Arnebrant, T.; T. Simonsson, *Acta Odontol. Scand.* **1991,** *49,* 281-288
3.   Vassilakos, N.; T. Arnebrant; J. Rundegren; P.-O. Glantz, *Acta Odontol. Scand.* **1992,** *50,* 179-188
4.   Attström, R.; P.-O. Glantz; B. Collaert; T. Simonsson, *Clinical and Biological Aspects of Dentifrises,* In Clinical and Biological Aspects of Dentifrises; G. Embery;Rölla, G. Oxford University: Oxford 1992; pp. 263-275.

5. Bohnert, J.L.; T.A. Horbett, *J. Colloid Interface Sci.* **1986,** *111,* 363-377
6. Rapoza, R.J.; T.A. Horbett, *J. Colloid Interface Sci.* **1990,** *136,* 480-493
7. Rapoza, R.J.; T.A. Horbett, *J. Biomed. Mat. Research.* **1990,** *24,* 1263-1282
8. Horbett, T.A.; J.L. Brash, *Proteins at Interfaces- Physicochemical and Biochemical Studies,* In Proteins at Interfaces- Physicochemical and Biochemical Studies; J.L. Brash;Horbett, T.A. Ed,ACS Symposium Series; American Chemical Society: Washington, DC 1987; 343, pp. 1-37.
9. Horbett T.;*Colloids Surf.* **1994,** in press
10. Horbett, T.;*Cardiovasc Pathol.* **1993,** *2,* 137s-148s
11. Lindman, B.; H. Wennerström, *Topics in Current Chem.* **1980,** *87,* 1-83
12. Scamehorn, J.F.; R.S. Schechter; W.H. Wade, *J. Colloid Interface Sci.* **1982,** *85,* 463-478
13. Fuerstenau, D.W., *Chemistry of Biosurfaces,* In Chemistry of Biosurfaces; D. Fuerstenau W 1971; 1, pp. 143-176.
14. Chander, S.; D.W. Fuerstenau; D. Stigter, *Adsorption from Solution,* In Adsorption from Solution; Otterwill;Rochester;Smith Academic Press: London 1983; pp. 197.
15. Arnebrant, T.; K. Bäckström; B. Jönsson; T. Nylander, *J. Colloid Interface Sci.* **1989,** *128,* 303-312
16. Somasundaran, P.; J.T. Kunjappu, *Colloids Surf.* **1989,** *37,* 245-268
17. Day, R.E.; F.G. Greenwood; G.D. Parfitt; *Proc. 4th Int Cong. on Surface Activity*; 1967; II, 1005.
18. Kronberg, K., *J. Colloid Interface Sci.* **1983,** *96,* 55-68
19. Han, K.N.; T.W. Healy; D.W. Fuerstenau, *J. Colloid Interface Sci.* **1973,** *44,* 407-414
20. Zhu, B.; T. Gu; X. Zhao, *J. Chem. Soc. Faraday Trans. 1.* **1989,** *85,* 3819-3824
21. Denoyel, R.; J. Rouquerol, *J. Colloid Interface Sci.* **1991,** *143,* 555-572
22. Tiberg, F., Ph.D. Dissertation *Self-Assembly of Nonionic Amphiphiles at Solid Surfaces.*; University of Lund, Sweden: 1994;
23. Wängnerud, P., Ph.D. Dissertation *Adsorption of Ionic Surfactants on Charged Solid Surfaces*; University of Lund, Sweden: 1994;
24. Pashley, R., M; J.N. Israelachvili, *Colloids Surf.* **1981,** *2,* 169-187
25. Wahlgren, M., Ph.D. Dissertation *Adsorption of Proteins and Interaction with Surfactants at Solid/Liquid Interfaces*; University of Lund, Sweden: 1992;
26. Tanford, C., *The Hydrophobic Effect: Formation of Micelles and Biological Membranes*; John Wiley & Sons, Inc: New York, 1980;
27. Ananthapadmanabhan, K.P., In *Interactions of Surfactants with Polymers and Proteins*; pp 319-366
28. Nozaki, Y.; J.A. Reynolds; C. Tanford, *J. Biol. Chem.* **1974,** *249,* 4452-4459
29. Subramanian, M.; B.S. Sheshadri; V.M. P., *J. Biosci.* **1986,** *10,* 359-371
30. Few,.A.V.; R.H. Ottewill; H.C. Parreira, *Biochem. Biophys. Acta.* **1955,** *18,* 136-137
31. Steinhardt, J.; N. Stocker; D. Carroll; K.S. Birdi, *Biochemistry.* **1974,** *13,* 4461-4468
32. Jones, M.N.; A. Wilkinson, *Biochem. J.* **1976,** *153,* 713-718
33. Kresheck, G.C.; W.A. Hargraves; D.C. Mann, *J. Phys. Chem.* **1977,** *81,* 532-537
34. Ericsson, B.; P.-O. Hegg, *Prog. Colloid & Polymer Sci.* **1985,** *70,* 92-95
35. Su, Y.-Y., T; B. Jirgensons, *Arch. Biochem. Biophys.* **1977,** *181,* 137-146
36. Nelson, C.A., *J. Biol. Chem.* **1971,** *246,* 3895-3901
37. Reynolds, J.A.; C. Tanford, *J. Biol. Chem.* **1970,** *245,* 5161-5165
38. Shirahama, K.; K. Tsujii; T. Takagi, *J. Biochem.* **1974,** *75,* 309

39. Lundahl, P.; E. Greijer; M. Sandberg; S. Cardell; K.O. Eriksson, *Biochim. Biophys. Acta.* **1986,** *873,* 20
40. Rapoza, R.J.; T.A. Horbett, *J. Biomater. Sci. Polymer Edn.* **1989,** *1,* 99-110
41. Ertel, S.I.; B.D. Ratner; T.A. Horbett, *J. Colloid Interface Sci.* **1991,** *147,* 433-442
42. McGuire, J.; M. Wahlgren; T. Arnebrant, *J. Colloid Interface Sci.* **1994,** in press
43. McGuire, J.; V. Krisdhasima; M. Wahlgren; T. Arnebrant, Separate contribution in this volume
44. Wahlgren, M.C.; M.A. Paulsson; T. Arnebrant, *Colloids Surf A.* **1993,** *70,* 139-149
45. Wahlgren, M.C.; T. Arnebrant, *J. Colloid Interface Sci.* **1991,** *142,* 503-511
46. Wahlgren, M.; T. Arnebrant; A. Askendal; S. Welin-Klintström, *Colloids Surf A.* **1993,** *70,* 151-158
47. Welin-Klintström, S.; A. Askendal; E. Elwing, *J. Colloid Interface Sci.* **1993,** *158,* 188-194
48. Wahlgren, M.; S. Welin-Klintström; T. Arnebrant; A. Askendal; H. Elwing, *Colloids Surf A.* **1994,** in press
49. Elwing, H.; C.-G. Gölander, *Adv. Colloid Interface Sci.* **1990,** *32,* 317-339
50. Gekko, K.; Y. Hasegawa, *Biochemistry.* **1986,** *25,* 6563-6571
51. Mitchell, D.J.; G.J.T. Tiddy; L. Waring; T. Bostock; M. McDonald P, *J. Chem. Soc. Faraday Trans. 1.* **1983,** *79,* 975-1000
52. Tilton, R.D.; E. Blomberg; P.M. Claesson, *Langmuir.* **1993,** *9,* 2102-2108
53. Blomberg, E., Ph.D. Dissertation *Surface Force Studies of Adsorbed Proteins;* Royal Institute of Technology, Sweden: 1993;
54. Elwing, H.; A. Askendal; I. Lundström, *J. Colloid Interface Sci.* **1989,** *128,* 296-300
55. Bäckström, K., Ph.D. Dissertation *Removal of Triglyceride and Protein Films from hard Surfaces by Aqueous Surfactant Solutions, an Ellipsometry Study;* University of Lund, Sweden: 1987;
56. Lindman, B.; S. Engström; K. Bäckström, *J. Surface Sci. Technol.* **1988,** *4,* 23-41
57. Malmsten, M.; B. Lindman, *Langmuir.* **1989,** *5,* 1105-1111
58. Rapoza, R.J.; T.A. Horbett, *Polymer Mater. Sci. Eng.* **1988,** *59,* 249-252
59. Courthaudon, J.-L.; E. Dickinson; D.G. Dalgleish, *J. Colloid Interface Sci.* **1991,** *145,* 390-395
60. Courthaudon, J.-L.; E. Dickinson; Y. Matsumura; D.C. Clark, *Colloids Surf.* **1991,** *56,* 293-300
61. Dickinson, E.; C.M. Woskett, *Food and Colloids,* In Food and Colloids; R.D. Bee Royal Society of Chem.: London 1989; pp. 74-96.
62. Wahlgren, M.; T. Arnebrant, *J. Colloid Interface Sci.* **1992,** *148,* 201-206
63. Slack, S.M.; T.A. Horbett, *J. Colloid Interface Sci.* **1988,** *124,* 535-551
64. Goddard, E.D.; K.P. Ananthapadmanabhan *Interactions of Surfactants with Polymers and Proteins;*, CRC Press: Boca Raton; Florida 1993

RECEIVED March 2, 1995

CONFORMATION AND ORIENTATION OF PROTEINS AT INTERFACES

Chapter 18

Calorimetric Observations of Protein Conformation at Solid–Liquid Interfaces

Guoying Yan, Jenq-Thun Li, Shao-Chie Huang, and Karin D. Caldwell[1]

Center for Biopolymers at Interfaces, University of Utah, Salt Lake City, UT 84112

Proteins of varying stability were adsorbed to polystyrene latex spheres, and the adsorption complexes were examined for structural transitions by means of microcalorimetry. Transition temperatures and enthalpy measurements were compared between given amounts of proteins in free and adsorbed form. From these measurements it appears that under the chosen adsorption conditions, serum albumin, fibrinogen and lysozyme may loose essentially all of their cooperatively folded structure, while proteins of greater structural stability, such as streptavidin, retain full structure and remain biologically active on the latex surface. Certain hydrogels of use as contact lens materials are known to adsorb lysozyme. From calorimetric observations on typical lens loads of lysozyme the enzyme appears to retain about half of its folded structure on the lens, although in a destabilized form to judge from the six degree reduction in transition temperature.

The adsorption of proteins and other highly structured macromolecules to solid surfaces frequently results in conformational changes within the adsorbing structure. Such changes are typically demonstrated by spectroscopic means, whereby the emission characteristics of some indigenous or artificially attached reporter group is monitored before and after the attachment. Although local rearrangements can be highlighted this way, the measurements provide no basis for an assessment of the overall structure of the protein on the surface.

Microcalorimetry has in recent years been developed to the point where accurate measurements of transition enthalpies can be obtained for milligram quantities of protein. Although several groups have contributed to this development (1-4), it is primarily Privalov and coworkers (5-8) who have led the way in refining this technique by showing that one can now reproducibly determine transition enthalpies

---

[1]Corresponding author

0097–6156/95/0602–0256$12.00/0

(ΔH) and transition temperatures for any protein of interest, provided it undergoes a significant cooperative unfolding in the temperature interval (typically 0-100°C) in which the solvent is free from phase transitions.

From the transition temperature (T_m), the width of the transition, and the area under the heat capacity curve, one can assign a relative stability to the selected protein in a particular solvent environment. In one of the earliest demonstrations of this point (5), heat capacity (C_p) curves were determined for hen lysozyme in buffers of different pH. Under the most acidic condition investigated (pH 2.0), the structural collapse of the protein occurred at the lowest temperature ($T_m = 56$°C) and the observed transition was relatively wide, indicative of a weakened cooperativity in the melting process. More importantly, the area under the peak, or ΔH, was the smallest, revealing a less structured form of the protein in this environment. As pH values increased, so did T_m. The transitions, meanwhile, showed signs of stronger cooperativity as they took place over narrower temperature intervals and required higher energy inputs to materialize.

Just as solvent environments may impact the structure of proteins, so may surfaces with which they come in contact. The thermodynamics of protein-surface interaction has been discussed extensively by Norde and coworkers (9-11). As a result of their detailed cataloguing of the fates of several carefully selected proteins, brought into contact with different surfaces under different conditions of pH and therefore charge, these authors conclude that adsorption, or lack thereof, is primarily a reflection of the stability of the protein. While proteins and surfaces of opposite charge always interact adsorptively, they found the converse not to be universally true. Although proteins, such as ribonuclease and lysozyme, tended to show low or no adsorption to hydrophilic surfaces with an overall charge of the same sign as the protein, there were other proteins, notably human serum albumin (HSA) and α-lactalbumin, that adsorbed even to surfaces from which they were coulombically repelled (9). The driving force for adsorption in this case is likely not enthalpic in origin, but rather of entropic nature and resulting, as far as concerns the protein, from its gain in configurational entropy upon unfolding. Based on these considerations, Norde introduced the terminology of "hard" and "soft" proteins to describe respectively those repelled by, and adsorbing to, hydrophilic surfaces of similar charge.

Despite the insight gained from calorimetric studies regarding protein stability, particularly as it is affected by different solvent environments, calorimetry has only rarely been used to analyze the structural contents of proteins on solid surfaces. In recent work (12), Middaugh and coworkers examined the structural condition of several proteins adsorbed to various modified silica particles, originally produced to serve as chromatographic supports. Despite its recognized "hardness", these authors found the enzyme lysozyme to be considerably destabilized, particularly on the hydrophobic surfaces. The observed loss of structure was accompanied by a loss of enzymatic activity. Clearly, the notion of lysozyme as a very hard protein must be somewhat modified. The same conclusion is drawn by Haynes and Norde in their contribution to this volume (13), in which they detail their recent calorimetric analyses of adsorption complexes involving this enzyme.

In the present study we will revisit the "hardness" question by examining the relationship between protein solution stability and structural retention upon adsorption for a select group of proteins and substrates of biomaterials and biotechnological importance.

Experimental

Materials. Polystyrene (PS) particles were purchased from Seradyn. Particles with diameters of 272 nm, 261 nm and 90 nm were used as indicated. The proteins Lysozyme, Bovine Fibrinogen (BFB), and Streptavidin (SA) were from Sigma Chemical Company, while HSA was obtained from Calbiochem Corporation. The buffer chemicals Tris, L-glycine, and phosphate buffered saline (PBS, pellets) were obtained from Sigma Chemical Co., while Na_2HPO_4 and KH_2PO_4 were purchased from Malinckrodt Specialty Chemicals, Inc. Vifilcon contact lenses were a gift from Ciba Vision Corporation.

Methods

Protein Adsorption to PS Latex. Lysozyme and HSA solutions were prepared with concentrations of 2 and 8 mg/mL, respectively, in PBS, pH 7.4. To 1 mL of protein solution was added 0.5 mL of a latex suspension containing 10 % (w/v) of solids. The mixture was incubated for 24 hours at room temperature, and the protein-particle mixture was then spun down in an Eppendorf table centrifuge (14,000 rpm), separated from the supernatant, and resuspended in PBS. This wash procedure was repeated three times, and the particles were then suspended in 700 μL PBS. Of this suspension, 600 μL was transferred to the calorimeter ampoule and 100 μL was submitted to amino acid analysis.

Solutions of streptavidin (2 mg/mL in deionized water) were prepared, and 300 μL aliquots were mixed with 58 μL of PS-261 latex particles in an Eppendorf tube. This mixture was determined to generate a monolayer surface concentration of SA (4 mg/m^2). The adsorption reaction was allowed to continue for 1 hour at 25°C with occasional vortexing. After separating the protein-coated particles from the supernatant, the adsorption complexes were washed with PBS followed by centrifugation; this procedure was repeated three times. Each sample was analyzed in duplicate.

Formation of the protein-particle complex for the titration experiments followed a slightly different protocol. Here, stock protein solutions of HSA (4.5 mg/mL in 10 mM phosphate buffer, pH 7.0) and BFB (2.5 mg/mL in 50 mM glycine, pH 8.5) were prepared and pipetted in 500 μL aliquots into the measuring ampoules. To the ampoules were then added different amounts of a 10% (w/v) particle suspension, followed by additional buffer to give a total sample weight of 600 μL. After sealing, the ampoules were left gently shaking on a shaker table for one hour prior to the calorimetric analysis.

Protein Adsorption to Vifilcon Lens. One mL portions of lysozyme solution, containing 2 mg protein/mL of 50 mM Tris buffer, pH 7.6, were incubated with three virgin Vifilcon lenses at room temperature. The whole lenses were immersed in the solution. After 24 hours, the protein-lens complexes were removed and washed carefully three times in Tris buffer. Calorimetry was then performed on two of the lenses (both in the same ampoule), while the third was submitted to amino acid analysis for quantification.

Calorimetry. Differential scanning calorimetry (DSC) was performed using a Hart Model 4207 instrument. The thermal characteristics of each one of its 1.2 mL ampoules were determined separately, using the appropriate buffer. In each case, the amount of sample subjected to analysis was determined by careful weighing of each ampoule before and after sample loading. The following steps constitute the operating protocol adhered to in this work: (a) scan down from 20°C to 15°C and stabilize the instrument at this temperature for 10 minutes; (b) scan up to 105°C at a rate of 1 degree per minute; (c) scan down to 20°C at a rate of 1.5 degrees per minute. All thermograms were background corrected by subtraction of either the buffer heat capacity or that of the buffer plus solid substrate, as appropriate. All runs were made at least in duplicates.

Protein Quantification. Whether in soluble or adsorbed form, the various proteins were quantified by means of an amino acid analysis procedure described in detail elsewhere (14). Following exhaustive hydrolysis in 6 M HCl for 20 hours at 105°C, the samples were evaporated to dryness and derivatized with phenyl isothiocyanate. The resulting mixtures of labeled amino acids were separated by reversed phase chromatography on an HP-1050 system from Hewlett-Packard, and the cumulative amount of amino acids was found by integration of the chromatogram.

For the calorimetric titration experiments the protein surface concentrations in the adsorption complexes were determined in separate batch adsorption experiments. Here the amount of protein adsorbed was mainly determined spectroscopically at 280 nm from the level of depletion in the supernatant with occasional verification by amino acid analysis.

Results and Discussion

Adsorption and Stability. A protein's adsorption from aqueous solution onto a hydrophobic solid surface is likely to qualitatively resemble its adsorption at the similarly hydrophobic air-water interface. In a model study involving six small, one-domain proteins, Wei (15) sought correlations between, on the one hand, their thermal stability and their stability towards denaturants such as urea and guanidinium hydrochloride (GuHCl) and, on the other hand, their observed tendencies to adsorb at the air-water interface from dilute solution and to unfold and spread at this interface. Not unexpectedly (16), there were clear correlations between calorimetric transition temperatures, T_m, and the concentrations of GuHCl bringing about unfolding of half the molecular population. In addition, lower T_m-values and $[GuHCl]_{1/2}$ -concentrations both correlated with the protein's tendency to unfold and spread at the air-water interface, to judge from the rate of change in surface tension.

It seems, therefore, that a case can be made for utilizing $[GuHCl]_{1/2}$ as a measure of the "hardness" of a protein in the Norde sense, at least for single domain proteins. The case of multi-domain proteins is less clear-cut, since the different domains may unfold uncooperatively. Which transition midpoint one observes is then entirely dependent on the method of detection. If, as is often the case, a spectroscopic method is selected, the overall protein stability will be judged by the unfolding of that particular domain in which the monitored chromophor resides. Mindful of these uncertainties, we will still attempt a hardness-ranking of the proteins whose interfacial behavior has been the focus of this study. Excluded from this ranking is bovine fibrinogen (BFB), due to its highly complex, multi-domain structure.

Beginning with the indisputably "soft", three-domain HSA molecule, this protein has the lowest reported $[GuHCl]_{1/2}$ of the set; the transition midpoint for the bovine analogue is reported to occur at a concentration of 1.6 M (17). The hen lysozyme, considered by Norde to be a "hard" protein (9), withstands significantly higher denaturant concentrations and unfolds at a $[GuHCl]_{1/2}$ of 4.3 M (15). Yet, this hardness appears modest in comparison with that seen for streptavidin (SA), a protein frequently used to immobilize ligands that have been selectively substituted with the small vitamin biotin, which it binds with an unusually high affinity ($K_a = 10^{15}$ M^{-1}). In our laboratory, SA has shown to remain folded at neutral pH even in 7 M GuHSCN (K. Dobaj and M. Adayanthoya, personal communication), a more potent denaturant than the chloride analogue due to the chaotropic character of the SCN-ion (18). Wilchek and coworkers (19) examined the ability of urea as well as of GuHSCN to unfold SA, and found no effect of a 6 M solution of the former, while a slow unfolding took place at that concentration of the latter, supporting the notion of a highly stable protein.

Calorimetry of Preformed Adsorption Complexes. In order to use microcalorimetry to evaluate the structural content of an adsorbed protein, two conditions must be met. First, the substrate must be sufficiently limited in volume to be accommodated by the 1.2 mL measuring cell, yet it should present enough surface area to allow measurements on about a mg of adsorbed protein. Second, the substrate should itself be free of significant thermal transitions in the temperature interval of interest.

Polystyrene (PS) nanoparticles are available in a variety of precisely controlled sizes. They are frequently used for diagnostic purposes in which case they may, for instance, be coated, either through adsorption (20) or through covalent attachment (21), with a specific IgG which causes the particles to aggregate measurably in response to the presence of antigen. Although these particles carry a negative charge (22) as a result of residual surfactant molecules trapped during the polymerization process, they are overwhelmingly hydrophobic as seen from their tendency to form essentially irreversible complexes in water with an array of uncharged polymers (23). Due to their regular shape and well defined size, these particles are easily quantified in suspension by a simple turbidity measurement (24), and the surface area available for protein adsorption is therefore easily determined. Fortuitously, this substrate shows no thermal transitions in the temperature interval of interest for protein calorimetry, as seen in Figure 1.

The protein HSA adsorbs rapidly and pseudo-irreversibly to PS latex particles. Following adsorption, the formed protein-particle complex can be extensively washed with protein-free buffer and characterized in terms of its residual protein load, as described in the Experimental section above. Given aliquots of this complex are then readily transferred to the measuring cell of the calorimeter for examination of possible structural changes in the protein that may have resulted from the adsorption. Figure 2 illustrates typical thermograms of comparable amounts of free and adsorbed HSA. From the striking absence of heat capacity features in the adsorption complex it is clear that all structure in the native protein which could be collapsed through the cooperative melting process had already been eliminated from the adsorbed form as a result of its interaction with the substrate. The notion of HSA as a "soft" protein is therefore fully supported by the calorimetric data.

The "hard" protein hen-egg lysozyme has been described (15) to unfold at an intermediate GuHCl concentration of 4.3 M, as discussed above. Whether or not this relatively high stability also manifests itself in the form of a substantial retention of structure upon adsorption to PS latex can be judged from Figure 3. Indeed, under the adsorption conditions used in this figure, also lysozyme appears to unfold extensively as it adapts to the PS surface. The protein surface concentration in this case is 1.74 mg/m^2, which is close to monolayer coverage. Since, as a rule, retention of biological activity will increase with increased protein close-packing (25), it is entirely possible that adsorption conditions may be found which allow the lysozyme to retain some amount of structure. In fact, the companion chapter by Haynes and Norde (13) demonstrates that under certain circumstances quantifyable (albeit small) amounts of structure are retained in lysozyme adsorbed to PS latex.

Although the behavior of lysozyme on PS may be of interest from a general protein science perspective, it is of limited practical significance. The contrary is true for the interaction of this enzyme with polymers of biomaterials interest, such as those

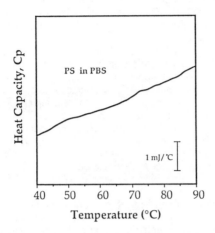

Figure 1. A typical thermogram of 42.86 mg PS latex spheres in PBS buffer, pH 7.4.

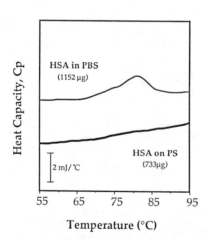

Figure 2. Thermal stability of HSA in free and adsorbed form. Amounts (from amino acid analysis) are indicated in the figure.

employed in contact lens manufacturing. Human lysozyme is the most abundant protein in tears (26); it adsorbs rapidly and extensively to hydrogels of FDA type IV, i.e., highly water-containing anionic polymer matrices (27). Despite the fact that human and hen-egg lysozyme are different molecular species, the latter, which is more readily available, is generally used in model tear experiments due to its similarities with the human analogue. In particular, the two proteins are comparable in size as well as in charge, and due to their basic nature they both bind coulombically to the negative type IV lens. As with any biomaterial, questions are continuously being asked regarding what influence the lens matrix might have upon the proteins with which it comes in contact.

In case of lysozyme, with its high adsorptivity to type IV materials, enough protein can be adsorbed by a lens to allow an *in situ* examination by microcalorimetry. This is illustrated in Figure 4 which compares hen egg lysozyme in solution with lysozyme pre-adsorbed to a couple of Vifilcon contact lenses. That amount of polymeric material is easily accommodated in the measuring cell and its protein load gives a readily measurable signal, as seen in the figure. Clearly, adsorption of the enzyme to this hydrogel occurs with less drastic consequences for the structure than in the case of PS. Although there are obvious signs of structural destabilization, indicated by a 6°C reduction in the transition temperature, a broadening of the heat capacity peak, and a reduction in the transition enthalpy, the effect is qualitatively similar to that seen by Privalov in his early work on pH induced destabilization of lysozyme (5).

In view of the destructive effect of PS surfaces on HSA and lysozyme, demonstrated in Figures 2 and 3, one may ask whether this substrate is at all suitable for adsorption of proteins whose biological function is to be maintained after immobilization. In earlier work on the adsorption of milk proteins to PS latex, we were able to calorimetrically detect a significant amount of structure in a β-lactoglobulin-particle complex (28). However, a more affirmative answer to this question is given by Figure 5, which shows thermograms for comparable amounts of the protein streptavidin in free and adsorbed form. Here, the presence of the substrate appears to have had little effect on the protein structure, to judge from T_m of the major transition as well as from the enthalpy of unfolding which both remain essentially unchanged by the adsorption. (It should be noted that we have seen some batch-to-batch differences in the transition behavior of SA, both in solution and in its adsorption complex with PS latex). In a separate study (24), the adsorbed streptavidin has been found to bind 0.4 moles per mole of its natural, high affinity ligand biotin. The reduced binding capacity, which is one tenth of that found in solution, is thought to result primarily from steric hindrance rather than from a significant structural change associated with the adsorption.

A Titration Approach to Calorimetry. The results described in the previous section were all obtained with protein-polymer complexes that had been allowed to form over extended periods of time (see Experimental) and had then been carefully washed to remove unadsorbed and loosely associated protein. Below, we will describe a somewhat different experimental design (29), developed to be less material consuming than the batch analysis described above, while allowing for the simultaneous examination of all protein in the cell, whether in solution, on the surface, or in a loose association with the complex.

The calorimetric titration experiment requires the availability of at least six well characterized measuring cells. Each ampoule is made to contain identical amounts of protein, delivered from a stock solution, to which is added different, well characterized amounts of particles. In this way, one can readily vary the ratio of protein to substrate surface area. The thermogram for each ampoule is then a

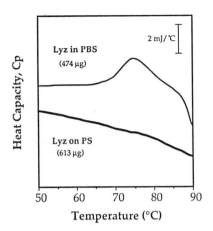

Figure 3. Thermal stability of lysozyme in free and adsorbed form. The amounts of protein (from amino acid analysis) are indicated in the figure.

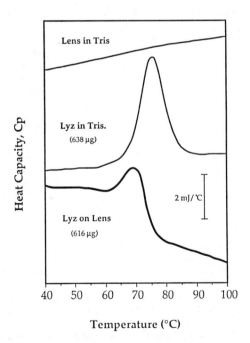

Figure 4. Lysozyme in free form and adsorbed to Vifilcon contact lens. Two lenses were accommodated in the cell; their combined protein load (from amino acid analysis) is indicated in the figure. Also indicated is the thermogram for two virgin lenses.

composite of the features displayed by all different forms of the protein in that cell.
The actual amounts of protein adsorbed in each case is estimated from a series of
separate batch experiments.

   This procedure is illustrated in Figure 6 for the adsorption of HSA to a small
(90 nm), high surface area PS latex. Here, the relative amounts of retained structure,
computed as the ratio of observed transition enthalpies for the protein-particle mixture
and the protein by itself, respectively, are plotted as a function of the particle:protein
ratio. One may examine the fractional loss, j, of structure upon adsorption by
expressing the observed transition enthalpy $\Delta H_{cal(particle-protein)}$ as a function of
the enthalpy for the protein itself, $\Delta H_{cal\ (protein)}$, the surface concentration Γ
generated in adsorption from that concentration, and the amount of surface area
available for adsorption. The latter is given by the product of specific surface area A
(m^2/mg particles), and mass m of particles added to the cell. Cast in terms of these
parameters, the loss factor j is easily identified:

$$\Delta H_{cal(part.-prot.)}\ /\Delta H_{cal\ (prot.)} = 1 - j\ \Gamma\ A\ m_{(part.)}/m_{(prot.)}$$

Here, a j-value of zero would imply no loss of structure upon adsorption, while a
value of unity would indicate a total loss of transition enthalpy by that portion of the
protein sample which is adsorbed to the particles. In addition to the measured data for
the relative retention of structure in the presence of different amounts of particles,
Figure 6 also includes a dotted line representing a j-value of unity under the prevailing
conditions of A = 63 m^2/1000 mg for PS 90 nm and Γ = 1.29 mg HSA/m^2. From the
thermograms of Figure 2, which showed a total loss of structure for HSA
preadsorbed to PS 272 nm, one could reasonably assume that the experimental data in
Figure 6 would fall close to the j = 1 line. Yet, they all lie significantly below this line
suggesting j-values in excess of 1. This finding, which is indicative of a more
extensive destabilization of the protein than just the complete unfolding of that portion
adsorbed to the surface, must be examined further. If supported by other
observations, it could mean that a certain fraction of the HSA molecules has
undergone a destabilizing collision with the substrate without irreversibly adhering to
it. Recent work by Norde (11) has shown that HSA, which has been brought back
into solution after an adsorption event and re-equilibrated with the starting buffer, has
a different helix content and adsorbs more readily than the starting material.
Presumably, once destabilized, this protein refolds only with great difficulty.

   The titration approach to microcalorimetry was also applied to bovine
fibrinogen (BFB), a highly complex multidomain protein. Its overall structure and
domain assignment was developed by Privalov and Medved (7), in part based on an
elaborate series of enzymatic digestions and calorimetric analyses of the fragments. A
representative thermogram of the intact molecule is shown in Figure 7. The two
transitions that are clearly distinguished were identified by Privalov and Medved as
the D (low melting) and E (high melting) domains.

   It was our original belief that the adsorption of fibrinogen to a surface of e.g.,
polystyrene would lead to a destabilization of one or the other of these domains. This
belief is not supported by the experimental observations summarized in Figure 8.
Instead of seeing a preferential loss of structure in one domain or the other, we find
both to have been reduced in the adsorption. In fact, the dotted line in Figure 8A,
commensurate with a j-value of unity and thus representing a 100% loss of structure
upon adsorption, appears to fit the experimental data remarkably well. This line was
computed based on a surface concentration Γ of 3.64 mg BFB/m^2; the value of A
remained the same as in Figure 6.

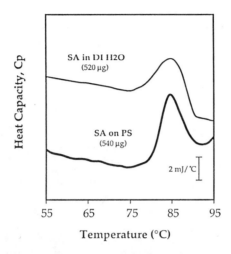

Figure 5. Streptavidin in solution (DI water) and adsorbed to PS 261 nm particles. The slight shift in T_m between the free and adsorbed form appeared to be batch specific for the protein, and was not observed in duplicate experiments, whereas the high temperature transition, seen for the adsorption complex, was a highly reproducible phenomenon.

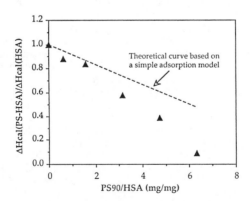

Figure 6. Relative retention of HSA structure upon addition of PS 90 nm particles. The surface concentration Γ of HSA at low particle load was 1.29 mg/m^2, and the particle area A available for adsorption was 63 m^2/g of solids. Initial protein concentration was 4.5 mg/mL in 10 mM phosphate, pH 7.0.

Bovine Fibrinogen in 50 mM Glycine Buffer (pH 8.5)		
	First transition	Second transition
Tm (°C)	54	96
ΔHcal (Kcal/mole)	805	318

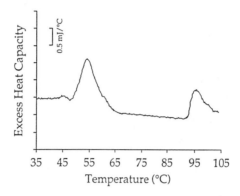

Figure 7. Thermogram of BFB in 50 mM glycine buffer, pH 8.5. The low T_m is associated with unfolding of the flanking D-fragments of the molecule (ref. 7), while the high T_m reflects unfolding of the central E-fragment.

Conclusion

As a complement to the more commonly used spectroscopic techniques for analysing protein conformations on surfaces, microcalorimetry has been found to offer a unique insight into the overall structural conditions of proteins in their association with solid substrates. The technique is tolerant to both the geometry and composition of the substrate, as long as it is free of major structural shifts in the temperature interval of interest, and as long as its surface-to-volume ratio is such that about a mg of adsorbed protein can be accommodated in the 1.2 mL measuring ampoule.

The calorimetric observations reported here suggest that certain "soft" proteins, so classified because of their unfolding in denaturant solutions of low to moderate concentration, undergo a more extensive loss of structure upon adsorption to hydrophobic surfaces than previously thought. This is seen to be true for small proteins, e.g., lysozyme that was previously thought of as moderately robust, as well as for the larger multidomain plasma proteins albumin and fibrinogen. In the present study we have focused our attention on one protein, streptavidin, which shows extreme "hardness", as well in its stability towards denaturants as in its retention of structure and biological activity upon adsorption to polystyrene beads. Other proteins, such as the various IgG species known to retain their antigen binding ability and to function well in a variety of particle based immunodiagnostic reagents, are likely to show similar calorimetric stability features.

Acknowledgments

This work has enjoyed support in part by NIH grant GM 38008-05 and in part by the Center for Biopolymers at Interfaces, University of Utah. GY gratefully acknowledges a fellowship from the Ciba Vision Corporation.

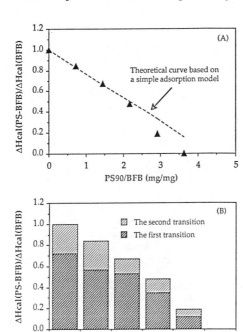

Figure 8. (A) Relative retention of BFB structure upon addition of PS 90 nm latex particles. The surface concentration of protein was determined separately as 3.64 mg/m^2, while particle area A was the same as in Figure 6. The initial protein concentration was 2.5 mg/mL of 50 mM glycine buffer, pH 8.5.
(B) The relative reduction of structure in the two peaks.

Literature Cited

1. Wadsö, I. In *New Techniques in Biophysics and Cell Biology*; Pain, R. H.; Smith, B. J., Eds.; John Wiley: New York, NY, 1974, Vol. 2; pp 85-126.
2. Jackson, W. M.; Brandts, J. F. *Biochemistry* **1970**, *9*, 2294.
3. Sturtevant, J. *Proc. Natl. Acad. Sci. USA* **1977**, *74*, 2236.
4. Biltonen, R.; Freire, E. *CRC Crit. Rev. Biochem.* **1978**, *5*, 85.
5. Privalov, P. L.; Khechinashvili, N. N. *J. Mol. Biol.* **1974**, *86*, 665.
6. Privalov, P. L. *Adv. Protein Chem.* **1979**, *33*, 167.
7. Privalov, P. L.; Medved, L. V. *J. Mol. Biol.* **1982**, *159*, 665.
8. Pfeil, W.; Privalov, P. L. *Biophys. Chem.* **1976**, *4*, 23.
9. Norde, W. *Clinical Materials* **1992**, *11*, 85.
10. Elgersma, A. V.; Zsom, R. L. J.; Norde, W.; Lyklema, J. *Colloids and Surfaces* **1991**, *54*, 89.
11. Norde, W.; Anusiem, A. C. I. *Colloids and Surfaces* **1992**, *66*, 73.
12. Steadman, B. L.; Thompson, K. C.; Middaugh, C. R.; Matsuno, K.; Vrona, S.; Lawson, E. Q.; Lewis, R. V. *Biotech. Bioeng.* **1992**, *40*, 8.
13. Haynes, C.; Norde, W., presented at Proteins at Interfaces, 1994, 207th ACS National Meeting, San Diego, CA.

14.   Yan, G.; Nyquist, G.; Caldwell, K. D.; Payor, R.; McCraw, E. *Inv. Ophthalmol. Vis. Sci.* **1993**, *34*, 1804.
15.   Wei, Ai-Ping *Surface Tension Kinetics of Model Proteins in Relation to their Structural Properties*; M.S. Thesis, University of Utah: Salt Lake City, UT, 1991.
16.   Creighton, T. E. *Proteins: Structures and Molecular Properties, 2nd Ed.*; W. H. Freeman and Company: New York, NY, 1993; pp 297.
17.   Saito, Y.; Wada, A. *Biopolymers* **1983**, *22*, 2123.
18.   Von Hippel, P. H.; Wong, K. Y. *J. Biol. Chem.* **1965**, *240*, 3909.
19.   Kurzban, G. P.; Bayer, E. A.; Wilchek, M.; Horowitz, P. M. *J. Biol. Chem.* **1991**, *266*, 14470.
20.   Fair, B. D.; Jamieson, A. M. *J. Colloid Interface Sci.* **1980**, *77*, 525.
21.   Quash, G.; Roch, A.; Niveleau, A.; Grange, J. *J. Immunol. Meth.* **1978**, *22*, 165.
22.   Prescott, J. H.; Shiau, S.-J.; Rowell, R. L. *Langmuir* **1993**, *9*, 2071.
23.   Li, J.-T.; Caldwell, K. D.; Rapoport, N. *Langmuir*, in press.
24.   Huang, S.-C.; Swerdlow, H.; Caldwell, K. D. *Anal. Biochem.*, submitted.
25.   Lewis, D.; Whatley, T. L. *Biomaterials* **1988**, *9*, 71.
26.   Berman, E. R. In *Biochemistry of the Eye*; Plenum Press: New York, NY, 1991; pp 76.
27.   Goldenberg, M. S.; Beekman, A. C. *Biomaterials* **1991**, *12*, 267.
28.   Caldwell, K. D.; Li, J.-M.; Li, J.-T.; Dalgleish, D. G. *J. Chromatogr.* **1992**, *604*, 63.
29.   Li, J.-T. *Plasma Protein Interactions with Copolymer Stabilized Colloids*; Ph.D. Dissertation, University of Utah: Salt Lake City, UT, 1993.

RECEIVED December 22, 1994

Chapter 19

Molecular Orientation in Adsorbed Cytochrome *c* Films by Planar Waveguide Linear Dichroism

John E. Lee and S. Scott Saavedra[1]

Department of Chemistry, University of Arizona, Tucson, AZ 85721

The heterogeneous distribution of amino acids present on the surface of a protein suggests that adsorption to materials of differing surface chemistry will produce different bound orientations. Investigating relationships between substrate surface chemistry and molecular orientation in hydrated protein films is hampered by a lack of suitable experimental techniques. Measuring absorption linear dichroism in a chromophore-containing protein film using planar waveguide ATR is one promising approach. This paper presents a theoretical and experimental description of this novel technique, and reports on measurements of molecular orientation in cytochrome *c* films adsorbed at hydrophilic and hydrophobic glass/buffer interfaces.

Non-specific protein adsorption occurs to some extent on virtually all insoluble materials of synthetic origin that are brought into contact with biological fluids (*1-4*). In comparison to specific binding between a receptor and its ligand, non-specific adsorption is generally thought to be a less localized interaction involving a larger fraction of a protein's amino acids. The larger contact region may be due to the multiple modes of binding that are possible between a substrate surface and the heterogeneous distribution of amino acids on the protein's external surface. If the conformational state of the protein is altered upon adsorption, formerly buried amino acids may be involved in the binding, further contributing to its multivalent nature. Furthermore, if the substrate surface itself is chemically microheterogeneous, the subset of amino acids involved in binding will vary substantially among individual protein molecules. Acting in concert, these effects may produce a film in which the geometric orientation of adsorbed molecules is isotropic with respect to the substrate surface.

However, in situations where a relatively high affinity binding mode is favored over competing modes, and the binding site on the protein is localized (which implies an asymmetric distribution of surface amino acids), a preferential orientation of adsorbed molecules may be produced. An example is provided by the work of Darst et al. (*5*). Using electron crystallography, they showed that *E. Coli* RNA polymerase holoenzyme forms two-dimensional, crystalline domains when electrostatically adsorbed to lipid layers at the air/water interface. The lipids lacked any specific ligands for binding the protein. The authors postulated that the negatively charged protein molecules "adsorb on the positively charged lipid surface in a predominant

[1]Corresponding author

0097–6156/95/0602–0269$12.00/0

orientation rather than at random, (and that) one region on the surface of the polymerase molecule may contain a relatively high local concentration of negatively charged amino acid residues."

Variations in surface coverage and bioactivity for a protein adsorbed on different surfaces have been observed by numerous other groups, and have been frequently attributed to differences in adsorbed molecular orientation. For example:

1) Lin et. al. (*6*) examined the specific activity of antibodies adsorbed to dichlorodimethylsilane (DDS)-treated glass. Partially denaturing the antibody prior to adsorption yielded a higher activity in the adsorbed state in comparison to adsorption of native antibody. The authors reasoned that partial denaturation exposed a hydrophobic patch of amino acids in the protein's Fc region. Hydrophobically driven adsorption produced an antibody layer with the antigenic binding sites preferentially oriented away from the DDS surface.

2) Lee and Belfort (*7*) reported that the enzymatic activity of ribonuclease A adsorbed to bare mica increased over a 24 hour period from 16% to 74% (relative to a solution phase activity of unity). Intermolecular force measurements between two adsorbed RNase A films showed that the rise in activity was accompanied by an increase in film thickness from 28 Å to 44 Å and an increase in adsorption density of 41%. The results were interpreted as follows: RNase A is initially adsorbed "side-on" with the positively charged active site facing the negatively charged mica surface. At long incubation times, the molecules reorient to an "end-on" geometry, which exposes the active site and reduces the mica surface area/molecule, yielding increased activity and adsorption density. The "end-on" orientation is likely stabilized by lateral electrostatic interactions between the RNase A molecules.

Although these studies support the premise that non-specific protein adsorption is not a geometrically random process, in contrast to Darst's work (*5*), orientation was not directly measured. They also suggest that adsorbed orientation is a function of adsorbent surface chemistry and can strongly influence biochemical function. A prerequisite to investigating these relationships is the capability to measure molecular orientation in submonolayer to monolayer hydrated protein films that are not crystalline (i.e., range from liquid crystalline to disordered). Polarized absorption techniques are an option but their application to monolayer protein films can be problematic from a sensitivity standpoint. Bohn and coworkers have addressed this difficulty by using a 150 μm thick glass coverslip to measure linear dichroism in an attenuated total reflection (ATR) geometry (*8,9*). This relatively thin internal reflection element supports a large number of internal reflections per unit length of beam propagation, which produces an enhanced optical pathlength. Using this approach, they determined mean molecular orientation in dried films of cytochrome b_5 mutants covalently attached to coverslips via disulfide bonding (*10*). An even greater sensitivity enhancement can be realized in the integrated optical limit (*8*). We have previously reported the use of planar integrated optical waveguide (IOW) spectrometry to measure attenuated total reflection in protein films at IOW/liquid interfaces (*11-14*). A theoretical and experimental description of the IOW-ATR technique and its application to molecular orientation measurements in hydrated heme protein films is provided here. Some preliminary results were reported previously (*14,15*).

Theory

We consider the case where a protein film of thickness d_f is formed by adsorption at the interface between a dielectric internal reflection element (IRE) and an aqueous solution containing dissolved protein. The refractive indices of the film, solution, and IRE are n_f, n_1, and n_2, respectively. The protein contains a bound chromophore that absorbs light at wavelength λ.

ATR Spectrometry of Bulk Samples and Thin Films. In Figure 1, light propagating in the IRE at angle θ_i is totally internally reflected at the waveguide/solution interface, which is the x-y plane. The chromophore concentration along the IRE surface normal (z-axis) is represented by a step function: *i)* in the film, which extends from $z=0$ to d_f, the concentration is c_f, and *ii)* in the bulk solution, which extends from $z=d_f$ to infinity, the concentration is c_b. When the absorption of evanescent energy per reflection in both the film and the bulk are relatively weak, the total absorption per reflection, A_t/N, is given by

$$A_t/N = A_b/N + A_f/N \tag{1a}$$

$$= \varepsilon_b c_b L_b + \varepsilon_f c_f L_f \tag{1b}$$

where A_b/N and A_f/N are the bulk and thin film absorbances per reflection, respectively, N is the total number of reflections at the IRE/solution interface, and ε_b and ε_f are the molar absorptivities for the bulk and thin film chromophores, respectively *(13)*. The terms L_b and L_f are the evanescent pathlengths in the bulk and thin film samples, respectively, and are given by

$$L_b = (I_e/I_i\,)d_p \tag{2}$$

$$L_f = (I_e/I_i\,)d_f \tag{3}$$

where d_p is the penetration depth of the evanescent wave, given by

$$d_p = (\lambda/4\pi n_2)[\sin^2\theta_i - (n_1/n_2)^2\,]^{-1/2} \tag{4}$$

and I_e/I_i is the interfacial transmitted intensity per unit incident beam intensity *(13)*. For transverse electric (TE) and transverse magnetic (TM) polarized light, I_e/I_i is given by

$$(I_e/I_i)_{TE} = \frac{n_1\,|E_y|^2}{n_2\,\cos\theta_i} \tag{5}$$

$$(I_e/I_i)_{TM} = \frac{n_1\left[\,|E_x|^2 + |E_z|^2\right]}{n_2\,\cos\theta_i} \tag{6}$$

where E_x, E_y, and E_z are the orthogonal components of the electric field vector of the evanescent wave, **E**, that extends into the solution phase.

$$|E_x|^2 = \frac{4\cos^2\theta_i\left(\sin^2\theta_i - (n_1/n_2)^2\right)}{\left(1 - (n_1/n_2)^2\right)\left[\left(1 + (n_1/n_2)^2\right)\sin^2\theta_i - (n_1/n_2)^2\right]} \tag{7}$$

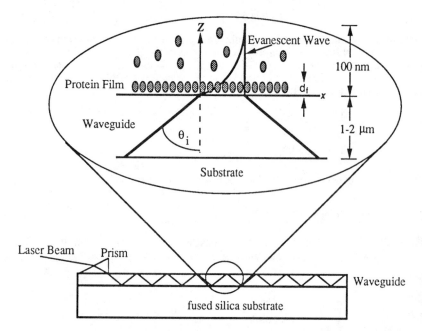

Figure 1. Schematic illustration of light propagating in an IRE at an internal waveguiding angle θ_i. The chromophore concentration along the IRE surface normal (z-axis) is represented by a step function. The drawing is not to scale. The film thickness, d_f, is less than 10% of the evanescent penetration depth. The lateral displacement of the beam along the x-axis upon reflection is the Goos-Hänchen shift.

$$|E_y|^2 = \frac{4\cos^2\theta_i}{1 - (n_1/n_2)^2} \tag{8}$$

$$|E_z|^2 = \frac{4\cos^2\theta_i \sin^2\theta_i}{\left(1 - (n_1/n_2)^2\right)\left[\left(1 + (n_1/n_2)^2\right)\sin^2\theta_i - (n_1/n_2)^2\right]} \tag{9}$$

As discussed in detail elsewhere (*16,17*), equations 1-6 are subject to some assumptions: *i)* the film thickness is less than 10% of the penetration depth (i.e., $d_f <$ 0.1 d_p), which is valid for a monolayer of small protein molecules; *ii)* the term I_e/I_i is not affected by the presence of the film, which is valid for a film thinner than 0.1 d_p with n_f that differs from n_1 by only a few percent.

ATR Linear Dichroism of a Heme Protein. Figure 2 diagrams the geometry of a linear dichroic ATR experiment in which the orientation of a protein-bound iron porphyrin at the IRE surface is determined. The general theoretical treatment and the notation of Thompson et al. (*18*) and Fraaije et al. (*19*) are used here. The iron porphyrin (heme) has two electric dipole transitions, μ_1 and μ_2, that lie orthogonal to one another in the heme plane, which is designated as the x'-z' plane. The quantity θ is the polar angle between the heme plane and the z-axis; ϕ is the azimuthal angle between the x and x' axes; and α is the angle between μ_1 and the z' axis. The dipole transitions interact with the electric field vector of the evanescent wave; the absorption intensity is proportional to $(\mu E)^2$. Two approximations can be made to simplify the analysis:

1) The heme dipole transitions are x, y polarized in the molecular plane, which means that $\mu_1 = \mu_2$ and absorption intensity is not a function of α. Eaton and coworkers have carried detailed polarized absorption studies on several heme protein crystals of known structure (*20-22*). Their work clearly demonstrates that the Q, B, and N heme absorption bands, extending over the 300-600 nm range, are x, y polarized.

2) There is no preferential ordering of molecules along either the x or y axes. In this azimuthally symmetric (uniaxial) geometry, all values of ϕ are equivalent and can be averaged over 2π (*9,19*). Under these conditions, the relative thin film absorbances along x, y, and z are given by (9, 19)

$$A_{f,x} = \frac{1}{2}|E_x|^2|\mu|^2\left(1 + \sin^2\theta\right) \tag{10}$$

$$A_{f,y} = \frac{1}{2}|E_y|^2|\mu|^2\left(1 + \sin^2\theta\right) \tag{11}$$

$$A_{f,z} = |E_z|^2|\mu|^2\left(\cos^2\theta\right) \tag{12}$$

Note that the trigonometric term in equations 10 and 11 accounts for the degeneracy of the in-plane heme dipole transitions (*19*). TE polarized light contains only an E_y component whereas TM polarized light is composed of only E_x and E_z components. The relationship between θ, averaged over a large ensemble of molecules, and the dichroic ratio of thin film absorbances (ρ) is then:

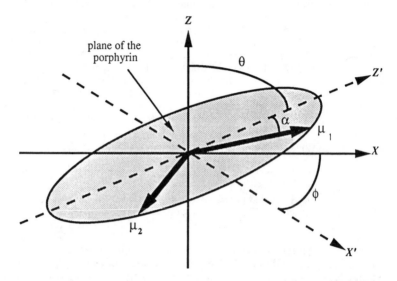

Figure 2. Schematic of the heme geometry relative to the laboratory coordinate system in a linear dichroic ATR experiment. The laboratory coordinate system is defined by the x, y, and z axes, with the origin at the point of reflection in the waveguide (x-y) plane. The heme plane lies at an angle θ from the IRE surface normal (z-axis). The heme electric dipole transitions μ_1 and μ_2 are perpendicular to one another in the molecular plane. The x', y', z' coordinate system defines the heme plane, with the origin at the center of the porphyrin ring. The angles α and ϕ are defined in the text.

$$\rho = \frac{A_{f,TM}}{A_{f,TE}} = \frac{A_{f,z} + A_{f,x}}{A_{f,y}} = \left[E_x^2 + \frac{2E_z^2 \cos^2\theta}{1 + \sin^2\theta} \right] \left[E_y^2 \right]^{-1} \quad (13)$$

IO Waveguiding. The reflection density of an IRE, N/D, is defined as the number of internal reflections per unit distance of beam propagation along the IRE/solution interface. Since N/D is inversely proportional to IRE thickness, the pathlength in an ATR experiment can be enhanced by using a thinner IRE. The ultimate in a thin IRE is a planar integrated optical waveguide (IOW), which is a substrate-supported IRE having a thickness on the order of λ. Interference between the multitude of totally reflecting light rays in a planar IOW restricts light propagation to a small number of discrete reflection angles (modes) that yield constructive interference. The theory that describes this behavior is well known (*23*). IOW-based spectroscopies have been applied to characterization of thin film and interfacial samples and reviews are available (*24*).

Lightguiding behavior in an IOW can be modeled using a simple ray optics approximation. Inter-modal comparisons of evanescent thin film pathlength in an IOW have shown this approximation to be valid (*13*). From a ray optics perspective, reflection density in an IOW is given by

$$N/D = (2d_2\tan\theta_i + \Delta_{21} + \Delta_{23})^{-1} \quad (14)$$

where d_2 is the IOW thickness, D is the distance over which the ATR measurement is made, and Δ_{21} and Δ_{23} are the Goos-Hänchen shifts at the IOW/solution and IOW/substrate interfaces, respectively (13). Expressions for Δ_{21} and Δ_{23} for a step-index, asymmetric planar IOW are given in reference 13. In the visible spectrum, values of N/D for IOWs range from several hundred to several thousand per cm. The enhanced optical pathlength makes it possible to measure ATR linear dichroism for a very weakly absorbing film of chromophores, specifically a heme protein film with a surface coverage on the order of 10^{-12} mol/cm^2 (14).

The dichroic ratio is determined using IOW-ATR by ratioing the thin film absorbances in TM and TE polarized modes. The thin film absorbances are recovered from the measured total absorbances by subtracting the bulk contribution (equation 1). A correction must also be applied if the reflection densities for the two modes are not equal. The dichroic ratio is given by

$$\rho = \frac{A_{t,TM} - A_{b,TM}}{A_{t,TE} - A_{b,TE}} = N_{TM}\left[E_x^2 + \frac{2E_z^2 \cos^2\theta}{1 + \sin^2\theta} \right]\left[N_{TE}\, E_y^2\right]^{-1} \quad (15)$$

where the subscripts TE and TM refer to the pair of modes under comparison.

Experimental

Reagents. Horse heart cytochrome *c* (cyt *c*, #C7752) and DDS were obtained from Sigma. Cyt *c* was purified by gel filtration on Sephadex G-25 (Sigma). It was eluted with 50 mM phosphate buffer, pH 7.2, containing 100 mM NaCl, and diluted to a final concentration of 35 μM (determined by absorbance measured at 514.5 nm).

Waveguides and Surface Preparation. Silicon-oxynitride planar IOWs were fabricated by plasma-enhanced chemical vapor deposition on fused silica substrates at the Microelectronics Center of North Carolina, as described previously (25). These waveguides are 1-1.5 μm thick and support 2-3 guided modes. Experiments were performed with bare (hydrophilic) glass and DDS-modified (hydrophobic) glass waveguides. Hydrophilic waveguides were prepared by soaking in PCC-54 surfactant (Pierce) for 24 hours, rinsing with deionized water, soaking in an 80°C chromic acid bath for 20 min, rinsing with deionized water, soaking in 1 M nitric acid for one hour, and rinsing again with deionized water. The surface was then blown dry with nitrogen and dried for two hours at 80°C. Hydrophobic glass waveguides were prepared by soaking hydrophilic waveguides in 2% DDS (v/v) in dry toluene (distilled over elemental sodium) for two hours. Nitrogen was bubbled through the solution during the silanization and the procedure was performed in a nitrogen-filled glove bag. The waveguide was then rinsed sequentially in toluene, toluene/ethanol, ethanol, ethanol/water, and water, followed by drying at 80°C for two hours. Wettability of waveguide surfaces was characterized by measuring static water contact angles with a goniometer. Contact angles on hydrophilic and hydrophobic surfaces were $10°\pm6.1°$ and $88°\pm1.2°$, respectively.

IOW-ATR Instrumentation. The instrumental arrangement was similar to previous descriptions (11,12,26). Waveguides were mounted in a flow cell that was installed on a rotary stage. The 514.5 nm line of a Coherent Innova 70 argon ion laser was prism coupled into TE and TM polarized waveguide modes by rotating the waveguide/flow cell assembly with respect to the stationary beam. Input polarization was selected with a Fresnel rhomb. The guided mode "streak" visible in the waveguide was photographed with a charge-coupled device (CCD) array cryogenically cooled to -102°C (26).

Data Collection and Analysis. Fifteen minutes after buffer was injected into the flow cell, images of TE and TM guided modes were recorded. Cyt c (35 μM) was then injected, allowed to equilibrate for 30 minutes, and guided modes were imaged again in both polarizations. Attenuation curves were generated by plotting the logarithm of the vertically integrated pixel intensity in the guided mode image against propagation distance. These curves were fit by least squares regression to $\log[I(D)] = -\alpha_T D + C$, where $I(D)$ is integrated pixel intensity as a function of D, the propagation distance in cm, α_T is the loss coefficient in cm^{-1}, and C is a constant.

Before protein is injected into the flow cell, $\alpha_T = \alpha_0$, the intrinsic propagation loss of the IOW. With protein present in the flow cell, $\alpha_T = \alpha_0 + \alpha_b + \alpha_f$, where α_b and α_f are the loss coefficients due to bulk and adsorbed protein, respectively. Since $\alpha_b D = A_b$, α_b can be calculated from knowledge of the physical parameters (N/D, n_1, n_2, n_3, θ_i) of the waveguide, using equations 1-6 and 14. The thin film loss coefficient is then recovered by difference ($\alpha_f = \alpha_T - \alpha_0 - \alpha_b$). Finally, since $\alpha_f D = A_f$, the dichroic ratio in eq 15 is simply the ratio of the TM and TE polarized α_f values.

Results and Discussion

Orientation in Cyt c Films. The heme group in cyt c is covalently bound to the polypeptide, and its geometric position relative to the three-dimensional protein conformation is known from the crystal structure (27). Thus in principle, the average molecular orientation of a uniaxial assembly of cyt c molecules can be determined relative to a laboratory-defined coordinate system by measuring the absorbance of the heme group at two orthogonal polarizations (22). The IOW-ATR technique solves the practical difficulty of detecting a statistically valid difference between two orthogonal absorbance measurements on a hydrated protein monolayer.

Isotherms for cyt c adsorption on hydrophilic and hydrophobic glass waveguides were measured using TE-polarized IOW-ATR spectrometry (data not shown). For both surfaces, the "plateau" region of the isotherm was observed at bulk protein concentrations >20 µM. Based on these data, a bulk cyt c concentration of 35 µM was selected to form adsorbed films for linear dichroism measurements. Thus in both cases, dichroic ratios were measured on films of maximum surface coverage. Although actual surface coverages were not determined independently, the measured thin film absorbances were consistent with approximately 50% monolayer coverage.

The mean tilt angle between the heme plane and the IOW surface normal in adsorbed cyt c films was calculated by rearranging eq 15 to solve for θ_i. On hydrophilic glass the tilt angle was $17°±2°$ (n=2), which means that the heme plane was oriented nearly vertical with respect to the waveguide (x-y) plane. Much different results were observed for films on hydrophobic waveguide surfaces. In this case, the tilt angle was $40°±7°$ (n=2) from the surface normal, indicating a mean orientation nearly equidistant between the waveguide plane and surface normal.

Two important points are evident from these data. First, the tilt angles on both surfaces are significantly different from $54.7°$, which is the mean angle that would be measured for a completely isotropic distribution of molecular orientations. Thus a geometrically anisotropic distribution of heme planes is present in the cyt c film on both surfaces. Second, the large difference between the tilt angles on hydrophilic and hydrophobic glasses demonstrate that the molecular orientation distribution is influenced by the chemical properties of the waveguide surface. It is well recognized that due to the asymmetric distribution of amino acids on the surface of a protein, one "face" of the molecule may be preferentially attracted to a given adsorbent material, while the other "faces" may be repelled (*1-4*). The attraction/repulsion geometry may be reversed at the surface of a second adsorbent having chemical properties that differ substantially from the first. To our knowledge, the data presented here are the first *direct* evidence supporting this hypothesis. However, as discussed below, limitations inherent in the method prevent us from deducing the mean orientation of the cyt c molecule at a waveguide/solution interface.

Several groups have investigated molecular orientation in adsorbed cyt c films, particularly on negatively charged, hydrophilic substrates. Given a pI of 10.0, the attraction to these surfaces is primarily electrostatic. Based on the electrochemical behavior of cyt c at a carboxy-derivatized gold surface, Tarlov and Bowden (*27*) proposed that adsorption takes place at the face where the edge of the heme plane is exposed to solvent. This geometry would place the heme plane approximately normal to the substrate surface plane, consistent our measurements on hydrophilic glass. Pachence et al. (*28*) and Kozarac et al. (*29*) used transmission linear dichroism and polarized reflection-absorption spectroscopies, respectively, to study the orientation of cyt c adsorbed to hydrated fatty acid and lipid films. Both groups concluded that the heme plane is approximately parallel to the adsorbent film plane. However, in both cases this conclusion was based on measuring a small difference between two very weak absorbances (on the order of 0.002 absorbance units). Using IOW-ATR, Walker et al. (*15*) measured a tilt angle of 48° for cyt c adsorbed to hydrophilic glass, substantially greater than reported here. However that measurement was made after flushing bulk dissolved protein from the flow cell, on a film that was ca. 25% of a monolayer, or about one half the surface coverage of the films examined here. It is possible that orientation may be affected by packing density (*7*), a concept that we are currently investigating.

Limitations of the IOW-ATR Technique. Using IOW-ATR measurements of absorption linear dichroism to examine orientation in protein films is subject to several limitations:

1) An internal reflection linear dichroism measurement on a uniaxial assembly of *x,y* polarized chromophores is rotationally degenerate about both the normal to the molecular plane of the chromophore (*y*-axis) and the normal to the waveguide plane (*z*-axis). Thus a specified tilt angle between the heme plane and the *z*-axis is consistent with a number of possible orientations for an adsorbed heme protein molecule. This of course does not imply that the adsorption process is geometrically random; it merely means that we cannot specify the contact region of the protein.

2) The tilt angle of the heme plane, not the protein molecule, is measured. If protein adsorption induces a conformational change that distorts the geometric relationship between the heme plane and the polypeptide, then heme orientation cannot be related to overall protein orientation. In the absence of supporting evidence, the maintenance of native conformation in the adsorbed state cannot be assumed. The use of the IOW-ATR technique to study orientation in protein films would therefore be complemented by concurrent studies of protein conformation.

3) An absorption linear dichroism measurement yields only the mean tilt angle of the dipoles in a molecular assembly. No information on the distribution of orientations is available. Thus we can detect the existence of anisotropy but cannot discern the degree of macroscopic ordering. In contrast, fluorescence emission anisotropy measurements are sensitive to both the mean and the distribution about the mean (*18*). The combined use of absorption linear dichroism and emission anisotropy techniques therefore appears to be a potentially powerful approach to characterizing order in a thin molecular assembly. Since the linear dichroism measurement specifies the mean tilt angle, the corresponding anisotropy measurement is sensitive only to the width of distribution..

Conclusions

We have described a sensitive planar waveguide ATR technique for measuring absorption linear dichroism in weakly absorbing organic films at the solid/liquid interface. This technique was used to demonstrate that the mean tilt angle of the heme in adsorbed cyt *c* films is dependent on the chemical properties of the adsorbent surface. We are presently extending this work to other heme proteins and adsorbent surface chemistries.

Acknowledgments

This research was supported by NIH Grant R03RR08097, NSF Grant CHE-9403896, and the donors of The Petroleum Research Fund, administered by the ACS. Helpful discussions with Dwight Walker and Monty Reichert of Duke University, and Lin Yang of the University of Arizona, are gratefully acknowledged.

Literature Cited

1. Andrade, J.D. In *Surface and Interfacial Aspects of Biomedical Polymers: Protein Adsorption*; J.D. Andrade, Ed.; Plenum Press: New York, **1985**; 2, Chapter 1.
2. Andrade, J.D.; Hlady, V. *Adv. Polym. Sci.*, **1986**, *79*, 1-63.
3. Norde, W. *Adv. Coll. Inter. Sci.*, **1986**, *25*, 267-340.
4. Horbett, T.A.; Brash, J.L. In *Proteins at Interfaces*; Horbett, T.A.; Brash, J.L., Eds.; American Chemical Society: Washington (D.C.), **1987**; Chapter 1.
5. Darst, S.A.; Ribi, H.O.; Pierce, D.W.; Kornberg, R.D. *J. Mol. Biol.*, **1988**, 203, 269-273.
6. Lin, J.N.; Andrade, J.D.; Chang, I.-N. *J. Immunol. Meth.*, **1989**, 125, 67-77.
7. Lee,C.-S.; Belfort, G. *Proc. Natl. Acad. Sci.*, **1989**, 86, 8392-8396.

8. Stephens, D.A.; Bohn, P.W. *Anal. Chem.*, **1989**, 61, 386-390.
9. Cropek, D.M.; Bohn, P.W. *J. Phys. Chem.*, **1990**, 94, 6452-6457.
10. Hong, H.G.; Bohn, P.W.; Sligar, S.G. *Anal. Chem.*, **1993**, 65, 1635-1638.
11. Saavedra, S.S.; Reichert, W.M. *Appl. Spectrosc.*, **1990**, 44, 1210-1217.
12. Saavedra, S.S.; Reichert, W.M. *Appl. Spectrosc.*, **1990**, 44, 1420-1423.
13. Saavedra, S.S.; Reichert, W.M. *Anal. Chem.*, **1991**, 62, 2251-2256.
14. Saavedra, S.S.; Reichert, W.M. *Langmuir*, **1991**, 7, 995-999.
15. Walker, D.S.; H.W. Hellinga; Saavedra, S.S.; Reichert, W.M. *J. Phys. Chem.*, **1993**, 97, 10217-10222.
16. Harrick, H.J. *Internal Reflection Spectroscopy*, 2nd ed.; Harrick Scientific: New York, **1979**.
17. Reichert, W.M. *Crit. Rev. Biocompat.*, **1989**, 5, 173-205.
18. Thompson, N.L.; McConnell, H.M.; Burghardt, T.P. *Biophys. J.*, **1984**, 46, 739-747.
19. Fraaije, J.G.E.M.; Kleijn, J.M.; van der Graaf, M.; Dijt, J.C. *Biophys. J.*, **1990** 57, 965-975.
20. Eaton, W.A.; Hochstrasser, R.M. *J. Chem. Phys.*, **1967**, 46, 2533-2539.
21. Makinen, M.W.; Eaton, W.A. *Ann. N. Y. Acad. Sci.*, **1974**, 206, 210-222.
22. Hofrichter, J.; Eaton, W.A. *Ann. Rev. of Biophys. Bioeng.*, **1976**, 5, 511-560.
23. Kogelnik, H. In *Integrated Optics*, 2nd ed., Springer Verlag: New York, **1979**; Chapter 2.
24. Bohn, P.W. *TRAC*, **1987**, 6, 223-233.
25. Walker, D.S.; Reichert, W.M.; Berry, C.J. *Appl. Spectrosc.*, **1992**, 46, 1437-1441.
26. Yang, L.; Saavedra, S.S.; Armstrong, N.R.; Hayes, J. *Anal. Chem.*, **1994**, 66, 1254.
27. Dickerson. R.E.; Takano, T.; Eisenberg, D.; Kallai, O.B.; Samsom, L.; Cooper, A.; Margoliash, E. *J. Biol. Chem.*, **1971**,5, 1511.
27. Tarlov, M.J.; Bowden, E.F. *J. Amer. Chem. Soc.*, **1991**, 113, 1847.
28. Pachence, J.M.; Amador, S.; Maniara, G.; Vanderkooi, J.; Dutton, P.L.; Blasie, J.K. *Biophys. J.*, **1990**, 58, 379.
29. Kozarac, Z.; Dhathathreyan, A.; Möbius, D. *FEBS Lett.*, **1988**, 229, 372-376.

RECEIVED February 1, 1995

Chapter 20

Human Serum Albumin Adsorption at Solid–Liquid Interface Monitored by Electron Spin Resonance Spectroscopy

R. Nicholov[1,2], N. Lum[2], R. P. N. Veregin[3], and F. DiCosmo[1,2]

[1]Centre for Plant Biotechnology, Department of Botany, University of Toronto, Toronto, Ontario M5S 3B2, Canada
[2]Institute of Biomedical Engineering, University of Toronto, Toronto, Ontario M5S 1A1, Canada
[3]Xerox Research Centre of Canada, 2660 Speakman Drive, Mississauga, Ontario L5K 2L1, Canada

Human serum albumin (HSA) was spin labelled with the 4-maleimido-tempo (MT) spin label. The experimental conditions favoured labelling of the free sulfhydryl group of HSA. The adsorption of HSA-MT on glass, polystyrene and polystyrene-butadiene beads at pH 4.5, 5.4 and 7.0 was studied. Our data show that the conformational status of HSA-MT adsorbed onto the glass surface differs from that in buffer. Changes in ESR spectral characteristics of HSA-MT display an additional immobilisation of the spin label due to the adsorption. ESR spectra report primarily on the conformational changes of HSA-MT in the vicinity of the spin label. There were no significant differences in ESR spectra of HSA-MT in buffer at pH 4.5, 5.4 and 7.0. However, ESR spectral characteristics of adsorbed HSA-MT were pH sensitive. The stronger immobilization of the MT was recorded when HSA was adsorbed onto polystyrene beads at the isoelectric point (pI) of the protein. This immobilization of MT corresponds to a bigger alteration of adsorbed HSA-MT near to the modified 34 cysteine residue. The most freely rotating MT was measured when HSA-MT was adsorbed onto the polystyrene-butadiene beads at pH 7.0; this corresponds to unfolding of adsorbed protein when attached to the negatively charged polystyrene-butadiene surface. The data suggests also that the point of attachment of HSA-MT onto the solid surface (glass, polystyreene and polystyrene-butadiene) during the process of unspecific adsorption is not in the vicinity of the spin label. ESR spectra of HSA-MT were recorded also at low temperature. By decreasing the temperature of HSA-MT in buffer from 25°C to -10°C we simulated the immobilization of the spin probe that resulted from the adsorption of spin labelled protein to the solid surface.

0097–6156/95/0602–0280$12.00/0
© 1995 American Chemical Society

The phenomenon of protein adsorption onto solid surfaces is important to many fields of study. Research has been done in pharmaceutical sciences, soils and agriculture, the silicon chip industry, water filtration, and protein processing. Adsorbed proteins affect the biocompatibility of artificial implants such as vascular grafts and soft tissue implants, cell adhesion for tissue growth, bacterial cell adhesion to urinary catheters, activation of the intrinsic and extrinsic systems in blood coagulation, complement activation, contact lens fouling and others [1].

Important parameters for protein adsorption include the amount of adsorbed protein, the adsorption-desorption kinetics and the conformation of the adsorbed protein. The details of conformational status of proteins at solid/liquid interfaces has been recognized as a important research topic by many authors [2,3] and has attracted increasing research efforts recently [3,4,5]. However the evaluation of the conformational status of the adsorbed protein molecule is a difficult task and the available information demonstrates convincingly that novel approaches are required if we are to develop a firm understanding of protein-surface interaction. In the majority of publications structural alterations of adsorbed proteins are evaluated using infrared spectroscopy and ellipsometry. ESR spectroscopy (ESR) offers other approach and direct measurements of the alteration of the protein structure. Narasimhan et al. [6] and Nicholov et al. [7] demonstrated that ESR has the capacity to monitor protein adhesion at a solid/liquid interface and is especially effective for direct evaluation of the conformational status of the adsorbed molecule. The spin label covalently bound to the protein molecule serves as a reporting group [8]. By modifying different protein moieties, at specific sites with spin labels, one can obtain specific information about the conformation of the protein in the vicinity of the spin label.

In this study, the ESR technique was applied to evaluate the conformational status of spin labelled human serum albumin (HSA-MT) adsorbed onto the surface of glass beads, polystyrene and polystyrene-butadiene beads. The adsorption was followed on the different surfaces at various pH values.

Materials and Methods

High grade crystalline human serum albumin was purchased from Sigma, (St. Louis, MO, Lot No. 127F9307). 4-maleimido-tempo (MT) spin label was purchased from Aldrich (Milwaukee, WI). Controlled pore glass beads were purchased from Sigma and polystyrene and polystyrene-butadiene beads were purchased from Polysciences, Inc. (Pasadena, CA).

Spin Labelling. HSA was labelled with 4-maleimido-tempo (MT) under conditions which preferably modified the free sulfhydryl group (Cys 34) [9]. Approximately 1.5 μmole of MT in methanol was transferred to a vial and evaporated to dryness under a nitrogen stream. A solution of HSA, 20 mg in 0.1 M phosphate buffered saline at the desired pH was added and stirred gently overnight at 4°C. Purification of the spin-labelled proteins from the free spin radicals was accomplished using a dialysis membrane (MW cutoff 12,000-14,000, Spectrum Medical Industries, Inc., Los Angeles, CA). The samples were carefully pipetted into the membrane and placed in 1:500 v/v 0.1 M PBS, which was changed after 1, 6, and 24 hours.

Protein Adsorption. Controlled pore glass beads, polystyrene, and polystyrene-butadiene beads were measured into three Ependorf tubes in quantities of approximately 42 mg, 45 mg, and 50 mg, respectively. To each set of surfaces was added one of the protein solutions, either at pH 4.5, pH 5.4, or pH 7.0. Vials were slowly shaken to ensure surface wetting of all the beads. The samples were than incubated for 24 hours before being washed with buffer. Tubes were gently mixed with buffer and then centrifuged for 2 minutes at 10,000 rpm. The buffer was removed, leaving enough liquid to prevent the beads from crossing the air/water interface. Washing was repeated three times. The ESR spectra were recorded after the washing. ESR spectra of adsorbed protein on to the glass or polymeric beads were recorded at the same conditions.

Stoichiometry of spin labelling was determined using methods as described by Nicholov et al [7]. The ratio of protein to spin label varied between 2 to 1 and 0.8 to 1. The size of the particles was of 1 μm and smaller.

ESR Measurements. ESR spectra were recorded with a Bruker ESR Spectrometer (Model 300, Germany) equipped with a variable temperature accessory. Magnetic field was 3375 G, sweep width was 100 G, and sweep time was 5.243 s. Time constant was 5.12 ms, modulation amplitude was 0.5 G, and microwave power was 10 mW. Number of scans ranged from 100 to 5000.

The mobility of the spin label (MT) and respectively the mobility of protein molecule in solution or adsorbed onto the solid surface was evaluated by rotational correlation time τ [7,10], hyperfine splitting $2T_{II}$ and semiquantitative parameters s/w and w/p. The last two parameters are the ratio of the amplitudes of the peaks s and w of ESR spectra as shown in Fig. 1

Results

ESR spectra of HSA-MT in buffer at room temperature. Figure 1A shows the ESR spectrum of HSA-MT in buffer (pH 7.0) at room temperature. The spectrum is broad and asymmetric with a considerably distorted line shape as compared with the spectrum of free MT in solution (Fig.1B). The MT spectrum in buffer is a characteristic of a molecule tumbling rapidly in aqueous phase with an rotational correlation time τ of $0.95.10^{-10}$s. The spin label covalently attached to the HSA molecule produced a completely different ESR spectrum. The low magnetic field resonance line was split into two lines (s and w Fig.1); the high field resonance line was also split into two relatively weak extrema. The distance (in Gauss) between the outer extrema marked in Fig.1 by arrows is a measure of hyperfine splitting $2T_{II}$. The ESR spectrum of HSA-MT in buffer has a hyperfine splitting $2T_{II}$ of (65.2 ± 0.7) G, which is 30.8 G bigger than that of MT in buffer. The increase of $2T_{II}$ points to an immobilization imposed onto the spin label when it is covalently attached to the HSA, although the protein molecule itself is moving freely in a solution at room temperature. The experimental values of $2T_{II}$ lie in the area typical for spin label with slow molecular motion. The hyperfine splitting $2T_{II}$ itself reflects the rotational mobility of MT relative to the point of attachment and the tumbling of the whole HSA molecule. The value of $2T_{II}$ demonstrates that MT is strongly immobilized when attached to HSA, suggesting that the nitroxyl moiety is

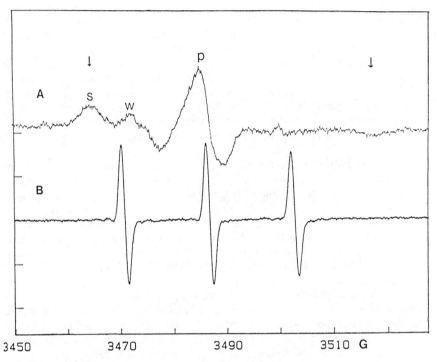

Figure 1. ESR spectra of human serum albumin (A), covalently labelled with 4-maleimido-tempo, in 0.1 M phosphate buffer saline at pH 7.0 and 25°C; (B) ESR spectra of 4-maleimido-tempo, in 01.M phosphate buffered saline at pH 7.0 and 25°C. Both spectra were recorded at magnetic field, 3370 G; sweep width, 100 G; sweep time, 5.243 s; time constant, 1.28 ms; modulation amplitude, 0.504 G; microwave power, 10 mW. The distance between the arrows corresponds to $2T_{\text{II}}$ in G.

not located on the surface of the protein, but rather is sterically hindered in a cleft of a protein fold. The spin probe is covalently attached to the protein molecule and because the attachment group is the shortest possible [11] we suppose that the mobility of the spin probe reflects also the mobility of the protein molecule in its entirety.

Rotational correlation time τ of HSA-MT in buffer is $\tau = 11.5$ ns calculated as shown in [12]. The increase of the rotational correlation time of MT-HSA shows also the restricted mobility of the spin label attached to the HSA. This parameter is a measure of the relative mobility of the spin label, including both the amplitude and the rate of motion.

The spectrum of MT-HSA in buffer has characteristics close to those of BSA labelled with maleimido tempo [7]. The similarity between amino acid sequences of HSA and BSA [13] and especially the number and the position of the free sulfhydryl groups most probably explains the similarity between ESR spectra of HSA-MT and BSA-MT. The most notable difference between both spectra is that MT is more strongly immobilized by the attachment to HSA than to BSA. This could be attributed to the differences in the tertiary structure of the molecules: HSA being heart-shaped [14] and BSA cylindrically-shaped [15].

HSA-MT in buffer at different pH values. The influence of pH on the process of protein adsorption results directly from the effect of pH on the net surface charges of the protein and consequently from unfolding of protein at a pH away from the isoelectric point [16]. The ability of HSA to bind a large variety of ions and the quantity of its surface charge groups [13] suggests the possibility of different patterns of unfolding of HSA at different pH values. By applying ESR spectroscopy to spin labelled HSA at different pH one can expect to obtain additional information on the conformational status of protein in buffer or adsorbed on a solid surface.

The isoelectric point of HSA is pH 4.5 [17] if the net surface charge of a protein at a given pH is defined as the difference between its own dissociated groups and bound ions. The isoelectric point of HSA determined on the basis of its own charge groups is at pH 5.4 [17]. At pH 7.0, HSA is negatively charged.

The ESR spectra recorded at pH 4.5, 5.4 and 7.0 pH are presented in Figure 2 and the ESR spectral parameters are shown in Table I. The comparison between these three spectra shows that at pH 4.5 and 5.4 $2T_{II}$, s/w and s/p are not significantly different and suggest restricted rotational mobility of the spin probe. At pH 7.0 the most freely rotating component of the spin label is better pronounced (Fig. 2, w) although the values of the ESR parameters are not very different. The immobilization of MT at pH 4.5 and 5.4 obviously reflects the fact that the globular proteins are most compact and stable near the isoelectric point. Above the isoelectric point at pH values greater than 6.0, the net charge of the protein is negative, the repulsion forces between identically charged groups provoke unfolding. Therefore one can expect that the ESR parameters of HSA-MT will change in buffer at pH 7.0. However, the ESR spectrum of HSA-MT at pH 7.0 is not significantly different from the spectrum recorded at the isoelectric point. Obviously, the repulsion forces between negatively charged groups are not strong enough to cause unfolding and a drastic increase in segmental motion in the vicinity

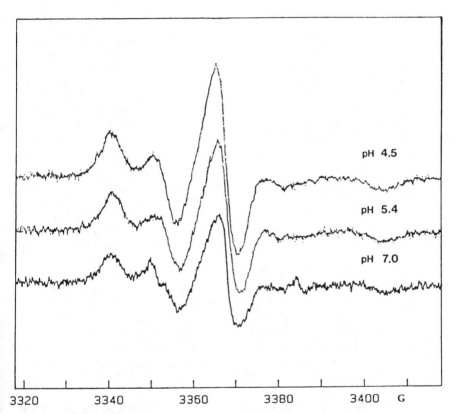

Figure 2. ESR spectra of human serum albumin spin labelled with 4-maleimido-tempo in 0.1 M phosphate buffer saline at pH 4.5, pH 5.4 and pH 7.0. The spectra recorded as in Figure 1.

of the spin label. Nevertheless, the ESR spectrum recorded at pH 7.0 shows greater mobility than the other spectra as indicated by the smaller s/w value.

HSA-MT at solid/liquid interface. ESR spectra of HSA-MT adsorbed on to the surface of controlled pore glass beads, polystyrene beads and polystyrene-butadiene beads were recorded in PBS at different pH values. The conformation of the protein molecule due to the adsorption and also the effect of pH on the conformational status of the adsorbed protein were monitored. The results are presented in Figures 3 and 4.

HSA-MT adsorption on controlled pore glass beads. ESR spectra of adsorbed HSA-MT on the surface of controlled pore glass beads were recorded at pH 4.5, 5.4 and 7.0. The sample of beads in buffer contained both adsorbed and desorbed proteins because of the process of spontaneous desorption of proteins. Special efforts were taken to remove the signal of desorbed protein molecules. The components of the ESR signals coming from the desorbed proteins were eliminated by asymptotical subtraction of the ESR spectrum of HSA-MT in solution from the recorded spectrum. The resulting spectra of only HSA-MT adsorbed onto the glass beads are presented in Fig.3. The portion of the ESR signal of the desorbed proteins was small (5%) because the desorption process is time dependent and the ESR spectra were recorded immediately after the washing. The data in Fig.3 shows that the ESR spectra of absorbed proteins are more immobilized than the HSA-MT spectra recorded in buffer. The additional immobilization of the spin label MT is a consequence of the adsorption. Because MT is not located on the surface of the protein, the modification of the ESR spectra revealed the conformation of the adsorbed molecule.

If the absorption is accomplish at different pH of the liquid phase the hyperfine splitting, $2T_{II}$, does not change significantly i.e. is not pH sensitive; however the semiquantitative parameters s/w and s/p do change with pH and are, therefore, pH sensitive (Table II). As a result of the adsorption on the glass surface, s/w values decrease with increasing pH, because the amplitude of the resonance line representing the more mobile component of the ESR spectra (w) increases with pH. Similarly, s/p values decrease with increasing pH as a result of the decrease in the amplitude of the s peak, which represents the slow motion component of the ESR spectrum.

Comparing the ESR spectra recorded at the three different pH values, we find the spectrum representing the most immobilized label was recorded at pH 4.5. If we consider the isoelectric point of HSA (determined on its own charges) one can see that the protein is positively charged at pH 4.5. This suggests a stronger attraction of the protein molecule to the glass surface. This result shows that the increased attraction of HSA-MT to the surface causes a stronger immobilization of the spin label.

An increase in the pH of the buffer led to an increase in the mobility of MT. The most freely rotating spin label was recorded at pH 7.0. At this pH, HSA molecules have a net negative charge; the glass surface is also negatively charged and the concentration of protons is relatively low. The immobilization of the MT

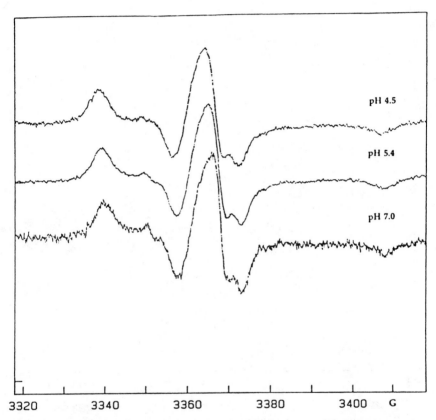

Figure 3. ESR spectra of human serum albumin spin labelled with 4-maleimido-tempo in 0.1 M phosphate buffer saline adsorbed to the surface of glass beads at pH 4.5, pH 5.4 and pH 7.0. The spectra of desorbed HSA-MT were subtracted, the resulted spectra are presented in the figure. The spectra were recorded as in Figure 1.

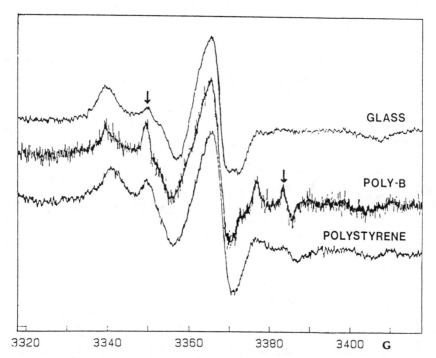

Figure 4. ESR spectra of human serum albumin spin labelled with 4-maleimido-tempo in 0.1 M phosphate buffer saline, pH 7.0, adsorbed on controlled pore glass, polystyrene-butadiene and polystyrene beads. The spectra were recorded as in Fig.1.

at pH 7.0 was less than at the lower pH value and consequently the attraction between protein and the surface is not as strong as at a lower pH (Fig. 3).

Table I. ESR spectral parameters of HSA-MT recorded in 0.1 M PBS buffer at different pH

pH	$2T_{II}$ (G)[*]	s/w	s/p
4.5	65.9 ± 1.3	0.64	0.27
5.4	64.5 ± 1.2	0.62	0.23
7.0	65.2 ± 1.4	0.56	0.28

[*]Values with root mean square error

   The experiments for all pH values were performed in 0.1 M PBS. The effect of ionic strength was not studied. Van Dulm et al. [18] proposed that there is a co-adsorption of cations in the inner region of the adsorbed layer. Assuming that this mechanism was involved and counterions reduced the attraction/ repulsion forces between the charged solid surface and the protein molecules, the data in Figure 3 and Table II show that this reduction did not completely obliterate the effect of pH. In addition our data shows that the structure of HSA-MT adsorbed on the glass surface differs from the conformation assumed in buffer, and that structural changes are pH sensitive. It is important once again to stress the fact that ESR data show conformational changes of HSA-MT only in the vicinity of the spin label and cannot be used to explain conformational modifications of the entire molecule.

HSA-MT adsorbed onto polystyrene beads. Table II presents the ESR spectral parameters for HSA-MT adsorbed on polystyrene and polystyrene-butadiene surfaces. The data shows that the adsorption of HSA-MT on the polystyrene surface modified the shape of the spectrum to a greater extent than the adsorption to the glass and polystyrene-butadiene beads. This distortion of the spectrum did not result from the paramagnetic centres of the polystyrene beads. The ESR spectrum of pure polystyrene beads was recorded under the same conditions as those used for the ESR spectra of adsorbed protein (data not shown) and did not reveal any ESR signals above the signal/noise ratio. The same procedure was applied for glass and polystyrene-butadiene beads.

   HSA-MT adsorbed on polystyrene beads also showed the greatest variation in ESR spectral parameters as a function of pH (Fig. 4). The hyperfine splitting constant, $2T_{II}$, decreased from 67.70 G at pH 4.5 to 65.05 G at pH 7.0. The semiquantitative parameter s/w changed from 0.86 at pH 4.5 to 0.56 at pH 7.0, and the parameter s/p changed from 0.34 at pH 4.5 to 0.23 at pH 7.0. The mobile component of the spectrum (peak w) increased with the increase of pH while the amplitude of the peak s decreased. The amplitude of the central peak p, is composed of both components: a mobile component and another with restricted mobility and the changes were not so visible. Taken together, alterations of the ESR spectra of adsorbed HSA-MT onto the polystyrene surface are more significant than those noted for the glass surface. Most probably, because the hydrophobicity of the polystyrene surface thr attraction/adsorption forces were stronger than those in the glass beads and this resulted in stronger immobilization of the spin probe and alteration of the ESR signal.

Table II. ESR spectral parameters of HSA-MT adsorbed onto different surfaces recorded in 0.1 M PBS at different pH values

pH and Surface	$2T_{II}(G)$	s/w	s/p
4.5			
Glass[*1]	68.13	0.76	0.29
Polystyrene[*2]	67.70	0.86	0.34
Polystyrene-Butadiene[*2]	66.20	0.67	0.26
5.4			
Glass	68.13	0.70	0.27
Polystyrene	66.50	1.23	0.34
Polystyrene-Butadiene	65.86	0.53	0.26
7.0			
Glass	68.13	0.67	0.21
Polystyrene	64.23	0.40	0.26
Polystyrene-Butadiene	65.05	0.56	0.23

[*1]The amount of desorbed HSA-MT was not subtracted from the ESR signal of absorbed HSA-MT as was done for the data presented in Figure 3.
[*2]The subtraction procedure was not applied to ESR spectra of HSA-MT on polystyrene and polystyrene-butadiene beads because of the low signal to noise ratio.

At pH 7.0, the rotational mobility of MT was significantly higher than those at pH 4.5 and pH 5.4. This could be attributed to the negative charges of the protein molecule at this pH.

HSA-MT adsorbed onto polystyrene-butadiene beads. The ESR spectra of HSA-MT adsorbed onto the polystyrene-butadiene beads (polystyrene beads specially treated to have negative charges on the surface) showed less immobilization than the polystyrene beads (Table II). As in the case of the glass beads, the hyperfine splitting constant, $2T_{II}$, does not increase significantly from pH 4.5 to pH 7.0 (Table II). However, the semiquantitative parameter s/w decreased from 0.67 at pH 4.5 to 0.56 at pH 7.0. The parameter s/p decreased from 0.26 at pH 4.5 to 0.23 at pH 7.0. The modifications of all three ESR spectral parameters show an increase of the spin probe mobility with an increase in pH away from the isoelectric point.

HSA-MT mobility at low temperature. By decreasing the temperature of HSA-MT in solution we induce an additional restriction to the rotational mobility of the spin probe. In this fashion we simulated the same immobilization of the spin probe that results from the adsorption of the protein to the solid surface.

The ESR spectra of HSA-MT in buffer at pH 7.0 were recorded over the temperature interval from 10° C to - 40° C (Fig. 5). The hyperfine splitting, $2T_{II}$, increased with the decrease of the temperature as it is shown in Fig.6 where it is plotted versus the absolute temperature. The ratio of resonance lines s/w increased also showing a progressive immobilization of the spin label with a temperature decrease.

The ESR spectrum recorded at -10°C and the ESR spectrum of HSA-MT adsorbed on the glass surface (Figure 3, pH 7.0) show similarities in shape, hyperfine splitting, and s/w ratio. Thus, the immobilization of MT in both cases can be considered comparable. The data shows that the decrease in temperature from 25°C to -10°C immobilizes the spin label to the same extent as the adsorption onto the glass surface.

Discussion

The covalent attachment of MT to HSA causes a considerable restriction of its mobility (comparison to MT in buffer) and produce a ESR spectrum with a distort line (Fig.1). This spectrum suggests that MT is positioned not on the surface of the molecule. Indeed, the SH group to which the spin label is bound is about 10 Å within the HSA [13] near the N terminus. The same spin probe (MT) bound to the SH group of BSA shows an almost identical ESR spectrum. The reason for this similarity is due to the position of the sulfhydryl group in both molecules. The MT submerged within the HSA molecule was used to report on its conformations during the adsorption.

The changes in ESR spectra generated by the adsorption of the protein on the solid surface [Fig.1 and Fig.4] were not as pronounced as the changes discussed in the previous paragraph but they are unquestionable. The differences between ESR spectra of adsorbed HSA-MT and ESR spectra of HSA-MT in buffer were related to the structural changes of the adsorbed protein in the vicinity of the MT. A spin probe positioned differently will most probably, produce a different modification of the ESR spectra, and a different extent of conformation change due to the adsorption. Our data support the reports [19,20] on conformational changes of the adsorbed proteins and additionally define the area of conformational alteration.

J.D. Andrade et al. [21] Norde et al. [22], and Barnabeu et al. [23] showed the pH dependence of adsorption isotherms, but the data does not provide information of the conformational status of the protein. Our data of adsorption performed at different pH values show that the defolding of the protein away of the isoelectric point (pH 7.0) is higher and conformational changes which restrict MT mobility smaller than at lower pH (pH 4.5 and 5.4). According to Kanal et al. [16], the denatured state of BSA is most pronounced at pH values below the isoelectric point. The ESR spectra presented here undoubtedly show that at pH 4.5

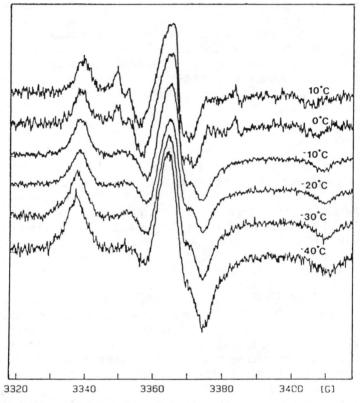

Figure 5. ESR spectra of human serum albumin spin labelled with 4-maleimido-tempo in 0.1 M phosphate buffer saline, pH 7.0, recorded over the temperature range of 10°C to -40°C. Magnetic field, 3370 G; sweep width, 100 G; sweep time, 5.243s; time constant, 1.28 ms; modulation amplitude, 0.504 G; microwave power, 10 mW.

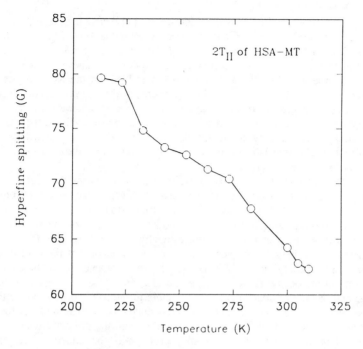

Figure 6. Hyperfine splitting constant, $2T_{II}$, in (G) plotted versus temperature in (K) at which the ESR spectrum of HSA-MT was recorded. ESR spectrometer conditions as in Figure 1.

the unfolding, if it occurs, does not effect significantly the packing of the polypeptide chain around the Cys34 residue. This difference is because the electrostatic repulsion, which causes denaturation of BSA occurs at a pH below 4.0, which was not studied herein. At pH values greater than 6.0, the net charge of the protein is negative, and according to Kanal et al. [16] the electrostatic repulsion is less strong and the protein denatures to an intermediate state which does not experience segmental motion. Assuming that this information is also relevant for HSA, our data suggests modest unfolding for MSA-MT at pH 7.0 and no increase of the segmental motion in close proximity to the 34 cysteine group where the spin label is attached.

Adsorption of HSA-MT at different surfaces showed that electrostatic attraction/repulsion forces effect the modification and conformation of the adsorbed molecule. However the adsorption on the hydrophobic polystyrene surfaces shows the strongest modification of the ESR spectra corresponding to the higher attraction. These findings definitively support the idea of the impact of hydrophobic attraction forces on the adsorption process [24].

The structure of the protein is considered more compact in buffer at, or near the pI. Unfolding occurs during the adsorption on the solid surface. This deviation from the most compact structure suggests an increase of the entropy of the system.

This presumption was developed by Norde et al. [2] and if accepted provides a good explanation of the spontaneous adsorption of the HSA-MT onto the glass beads. Our data show the modification of the structure of HSA around the spin probe during the process of adsorption and suppports the above concept.

Norde et al.[2,22] and Koutsoukos et al.[25], on the basis of adsorption isotherms, analyzed the changes in the free energy during the process of adsorption, and showed that spontaneous adsorption occurs concomitant with increases in the entropy of the system. In our present study an attempt to evaluate the changes in free energy was made by simulating the immobilization of the spin label due to adsorption of the spin labelled protein by imposing a restriction in the spin label mobility by low temperatures.

This simulation allows us to evaluate the free energy associated with the adsorption. Thus, the adsorption free energy can be calculated to be between 30 and 35 kT. Norde et al. [2] discussed the adsorption free energy of a ten segment polymer molecule to be in the range of 10 kT. This value of the adsorption free energy is in good agreement with the estimation of the HSA-MT free energy of adsorption onto the glass surface. However, the evaluation of the conformational adsorption free energy using ESR data is relative. It depends on the position of the MT spin probe on the HSA molecule and also on the distance between the spin probe and the site of attachment of the protein to the solid surface. Since the spin probe only gives information about its immediate environment and not the entire molecule, the adsorption free energy associated with the protein molecule outside of the immediate vicinity of the spin probe must be added. Therefore, the adsorption free energy of the HSA-MT will be much higher than the value estimated by the MT spin label. By varying the location of the spin label along the protein molecule, conformational changes of different segments of the molecule and their associated adsorption free energies can be evaluated.

Although the evaluation of the free energy may be considered fairly speculative, the data presented in this study show that a decrease of the temperature from 25°C to -10°C modified the ESR spectra and immobilized the spin label of HSA-MT in buffer to the same extent as did the adsorption onto the glass surface at room temperature.

Acknowledgments

Research in our laboratory is funded by a Medical Research Council of Canada grant. The ESR facility was made available by Xerox Research Centre of Canada.

Literature Cited

1.   Horbett, T.A. In *Biomaterials: Interfacial Phenomena and Applications*; Cooper, S.L.; Peppars, N.A., Ed.; American Chemical Society: Washington, DC, **1982**; pp 233-244.
2.   Norde, W.; Fraaye, J.G.E.M.; Lyklema, J. In *Proteins at Interfaces; Physical and Biochemical Studies* **1987**; pp 36-45.
3.   Morrisey, B.W.; Stromberg, R.R. *J. Coll. Interface Sci.* **1974** *46*, 152-164.

4. Leininger, R.I; Huston, T.B.; Jakobsen, R.J. In *Blood in Contact With Natural and Artificial Surfaces;* Annals New York Acad. Sci. **1987**; Vol. 516, 173-183.
5. Jakobsen, R.J.; Wasacz, F.M. In *Proteins at Interfaces: Physicochemical and Biochemical Studies.* Brash, J.L.; Horbett, T.A., Eds.; American Chemical Society: Washington, DC, **1989**; pp 339-361.
6. Narasimhan, C.; Lai, C.-S. *Biochemistry* **1989**, *28*, 5041-5046.
7. Nicholov, R.; Veregin, R.P.; Neumann, A.W.; DiCosmo, F. *Colloids and Surfaces A* **1993**, *7*, 159-166.
8. Likhtenshtein, G.I. In *Spin Labelling Methods in Molecular Biology*: Wiley & Sons, NY, **1974**; pp 125-149.
9. Fung, L.W-M. *Biophys J.* **1981**, *33*, 252-262.
10. Knowles, P.F.; Marsh, D.; Rattle, H.W. *Magnetic Resonance of Biomolecules*, Wiley & Sons: London, **1976**; pp 168-207.
11. Lai, C.-S.; Tooney, N.M.; Ankel,E.G. *FEBS Letters* **1984**, *173*, 283-286.
12. Freed,J.H. In *Spin Labelling, theory and application.* Berliner, L.J.(Ed.), N.Y., **1974**, 1, pp 53-133.
13. Brown, J.R. In *Albumin Structure, Function and Uses, Pergamon*; Rosenoer, V.M.; Oratz, M.; Rothschild, M.A., Eds.; New York, NY, **1977**; pp 27-52.
14. He, X.M.; Carter, D.C. *Nature* **1992**, *358*, 209-214.
15. Theodire, P. Jr. In *Chemistry and Physiology of the Human Plasma Proteins*, Bing, D.N.H., Ed.; Pergamon: New York, NY, **1979**; pp 161-245.
16. Kanal, K.M.; Fulerton, G.D.; Cameron, I.L. *Biophysical J.* **1994**, *66*, 153-160.
17. Fogh-Andersen, N.; Bjerrum, P.J.; Siggaard-Andersen, O. *Clin. Chem.* **1993**, *39*, 48-52.17.
18. Van Dulm, P.; Norde, W.; Lyklema, J. *J. Coll. Interface Sci.* **1981**, *82*, 77-82.
19. Nygren, H.; Stenberg,M. *J. Biomed. Mater. Res.* **1988**, *22*, 1-11.
20. Hlady, J.D.; Van Wagenen, R.A.; Andrede, J.D. In *Surface and Interfacial Aspects of Biomedical Polymers*, Andrede,J.D.(Ed.) N.Y., **1985;** pp 81-119.
21. Andrade, J.D.; Hlady, V.; Wei, A-P.; Ho, C-H.; Lea, A.S.; Jeon, S.I.; Lin, Y.S.; Stroup, E. *Clinical Materials,* **1992**, *11*, 67-84.
22. Norde, W.; Lyklema, J. *J. Colloid Interface Sci.* **1978**, *66*, 257-261.
23. Bernabeu, P.; Caprani, A. *Biomaterials* **1990**, *11*, 258-264.
24. Brynda,E; Cepalova, N.A.; Stol, M.; *J.Biomed. Mater. Res.* **1984**, 685-693.
25. Koutsoukos, P.G.; Norde, W; Lyklema, J. *J. Colloid Interface Sci.* **1983**, *95*, 385-392.

RECEIVED March 24, 1995

Chapter 21

Proteins at Surfaces Studied with the Surface Force Technique

Eva Blomberg and Per M. Claesson

Laboratory for Chemical Surface Science, Department of Chemistry and Physical Chemistry, Royal Institute of Technology, S–100 44 Stockholm, Sweden, and
Institute for Surface Chemistry, Box 5607, S–114 86 Stockholm, Sweden

Some results obtained by using the interferometric surface force technique for studying the interactions between adsorbed protein layers and between such layers and surfaces are presented. We have chosen to report results obtained for several types of proteins in order to emphasize differences and similarities in the behaviour. In this chapter we have also included sections describing the normal experimental procedure as well as some common difficulties which we have encountered during our studies of proteins with the surface force technique. It is hoped that these sections can be of use for readers that has no or a limited experience with this technique.

The surface of proteins are generally very heterogeneous with positive and negative charges, groups with hydrogen bonding abilities, as well as non polar hydrophobic regions. This complexity of the protein surface means that each type of protein can interact with other molecules and surfaces in a great number of ways. There are possibilities for ionic interactions (both repulsive and attractive), hydrogen bonding, hydrophobic interaction, hydration forces, acid-base interactions, and, of course, the always present van der Waals force. The most important driving forces for protein adsorption are often regarded to be hydrophobic interaction and ionic interactions, combined with an entropy gain caused by conformational changes of the protein during the adsorption (1, 2). Due to the assymetry of the protein surface, the importance of all of these interactions during an adsorption event will depend on the orientation of the molecule. Considering the large difference in overall size, shape and flexibility of different types of proteins (see e.g. Figure 1) it is of course not possible to treat proteins as a homogeneous group of molecules. Instead, a large flexible glycoprotein such as mucin will have completely different solution and interfacial properties than a small compact protein like lysozyme. For this reason one should be very cautious about drawing general conclusions about the behaviour of proteins at interfaces based on studies of only a few types of proteins. Nevertheless, the need of science (and scientists) to be able to predict the behaviour of as yet not studied systems is ever present. In this chapter we have grouped the proteins studied into four classes, compact globular, soft globular, amphiphilic and random coil like proteins. This distinction is of course not strict but we feel that different proteins which fall into the same category will have at least some aspects

0097–6156/95/0602–0296$12.00/0

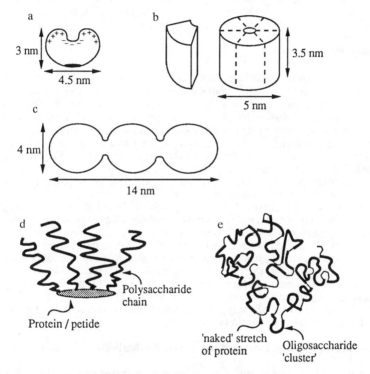

Figure 1. A schematic picture showing the overall structure of a) lysozyme ($M_w \approx$ 14 500, i.e.p = 11-11.5), b) insulin ($M_w \approx$ 5 808, i.e.p = 5.5), c) human serum albumin ($M_w \approx$ 66 000, i.e.p = 4.7-5.2), d) proteoheparan sulphate ($M_w \approx$ 0.175 x 10^6, protein core ($M_r \approx$ 38 kDa), 3-4 heparan sulfate side chains ($M_r \approx$ 35 kDa)), e) mucin ($M_w \approx$ 5-25 x 10^6, radius of gyration $\approx$ 190 nm).

of their interfacial behaviour in common, even though they certainly also will behave differently in many ways.

A number of review articles on somewhat similar subjects to what is presented here has recently been published. Among these Leckband and Israelachvili (3) focus mainly on molecular recognition, whereas Luckham and Hartley in a review on interactions between biosurfaces have included a section discussing protein interactions (4). We hope that this chapter will serve as a complement to their presentations. In the limited space of the chapter we have chosen to discuss mostly data obtained from our own research. Hence, this chapter is not a review of the existing knowledge in the field of surface forces and proteins. However, in the near future we intend to publish a review on the subject that also discusses the many important contributions from other research groups.

Surface Force Technique

With the interferometric surface force technique (5, 6) the total force acting between two macroscopic (A ≈ 1 cm^2) molecularly smooth surfaces in a crossed cylinder configuration is measured as a function of surface separation. This choice of geometry is made due to its experimental convenience. There will be only one contact position, which easily can be changed by moving the surfaces, and there is no problem with aligning the surfaces. One of the surfaces are mounted on a piezo electric tube that is employed to change the surface separation. The other surface is mounted on the force measuring spring.

Muscovite mica, a layered aluminosilicate mineral, is the preferred substrate in these measurements due to the ease with which large molecularly smooth thin sheets can be obtained. The surface chemical properties of these surfaces can be modified in a number of ways including adsorption from solution, Langmuir-Blodgett deposition, and plasma treatment (in some cases followed by reactions with chlorosilanes). The mica sheets are silvered on their back-side and glued onto half-cylindrical silica discs (silvered side down), normally using an epoxy glue. When white light is introduced perpendicular to the surfaces an optical cavity is formed between the silvered backsides of the mica surfaces. From the wavelengths of the standing waves produced the surface separation can be measured to within 0.1-0.2 nm. The deflection of the force measuring spring can also be determined interferometrically, and the force is calculated from Hooke's law.

Typical Experimental Procedure and Data Evaluation

The procedure used when studying the interactions between protein layers adsorbed to solid surfaces is briefly described below. The experimental results are sensitive to contaminants, particularly particles adsorbed to the surfaces, and therefore it is advisable to always make some control experiments before the actual research is started.

Once the two surfaces are mounted in the surface force apparatus they are brought into contact in dry air. When the surfaces are uncontaminated the long-range force is due to the attractive van der Waals force and the adhesion between the surfaces is high. In the next step, the measuring chamber is filled with pure water or a weak aqueous electrolyte solution. Under these conditions it is well established that the long-range force is dominated by a repulsive electrostatic double-layer force and the short-range interaction by a van der Waals attraction, and that the measured forces are consistent with theoretical predictions based on the DLVO theory. Hence, it is easy to establish that the measured forces at this stage are as expected in an uncontaminated system. The adhesive contact between the surfaces in water (i.e. the wavelengths of the standing waves with the surfaces in contact) defines the zero surface separation. When everything appears as normal the salt concentration and pH is changed to the values preferred under the experiment. The forces are measured a few times and the results are checked for reproducibility and compared with previous results (if available). It is only at this stage that the protein is introduced into the measuring chamber and the true research can be started.

The force is not measured between individual molecules but rather all molecules over an area of tens to hundreds μm^2 contribute to the force. Hence, the measured total force is an average force that depends on the orientation of the molecules. When the driving force for adsorption of one part of the molecule is very different to that of the other part, e.g. surfactants on a hydrophobic surface, one orientation of the molecules will dominate. In other cases, like for many proteins on surfaces one can expect that different orientations on the surface are possible and the force measured is then an average over these orientations.

The force (F_c) is measured between crossed cylinders with a geometric mean radius R, which is related to the free energy of interaction between flat surfaces per unit area (W) (7, 8):

$$\frac{F_c}{R} = 2\pi W = 2\pi \sum_i W_i \tag{1}$$

where W_i represents the different types of forces acting between the surfaces. This relation is valid provided that R (about 2 cm) is much larger than the range of the forces between the surfaces (typically 10^{-5} cm or less). A requirement fulfilled in this experimental set-up. The local radius can be measured from the shape of the standing wave pattern. Experimentally it is found that under the influence of strong forces, particularly when the forces vary rapidly with surface separation, the local radius will change as the surfaces are pushed firmly together (9, 10). This is due to the flattening of the glue supporting the mica surfaces. (Note that the measured surface separation is not influenced by the deformation of the glue since the compressed glue is located outside the optical cavity). When the shape of the surfaces changes with the separation the Derjaguin approximation is no longer valid.

An increased understanding of the results can often be obtained by interpreting the measured total force in terms of various force contributions. However, it should be stressed that this procedure in itself is an approximation since all forces are interrelated and not strictly independent. Nevertheless, it is often a useful approximation. The most common procedure is to calculate the electrostatic double-layer force and the van der Waals force for the system under investigation. The discrepancy between measured and calculated forces is interpreted in terms of other force contributions. A complication in this respect is that in general the location of the protein layer – solution interface is not well defined. This means that the location of the plane of charge, from which the double-layer force acts, is not well defined. A further complication is that the dielectric properties of the adsorbed protein layer is not well known and therefore the van der Waals interaction is hard to calculate with a good precision. In practise we have in our analysis assumed the plane of charge to be at the distance where the short-range force due to the compression of the protein becomes dominant. This is a reasonable choice since we know that at smaller distances the Poisson-Boltzmann treatment of the double-layer certainly does not hold. When highly charged proteins are present in solution it has been shown that they make a significant contribution to the electrostatic decay-length in low ionic strength solutions (11). For the results presented here, this effect is not important. For the van der Waals force we have chosen the same plane of origin as for the double-layer force. Two extreme cases for the van der Waals force were considered, using either the Hamaker constant A = 2.2 x 10^{-20} J, equal to that for mica interacting across water, or A = 0, corresponding to no van der Waals force between the hydrated protein layers. From this discussion it should be clear that the comparison between theoretical and experimental force curves is not straightforward and that only large discrepancies should be regarded as significant.

Compact Globular Proteins

Lysozyme is a small compact protein with dimensions approximately 4.5 x 3 x 3 nm. It is positively charged at pH 5.6, the pH used in this study. The positive charges on the molecule are located in such away that it is difficult to use electrostatic considerations to predict the orientation of lysozyme on negatively charged surfaces. The forces acting between mica surfaces across a 10^{-3} M NaCl solution containing various amounts of lysozyme are shown in Figure 2. Without any lysozyme in solution the forces are well described by DLVO theory with a repulsive double-layer force dominating at large separations (apparent surface potential -85 mV) and an attractive van der Waals interaction predominating at distances below 2 nm. When lysozyme is introduced into the surface force apparatus to a concentration of 0.002 mg/ml the forces changes dramatically (Figure 2). The repulsive double layer force is much reduced (magnitude of the apparent interfacial potential 16 mV) and a steric repulsion due to the adsorbed layer is present at distances below about 6 nm. This corresponds to contact between side-on oriented lysozyme on each surface. A further increase in lysozyme concentration to 0.02 mg/ml does hardly affect the interfacial potential. However, the thickness of the adsorbed layer has increased significantly to about 9-10 nm. This corresponds to contact between end-on oriented molecules. A further increase in lysozyme concentration to 0.2 mg/ml hardly affects the measured forces. We can draw three conclusions from these results. First, when the lysozyme concentration is increased some molecules adsorb in an end-on orientation and we most likely end up with a layer composed of proteins adsorbed end-on and side-on. Second, the additional adsorption of the charged proteins hardly affects the interfacial charge. Hence, due to the low dielectric constant of the adsorbed layer free charges are not easily incorporated, and upon adsorption acid-base equilibria are shifted towards the uncharged state and small counterions co-adsorb. This confirms the conclusion made by Norde based on electrophoretic mobility measurements and electron spin resonance (ESR) spectroscopy (*12, 13*). Third, the layer thicknesses obtained are consistent with the size of the molecules in solution. Hence, no global changes in the protein structure occur upon adsorption or upon applying an external compressive force. In this respect lysozyme behaves like other small compact proteins like insulin, (*11, 14*), cytochrome C (*15-17*), RNAse (*18*), and myelin basic protein (*16, 17*). It should be noted, however, that small changes in the protein structure do take place upon adsorption of lysozyme onto silica. This can be seen clearly using e.g. CD-spectroscopy (*19*). Such small changes in the structure of the protein is not easily detected with the surface force technique.

At the higher protein concentrations a weak non-electrostatic repulsion is observed at separations between 12 and 10 nm. The force in this distance regime is more repulsive on approach than on separation. These findings indicate that a second layer of lysozyme is weakly associated with the firmly bound one, most likely through association via the hydrophobic patch located opposite to the active cleft. It is the work needed to remove this layer from the contact zone that gives rise to the repulsion observed between 12 and 10 nm. Clearly, the layer does not reform during the time it takes to measure the force profile. However, the outer layer does reform when the surfaces are left apart. The formation of an outer layer is consistent with the fact that there is an attractive force between the adsorbed (mono) layers (Figure 2) and that lysozyme molecules form dimers in solution at high enough concentrations (*20*). It should be emphasized that the surface force technique readily detects the presence of any weakly adsorbed outer layer on weakly charged surfaces since the presence of this layer affects the weak long-range forces. A similar layer is not so easily detected by e.g. ellipsometry which primarily measures the adsorbed amount, a quantity which hardly is affected by a few molecules adsorbed in an outer layer. More information about the forces acting between lysozyme coated surfaces can be found in *21, 22*

Insulin is also a small compact protein. It readily associates in solution to form dimers, tetramers and hexamers (*23, 24*). The overall shape of some of the different solution species is shown in Figure 1. The adsorption of insulin on mica (*11*) and hydrophobized mica (*14*) and the resulting interaction forces have been studied with the interferometric surface force technique. On hydrophobized mica it was possible to follow the build-up of the adsorbed layer. The forces measured across hydrophobized mica (i.e. mica coated with a Langmuir Blodgett layer of dimethyldioctadecyl ammonium bromide) across a solution containing 1.9 mg/ml insulin crystals at pH 7.3 are shown in Figure 3. Under this solution condition the fraction of monomers, dimers, tetramers and hexamers are 0.09, 0.65, 0.26, and 0.01, respectively. As the adsorption proceeds the layer thickness and the repulsive double-layer force, dominating at D > 12 nm, increases. The results are consistent with an initial adsorption of monomers or dimers in side-on conformation. The layer thickness also increases with the adsorption time indicating formation of hexamers on the surface. Hence, a more dramatic build-up of the protein layer is observed for insulin than for lysozyme. It seems plausible to assume that whenever there is possibilities for attractive interactions between proteins, more than one layer of proteins may eventually adsorb. This has particularly clearly been demonstrated for adsorption of cytochrome C at the isoelectric point on mica (*15*).

Soft Globular Proteins

Human serum albumin (HSA) is a globular but rather flexible protein with an overall dimension of 14 x 4 x 4 nm. It consists of three globular units held together by short flexible regions. This structure was initially confirmed for fatty acid free HSA crystallized from polyethylene glycol ($M_w \approx 400$ g/mol) by X-ray diffraction studies using multiple isomorphous replacement (MIR) (*25*). However, more recent more accurate MIR X-ray diffraction studies of the same protein instead indicate that the overall shape can be approximated with an equilateral triangle with sides of about 8 nm and average thickness of 3 nm (*26*). Considering, the rather flexible regions linking the globular parts of HSA it is not obvious which overall shape adsorbed HSA will adopt, which complicates the interpretation of the force data in terms of protein orientation and monolayer adsorption versus multilayer adsorption. The adsorption of fatty acid containing HSA at a low ionic strength (10^{-3} M NaCl) at pH 5.5, where the protein is weakly negatively charged, onto negatively charged mica and the resulting surface forces have been investigated (*27, 28*).

The forces measured depend strongly on the HSA concentration in bulk solution as illustrated in Figure 4. In all cases the long-range force is dominated by a repulsive double-layer force originating from the charges on the mica surface and on the HSA. When the HSA concentration is 0.001 mg/ml a hard wall repulsion is encountered at a separation of about 2 nm. This distance is less than the smallest cross-section of HSA. Hence, it is clear that when the number of molecules adsorbed on the surface is small the conformation of the protein can easily be changed by an external compressive force, indicating a rather limited structural stability. On separation, an attractive minimum is observed at a distance of about 4-5 nm. This corresponds to the expected thickness of one monolayer of HSA adsorbed side-on. At higher concentration the data do not provide any evidence for surface denaturation. This,

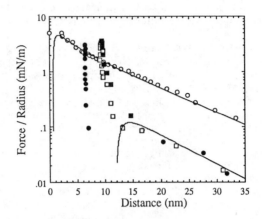

Figure 2. The forces measured between mica surfaces across a 10-3 M NaCl solution at pH 5.6. The forces were measured as a function of lysozyme concentration, 0.002 mg/ml (filled circles), 0.02 mg/ml (open squares), and 0.2 mg/ml (filled squares). Open circles represent the force measured in a pure 10^{-3} M NaCl solution at pH 5.6. The solid curves represent the forces calculated from the DLVO theory (A = 2.2 x 10^{-20} J and κ^{-1} = 9.6 nm).

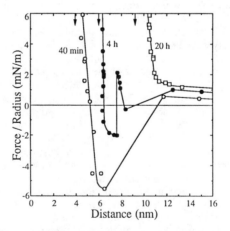

Figure 3. The forces as a function of separation between mica surfaces immersed in a solution containing 1.9 mg/ml insulin at pH 7.3. The forces were measured after different adsorption times. The arrows indicate the layer thickness under a high compressive load. The dashed lines are guides for the eye that serve to connect different parts of the force curve.

however, does not prove that no changes in conformation occur upon adsorption. For instance at a concentration of 0.01 mg/ml the range of the non-DLVO force increases to about 7-8 nm, close to the expected thickness for one side-on monolayer adsorbed on each surface. These results do not show conclusively that no molecules are adsorbed end-on. Rather, the correct conclusion is that if some end-on oriented molecules exist on the surface then the force needed to change the orientation from end-on to side-on is small compared to the electrostatic double layer force. When a high compressive force is applied the thickness decreases to about 5-6 nm. No, or a weak, adhesion force is observed between the layers. When the HSA concentration is even larger, 1 mg/ml, the range of the non-DLVO repulsion is 12-15 nm. This is considerably larger than expected for a side-on monolayer on each surface, demonstrating that some molecules are adsorbed at an angle to the surface. Under the action of a high compressive force the layer thickness is reduced to about 8-9 nm. No adhesion between the surfaces is observed under these conditions.

We note that the protein do adsorb onto mica despite that it has the same net sign on its charge as the surface. This means that either some positively charged regions adsorb to the surface or that the driving force for adsorption is not due to charge-charge interactions. It has been concluded by Norde (*29*) that one important reason for protein adsorption is an increase in entropy due to structural changes. It seems likely that this is important for the case of HSA on mica since the adsorbed layer is considerably more compressible than that formed by e.g. lysozyme, and that the HSA layer thickness at low adsorption densities is not consistent with the dimension of the protein in solution.

At high ionic strength, 0.15 M NaCl, it was found that a significant fraction of the proteins initially adsorb end-on. This is illustrated in Figure 5, that displays the forces measured during the first and a subsequent approach of the surfaces. On the first approach a repulsive force is experienced at distances below about 37 nm, corresponding roughly to the size of end-on molecules interacting with an electrostatic double-layer force. Upon further compression the increase in repulsion with decreasing separation is very limited until the surfaces reach a separation of slightly below 10 nm. The final position of the surfaces under a high compression is 4.4 nm. We interpret these findings as evidence for a pressure induced change in orientation of the initially end-on oriented fraction to side-on. When the surfaces are kept apart only some HSA molecules do change back to an end-on orientation as evidenced by the considerably less long ranged forces observed upon a subsequent approach (Figure 5).

The forces operating between mica surfaces in the presence of bovine serum albumin (*30*) have been reported to be considerably more long-ranged than those acting in the presence of HSA. At present we do not have any explanation for this difference, but the subject requires further investigations.

Amphiphilic Proteins

Some proteins have an amphiphilic character with one large hydrophobic region separated from a large hydrophilic region. These proteins are often membrane bound, or, like casein, act as stabilisers for hydrophobic emulsions. In many cases this type of protein can be expected to have a strong preferential orientation on hydrophobic surfaces. This has been shown to be the case for *proteoheparan sulfate*, a glycoprotein with a rather hydrophobic peptide region with attached strongly negatively charged hydrophilic glycosaminoglycan side chains. For this protein the peptide chain adsorb strongly to hydrophobic surfaces whereas the polysaccharide chains do not adsorb (*31*).

The forces measured between proteoheparan sulfate coated hydrophobic surfaces across a protein-free aqueous salt solution is shown in Figure 6. In a 0.1 mM

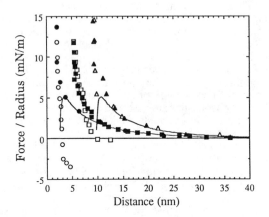

Figure 4. The forces as a function of separation between mica surfaces immersed in 10^{-3} M NaCl at pH 5.6. The solution also contained HSA at concentrations of 0.001 mg/ml (circles), 0.01 mg/ml (squares) and 1 mg/ml (triangles). Filled and unfilled symbols represent the force measured on compression and decompression, respectively. The solid curves represent the forces calculated from the DLVO theory (A = 2.2 x 10^{-20} J and κ^{-1} = 8.5 nm).

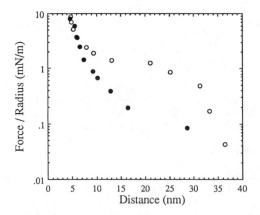

Figure 5. The forces as a function of separation between mica surfaces immersed in a solution containing 0.01 mg/ml HSA and 0.15 M NaCl at pH 5.6. Forces measured on the first approach is represented by unfilled circles, and forces measured on the second approach by filled circles.

NaCl solution a repulsive double-layer force dominates the long-range interaction. At a separation below 9 nm a steric repulsion due to compression of the glycoprotein is predominating. As the surfaces are separated, a small attractive minimum is noted at a separation of 13 nm. An addition of $CaCl_2$ to a concentration of 1.25 mM results in a significant decrease in the long range repulsive force. However, after addition of $CaCl_2$ the decay length of the force is no longer consistent with an electrostatic force. Instead the force is of steric origin and most likely due to interactions between the carbohydrate chains. We note that the compressed layer thickness also decreases somewhat upon addition of $CaCl_2$. On separation a comparatively strong attraction is observed at a separation of 10 nm. The increase in adhesion observed upon addition of $CaCl_2$ can be rationalized as resulting from a decreased repulsive interaction rather than invoking a calcium dependent attractive force. A further addition of $CaCl_2$ to 2.5 mM results in a reduction of the long-range steric force but no further decrease in the compressed layer thickness. These observations can be rationalized in terms of the decreased electrostatic repulsion between charged segments within the carbohydrate chains resulting in an entropically driven chain contraction and thus a less long ranged steric repulsion. Note that the strong electrostatic repulsion observed before addition of $CaCl_2$ precludes any determination of the range of the long range component of the steric force under this condition. For further information see *31*.

Random Coil-Like Proteins

Mucin is a very large linear and flexible glycoprotein. About 80% of the weight is due to oligosaccharides that are clustered in regions flanked by stretches composed predominantly of amino acids. It is an important type of glycoprotein since it covers many internal surfaces in the body and thus will be one of the primary molecules that e.g. drugs will interact with. The forces between hydrophobic surfaces precoated with a layer of rat gastric mucin (RGM), a weakly charged mucin, have been investigated (*32*). In this case it was found that the forces were predominantly of steric origin as evidenced by the very weak salt dependence (Figure 7). It was also noted that for the more highly charged pig gastric mucin it was very difficult to measure (quasi)equilibrium forces due to a very slow relaxation of the adsorbed layer (*32*), indicating that viscous forces may be of importance for the protective function of mucins. The interaction forces between mucin coated surfaces have also been studied by Perez (*33*).

Several theoretical models for the forces operating between polymer-coated surfaces have been developed for quasi-equilibrium (restricted equilibrium) situations (*34–36*). They predict that the same forces should be measured on approach and on separation. This is often the case when forces between surfaces coated with homopolymers are measured under poor solvency conditions (*37*). Quasi-equilibrium forces have also been observed for surfaces coated with heterogeneous polymers which adsorb strongly via specific anchor groups utilizing electrostatic forces (*38*) or hydrophobic interactions as for proteoheparan sulfate or ethylhydroxyethyl cellulose (*39*). The nonadsorbing segments may then experience good solvency conditions and for sufficiently high adsorbed amounts the polymer-coated surfaces will repel each other (*38*). However, when the polymer-surface interaction is neither very strong nor very weak nonequilibrium forces are experienced as a rule rather than as an exception. This implies that the displacement of polymers from between the surfaces and/or slow conformational changes occur. Clearly, relaxation effects in adsorbed polymer layers, including mucin, and their consequences for the forces acting between polymer-coated surfaces are very important.

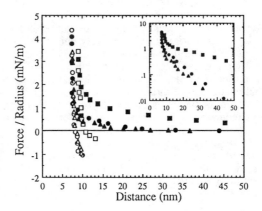

Figure 6. Comparison of the forces obtained between proteoheparan sulphate layers adsorbed at hydrophobized mica as a function of the distance of separation at different excess electrolyte concentrations. The forces were measured on approach and on separation in 0.1 mM NaCl (squares), 1.25 mM $CaCl_2$ (circles) and 2.5 mM $CaCl_2$ (triangles). Filled symbols represent forces measured on approach and unfilled symbols forces measured on separation.

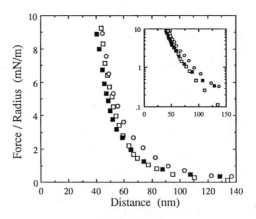

Figure 7. Forces as a function of separation after adsorption for 24 hours from a 0.1 mg/ml RGM solution, followed by dilution by a factor of 3000. The forces were measured on approach. The NaCl concentration was 10^{-4} M (open squares), 10^{-2} M (filled squares) and 0.15 M (open circles).

Protein-Surface Interactions

The desorption of proteins, like for many other polymers, is very slow. This makes it possible to first replace the protein solution with a protein-free aqueous solution and then remove one of the protein coated surfaces and instead insert one bare mica surface. At this stage it is possible to measure the forces between one protein coated surface and one bare surface. Results from such measurements for lysozyme and for human serum albumin are shown in Figure 8. As expected, replacing one of the protein coated surfaces with bare mica results in a halving of the contact separation. More interesting is that the adhesion force between the mica surface and the protein coated surface is high, considerably higher than that between two protein coated surfaces. We also note that the adhesion force is higher between negatively charged mica and HSA than between mica and lysozyme despite that HSA is weakly negatively charged and lysozyme positively charged. Again, this points to a rather limited importance of the net charge of the protein in comparison with the protein flexibility for the protein-surface interaction. Despite the strongly attractive force measured it was possible to separate the surfaces and remeasure the same force on a subsequent approach. This indicates that the proteins are most strongly bound to the surface that they initially adsorb on, and that only a limited (if any) material transfer between the two surfaces takes place. This is not very surprising since the proteins are asymmetric and will orient in such a way that they interact most favourably with the surface they adsorb on. When a second surface is brought into contact with the protein coated surface it will interact less favourably with the opposite side of the protein. It is, however, interesting that even during several minutes in contact the proteins do not reorient to interact equally favourably with both surfaces.

Pressure Induced Changes in Adsorbed Layers.

A practical problem that has been observed when studying proteins on surfaces with the surface force technique is that irreversible changes in the adsorbed layer may take place when a strong compressive force is applied. This has to do with the fact that many proteins are neither very strongly or very weakly bound to the surface causing them to be pushed out from between the surfaces only at such high forces that the surfaces have started to flatten (due to the deformation of the supporting glue). Under such conditions molecules at the edge of the flat region can leave the contact zone whereas those in the middle of the flat region will be trapped and pushed together (*40*). (This phenomenon is related to elastohydrodynamic lubrication (*41*)). When the surfaces are separated again the "lump" of proteins that has formed in the middle of the contact region will often remain at the contact position for a very long time giving rise to long-range repulsive forces. A comparison between the forces observed before and after such pressure induced changes in the layer has occurred is seen in Figure 9.

Hence, when studying proteins one should always determine how readily such changes in the adsorbed layer occur, and never apply such strong forces. As a comparison weakly bound molecules will be removed from between the surfaces already when a weak force is applied (e.g. the outer protein layer of lysozyme in Figure 2), whereas strongly bound molecules always will remain at the same position on the surface.

Conclusions

From this study it is shown that the surface force technique is suitable for the study of several aspects of proteins on surfaces, such as long-range forces, contact forces, molecular orientation and compressibility. It was found that small compact proteins

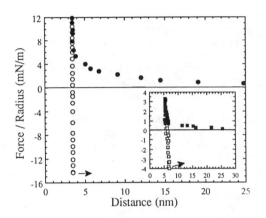

Figure 8. Forces between one protein coated surface and one bare mica surface across a 10^{-3} M NaCl solution. The forces measured in the case of HSA are represented by circles. Filled symbols represent forces measured on approach and unfilled symbols forces measured on separation. The insert show the forces in the case of lysozyme, represented by squares.

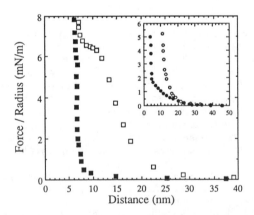

Figure 9. Forces as a function of separation between lysozyme coated surfaces (squares) before (filled symbols) and after (unfilled symbols) irreversible pressure induced changes have been introduced by applying a strong compressive force. The insert show the forces between HSA coated surfaces (circles) before (filled) and after (unfilled) pressure induced changes in the layer has taken place.

like lysozyme and insulin did not show any large conformational changes upon adsorption. When the adsorption had reached equilibrium the adsorbed lysozyme nearly neutralised the surface charge of the mica substrate. When separating the lysozyme or insulin monolayers from contact, an adhesion force is present, demonstrating the existence of an inter-protein attraction that is involved in the adsorption of an outer layer. It was found that negatively charged human serum albumin adsorbs onto negatively charged mica. Small surface induced structural changes take place upon adsorption and further structural changes can be induced by applying an external compressive force. Hence, it seems plausible to believe that compact globular proteins do not change conformation to the same degree as more soft/flexible proteins do. The forces between proteoheparan sulfate and between mucin layers, that expose polymer chains towards the solution, contain an important steric component, similar to that acting between surfaces coated with nonbiological polymers.

Literature Cited

1. Norde, W. *Adv. Colloid Interface Sci.* **1986**, *25*, 267.
2. Wahlgren, M. *Adsorption of Proteins and Interactions with Surfactants at the Solid/Liquid Interface*, PhD-thesis, University of Lund, 1992.
3. Leckband, D.; Israelachvili, J. N. *Enzyme Microb. Technol.* **1993**, *15*, 450.
4. Luckham, P. F.; Hartley, P. G. *Adv. Colloid Interface Sci.* **1994**, *49*, 341.
5. Israelachvili, J. N.; Adams, G. E. *J. Chem. Soc. Faraday Trans. I.* **1978**, *74*, 975.
6. Parker J. L.; Christenson H. K.; Ninham B. W. *Rev. Sci. Instrum.* **1989**, *60*, 3135.
7. Derjaguin, B. W. *Koll. Z. Z.* **1934**, *69*, 155.
8. Israelachvili J. N. *Intermolecular and Surface Forces - with applications to colloidal and biological systems*, Academic Press, London, 1991, 2nd edition.
9. Pashley, R. M. *J. Colloid Interface Sci.* **1981**, *80*, 153.
10. Horn, R. G.; Israelachvili, J. N.; Pribac, F. *J. Colloid Interface Sci.* **1987**, *115*, 480.
11. Nylander, T.; Kékicheff, P.; Ninham, B. *J. Colloid Interface Sci.* **1994**, *164*, 136.
12. Norde, W.; Lyklema, J. *J. Colloid Interface Sci.* **1978**, *66*, 285.
13. van Dulm, P.; Norde, W.; Lyklema, J. *J. Colloid Interface Sci.* **1981**, *82*, 77.
14. Claesson, P. M.; Arnebrant, T.; Bergenståhl, B.; Nylander, T. *J. Colloid Interface Sci.* **1989**, *130*, 130.
15. Kékicheff, P.; Ducker, W. A.; Ninham, B. W.; Pileni, M. P. *Langmuir* **1990**, *6*, 1704.
16. Afshar-Rad, T.; Bailey, A. I.; Luckham, P. F.; MacNaughtan, W.; Chapman, D. *Colloids Surf.* **1988**, *31*, 125.
17. Luckham, P. F.; Ansarifar, M. A. *British Polymer J.* **1990**, *22*, 233.
18. Lee, C.-S.; Belfort, G. *Proc. Natl. Acad. Sci. USA* **1989**, *86*, 8392.
19. Norde, W.; Favier, J. P. *Colloids Surf.* **1992**, *64*, 87.
20. Deonier, R. C.; Williams, J. W. *Biochemistry* **1970**, *9*, 4260.
21. Tilton, R. D.; Blomberg, E.; Claesson, P. M. *Langmuir* **1993**, *9*, 2102.
22. Blomberg, E.; Claesson P. M.; Fröberg, J. C.; Tilton, R. D. *Langmuir* **1994**, *10*, 2325.
23. Pekar, H. K.; Frank, B. H. *Biochemistry* **1972**, *11*, 4013.
24. Holloday, L. A.; Ascoli, M.; Puett, D. *Biochimi. Biophys. Acta* **1977**, *494*, 245.
25. Carter, D. C.; He, X.-M.; Munson, S. B.; Twigg, P. D.; Gernert, K. M.; Broom, M. B.; Miller, T. Y., *Science* **1989**, *244*, 1195.
26. He, X.-M.; Carter, D. C., *Nature* **1992**, *358*, 209.

27. Blomberg, E.; Claesson, P. M.; Gölander, C.-G. *J. Dispersion Sci. Technol.* **1991**, *12*, 179.
28. Blomberg, E.; Claesson P. M.; Tilton, R. D. *J. Colloid Interface Sci.* in press.
29. Arai, T.; Norde, W. *Colloids Surf.* **1990**, *51*, 1.
30. Fitzpatrick, H.; Luckham, P. F.; Eriksen, S.; Hammond, K. *Colloids Surf.* **1992**, *65*, 43.
31. Malmsten, M.; Claesson, P.M.; Siegel, G. *Langmuir* **1994**, *10*, 1274.
32. Malmsten, M.; Blomberg, E.; Claesson, P. M.; Carlstedt, I.; Ljusegren, I. *J. Colloid Interface Sci.* **1992**, *151*, 579.
33. Perez, E.; Proust, J. E. *J. Colloid Interface Sci.* **1987**, *118*, 182.
34. de Gennes, P. G. *Adv. Colloid Polymer Sci.* **1987**, *27*, 189.
35. Scheutjens, J. M. H. M.; Fleer, G. J. *Macromolecules* **1985**, *18*, 1882.
36. Klein, J.; Pincus, P. *Macromoleules* **1982**, *15*, 1129.
37. Patel, S. S.; Tirrell, M. *Annu. Rev. Phys. Chem.* **1989**, *40*, 597.
38. Hadziioannou, G., Patel, S., Granick, S., and Tirrell, M. J. Am. Chem. Soc. **1986**, 108, 2869.
39. Malmsten, M.; Claesson, P.M.; Pezron, E.; Pezron, I. *Langmuir* **1990**, *6*, 1572.
40. Blomberg, E.; Claesson, P. M.; Christenson, H. K. *J. Colloid Interface Sci.* **1990**, *138*, 291.
41. Roberts, A. D.; Tabor, D. *Proc. R. Soc. London* **1971**, *A325*, 323.

RECEIVED March 1, 1995

Chapter 22

Neutron Reflectivity of Adsorbed Protein Films

Peter J. Atkinson[1], Eric Dickinson[1], David S. Horne[2,4], and
Robert M. Richardson[3]

[1]Procter Department of Food Science, University of Leeds,
Leeds LS2 9JT, United Kingdom
[2]Hannah Research Institute, Ayr KA6 5HL, Scotland
[3]School of Chemistry, University of Bristol, Bristol BS8 1TS,
United Kingdom

Direct information on the thickness and structure of adsorbed milk
protein layers at oil/water and air/water interfaces has been obtained
using the technique of specular neutron reflection on the CRISP
reflectometer at the Rutherford-Appleton Laboratory. Segment density
profiles are presented for β-casein adsorbed at the air/water interface
as a function of bulk solution pH and protein concentration. The
results are compared with film properties derived using other
techniques. Data from a neutron reflectivity study of the competitive
adsorption of β-casein and the nonionic surfactant, $C_{12}E_6$, are also
included.

By virtue of their amphiphilic nature, proteins are very surface active and adsorb
readily at interfaces. This adsorption behavior is of great practical importance in
fields as diverse as the pharmaceutical, biomedical, cosmetic and food industries.
Our particular area of interest is the food industry, where proteins, especially dairy
proteins, are employed in the stabilization of oil-in-water emulsions. Here the protein
layer protects the oil droplets against immediate coalescence and provides long-term
stability against flocculation and eventual coalescence.

Despite much research on protein adsorption to various interfaces, there is
still relatively little knowledge of the actual conformations adopted by the adsorbed
protein molecules. In this paper, we describe the technique of neutron reflection
which yields such information directly as the density distribution normal to the
interface. The protein we employ is the milk protein, β-casein, measuring the
conformational profile at the air/water interface as a function of solution pH and
protein concentration.

Many biocolloidal systems of technical and commercial importance contain
small molecule surfactants as well as proteins. The distribution of these molecules
between the surface and the bulk phase affects the properties of the systems
(stability, rheology etc.)(1). While the competitive displacement of milk proteins by

[4]Corresponding author

0097–6156/95/0602–0311$12.00/0
© 1995 American Chemical Society

nonionic and anionic surfactants has been demonstrated experimentally at the planar oil/water interface (2) and at the emulsion droplet surface (3) and the behaviour modelled by Monte Carlo (4) and molecular dynamics (5) simulation, the detailed structure of the mixed protein/surfactant layers remains unknown. Preliminary measurements of the competitive displacement of β-casein by the nonionic surfactant, $C_{12}E_6$ (hexaoxyethylene dodecyl ether) are detailed here, together with a description of a strategy for locating the surfactant in the mixed layer.

Neutron Reflection

A general description of the neutron reflection technique has been provided by Richards and Penfold(6). Its application to the study of the conformation of proteins adsorbed at interfaces has been described in detail previously(7). In many of their interactions with matter, the behaviour of slow neutrons can be expressed in theoretical terms based on analogies with light. Thus scattering, interference, diffraction and reflection all have their counterparts in neutron behavior and follow very similar laws to electromagnetic radiation. The specular reflection of neutrons from an interface thus depends quantitatively on the refractive index profile perpendicular to the interface, which is related to the coherent scattering-length density. In turn this scattering-length density profile depends on the chemical composition and number density of scattering species. Hence detailed information regarding the interfacial region is contained in the reflectivity profile. A further advantage of neutrons is their short wavelengths, some two to three orders of magnitude smaller than light, so that interfacial layers are probed on the molecular scale. Neutrons also possess the further distinct advantage that isotopes of the same element may manifest different scattering lengths, hydrogen and deuterium being the most advantageous and most widely used. This is exploited in two ways here, first by contrast-matching the solvent water to air or other liquid, so that only the adsorbed layer is "seen" and the protein profile generates the reflectivity observed and secondly by employing both deuterated and hydrogenated versions of the nonionic surfactant in the competitve adsorption studies.

The essence of a neutron reflection experiment is to measure the neutron reflectivity as a function of the momentum transfer vector perpendicular to the reflecting surface. In principle, it should be possible to transform the measured reflectivity data to obtain the adsorbed segment density profile directly. In practice, as is often the case in attempting to implement a mathematical inversion procedure, the lack of high quality data in ranges of wave vector largely inaccessible due to experimental limitations precludes this approach. Detailed analysis has therefore been accomplished by adopting a two-layer model function for the profiles and seeking a best-fit to the data.

Materials and Methods

The β-casein (genetic variant B, obtained from a single cow homozygous for this variant) was prepared from acid casein precipitated from fresh skim milk. Separation from the other caseins was achieved by ion-exchange chromatography on a Sepharose Q column. The fraction corresponding to β-casein was dialysed exhaustively against distilled water and freeze-dried. Polyacrylamide gel electrophoresis and fast protein liquid chromatography demonstrated its purity to be greater than 99%. Research grade $C_{12}E_6$ was purchased from Sigma Chemicals and used as supplied (stated purity 99%). The deuterated surfactant, denoted d-$C_{12}E_6$, was a gift from Unilever Research, Port Sunlight, Cheshire, U.K. D_2O was obtained from MSD Isotopes Ltd. All solutions were prepared using Milli-Q ultrapure water with Analar grade reagents as buffering agents.

The reflectivity profiles were measured using the CRISP instrument at the Rutherford-Appleton Laboratory (Chilton, Oxfordshire, U.K.) (8). This instrument uses a pulsed polychromatic beam of neutrons and records their time of flight so as to vary wavelength λ at the selected fixed grazing incidence angle θ. In a typical experiment a buffered solution (20 mmol/dm³ imidazole/HCl, pH 7.0) containing β-casein and a variable amount of $C_{12}E_6$ all dissolved in air-contrast-matched-water (CMW) (8%D_2O) was carefully poured into a clean Teflon trough, approximately 60 cm³ of solution being required to give a proud meniscus above the edge of the trough. Further details of the apparatus and procedure may be found elsewhere (7,9,10).

Results and Discussion

We first consider results for β-casein adsorbed to air-contrast-matched water in the absence of surfactant. Preliminary analysis of the reflectivity profiles is best accomplished by application of the Guinier approximation to the data obtained at low scattering vector, q $(q=(4\pi/\lambda)\sin \theta)(7,11)$. For an adsorbed layer on top of a homogeneous subphase the reflectivity can be written as a sum of three terms in the kinematic approximation(12).

$$R(q) \approx R_0(q)\Delta\rho^2 - R_1\, \Delta\rho + R_2(q) \qquad (1)$$

The first term is the Fresnel reflectivity from a clean sharp interface (no adsorbed layer):

$$R_F(q) = R_0(q)\Delta\rho^2 = 16\pi^2\, \Delta\rho^2/q^4 \qquad (2)$$

The quantity $\Delta\rho$ is the difference in scattering length density between the incident medium and the subphase. When the subphase is contrast-matched to the air, $\Delta\rho = 0$. Both the first term (R_F) and the second term ($R_1(q)\, \Delta\rho$) disappear and the reflectivity is simply given by

$$R(q) = R_2(q) \qquad (3)$$

In practice, when $\Delta\rho = 0$ the reflectivity is weak and so the kinematic approximation holds very well. Applying a Guinier-type approximation, the behaviour of the function $R_2(q)$ at low q can be written as

$$R_2(q) = (16\pi^2/q^2)m^2\exp(-q^2\sigma^2) \qquad (4)$$

where σ is the second moment of the adsorbate distribution function normal to the interface

$$\sigma = (\langle z^2 \rangle - \langle z \rangle^2)^{1/2} \tag{5}$$

and m is the scattering length density integrated over the adsorbed layer:

$$m = \int \rho (z) \, dz \tag{6}$$

A Guinier plot of $ln\ q^2R(q)$ against q^2 thus gives a straight line of slope equal to $-\sigma^2$ and intercept equal to $ln[16\pi^2 m^2]$. Hence the slope gives a measure of the thickness of the adsorbed layer and the intercept gives the quantity of material adsorbed at the interface. The integrated scattering length density can be related to the adsorbed amount in units of mass per unit area, Γ, knowing the atomic composition of the adsorbate and using the equation

$$\Gamma = (M_w m/N_A) \, [\Sigma \, b_i]^{-1} \tag{7}$$

where M_w is the molecular weight of the adsorbate, N_A is Avogadro's number, and b_i is the scattering length of the ith atom in the adsorbate molecule. Table I lists the values calculated for surface coverage and the square root of the second moment of the adsorbed layer profile from profiles measured as a function of buffer pH and bulk protein concentration.

Table I. Derived Guinier plot parameters for pure β-casein adsorbed at air/contrast-matched-water interface

pH	Bulk Protein Conc. (wt %)	Γ (mg/m^2)	σ (nm)
7.0	5×10^{-2}	2.7 ± 0.1	1.80 ± 0.06
7.0	5×10^{-3}	2.05 ± 0.1	1.65 ± 0.07
7.0	5×10^{-4}	1.91 ± 0.1	1.81 ± 0.07
6.0	5×10^{-3}	2.85 ± 0.1	2.39 ± 0.08
5.4	5×10^{-3}	3.90 ± 0.15	2.57 ± 0.08

The bulk concentration of 5×10^{-3} wt% lies in the expected plateau region of the adsorption isotherm of β-casein at both the air/water and oil/water interfaces (13). The calculated surface coverage of $2.05 +/- 0.10$ mg/m^2 at pH 7.0 is close to that reported by Graham and Phillips (14) ($\Gamma \approx 2.5$ mg/m^2). The observed increase in surface coverage inferred from the Guinier plots as bulk protein concentration is increased to 5×10^{-2} wt% is also consistent with their findings as is our result that reducing the protein content to 5×10^{-4} wt% gives little change in surface coverage

(Table I). Layer thickness, calculated from the slopes of the Guinier plots, remains almost constant across the protein concentration range, with possibly a marginal increase at higher surface coverage but a similar slight increase at the lowest concentration employed. The increase at the higher concentration would be expected if secondary layer coverage were being initiated, but the behaviour at low concentrations is not and merits further study to determine if these variations are significant.

We observe a considerable increase in reflectivity as the solution pH is decreased. The layer thickness at pH 5.4 derived from the Guinier plot is some 50% greater than the pH 7.0 value and the adsorbed amount almost 100% larger. This behaviour is consistent with the increased aggregation of β-casein molecules in solution as the pH approaches the isoelectric point pI (net molecular charge tends to zero), since one expects increased protein surface coverage under conditions where the protein solution is close to phase separation or precipitation. The values calculated here for Γ and σ at pH 5.4 are almost exactly the same as those reported previously (7) for β-casein adsorbing from unbuffered distilled water at air/water and oil/water interfaces. We conclude from this that the effective pH in that preliminary experiment was close to pI, partially due to the buffering action of the protein itself, and partially to the additional acidity resulting from dissolved carbon dioxide.

The preceding observations are completely independent of any model assumed for the structure of the adsorbed protein layer. Further analysis of the reflectivity data was accomplished by model fitting, as described in our earlier publication (7), using the matrix method of Abélès as summarised in ref 15 to calculate the function $R_2(q)$ from an assumed model for the scattering length density function $\rho(z)$. Our model divides the interface into a number of uniform layers. The number of layers, their thickness, scattering length density and a roughness parameter may all be adjusted to achieve an optimum fit to the data, using a nonlinear least-squares fitting routine. The scattering length density of the adsorbed layers is closely related to the protein volume fraction, ϕ, by the formula

$$\rho(z) = \phi(z)\rho_P + (1 - \phi(z))\rho_S \qquad (8)$$

where ρ_P is the scattering length density of the pure protein, ρ_S is that of the solvent, and $\phi(z)$ is the volume fraction profile of the protein, z being the distance measured normal from the interface. Best fits were achieved with a two layer model exemplified by the plot shown in Figure 1 for the adsorbed β-casein profile at pH 7.0 at a bulk concentration of 5×10^{-3} wt%. The calculated segment density profile indicates a dense inner layer of volume fraction (ϕ_1) 0.9 and thickness (d_1) approx. 1 nm immediately adjacent to the interface and beyond that extending into the aqueous phase a more tenuous and extensive outer layer, 4-5 nm thick (d_2) and of volume fraction (ϕ_2), 0.14. Such layer thicknesses agree well with hydrodynamic layer thicknesses inferred in studies of β-casein adsorption to polystyrene latex particles from dynamic light scattering measurements (16,17), bearing in mind the known sensitivity of hydrodynamic measurements for segment density at the periphery of the adsorbed layers.

The calculated variations in the fitted parameters with solution pH are plotted in Figures 2a and 2b. The volume fraction of the inner layer (ϕ_1) remains high at 0.9

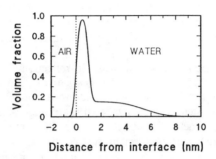

Figure 1. Two-layer model fit of segment density distribution calculated for β-casein (5×10^{-3} wt%) adsorbed at air/ water interface at pH 7.0.

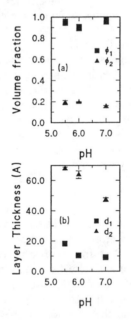

Figure 2. Results of fitting two-layer model to neutron reflectivity data obtained for β-casein (5×10^{-3} wt%) adsorbed at the air/water interface with subphase buffered to pH values indicated using imidazole (20mM) and HCl. (a) Variation of volume fraction parameters (ϕ_1 inner and ϕ_2 outer) with pH. (b) Variation of derived layer thicknesses (d_1 inner and d_2 outer) with pH.

to 0.95 as the solution pH is decreased. It is difficult to see how it could increase further. That of the outer layer (ϕ_2) increases by about a third from 0.14 to 0.19 as the pH is decreased from 7.0 to 5.4. The thickness of the inner layer (d_1) is almost doubled from just under 10 Å at pH 7.0 to 18 Å at pH 5.4. The outer layer thickness increases even more from 48 to 68 Å over the same pH range. It seems rather likely therefore that a second layer of protein forms on approaching pI, as schematically described in Figure 3. The train-and-tail model for the adsorbed protein molecule at pH 7.0 has been inferred from our earlier studies and fits well the two-layer structure derived from neutron reflectance measurements. Adding a second layer of β-casein should approximately double d_1 whilst ϕ_1 should remain fairly constant. The volume fraction, ϕ_2, should increase and the outer layer thickness, d_2, should also increase as indicated in the simple model in the diagram. The data for pH 6.0 seem to suggest the additional protein packs loosely into the outer diffuse layer. The increase in d_2 is quite large and ϕ_2 is increased, but d_1 and ϕ_1 remain about the same. If the extra protein sat neatly on the first layer, then d_2 would increase by as much as d_1. That it should be observed to be much greater is an indication that the additional protein is perhaps adopting a more extended or simply different conformation from the molecular layer nearest the interface.

When the same model analysis is applied to the reflectivity data obtained as a function of bulk protein concentration at pH 7.0, a similar picture emerges. Again the best-fits are obtained with a two-layer model for the adsorbed protein molecule. Table II lists the parameters determined in these fitting procedures. Increasing the protein concentration by a factor of 10 from the standard bulk concentration of 5×10^{-3} wt% increased the surface coverage to 2.7 mg/m^2 (Table I). Fitting indicates that this produces no increase in the volume fraction of the inner layer, but an increase in its thickness from 10.1 Å to 13.4 Å, almost exactly compensating for the increase in protein loading. No changes are found in either the thickness or volume fraction of the outer layer as a result of this increase in Γ. Apparently the extra protein is accommodated by the molecules compacting closer together in this instance rather than forming a second molecular layer as the behaviour observed on lowering the solution pH suggests. Decreasing the bulk protein concentration to 5×10^{-4} wt% produces little change in surface coverage, as calculated from the intercept of the Guinier plot. The fitting to a model profile gives an inner layer marginally thinner than the 5×10^{-3} wt% profile at 9.5 Å and of a lower volume fraction than previously calculated for this protein ($\phi_1 = 0.85$). The outer layer features an unchanged volume fraction of 0.14 but now appears thinner at 42 Å. At the lower protein concentration, the adsorbed molecule has apparently more space to move around on the surface and the tail extending out into the aqueous phase is more flexible and less constrained by its neighbours.

These neutron reflectivity studies of the adsorption of β-casein at the air/water interface are providing a detailed picture of the behaviour of the adsorbed molecule, largely substantiating the scenario inferred from hydrodynamic measurements of adsorbed layer thickness and changes in this parameter consequent on enzymic digestion of the adsorbed molecule (*16-19*).

Table II. Fitted parameters for segment density profiles derived from reflectivity data for β-casein adsorbed at air/contrast-matched-water interface (pH 7.0) from different bulk protein concentrations

Bulk Protein Concentration (wt%)	d_1 (Å)	ϕ_1	d_2 (Å)	ϕ_2	Γ (mg. m^{-2})
5×10^{-4}	9.5	0.85	42	0.14	1.91
5×10^{-3}	10.1	0.90	49	0.14	2.05
5×10^{-2}	13.4	0.92	49	0.14	2.70

The addition of hydrogenated nonioinic surfactant, $C_{12}E_6$, to the aqueous phase (pH 7.0) at concentrations of the order of 1×10^{-4} wt% and above leads to a substantial reduction in the reflectivity of the β-casein layer adsorbed from a bulk protein solution of 5×10^{-3} wt%. Figure 4 shows plots of the logarithm of the measured reflectivity (R) against scattering vector (q) for four different surfactant concentrations. The especially large fall in reflectivity between surfactant concentrations of 2.5×10^{-4} wt% and 5×10^{-4} wt% is suggestive of substantial displacement of protein from the air/water interface in this surfactant concentration range. Corkhill and co-workers (20) determined the cmc in water to be 5×10^{-3} wt% for this surfactant, so that the observed protein displacement is taking place predominantly below this critical concentration. The molecular scattering length for hydrogenated $C_{12}E_6$ calculated on the basis of its chemical composition is some 400 times smaller than for β-casein. Allowing for the 50-fold difference in molecular weight, a complete layer of this surfactant is still calculated to be approximately 8 times less reflective than a monolayer of protein. The low reflectivities observed at surfactant concentrations of 1×10^{-2} wt% are consistent with these calculations and indicate total displacement of protein from the air/water interface at these surfactant levels at neutral pH. Because the hydrogenated surfactant is such a low-power reflector of neutrons relative to the β-casein, it is possible to calculate from the Guinier parameters derived from the reflectivity data a titration curve for the displacement of protein by surfactant (11). Such plots reproduce the behaviour seen in more conventional chemical laboratory studies of the competitive adsorption between β-casein and nonionic surfactant (21). With the technique of neutron reflection relying heavily on model fitting for data interpretation, it is gratifying it can also reproduce the results of the simpler, less esoteric methods. This imparts greater faith to the reliability and acceptability of parameters derived from the more complicated analytical procedures required for full interpretation of the data. Thus we anticipate that a full analysis of the reflectivity data will provide a detailed picture of the composition and structure of the mixed protein/surfactant layer. To exploit fully the advantage provided in neutron reflection of the varying contrasts of different isotopes, we have recently performed experiments involving deuterated surfactant and mixtures of deuterated and hydrogenated surfactant to provide a range

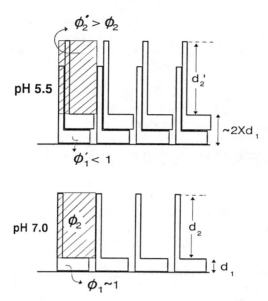

Figure 3. Highly idealised schematic representation of adsorption of β-casein at the air/water interface showing single layer at pH 7.0 and stacking of layers at pH 5.4 with symbols as defined for Figure 2 and primes distinguishing parameters at pH 5.4.

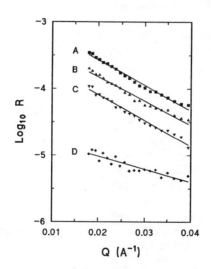

Figure 4. Influence of non-ionic hydrogenated surfactant $C_{12}E_6$ on reflectivity of β-casein at air/contrast-matched-water interface (5×10^{-3} wt% protein pH 7.0). The logarithm of the reflectivity is plotted against wave transfer vector, q, for various concentrations of hydrogenated surfactant: A, 0 wt%; B, 1×10^{-4} wt%; C, 2.5×10^{-4} wt%; D, 5×10^{-4} wt%.

of contrasts on a contrast matched subphase which will allow the position of the surfactant in the mixed layer to be located unambiguously. Analysis of this data is now underway and a full interpretation will be presented in a subsequent paper.

Acknowledgments

We thank Drs. Jeff Penfold and John Webster of the Rutherford-Appleton Laboratory for their technical assistance and enthusiatic support of this research. Financial support from SERC, AFRC through their LINK Scheme and The Scottish Office Agricultural and Fisheries Department is acknowledged.

Literature Cited

1. Dickinson, E. *An Introduction to Food Colloids*; Oxford University Press: Oxford, 1992.
2. Dickinson, E.; Euston, S.R.; Woskett, C.M. *Prog. Colloid Polym. Sci.* **1990**, *82*, 65.
3. Courthaudon, J.-L.; Dickinson, E.; Dalgleish, D.G. *J. Colloid Interface Sci.* **1991**, *145*, 390.
4. Dickinson, E.; Euston, S.R. *Molec. Phys.* **1989**, *68*, 407.
5. Dickinson, E.; Pelan, E.G. *J. Chem. Soc. Faraday Trans.* **1993**, *89*, 3453.
6. Richards, R.W.; Penfold, J. *Trends Polym. Sci.* **1994**, *2*, 242.
7. Dickinson, E.; Horne, D.S.; Phipps, J.S.; Richardson, R.M. *Langmuir* **1993**, *9*, 242.
8. Penfold, J.; Ward, R.C.; Williams, W.G. *J. Phys. E, Sci. Instrum.* **1987**, *20*, 1411.
9. Penfold, J.; Thomas, R.K. *J. Phys. Condensed Matter* **1990**, *2*, 1369.
10. Cosgrove, T.; Phipps, J.S.; Richardson, R.M. *Colloids Surf.* **1992**, *62*, 199.
11. Dickinson, E.; Horne, D.S.; Richardson, R.M. *Food Hydrocolloids* **1993**, *7*, 497.
12. Crowley, T.L. *D. Phil. Thesis* Oxford Univ. 1984.
13. Benjamins, J.; de Feijter, J.A.; Evans, M.J.A.; Graham, D.E.; Phillips, M.C. Faraday Disc. Chem. Soc. **1975**, *59*, 218.
14. Graham, D.E.; Phillips, M.C. *J. Colloid Interface Sci.* **1979**, *70*, 403.
15. Heavens, O.S. *Optical Properties of Thin Films;* Butterworths: London, 1955.
16. Dalgleish, D.G. Colloids Surf. **1990**, *46*, 141.
17. Brooksbank, D.V.; Davidson, C.M.; Horne, D.S.; Leaver, J. *J. Chem Soc. Faraday Trans.* **1993**, *89*, 3419.
18. Leaver, J.; Dalgleish, D.G. *Biochim. Biophys. Acta* **1990**, *1041*, 217.
19. Dalgleish, D.G.; Leaver, J. *J. Colloid Interface Sci.* **1991**, *141*, 228.
20. Corkhill, J.M.; Goodman, J.F.; Ottewill, R.H. *Trans Faraday Soc.* **1961**, *57*, 1627.
21. Dickinson, E.; Tanai, S. *J. Agric. Food Chem.* **1992**, *40*, 179.

RECEIVED February 10, 1995

Chapter 23

Mechanisms and Consequences of Protein Adsorption on Soil Mineral Surfaces

H. Quiquampoix[1], J. Abadie[1], M. H. Baron[2], F. Leprince[1],
P. T. Matumoto-Pintro[1], R. G. Ratcliffe[3], and S. Staunton[1]

[1]Laboratoire de Science du Sol, Institut National de la Recherche Agronomique, 2 place Pierre Viala, 34060 Montpellier, France
[2]Laboratoire de Spectrochimie Infrarouge et Raman, Centre National de la Recherche Scientifique, 8 rue Henry Dunant, 94320 Thiais, France
[3]Department of Plant Sciences, University of Oxford, South Parks Road, Oxford OX1 3RB, United Kingdom

Both the catalytic activity of extracellular enzymes secreted by microorganisms for the degradation of soil organic matter (C, N, P and S cycles) and the rate of biodegradation of proteins considered as substrates in the N cycle are influenced by adsorption on soil mineral surfaces. The main consequence of adsorption is a pH-dependent modification of the macromolecular conformation. This phenomenon has been followed for albumin adsorption on montmorillonite by its effect on the modification of the secondary structure using FTIR and on the modification of the specific interfacial area using NMR. The relative importance of electrostatic and hydrophobic interactions has been deduced from co-adsorption experiments of albumin with its methylated derivative. The results obtained are used to interpret the effects of several surfaces (clay minerals, oxyhydroxides, clay-humic complexes) on the catalytic activities of different enzymes (proteases, glucosidases, phosphatases, phytases).

By far the greatest incentive for research in the domain of protein adsorption on solid surfaces arises from the numerous technological and medical applications that depend on this phenomenon (1). These include immobilized enzyme reactors, protein chromatographic separation, solid phase immunoassays, biocompatibility of medical implants and prostheses and fouling of contact lenses. However it is often forgotten that protein adsorption is also an important natural phenomenon in terrestrial and aquatic ecosystems (2-4). Bacteria and fungi involved in the biodegradation of the organic matter found in these natural environments secrete extracellular enzymes. High molecular weight polymers cannot cross the membranes of the microorganisms and extracellular enzymes are necessary to cleave these polymers into their constituent monomers, solubilizing the organic matter and thereby rendering it easily transportable by the membrane permeases. A characteristic of the soil environment is the high ratio of the surface area of the solid to the volume of the liquid filled pore space. Together with the high adsorptive properties of the soil clay minerals this indicates that the interaction of extracellular enzymes with mineral surfaces is likely to be important. The process also occurs in aquatic environments where colloidal matter is in suspension, but in a more dilute regime. The adsorption of enzymes on negatively charged surfaces is known to shift the optimum pH of the catalytic activity towards alkaline values (5-8) and, more generally, to reduce this overall activity (9). Until now no general agreement has been reached on the mechanisms responsible for the optimum pH shift despite the practical importance of this phenomenon in ecological and biotechnological processes. The work presented in this article should help to elucidate the mechanisms responsible for these perturbations of enzyme activity in the adsorbed state.

Proteins may also act as substrates for enzyme action in soils. The process known as ammonification in the nitrogen cycle, which results in ammonium formation, is largely controlled by proteases. The interaction of proteins with mineral surfaces can protect them from the proteolytic action (formation of interlayer complexes with clays) or, conversely, can make them more susceptible to proteolytic attack by rendering the peptide bonds more accessible (denaturation on the surfaces). Thus, either by an effect on the proteases or by an effect on the substrates, mineral surfaces can affect the N cycle in the soil in a complex way.

However the importance of the study of the interaction between soil clay mineral surfaces and enzymes extends beyond the limits of soil science. Clays are a subdivision of the phyllosilicate mineral class and therefore have many properties in common with other members of the class. Phyllosilicates are increasingly studied in surface science and related fields. For example muscovite, a phyllosilicate from the mica group, is the only mineral that has a surface which is sufficiently smooth at a molecular level over macroscopic dimensions to be usable for the surface force apparatus. This property also makes it a support surface of choice for atomic force microscopy. For these reasons the surface forces of phyllosilicates are probably amongst the better understood. Another interesting property results from the origin of the electronegative charge of the basal surface of clays, namely isomorphic substitutions in the crystal lattice. A range of phyllosilicates can therefore exist with differing surface electrical charge but the same siloxane surface, according to the extent of these substitutions. This situation differs considerably from the polymer surfaces where a change in the electrical charge implies a change in the chemistry of the interface (nature and density of the surface monomers).

Experimental

The adsorption of proteins on various minerals was studied in dilute aqueous suspension. The following reference minerals were used: Wyoming montmorillonite, saturated in sodium; goethite, synthesized according to the method of Schwertmann (10); talc, obtained from Luzenac. The clay fractions of soils were prepared by granulometric separation. The adsorption isotherms of non-enzymatic proteins were obtained by a depletion method and the protein concentration measured by UV spectroscopy.

Nuclear Magnetic Resonance Spectroscopy. The adsorption of proteins on clays displaces the charge compensating cations from the surface. This phenomenon has been used to follow the surface coverage of montmorillonite (1 g dm^{-3}) by bovine serum albumin (BSA) (11). A small quantity of a paramagnetic cation, Mn^{2+}, was added to the clay suspended in a phosphate buffer. The release of Mn^{2+} from the surface caused by protein adsorption was followed by its interaction with the orthophosphate ions using ^{31}P NMR spectroscopy. This interaction results in a line broadening of the ^{31}P NMR signal of orthophosphate due to a decrease of its spin-spin relaxation time. The line broadening effect is proportional to the concentration of released Mn^{2+}, with no contribution from surface adsorbed Mn^{2+}. The orthophosphate signal was recorded at 121.49 MHz on a Bruker CXP 300 spectrometer.

Fourier Transform Infrared Spectroscopy. Transmission spectra of BSA (30 g dm^{-3}) in solution or in presence of montmorillonite (57 g dm^{-3}) were recorded on a FTIR Perkin-Elmer 1720 spectrometer between 1350 and 1800 cm^{-1} in ^2H_2O at different p^{2}H and after 4 h (*12*). The decomposition of the amide I signal was performed by (i) acquisition of the frequencies of the different components from the second derivative and deconvolution of the spectra, (ii) selection of a line shape (0.25 Gaussian / 0.75 Lorentzian) and a band width (12 cm^{-1}) fitting the experimental spectra, and (iii) calculation of the intensity of the components by a least squares iterative curve fitting.

Enzyme Activity Measurements. The catalytic activity of enzymes was measured using three procedures (*13*). Procedure A is a measurement of the enzyme activity without adsorbent surfaces and acts as a control. Procedure B measures the enzyme activity in the presence of adsorbent surfaces, but all the enzyme is not necessarily adsorbed. Procedure C measures the activity of the supernatant after centrifugation of B and represents the contribution of non-adsorbed enzyme to the overall activity measured by procedure B. From these primary data, more refined parameters can be calculated. The relative activity R of the adsorbed enzyme is the ratio of the activity due to the fraction adsorbed (B-C) and that of an equal quantity of enzyme in solution (A-C), namely R = 100 (B-C) / (A-C). The proportion of non-adsorbed enzyme is F = 100 C / A.

When the effect of an initial pH of adsorption on the catalytic activity at another pH was measured, the enzyme and the mineral were firstly incubated for 2h at the adsorption pH, then the pH was adjusted to the reaction pH and incubated for another 2h period, and after that the catalytic activity was measured.

Discussion of the Interaction between BSA and Montmorillonite

The proteins most relevant to this study are the extracellular enzymes secreted by soil bacteria and fungi or plant roots. However, the major limitation to physicochemical studies is the availability of a sufficient amount of well purified enzymes, although in the near future progress in molecular biology will improve this pervasive problem in biochemistry. In particular adsorption isotherms and low sensitivity spectroscopical methods such as NMR and FTIR require large amounts of proteins. For this reason, we have chosen BSA as a model protein for our studies. This protein has the two advantages it is commercially available in a well purified state at a reasonable price, and it is among the most studied proteins for their adsorption properties, due to the problem of blood thrombogenicity on the artificial surfaces of prostheses (*14*).

Models of Protein Adsorption *vs* pH. It has been often observed that the maximum adsorption on surfaces bearing an electrical charge occurs at the isoelectric point of the protein (*15*). Most of the hypotheses currently put forward on the mechanisms responsible for this phenomenon favour symmetric models. By symmetric model we mean that the same fundamental reason is given for the decreasing adsorption above and below the i.e.p.. Either an increasing electrostatic lateral repulsion between like-charged (positive below the i.e.p., negative above) proteins arises at pH distant from the i.e.p. range making the surface less densely covered (*16*), or alternatively this electric charge weakens the structural stability of the protein by intramolecular repulsions and the protein spreads out on the surface, requiring thus a lesser amount to saturate it (*17*).

But the main weakness of these assumptions is that no independent measurements of the surface coverage and of the number of adsorbed layers can be made to support them, since it is clear that the following parameters are interrelated: quantity adsorbed, number of adsorbed layers, total surface coverage of the solid and specific area of solid occupied by a single protein molecule. It is necessary to know the first three of them to calculate the fourth, which is particularly important since it is conformation related. This is of course essential with flexible macromolecules, like proteins, since the adsorption could (i) lead to the formation of a monolayer or a multilayer, (ii) involve native or unfolded forms of the polymer, and (iii) be "side-on" or "end-on", even if the polymer conserves the same conformation than in solution.

Maximum Adsorption and Layer Structure. As shown in Figure 1a, adsorption of BSA on montmorillonite follows the usual trend with a maximum occuring around the i.e.p. which is at pH 4.7. Both methods used in this study, namely the depletion method (centrifugation, then determination of the concentration of non-adsorbed protein by UV absorbance) and the NMR method (determination of the quantity of protein above which no more release of paramagnetic cation is observed) give the same result (11). This agreement goes further than a simple confirmation by two independent methods. It is important to note that the NMR method does not include a centrifugation step and that it detects the direct contact between the first layer of protein and the mineral surface. Thus we can deduce (i) that the maximum adsorption is not an artefact of the depletion method resulting from the lower solubility of proteins at their i.e.p., with protein and clay co-elimination in the pellet after centrifugation, and (ii) that no more than a monolayer is adsorbed on the clay surface, otherwise the NMR method should give lower values than the depletion method.

Surface Coverage of the Solid. The cation exchange properties of clay minerals have been used in this study to obtain the surface coverage with NMR (11). The extent of Mn^{2+} exchange following adsorption has been taken as a measurement of this parameter. Indeed, in the pH range studied, the side chains of lysine and arginine remain positively charged and can replace the charge compensating cations of the clay surface. On the other hand the carboxyl groups of the side chains of glutamic and aspartic acids are mainly in a protonated state for BSA adsorbed on montmorillonite as has been shown by FTIR spectroscopy (12). Figure 1c shows that BSA covers a constant value of about 80% of the montmorillonite surface below the i.e.p. but the surface coverage decreases above, reaching a zero value above pH 7. Thus, this parameter is strongly asymmetric with respect to the i.e.p. and indicates the occurrence of strong electrostatic repulsions above the i.e.p. between the negatively charged protein and clay surface.

Unfolding of the Protein. The knowledge of the quantity of BSA adsorbed, of the monolayer character of its adsorption and of the surface coverage allows the calculation of the specific interfacial area of the contact between the protein and the mineral surface (12). For example, if no more than a monolayer is adsorbed, the specific interfacial area can be deduced from the ratio of the surface coverage and the quantity of protein adsorbed. Figure 1b shows the result of such a calculation which indicates that at and above the i.e.p. the specific interfacial area is constant with the value of about 60 nm^2, but below the i.e.p. this area increases with decreasing pH, reaching 120 nm^2 near pH 2.5. We can deduce that at and above the i.e.p. the BSA is adsorbed side-on since the measured area is near the value which can be calculated from the molecular dimensions of the molecule (18), 14 nm x 4 nm x 4 nm, in this configuration. On the other hand, below the i.e.p. the calculated area greatly exceeds the molecular dimensions of the native BSA. Therefore as the pH decreases the positive electric charge of the protein increases and gives rise to strong attractive electrostatic interactions with the negatively charged clay surface which results in a spreading out of the unfolded BSA. This process of unfolding is not at all symmetric with respect to the i.e.p.

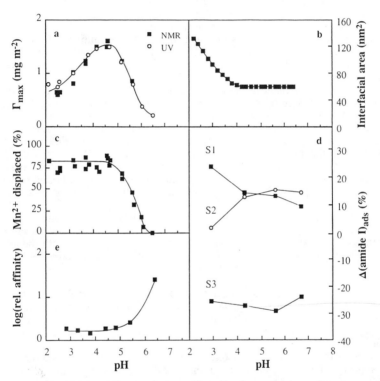

Figure 1.   The pH dependence of the interaction of BSA with montmorillonite. (a) Maximum amount of BSA adsorbed deduced from UV absorption on the supernatant after centrifugation and from NMR spectroscopy without centrifugation. (b) Calculated specific interfacial area of adsorbed BSA. (c) Fraction of Mn^{2+} displaced by BSA from montmorillonite deduced from NMR spectroscopy. (d) Evolution of different amide I components of FTIR spectra of BSA on adsorption (% in the adsorbed state - % in solution). (e) Affinity for the montmorillonite of methylated BSA relative to BSA.

Electrostatic *vs* Hydrophobic Interactions. From a qualitative point of view, it is much easier to show the occurence of electrostatic interactions than hydrophobic interactions, due to the dependence of the former on pH. Nevertheless the hydrophobic interactions are suspected to play an important role in protein adsorption (*15*). We have obtained evidence of this type of interaction by studying the coadsorption of BSA and its methylated derivative on montmorillonite (*19*). Since reductive methylation of lysine groups does not change their pK_a, the electric charge of both proteins should be the same at all pH values and only their hydrophobicities should differ. Figure 1e shows that the relative affinity of the methylated form is higher on the entire range of pH studied, by a constant factor below the i.e.p. and increasing strongly above the i.e.p.. The reason for the increase in the affinity of the methylated form over the untreated form above the i.e.p. is related to the fact that in this pH range the proteins and the mineral surface are both negatively charged and thus only hydrophobic interactions can overcome this electrostatic barrier.

The exact mechanism of these hydrophobic interactions is not immediately apparent since montmorillonite is a very hydrophilic surface, swelling in presence of water (smectite class of the clay minerals). But in fact the charge compensating cations such as Na^+ in this case, which are Lewis acids, are therefore the true hydrophilic centers. Since these cations are exchanged on adsorption, the surface in contact with the protein is the siloxane surface of the silica tetrahedral sheet. Several experimental and theoretical studies have shown that siloxane surfaces are hydrophobic (*20-22*). Thus it is not surprising that the montmorillonite surface can give rise to hydrophobic interactions when cation exchange occurs at its surface.

Secondary Structure of the Adsorbed Protein. The decomposition of the amide I band led to 5 components common to BSA in solution and adsorbed on montmorillonite, and to 2 additional components appearing only with the adsorbed BSA. Figure 1d shows the evolution of these components, regrouped in 3 classes: S_1 (1662 cm^{-1} + 1619 cm^{-1}), S_2 (1681 cm^{-1} + 1672 cm^{-1} + 1639 cm^{-1}), and S_3 (1655 cm^{-1} + 1633 cm^{-1}).

S_1 is the sum of the components appearing only for the adsorbed BSA: 1662 cm^{-1} is attributed to peptide units engaged in slightly disordered α-helices arranged in bundles at the clay surface and 1619 cm^{-1} to extended β-strands associated by intermolecular hydrogen bonds between adjacent molecules on the surface.

S_2 is the sum of the components corresponding to non-ordered peptide units and appearing both in solution and in the adsorbed state: 1681 cm^{-1} is assigned to non-hydrogen bonded peptide units situated in apolar domains and 1672 cm^{-1} to the same type of units, but in a more polar environment; 1639 cm^{-1} corresponds to non-ordered peptide units associated to water.

S_3 is the sum of the components corresponding to classical ordered secondary structures: 1655 cm^{-1} is attributed to α-helices and 1633 cm^{-1} to extended strands forming intramolecular β-domains.

It is apparent from Figure 1d that there is an important loss of ordered secondary structures on adsorption, between 24 and 28% of the amino-acids. This confirms other spectroscopic analysis on desorbed proteins (*23*) or directly on adsorbed proteins (*24, 25*). The data in Figure 1d exhibit no marked influence of the pH on the disappearance of the ordered secondary structures. However the classes of structures into which the disappearing α-helices and β-structures (S_3) are transformed (S_1 and S_2), differ according to the pH. Below the i.e.p. the decrease in S_3 is matched by the increase in S_1. At the i.e.p. and above, the decrease in S_3 is equilibrated by an approximatively equal increase in S_1 and S_2.

The S_2 fraction represents less ordered secondary structures, according to the assignments currently made on proteins in solution. However as it is mainly related to hydration of the protein (*26*), it does not necessarily imply protein denaturation. There are some indications that the S_1 fraction represents a more unfolded molecular structure: (i) the fact that the S_1 fraction is strictly associated with the presence of clay mineral, (ii) the fact that the intermolecular β-structures (1619 cm^{-1}) increases at low pH when the unfolding of the protein on the surface increases the probability of intermolecular contact (*11*), and (iii) the fact that the 1619 cm^{-1} component is always associated with denaturation or aggregation of proteins both in solution and in the adsorbed state (*27*).

Application to the Interpretation of Enzyme Activity in Soil

Although the results described above were obtained from a protein with no catalytic function, the conclusions are helpful in understanding the influence of mineral and organo-mineral surfaces on extracellular enzyme activity. This can be illustrated with recent work from our laboratory on some of the enzymes involved in the biogeochemical cycles of N, C and P in soils.

Inadequacy of the Surface pH Hypothesis. A survey of the literature shows that the alkaline pH shift in the catalytic activity of enzymes adsorbed on electronegative surfaces has often been interpreted in terms of a local pH effect. In this interpretation it is supposed that the enzyme retains its native conformation on adsorption and that the pH in the vicinity of the catalytic active site is lower than the bulk solution pH measured with a glass electrode, due to the formation of a diffuse double layer at the mineral surface (*5-9*).

Most of the observations made on fixed enzymes in biotechnology and on the effect of mineral surfaces on extracellular enzyme activity in soil science have been interpreted according to this model. An important consequence of the local pH effect mechanism is that the catalytic activity for the adsorbed enzyme should be higher than that of the enzyme in solution in the alkaline pH range. However an extensive survey of the literature for the last 30 years shows that when the data are given in absolute values, and not normalized on their maximum, the catalytic activity of an adsorbed enzyme is always lower or equal to that of the enzyme in solution. In fact a misleading representation of the catalytic activity of adsorbed enzymes is probably partly responsible for masking this result and this can be illustrated by considering the pH dependence of the interaction between bovine α-chymotrypsin and montmorillonite. Figure 2a shows the experimental data directly measured according to the protocoles previously defined: the catalytic activity in solution (A), in a montmorillonite suspension (B) and in the supernatant of B (C). It is clear that the activity of the adsorbed enzyme at any pH is never higher than the activity of the control in solution and the best way to describe the observed effect is a strong inhibition at low pH rather than an optimum pH shift. But if both curves A and B are normalized to their respective optimum pH as in Figure 2b, a representation often found in the literature on adsorbed enzymes (*5-7*), the pH shift becomes more apparent to the detriment of the inhibitory effect.

The interaction of an *Aspergillus ficuum* phytase with montmorillonite, shown in Figure 2c, is another clear example of the inadequacy of the local pH hypothesis. This enzyme has the interesting property of having two optimum pH values for catalytic activity, one at pH 2.5 and the other at pH 5.5. According to the local pH hypothesis both maxima should be shifted. But Figure 2c shows rather a complete inhibition of the pH 2.5 peak and a pH 5.5 peak practically unaffected.

Further evidence against the surface pH hypothesis is given in the Figure 2d which shows the effect of adsorption on montmorillonite at particular pH values on the subsequent pH dependence of the activity of sweet almond β-D-glucosidase (*13*). At pH values 3.6, 4.4 and 5.7, the enzyme incubated in the absence of clay (-M) was stable. However, when the enzyme was adsorbed on montmorillonite

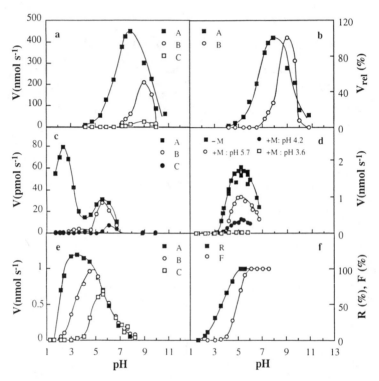

Figure 2. The pH dependence of the adsorption and catalytic activity of various enzymes on montmorillonite. (a) Catalytic activity of bovine α-chymotrypsin measured according to the procedures A (control in solution), B (in presence of clay) and C (supernatant of B). (b) same results as in (a) but with the curves A and B normalized to their maximum. (c) Catalytic activity of *Aspergillus ficuum* phytase measured according to the procedures A, B and C. (d) Effect of the pH of adsorption on the catalytic activity of sweet almond β-D-glucosidase; -M: control without montmorillonite; +M: with montmorillonite at different pH of adsorption. (e) Catalytic activity of *Aspergillus niger* β-D-glucosidase measured according to the procedures A, B and C. (f) Relative catalytic activity in the adsorbed state R and non-adsorbed fraction F of *A. niger* β-D-glucosidase calculated from (e).

(+M) at these pH values and then its activity measured across the full range of pH, the maximum values obtained at pH 5.5 showed a decrease in activity as the initial pH was lower. This result cannot be explained by a local pH effect since a modification of the physicochemical composition of the bulk solution should instantaneously modify the structure of the diffuse double layer and no lasting effects should be observable 2 h after this pH change.

Finally, even if it is true that the proton activity is higher in the vicinity of an electronegative surface, the tendency of a proton to react with the active site of the enzyme is not given by the proton activity but by its molar free energy, namely its electrochemical potential, $\mu = \mu_0 + RT \ln [H^+] + F\Psi$. The system being in a thermodynamic equilibrium, the molar free energy of the proton is the same in the bulk of the solution and at the surface $(2, 28)$. Thus it is not clear why a diffuse double layer effect should shift the optimum pH.

The pH-Dependent Modification of Conformation Hypothesis.   In contrast to the preceding model, which assumes that the enzyme adopts the same conformation in solution and in the adsorbed state, another mechanism may be postulated where there is a pH dependent unfolding of the enzyme in the same way as we have demonstrated for the interaction of BSA with montmorillonite. The data obtained for β-D-glucosidase of *Aspergillus niger*, which has a i.e.p. of 4.0, provide a good example (29). Figure 2e shows once again that the interaction with montmorillonite does not lead to a catalytic activity which is higher than that of the control at alkaline pH. Figure 2f illustrates the importance of electrostatic interactions in its reaction with montmorillonite. Above the i.e.p. the non-adsorbed fraction, F, of the enzyme increases progressively with pH until no enzyme is adsorbed. As pH increases above the i.e.p. the charge on the enzyme increases, thereby increasing the electrostatic repulsion from the electronegative surface of the clay. These electrostatic interactions are also responsible for the decrease in the relative catalytic activity, R, with decreasing pH since the increase in the positive charge causes a gradual unfolding of the enzyme on the electronegative surface.

Bovine α-chymotrypsin and *A. ficuum* phytase, which have i.e.p. of 8.6 and 5.0-5.4 respectively, confirm the model of a pH dependent modification of conformation since, as shown in Figures 2a and 2c, their catalytic activities at pH below their i.e.p. are drastically reduced.

Finally, the irreversibility of the effects of adsorption at acid pH on the catalytic activity subsequently measured at higher pH (Figure 2d) is readily explained using this model. The unfolding of the enzyme at acid pH increases the number of points of contact with the clay surface. An increase in pH does not necessarily enable the enzyme to adopt the conformation that it would have had in solution at that pH, as this would require an activation energy equivalent to the adsorption energy of all the additional amino-acids which have been brought into contact with the surface. This activation energy may be greater that the available thermal energy which explains the observed irreversibility of the adsorbed conformation (15).

Protective Effect of the Soil Organic Matter.   In the preceding discussion we have considered the effect of bare clay surfaces. However clays present in soils do not have clean surfaces since they associate with organic matter to form clay-humic complexes. Figure 3a shows the interaction of a phosphatase from the ectomycorrhizal fungus *Hebeloma cylindrosporum* with a suspension of a ferrallitic soil, using the experimental procedures A, B and C. Figure 3b shows the interaction of the same enzyme with the same soil after a pretreatment to destroy the organic matter (6 % H_2O_2 at 80°C for 12 h). It is evident that the destruction of the organic matter increases the adsorption of the enzyme and decreases the catalytic activity in the adsorbed state since the values of the curves C and B are

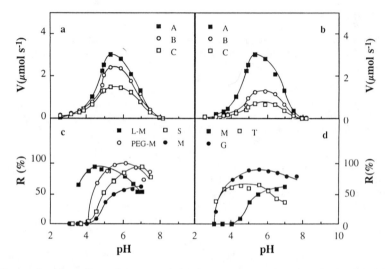

Figure 3. The effect of an organic matter coating and surface hydrophobicity / hydrophilicity on the pH dependence of the catalytic activity of an adsorbed enzyme. (a, b) Interaction of *Hebeloma cylindrosporum* acid phosphatase with a ferrallitic soil: (a) without treatment and (b) with destruction of the organic matter; catalytic activities measured according to the procedures A (control in solution), B (in presence of soil fraction) and C (supernatant of B). (c, d) Relative catalytic activity R of sweet almond β-D-glucosidase adsorbed (c) on different clay-organic complexes and (d) on different mineral surfaces; soil fraction rich in organic matter (S), polyethylene glycol-montmorillonite complex (PEG-M), lysozyme-montmorillonite complex (L-M), montmorillonite (M), goethite (G), talc (T).

lower than those measured without the pretreatment.   In order to elucidate the mechanisms responsible for the protective effect of the organic matter the catalytic activity of sweet almond β-D-glucosidase was investigated on a range of coated clay mineral surfaces (*30*).   Figure 3c shows the effect of a bare surface of montmorillonite (M) on the relative catalytic activity, R, of the enzyme, with denaturation of the enzyme on the surface below pH 4.   A soil fraction, rich in organic matter (S) illustrates the protective effect of the latter since values of R are higher above pH 4, but below this pH the denaturation of the enzyme remains complete.   An artificial montmorillonite-polyethylene glycol complex (PEG-M) has a similar effect to that of the natural soil fraction, namely an increase in the relative activity above pH 4 but zero residual activity below pH 4.   In contrast a lysozyme-montmorillonite complex (L-M)) exhibits a different effect, in this case no inhibition of enzymatic activity is observed at pH 4.   Since lysozyme is a protein with a high i.e.p., 11.7, it binds tightly to the negatively charged surface and renders it electropositive.   The catalytic activity is thus maintained at pH 4 because sweet almond β-D-glucosidase is positively charged at this pH and the lysozyme coating suppresses the attractive electrostatic interactions which would have existed between the β-D-glucosidase and the bare montmorillonite surface.   The fact that the coating of polyethylene glycol, which is a neutral polymer, affords no protection to the activity of the adsorbed enzyme at pH 4 may be due to an exchange of polyethylene glycol with the positively charged enzyme at this low pH. However when the pH is increased the enzyme becomes progressively negatively charged and therefore loses its capacity to displace the polyethylene glycol.   A layer of hydrophilic polymer therefore protects the enzyme from the hydrophobic interactions with the surface at these higher pH.   The similarity of the effects of the montmorillonite-polyethylene glycol complex and the soil fraction suggests that although certain forms of humic substances may be exchanged by enzymes at low pH, they are not exchanged at higher pH where they therefore play a protective role.

Hydrophilic and Hydrophobic Surfaces.   Although the electrostatic interactions between proteins and surfaces explain most of the observed conformational changes, they cannot completely account for the observed phenomena in every case. Figure 3d compares the effects of goethite (G), talc (T) and montmorillonite (M) on the relative catalytic activity of sweet almond β-D-glucosidase (*30*).   The goethite surface has the least destabilizing effect on the structure of the enzyme in the range of pH studied.   The surface of this iron oxyhydroxide is hydrophilic and its surface charge is low under the experimental conditions used because of the complexation between the surface hydroxyl groups and citrate which was used as a buffer.   The basal surface of talc is hydrophobic and possesses no electric charge.   However this hydrophobic surface nevertheless affects the catalytic activity to a greater extend that does the hydrophilic goethite surface.   This confirms the findings obtained on artifical organic surfaces which demonstrate the intervention of hydrophobic forces. This also agrees well with our demonstration that methylated BSA has a greater affinity for a montmorillonite surface than does native BSA.   Talc, like montmorillonite, has a surface composed of siloxane groups.   The fact that only montmorillonite causes a complete inhibition of the catalytic activity of sweet almond β-D-glucosidase at low pH clearly demonstrates that even if hydrophobic surfaces are more destabilizing than hydrophilic surfaces, electrostatic interactions play the most important role.

Conclusions

The adsorption of BSA on montmorillonite has been studied using various techniques; adsorption isotherms, modification of the hydrophobicity of the protein by methylation, degree of coverage of the mineral surface by NMR, secondary structure by FTIR. This investigation has led to the formulation of a coherent interaction model which explains the pH dependent effect of clays on the catalytic activity of the extracellular enzymes which are released by microorganisms and plant roots into soil. Thus (i) below the i.e.p. the electrostatic interactions between the clay surface and the positively charged enzyme lead to an unfolding of the latter as the pH is lowered which is responsible for the observed lowering of catalytic activity; (ii) near the i.e.p. the electrostatic interactions are minimal, the perturbation of the tertiary structure of the enzyme is slight and its activity is therefore maintained; (iii) above the i.e.p. the electrostatic interactions are repulsive since the enzyme is negatively charged whereas the surface remains electronegative. In that case the enzyme is in solution and free to diffuse in the water-filled soil pore network.

We conclude that the model hitherto invoked to explain the perturbation of enzyme activity by a surface pH effect is incompatible with experimental results.

Acknowledgments

Part of this work was funded by doctoral studentships from the Région Languedoc-Roussillon (F.L.) and Brazil (P.T.M.P). The assistance of François Mazella in the preparation of the figures is greatly acknowledged.

Literature Cited

1   *Proteins at Interfaces. Physicochemical and Biochemical Studies*; Brash, J.L.; Horbett, T.A., Eds.; ACS Symposium Series 343; American Chemical Society: Washington, DC, 1987.
2   Fletcher M. *Adv. Microbial Physiol.* **1991**, *32*, 53-85.
3   Burns, R.G. In *Soil Enzymes*; Burns, R.G., Ed.; Academic Press: London, UK, 1978, pp 295-340.
4   Theng, B.K.G. *Formation and Properties of Clay-Polymer Complexes*; Chapter 7; Elsevier: Amsterdam, NL, 1979; pp. 157-226.
5   Douzou, P.; Petsko, G.A. *Adv. Protein Chem.* **1984**, *36*, 245- 361.
6   Katchalski., E.; Silman, I.; Goldman,G. *Adv. Enzymol.* **1970**, *33*, 445-536.
7   Mc Laren, A.D.; Estermann E.F. *Arch. Biochem. Biophys.* **1957**, *68*, 157-160.
8   Mc Laren, A.D.; Packer L. *Adv. Enzymol.* **1970**, *33*, 245-308.
9   Aliev, R.A.; Gusev, V.S.; Zvyagintsev, D.G. *Vestn. Mosk. Univ. Biol. Pochvoved.* **1976**, *31*, 67-70.
10 Schwertmann, U.; Cambier P.; Murat, E. *Clays Clay Miner.* **1985**, *33*, 369-378.
11 Quiquampoix, H.; Ratcliffe, R.G. *J. Colloid Interface Sci.* **1992**, *148*, 343-352.
12 Quiquampoix H.; Staunton S.; Baron M.-H.; Ratcliffe R.G. *Colloids Surf.* **1993**, *75*, 85-93.
13 Quiquampoix, H. *Biochimie* **1987**, *69*, 753-763.
14 Brash, J.L., In *Proteins at Interfaces. Physicochemical and Biochemical Studies*; Brash, J.L.; Horbett, T.A. Eds.; ACS Symposium Series 343; American Chemical Society: Washington, DC, 1987; 490-506.
15 Norde, W. *Adv. Colloid Interface Sci.* **1986**, *25*, 267-340.
16 Eirich, F.R. *J. Colloid Interface Sci.* **1977**, *58*, 423-436.

17 Norde, W.; Lyklema, J. *J. Colloid Interface Sci.* **1978**, *66*, 257-265.
18 Squire, P.G.; Moser, P.; O'Konski, C.T. *Biochemistry* **1968**, *7*, 4261-4272.
19 Staunton, S.; Quiquampoix, H. *J. Colloid Interface Sci.* **1994**, *166*, 89-94.
20 Bleam, F. *Clays Clay Miner.* **1990**, *38*, 527-536.
21 Jaynes,W.F.; Boyd, S.A. *Clays Clay Miner.* **1991**, *39*, 428--436.
22 Skipper, N.T.; Refson, K.; Mc Connell, J.D.C. *Clay Miner.* **1989**, *24*, 411-425 .
23 Norde, W.; MacRitchie, F.; Nowicka, G.; Lyklema, J. *J. Colloid Interface Sci.* **1986**, *112*, 447-456.
24 Jakobsen R.J.; Wasacz F.M. *Appl. Spectrosc.* **1990**, *44*, 1478-1490.
25 Kondo, A.; Oku, S.; Higashitani K. *J. Colloid Interface Sci.* **1991**, *143*, 214-221.
26 Wantyghem, J.; Baron, M.H.; Picquart, M.; Lavialle, F. *Biochemistry* **1990**, *29*, 6600-6609.
27 R.J. Jakobsen and F.M. Wasacz, In *Proteins at Interfaces. Physicochemical and Biochemical Studies*; Brash, J.L.; Horbett, T.A. Ed.; ACS Symposium Series 343; American Chemical Society: Washington, DC, 1987; 339-361.
28 Rouxhet P.G. In *Enzyme Engineering*; Okada H.; Tanaka A.; Blanch H.W., Eds.; Annals of the New York Academy of Sciences, 613; The New York Academy of Sciences: New York, NY, 1990, Vol. 10; pp 265-278.
29 Quiquampoix, H.; Chassin, P.; Ratcliffe, R.G. *Progr. Colloid Polym. Sci.* **1989**, *79*, 59-63.
30 Quiquampoix, H. *Biochimie* **1987**, *69*, 765-771.

RECEIVED March 1, 1995

Chapter 24

Reactivity of Antibodies on Antigens Adsorbed on Solid Surfaces

P. Huetz[1,5], P. Schaaf[2,3], J.-C. Voegel[1], E. K. Mann[2], B. Miras[2], V. Ball[1], M. Freund[4], and J.-P. Cazenave[4]

[1]Centre de Recherches Odontologiques, Institut National de la Santé et de la Recherche Médicale, Contrat Jeune Formation 92−04, 1 Place de l'Hôpital, 67000 Strasbourg, France
[2]Institut Charles Sadron, Centre National de la Recherche Scientifique, Université Louis Pasteur, 6 rue Boussingault, 67083 Strasbourg Cedex, France
[3]Ecole Européenne des Hautes Etudes des Industries Chimiques de Strasbourg, 1 rue Blaise Pascal, B.P. 296F, 67008 Strasbourg Cedex, France
[4]Centre Régional de Transfusion Sanguine, Institut National de la Santé et de la Recherche Médicale U311, 10 rue Spielmann, 67085 Strasbourg Cedex, France

The aim of this study was to analyze the antigen/antibody reactivity after antigen adsorption onto a solid surface. Three systems were evaluated: (i) the IgG/IgY, (ii) the IgG/anti-IgG and the Fib/anti-Fib systems, using either an optical technique (scanning angle reflectometry) or a radiolabeling approach which allows quantifying precisely such side effects as desorption or exchange reactions. With the optical technique, antibody/antigen molar reactivity ratios of 0.8, 1 and ≈6 were found for the three systems respectively. The behaviour of the Fib/anti-Fib system was quantitatively analyzed using the radiolabeling technique: desorption of the adsorbed antigen molecules and/or exchange reactions by incoming antibody molecules were precisely evaluated. The reactivity ratios varied strongly, from ≈1.5 down to zero, between anti-Fib solutions which were prepared slightly differently, indicating a problem linked to the radiolabeling.

Biological fluids are complex ionic aqueous mixtures of various macromolecules, cells and microorganisms. If we compare the sizes of a protein and a cell for instance, the difference of two or three orders of magnitude between them leads to diffusion coefficients and characteristic transport times which are significantly different. This implies that the first process which takes place when a biological fluid comes into

[5]Current address: Department of Biophysical Chemistry, Groningen Biomolecular Sciences and Biotechnology Institute, University of Groningen, Nijenborgh 4, 9747 AG Groningen, Netherlands

contact with a solid surface is the adsorption of different proteins onto this surface. But even this apparently simple problem is fairly complicated, so that researchers have for years addressed the problem of protein adsorption from single component aqueous solutions. From these numerous investigations, a few general rules emerged: (i) adsorption of proteins onto solid surfaces is not fully reversible, so that thermodynamic laws can seldom be applied and only with great care (*1*), (ii) the affinity of proteins for a given surface usually increases with the mass of the macromolecules (*2*), and (iii) for a given protein/surface system, the affinity for the surface is the highest at a pH value close to the isoelectric point of the protein (*3*).

After having concentrated on single component solutions, the investigations moved towards the understanding of the interactions between a solid surface and a solution containing a mixture of proteins. The additional fundamental process that occurs, beyond those just described, is an exchange mechanism between the adsorbed proteins and the free macromolecules of the solution, a phenomenon which has been clearly demonstrated by different radiolabeling experiments (*4, 5*). Out of the reported results, it appears that low molecular weight proteins adsorbed on surfaces are more easily replaced by high molecular weight proteins than the contrary (*6, 7*). Such an observation has also been reported for synthetic polymers (*8, 9*).

In this work, we focus on the evaluation of the reactivity of adsorbed antigens with their respective antibodies by two complementary approaches: scanning angle reflectometry and a radiolabeling technique. Particular attention was devoted to a quantitative evaluation of all the side effects influencing the determination of the molar reactivity ratio.

We selected different immunologic systems: (i) antibodies from hen eggs (IgY) directed against human γ-immunoglobulins (IgG), (ii) rabbit antibodies (anti-IgG) directed against IgG, and (iii) rabbit antibodies (anti-Fib) directed against human fibrinogen (Fib). All the antibodies are polyclonal.

Using the radiolabeling technique, we also determined the IgG and Fib isotherms onto the bare surface, for the same hydrodynamical conditions under which we tested the antigenic reaction.

Experimental methods and materials

Three different antigen-antibody pairs were studied. Lyophilized human polyclonal IgG (Mw ≈ 150000) was prepared as described in (*10*) and isolated from blood given by about 1000 donors. The reactivity against this molecule of two different antibodies was considered: purified polyclonal IgY (Mw ≈ 170000) isolated from the egg yolk of hens immunized against human IgG, and a solution of purified rabbit anti-IgG (Mw ≈ 150000). Human fibrinogen (Mw ≈ 340000), of grade L and coagulability >90%, was purchased from Kabi Vitrum (Stockholm, Sweden). We tested the reactivity of a purified rabbit γIgG (anti-Fib) solution directed against this human Fib. IgG, IgY, anti-IgG and anti-Fib were provided by the Centre Régional de Transfusion Sanguine (Strasbourg).

All protein solutions were prepared in PBS buffer solutions, type 1 or type 2, as indicated in the text. Type 1 PBS buffer was prepared by dissolving 10 mmol of $NaH_2PO_4 \cdot 2H_2O$, 100 mmol of NaCl and 1 mmol of NaN_3 per liter of deionized water (Super Q, Millipore), the pH being adjusted to 7.8 with concentrated NaOH. Type 2 PBS buffer was prepared with 50mM $Na_2HPO_4 \cdot 12H_2O$ and 0.15M NaCl, the pH adjusted to 7.5 with 50mM $NaH_2PO_4 \cdot 2H_2O$. In all experiments, the buffer was filtered with a Millex GV-0.22μm filter, and degassed before use. The same buffer was used throughout an experiment for the rinsing steps and to dilute the antigen and antibody solutions.

Silica provided the adsorbing surface in all experiments: one flat surface of a prism made of Herasyl (Heraus, refractive index n=1.45718) for the reflectivity experiments, circular tubing for the radiotracer experiments. In all cases these surfaces were cleaned with sulphochromic acid, and the hydrophilicity verified. Note that adsorption took place under flow conditions for the radiotracer experiments, but non-flow, non-stir conditions for the reflectivity experiments.

The experimental apparatus used in the radiotracer experiments has been described by Boumaza et al. (11) and consists of a silica tube of inner diameter 0.17 cm and length 10 cm or 20 cm. Antigens and antibodies were radiolabeled (^{125}I) using a technique adapted from the method of McFarlane (12) in which iodine monochloride is the iodinating agent. Proteins were stored in a concentrated form (0.4 to 1% (w/w)) at -20°C. Just before use, they were quickly thawed at 37°C and diluted in type 2 PBS buffer. Concentrations were determined by absorbance of the solutions at 280 nm (spectrophotometer Beckman, model 34, Inst. Inc. Fullerton, USA) and specific activities by γ counting (Minimaxi γ, United Technologies, Packard Instrument, USA). The way we calculated the adsorbed protein amounts will be discussed in a precise manner in the section dealing with the Fib/anti-Fib reaction.

Structural information on the interfacial adsorbed layer was evaluated with the help of Scanning Angle Reflectometry (13). This technique is based on the variation, after adsorption, of the reflection coefficient of a light wave polarized in the plane of incidence (p-wave) around the Brewster angle (θ_B). At this angle, about 42.5° for the silica-water interface, the reflectivity of a p-wave would be null for a perfectly abrupt, planar interface (Fresnel interface) (14). The signal here is thus very sensitive to deviations from this interface: for example, any adsorbed film. The light source is a 5 mW He-Ne laser (λ = 632.8 nm). This light, polarized in the plane of incidence, passes through a silica prism to reflect off the inner, optically polished interface, which forms one face of the experimental cell. The reflected angle is selected with an angular precision of ± 0.01%. The reflected intensity was recorded at various angles θ around θ_B, after previous equilibration of the silica surface with PBS buffer. These data constituted the reference signal (Figure 1). In order to study the reactivity of the antigen/antibody systems, the antigen solution was then injected, over five minutes, into the cell to replace the buffer. The protein solution was then left in contact with the surface, and the measurement of the reflectivity curve I(θ) repeated (Figure 1). Each curve required approximately 5 minutes to measure, limiting the adsorption kinetics which can be followed. As adsorption proceeded at the interface, the reflectivity at θ_B increased; the shape of the curve and the position of its minimum also changed (Figure 1). These changes depend on both the thickness and the density of the adsorbed layer, which can therefore be deduced. This evolution was followed for several hours. The solution was then replaced by buffer. The antibody solution was injected similarly. Typical reflectivity curves, taken at various points during the successive adsorption of antigen and antibody, are shown in Figure 1, along with the fits to these curves given by a simple model of the interface.

We adopted for the interface the common model of a homogeneous, isotropic layer defined by a thickness L_0 and a mean refractive index n (14). The actual protein layer is not homogeneous. In particular, protein molecules may adsorb in both the "end-on" and "side-on" positions, with possible conformational changes upon or after adsorption. L_0 is thus an optical average over the different molecular configurations occuring in the layer. The mean refractive index n of the layer is related to the protein concentration c within this layer by the known refractive index increment dn/dc for the protein with respect to the solvent (dn/dc = 0.18 cm^3/g for Fib, (*15*)). The total adsorbed quantity is then given by the product $\Delta n \cdot L_0$ (with Δn the refractive index difference between layer and solvent).

All experiments were performed at room temperature (22±2°C).

Results and discussion

Layer Structure Determination by Scanning Angle Reflectometry (SAR). The concentration of the antigen was set at 0.05% (w/w), unless otherwise indicated, in order to attain saturation of the surface (see isotherms below), whereas the antibody concentration was set at 0.01% (w/w). The total adsorbed quantity is determined with a precision of the order of 10%. The thickness L_0 is reproducible within about 20% from one experiment to another. On the other hand, the total adsorbed quantity, $\Delta n \cdot L_0$, varies. This quantity adsorbed is known to depend sensitively on the state of the surface (*16*). The results presented here are preliminary in the sense that only one or two trials have been performed for each antigen/antibody pair.

The IgG/IgY System. Human polyclonal IgG was reconstituted in type 1 buffer and injected. The reflected intensity led to a value $L_0 = 20$ nm and a refractive index difference $\Delta n = 2.2 \cdot 10^{-2}$. The amount of adsorbed IgG was then calculated by assuming the refractive index increment dn/dc to be the same as for fibrinogen (0.18 cm^3/g). The adsorbed amount was found to be equal to 0.25 µg/cm^2 (Exp. 1a, Table I). The width of the layer is close to the " end-on " dimension of the IgG molecule (of dimensions 23.5x4.4x4.4 nm^3, (*17*)). The surface concentration corresponds to total coverage of 15% if every molecule is in this end-on position, or 80% for a " side-on " position, which should be compared to the maximum, or " jamming-limit ", coverage of about 55% predicted by simple adsorption models like " Random Sequential Adsorption " (RSA) model (*18*).

We then investigated the reactivity of the adsorbed IgG molecules with purified polyclonal IgY (Mw ≈ 170000). We found that a significant amount of IgY from the 0.01% (w/w) IgY solution reacted with the IgG layer: the measurement of the reflectivity led to total protein layer of 0.48 µg/cm^2. Withdrawing the adsorbed amount relative to the simple IgG layer, one obtains that IgY molecules are fixed at a surface concentration $\Delta\Gamma_{IgY} = \Gamma_{Tot} - \Gamma_{IgG} = 0.23$ µg/cm^2. Taking into account the molecular weight of the two species, the IgG/IgY reactivity is in a ratio of 1:0.8.

The IgG/anti-IgG System. The human IgG solution was prepared in type 2 PBS buffer. We again obtained $L_0 = 20$ nm for the IgG layer, but a somewhat higher density for this trial, with $\Delta n = 4.0 \cdot 10^{-2}$. This implies protein adsorption of 0.45 µg/cm^2 (Exp. 1b, Table I). This layer was put in contact with a 0.01% (w/w) solution of purified rabbit anti-IgG (Mw ≈ 150000). The amount of adsorbed anti-IgG was 0.40 µg/cm^2; again, the antibody reacts in an approximately 1:1 ratio with the IgG on the surface.

An IgG solution, at 0.01% (w/w), was again introduced into the cell after rinsing. The total layer thickness increases immediately by 4 nm, suggesting that accessible reactive sites remain on the antibody after adsorption. The relatively small additional layer thickness suggests that the additional antigen is either predominately in the side-on position or considerably intercalated in the antibody layer. The quantity of IgG adsorbed at the surface increases slowly, over about 15 hours, to 0.05 $\mu g/cm^2$ this corresponds to 0.13 antigen per antibody already adsorbed.

Table I. Amounts of adsorbed antigen (Γ) and immobilized rabbit antibodies ($\Delta\Gamma$, raw signals)

Exp.	Antigen	Bulk conc. (% (w/w))	L_0 (nm)	Γ ($\mu g \cdot cm^{-2}$)	Antibody (0.01% (w/w))	ΔL_0 (nm)	$\Delta\Gamma$ ($\mu g \cdot cm^{-2}$)
1a	IgG	0.050	20	0.25	IgY	---	0.23
1b	IgG	0.050	20	0.45	anti-IgG	13	0.40
1c	Fib	0.050	22	0.25	anti-Fib	63	0.52
2a	Fib*	0.010		0.32	anti-Fib$_1$*		0.30
2b	Fib*	0.010		0.40	anti-Fib$_1$*		0.27
2c	Fib	0.010		---	anti-IgG*		0.084
2d	Fib	0.010		---	anti-Fib$_2$*		0.076
2e	Fib	0.057		---	anti-Fib$_2$*		0.046
2f	Fib	0.056		---	anti-Fib$_2$*		0.040
2g	Fib	0.010		---	anti-Fib$_3$*		0.16
2h	Fib	0.050		---	anti-Fib$_3$*		0.11
2i	Fib*	0.011		0.29	anti-Fib$_3$		-0.03

Experiments 1a-c: SAR, experiments 2a-i: radiotracer technique. Corresponding thicknesses are given for SAR measurements (L_0, ΔL_0). For experiment 2i, the quantity of Fib desorbed by unlabeled anti-Fib molecules was evaluated (thus the negative value). Indices of the different anti-Fib are related to their origin (see text explanations). (*) means that corresponding proteins were labeled.

The Fib/anti-Fib System. A 0.05% (w/w) solution of Fib prepared in type buffer was injected. The layer thickness, refractive index, and total adsorbed quantity as they evolve in time after introduction of the protein are given in Figures 2a-c for typical experiment. The values saturate at L_0 = 22 nm and $\Delta n = 1.8 \cdot 10^{-2}$ for a total adsorbed amount of 0.25 $\mu g/cm^2$ (Exp. 1c, Table I). The value of L_0 is intermediate between the end-on and side-on dimensions of the molecule (of dimensions 45x9x nm^3, (19)); probably both configurations are present in the layer, to give the average 22 nm value. This value is very close to that found in earlier experiments (16) at similar protein concentrations. The density would correspond to a coverage of about 20% in the "end-on" position and 120% in the "side-on" position. Both the thickness and the density imply that a significant fraction of the protein molecules approach the "end-on"

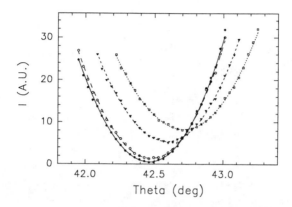

Figure 1. SAR: Reflected intensity (arbitrary units) as a function of angle, taken at various points during the successive adsorption of fibrinogen and its antibody (sample curves from a single experiment).

• : Before adsorption (Fresnel curve); ○ : after 100 min of 0.05% (w/w) fibrinogen adsorption; ▽ : after 60 min of 0.01% (w/w) anti-fibrinogen reaction; □ : after 800 min of anti-Fib reaction.

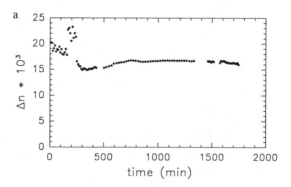

Figure 2a. SAR: Variation of the refraction indice (Δn) of the layer with time (0-160 min: Fib (0.05% (w/w)), 160-230 min: PBS, 230-1800 min: anti-Fib (0.01% (w/w))).

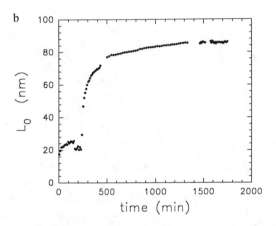

Figure 2b.   SAR: Variation of the layer thickness (L_0) with time (same conditions as in   Figure 2a).

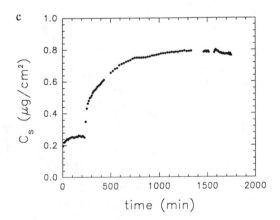

Figure 2c.   SAR: Variation of the surface concentration ($\Delta n \cdot L_0$) with time (same conditions as in Figure 2a).

position under the conditions of this experiment. A range of different layer thicknesses have been observed depending on the substrate (*16, 20*).

The surface was then rinsed with buffer solution. As seen in Figures 2a-c, the protein layer appears to rearrange to become more compact (both thinner and more dense, while the total adsorbed quantity remains unchanged) during this process. This compactification was observed consistantly in adsorbed protein layers. It may represent a shift in the proportions of " end-on " to " side-on " positions for the molecules adsorbed.

A purified rabbit γIgG (anti-Fib) solution directed against human fibrinogen was then injected, and led to the values $L_0 = 85$ nm, $\Delta n = 1.6 \cdot 10^{-2}$ for both layers, i.e. $\Delta\Gamma = 0.52$ $\mu g/cm^2$ of anti-Fib. Considering the molecular weight of the two species, the reactivity Fib/anti-Fib is thus about 1:5. A second similar experiment found a ratio of 1:8. Such large ratios suggest aggregation of the anti-Fib molecules in solution.

We have seen that the method of scanning angle reflectometry can give information about the total quantity of adsorbed molecules and the layer thickness. This can be used to estimate the reaction between the antibody and its antigen adsorbed on the surface. Note however that it cannot distinguish between the different species. In particular, any departure of the initially adsorbed antigen will remain unaccounted for.

Radiotracer Experiments.

Isotherms of Human Fibrinogen and γIgG. We present the isotherms of human Fib and IgG in Figures 3 and 4.

The Fib isotherm was realized by injection of labeled Fib, in type 2 PBS buffer, at a flow rate of 2.4 $cm^3 \cdot h^{-1}$ during 3.45 h, followed by a PBS rinse of 1.30 h at 4.8 $cm^3 \cdot h^{-1}$ (same conditions as those used in the Fib/anti-Fib reaction studies).

The plateau value (≈ 0.50 $\mu g \cdot cm^{-2}$) would correspond to a coverage of about 45% for an "end-on" and of 250% for a "side-on" configuration if one considers a parallelepiped of dimensions $45 \times 9 \times 6$ nm^3 for the Fib molecule. One would thus conjecture that the molecules are mainly (if not exclusively) fixed in an "end-on" configuration. The coverage in this position is in reasonable agreement with that expected in the jamming limit for the RSA model.

The γIgG isotherm was realized under slightly different conditions than for the Fib one: the γIgG molecules were adsorbed on the tube at the same $fr = 2.4$ $cm^3 \cdot h^{-1}$ but during 4.2 h, followed by a PBS rinse of 1.23 h at 4.8 $cm^3 \cdot h^{-1}$.

The plateau was reached at a bulk concentration of about 0.050% (w/w) with a surface concentration of ≈ 0.8 $\mu g \cdot cm^{-2}$. This corresponds to a coverage of $\approx 49\%$ for an "end-on" and 260% for a "side-on" configuration for ellipsoïd dimensions for the IgG molecule of $23.5 \times 4.4 \times 4.4$ nm^3. Thus, the same conclusions can be drawn as for the Fib molecules.

If we compare the values of the Γ obtained on the prism and the tube surfaces, we notice that, for the Fib as well as for the IgG molecules (see Table I and isotherms), they vary by a factor of two for identical bulk concentrations. We see that it is very difficult to obtain reproducible values for adsorbed quantities and to compare them between two different surfaces (here both made of hydrophilic silica): hence, the state of the surface influences crucially the adsorption.

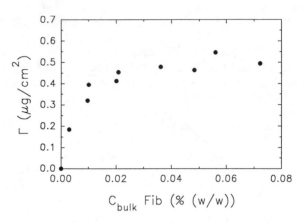

Figure 3. Adsorption isotherm of human Fib onto silica at a flow rate of 2.4 $cm^3 \cdot h^{-1}$ during 3.45 h (followed by a PBS rinse of 1.30 h at 4.8 $cm^3 \cdot h^{-1}$). Each point represents an independent experiment.

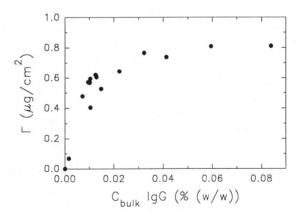

Figure 4. Adsorption isotherm of human IgG onto silica at a flow rate of 2.4 $cm^3 \cdot h^{-1}$ during 4.2 h (followed by a PBS rinse of 1.23 h at 4.8 $cm^3 \cdot h^{-1}$).

If the Γ at low bulk concentration is assumed to be limited, at a given time (here $\approx$ 4 hours), by diffusion of molecules to the surface, the diffusion coefficients of Fib and IgG may be evaluated, using the equation *(21)*: $D = (\Gamma/(2C))^2.\pi/t$, where C is the solution concentration. This leads to values of $\approx$ 2 orders of magnitude lower than the corresponding literature values $(2.0 \cdot 10^{-7}$ cm$^2 \cdot$s^{-1} and $4 \cdot 10^{-7}$ cm$^2 \cdot$s^{-1} for Fib and γ IgG, respectively). These depressed values may be related to boundary layer effects. However, even the saturation value of Γ appears to depend on the bulk concentration, implying that the limitations are not merely diffusive.

Study of the Fib/anti-Fib Reaction.   The concentration of the antigen solution was fixed at either 0.05 or 0.01% (w/w), yielding surface concentrations near saturation and at about three-fourths saturation respectively. The antibody solution was constantly 0.01% (w/w). An experiment starts (step 1, Figure 5) with injection of type 2 PBS buffer into the tube previously cleaned with sulphochromic acid. This step is followed (step 2, Figure 5) by injection of ^{125}I labeled Fib until the radioactivity stays approximately constant over a period of 30 min, indicating that the system has reached equilibrium. We chose for this purpose an adsorption time (t) of 3.45 h at a flow rate of 2.4 cm$^3 \cdot$h^{-1}. In step 3, the solution is replaced by buffer (rinse of 1.30 h at a flow rate of 4.8 cm$^3 \cdot$h^{-1}) and the increase (Δh) in activity at the end of this step is directly related to the amount of adsorbed antigen. The reactivity of the anti-Fib on the adsorbed molecules was then evaluated by flowing the labeled antibody solution (t = 2.45 h, flow rate = 2.4 cm$^3 \cdot$h^{-1}). A rough estimate of the protein uptake due to the antigen/antibody reaction was evaluated by Δh' - Δh after a PBS rinse of t = 2.35 h at a flow rate of 2.4 cm$^3 \cdot$h^{-1} (steps 4 and 5, Figure 5).

Quantitative analysis requires previous *in situ* calibration in order to relate the observed counts to the quantity of adsorbed protein. Two methods can be used. Method 1 starts with the passivation of the inner tube surface: a solution of concentrated fibrinogen (0.1% (w/w)) is injected and adsorption is allowed to proceed during $\approx$ 1 hour. The solution is then replaced by PBS buffer and the tube is heated during 10 min at $\approx$ 60°C, the temperature at which the Fib undergoes a thermal denaturation *(22)*. A radiolabeled protein solution of known concentration and specific activity is then injected and, since under these conditions no adsorption occurs, the radioactivity at the end of step 3 is equal to that of step 1 (background noise), and the increase in activity in step 2 is due only to the known amount of the protein present in the bulk solution. In method 2, for each adsorption experiment one measures the difference of the signal between step 2 (just before the rinsing step) and step 3 (at the end of the PBS rinse); this corresponds approximately to the radioactivity due to the protein amount present in the bulk solution *(11)*. This method has the disadvantage of being affected by the desorption of any weakly bound protein layer during the rinsing step, unlike method 1. On the other hand, it presents the advantage that calibration is made for each individual experiment, allowing for such intrinsic apparatus variations as those linked to the tube removal and replacement for cleaning. Both methods lead to an effective counting tube length (L_{eff}), calculated by writing the equality between the ratio of the activity due to the bulk protein to the specific activity at the same reference date (which is the amount effectively " seen " by the NaI detector), and the product $c\pi r^2 L_{eff}$, where c is the protein solution concentration determined by absorbance at 280 nm and r the tube radius (r = 0.085 cm). L_{eff} varied between 4 and 5 cm, and the

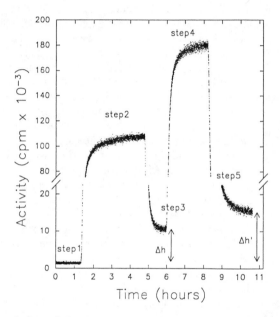

Figure 5. Evolution of the detected radioactivity versus time in a typical Fib adsorption experiment, followed by the reaction with anti-Fib molecules. Step 1: PBS buffer injection; step 2: labeled Fib injection; step 3: replacement of the protein solution by pure PBS buffer; step 4: labeled anti-Fib injection; step 5: replacement of the anti-Fib solution by pure PBS buffer; Δh: increase in activity related to the amount of adsorbed Fib; Δh': increase in activity proportional to the amount of adsorbed Fib and anti-Fib.

precision of this limits the precision of the estimation of the surface concentration (Γ) of the adsorbed protein layer. We compared the Γ of the adsorbed protein amounts calculated with the help of L_{eff} obtained by both methods. The approximately 0.03 $\mu g \cdot cm^{-2}$ difference between these values was nearly constant; but molar reactivity ratios were unaffected. With no obvious difference in the quality of the two methods, we chose to give all the values evaluated with method 1.

The labeled Fib, and later the labeled anti-Fib, were injected at a bulk concentration of 0.010% (w/w). The immobilized quantities are given in Table I. It is natural to attribute the change in the signal ($\Delta\Gamma = \Delta h' - \Delta h$) to the reaction of the adsorbed Fib molecules with the incoming anti-Fib ones. However, two other processes must be taken into account: (i) the desorption of labeled Fib molecules, whether directly or in exchange for the incoming anti-Fib molecules or as Fib/anti-Fib complexes and (ii) direct adsorption and/or adsorption by exchange of the anti-Fib molecules. The consequence of the first process is a decrease of the interfacial concentration of the initial adsorbed Fib layer, whereas the second implies that only some fraction of the anti-Fib molecules present at the surface are actually in direct reaction with the first adsorbed layer.

Two separate types of experiments, realized under the same experimental conditions as before but using selective labeling, quantifies these two contributions. To test for contribution (i), only the Fib molecules were labeled (Exp. 2i, Table I). The Fib concentration was found to decrease by $\Delta\Gamma = -0.03$ $\mu g \cdot cm^{-2}$ after the unlabeled anti-Fib was introduced. To test for contribution (ii), the adsorption of unlabeled Fib was followed by the injection of labeled rabbit γ-IgG which did not react with the Fib (anti-IgG*, Exp. 2c, Table I). The observed value of $\Delta\Gamma$, ≈ 0.08 $\mu g \cdot cm^{-2}$, was then due to other processes, whether direct adsorption or an exchange process.

If one takes into account these two contributions, molar reactivity ratios of 1.7 and 1.2 were found for experiments 2a and 2b respectively. Ratios above unity are not unexpected, since the polyclonal antibodies may react with different epitopes of the same antigen molecule.

In order to verify the reproducibility of these molar reactivity ratios, we repeated the experiments with two other stock solutions of purified anti-Fib. The results of these experiments, with what will be called solutions 2 and 3, as well as those with solution 1 as already discussed, are presented in Table I (with the preparation of anti-Fib molecules denoted by the indice). These three solutions were issued from three different purifications of the same crude plasma of rabbits immunized against human Fib. The proportion of labeled proteins was increased in solution 2, by doubling the quantity of NaI* added to the protein in the reaction medium, in an attempt to increase the signal at the interface (the specific activity was $\approx 20 \cdot 10^6$ cpm/mg compared to $\approx$ $7.5 \cdot 10^6$ and $4.5 \cdot 10^6$ cpm/mg for solutions 1 and 3 respectively). However, ELISA tests showed that while solutions 1 and 3 had about the same titer, i.e. the same reactivity against Fib, the reactivity for solution 2 was two to three times lower.

The results are as follows: (i) with anti-Fib*$_2$ molecules, we see (exp. 2d, Table I) that the signal is of the same order of magnitude as that obtained in experiment 2c, with anti-IgG* molecules. No significant reaction occured in this case. This was confirmed by two similar experiments for which the Fib solution was injected at higher bulk concentrations ($\approx 0.056\%$ (w/w), exps. 2e, 2f, Table I). If the anti-Fib*$_2$ molecules (same bulk concentration) had reacted proportionally with the adsorbed Fib, the signal would have been significantly higher. The fixed amounts were in fact slightly lower than for experiment 2e: when the Fib is near saturation, direct adsorption of antibodies is less important.

(ii) With anti-Fib*$_3$ molecules, we performed two experiments in which the (unlabeled) Fib solution was injected at 0.010% (w/w) (exp. 2g) and 0.050% (w/w) (exp. 2h, Table I). The anti-Fib*$_3$ signals led to $\Delta\Gamma$ of 0.16 and 0.11 μg·cm^{-2}, respectively.

The advantage of the use of unlabeled antigen molecules is that all subsequent signal is entirely due to antibody adsorption or reaction; the desorption of labeled antigen molecules can, and has, led to lower total signals after than before antibody adsorption. However, without labeling, the surface concentration of the antigen molecule must be taken from the isotherm, and the value is less precise than the one directly obtained in the same experiment with both radiolabeled antigen and antibody molecules (from this isotherm, Figure 3, bulk solutions of 0.010% and 0.050% correspond to a Γ of 0.35 and 0.50 μg·cm^{-2}, respectively). On the other hand, if both molecules are labeled, it is important to have an estimation of the $\Delta\Gamma$ of the antigen at the end of step 5 in order to appreciate this contribution in the resulting signal, even if the conclusion is that it is negligeable.

Again, the effect of adsorption processes unrelated to the direct antigen-antibody reaction must be taken into account. For experiment 2g, this is given by experiments 2c, 2d and 2i (Fib at a bulk concentration of 0.01%), and for experiment 2h, by experiments 2e and 2f (higher Fib bulk concentration). Molar reactivity ratios of mean values 0.56 and 0.30 were then found for experiments 2g and 2h respectively, quite different from the results obtained with the first anti-Fib preparation (exps. 2a and 2b).

These results indicate that a real problem exists linked to the radiolabeling itself. We have shown that for the anti-Fib*$_1$ and anti-Fib*$_3$ molecules, prepared from solutions 1 and 3 exhibiting the same reactivity levels in ELISA tests and with identical ^{125}I activities, the molar reactivity ratio is about three times higher for anti-Fib*$_1$ compared to anti-Fib*$_3$ molecules, whereas for the anti-Fib*$_2$, 2 to 3 times less reactive (ELISA) and with a labeling $\approx$ 4 times stronger, a total absence of reactivity could be shown. The influence of the radiolabeling on the activity of rabbit antibodies was qualitatively confirmed by batch immunologic experiments performed on the systems Fib/anti-Fib$_2$ and Fib/anti-Fib$_3$. These experiments were liquid phase immunoprecipitation tests. In one set of these tests, the antigen was labeled whereas the antibody was not, inversing the labeling for the other test. We then varied the molar ratios Fib/anti-Fib and incubated the mixtures at 37°C during 1 hour. If the molecules react, they form an "immunologic network" that precipitates when antigens and antibodies are at equivalence. After one or two days at 4°C, the solutions were then centrifuged and the radioactivity of the precipitate and the supernatant measured. The labeled molecules were of course the same as those used preceedingly in the tube experiments.

We found that for the system Fib/anti-Fib$_2$, with the anti-Fib$_2$ labeled, no radioactivity appears in the precipitate, whereas all the initial activity remains in the supernatant. This means that in the precipitate, we have only unlabeled Fib/anti-Fib$_2$ complexes, the labeled/unlabeled ratio of the anti-Fib$_2$ molecules being here of about 1:1200. For the system Fib/anti-Fib$_3$, about 5% of the initial anti-Fib*$_3$ is found in the precipitate, and correlatively disappears in the supernatant. For both systems, when the Fib is labeled and the anti-Fib not, all the initial radioactivity appears in the precipitate and disappears in the supernatant, the reaction being then total between the antigen and antibody molecules.

These results are in good agreement with those observed in the tube experiments. They suggest that the rabbit antibody is very sensitive to the radiolabeling method employed, and from one labeling to another the reactivity is more or less modified.

This observation is supported by the works of Pressman and Sternberger (*23*), and Koshland (*24, 25*), who demonstrated that extensive iodination (by the same McFarlane- derived method) destroys the precipitating capacity of a rabbit antibody, whose active site contains an iodine-reactive residue. Iodination thus reduces the binding capacity by a direct attack at the active site, by substitution or disubstitution in the phenolic group of a tyrosyl residue, which is the more sensitive to this attack, or on both the nitrogen and carbon atoms of the imidazole ring of a histidyl residue. Other groups may also be subjected in a γ-globulin to a chemical modification which could affect its reactivity, such as oxidation of cysteinyl and cystinyl residues to cysteic acid, methionyl residues to the corresponding sulfoxide or sulfone, or tryptophanyl residues. These changes could enhance the rigidity of the arms of the γIgG by inducing a steric hindrance at the angle region of the Fab fragments, or result in a local rupture of its 3-dimensional structure, resulting in a partial denaturation and a correlated inactivation.

Another indication of this effect due to the ^{125}I radiolabeling is brought by a complementary study concerning the reactivity of the system IgG/anti-IgG with a latex particle test. Unlabeled IgG were adsorbed under saturation conditions (Γ_{max} = 0.54 μg·cm^{-2}) on latex particles 790 nm in diameter. After elimination of the free IgG molecules in solution by successive centrifugations and redispersions with buffer, a solution of ≈ 0.030% (w/w) labeled anti-IgG (of high specific activity ≈ 12.5·10^6 cpm/mg) was brought into contact with the latex-IgG complexes and the reaction allowed to take place during 2 hours under gentle agitation. The reaction was stopped by centrifugation, and a measure of the activity of the supernatant led, by depletion, to the amount of anti-IgG* fixed on the surface. This quantity (0.11 μg·cm^{-2}) was found to be equal to the quantity of IgG adsorbed on these latex particles in experiments where we studied the homogeneous exchange mechanism IgG/IgG* under the same experimental conditions (Γ_{IgG*} = 0.13 μg·cm^{-2}) (*26*). Such an observation cannot be imputed to an eventual bad orientation of the antigen epitopes, as the antibody is polyclonal and recognizes a variety of determinants on the antigen molecule. Thus, this demonstrates that no reaction between IgG and the labeled anti-IgG occured. Again, an effect of the radiolabeling on the activity of the anti-IgG may be implied.

However, these effects do not concern the adsorption properties of the protein molecules on which we worked. Indeed, this has been demonstrated by Schmitt *et al.* (*27*), and is true in so far as less than one atom of iodine is statistically present per molecule, which was always the case in the radiolabelings we made (never higher as ≈ 1 per 200 molecules).

Conclusion

The Scanning Angle Reflectometry technique offers the possibility of analyzing the thickness variations of an adsorbed protein layer, and is thus well adapted to a qualitative evaluation of the reactivity between biological macromolecules. Hence, if an antibody reacts with an antigen, the evolution of the signal demonstrates it clearly and unambiguously. Such qualitative information is not immediately accessible with the radiotracer technique. However, to quantify the different side reactions (desorption of Fib and direct adsorption or exchange mechanisms of anti-Fib), which cannot be omitted in a precise estimation of the molar reactivity ratios, the radiotracer experiments are perfectly adapted and the adsorbed quantities are well evaluated.

The radiolabeling technique highlighted the unforseen problem of the ^{125}I labeling method used in this work with regard to the reactivity of rabbit antibodies. As this problem is linked to the presence of aminoacids which react covalently with ^{125}I atoms on the Fab fragment of the antibody, i.e. the epitope binding site, an antibody which belongs to a different species may not be affected by this kind of labeling technique. Nevertheless, a verification of the reactive ability of the labeled antibody towards its antigen in solution, e.g. with a liquid phase immunoprecipitation test, is essential whatever the chosen labeling technique or the origin of the antibody. Such an information is not given by an ELISA test, since the proportion of labeled molecules is generally low compared to the unlabeled ones: only the " cold " antibodies which react are detected.

From this point of view, reflectometry has the appreciable advantage that one can work with unaltered native molecules.

The tube geometry is a good model to mimic what occurs in biological systems, i.e. blood vessels. The sensitivity could be enhanced by increasing its surface, e.g. with a system composed of hollow silica capillaries (28). The possibility of studying molecular mechanisms under flow, i.e. precisely defined hydrodynamic conditions, is an appreciable advantage. This is in contrast with the SAR experiments as performed here, under static conditions. The kinetics of adsorption are certainly different: such factors as the relative frequency of side-on and end-on adsorption may be affected.

By combining the SAR and radiotracer techniques, we obtain different kinds of information that lead to a better comprehension and evaluation of the reactivity phenomena occuring between biological macromolecules. The SAR technique gives additional structural information, while the selective labeling of the radiotracer techniques allows one to distinguish between processes involving the different molecules.

Our study is a basis for further studies of the reactivity of antibodies with antigens adsorbed on an existing protein layer. Lutanie et al. (29) have observed a complete loss of reactivity of IgY with IgG adsorbed on a human albumin layer. To be generalized, this result should be verified on different types of surfaces and with a variety of immunological systems. The nature of the first adsorbed layer certainly plays a major role in this mechanism. As exchange processes cannot be detected by SAR, the radiotracer technique is of great utility for such an investigation. Using the controls described here, one can avoid misleading results.

Acknowledgments

One of the authors, Ph. H., thanks the Faculty of Odontology of Strasbourg for financial support.

Literature Cited

1.   Lyklema, J. *Colloids and Surfaces* **1984**, *10*, 33.

2. Wojciechowski, P.; ten Hove, P.; Brash, J.L. *J. Colloid Interface Sci.* **1986**, *111*, 455.
3. Norde, W.; Lyklema, J. *J. Colloid Interface Sci.* **1978**, *66*, 257.
4. Brash, J.L.; Uniyal, S.; Pusineri, C.; Schmitt, A. *J. Colloid Interface Sci.* **1983**, *95*, 28.
5. Voegel, J.-C.; de Baillou, N.; Schmitt, A. *Colloids and Surfaces* **1985**, *16*, 289.
6. Absolom, D.R.; Zingg, W.; Neumann, A.W. *J. Biomed. Mat. Res.* **1987**, *21*, 161.
7. Norde, W.; Favier, J.P. *Colloids and Surfaces* **1992**, *65*, 17.
8. Pefferkorn, E.; Carroy, A.; Varoqui, R. *J. Polym. Sci. Polym. Phys. Ed.* **1983**, *23*, 1997.
9. De Gennes, P.-G. *C. R. Acad. Sci. Paris II* **1985**, *301*, 1399.
10. McKinney, M.M.; Parkinson, A. *J. Immunol. Methods* **1987**, *96*, 271.
11. Boumaza, F.; Déjardin, Ph.; Yan, F.; Bauduin, F.; Holl, Y. *Biophys. Chem.* **1992**, *42*, 87.
12. McFarlane, A.S. *Nature* **1958**, *182*, 53.
13. Schaaf, P.; Déjardin, Ph.; Schmitt, A. *Rev. Phys. Appl.* **1986**, *21*, 741.
14. Azzam, R.M.A.; Bashara, N.M.; in *Ellipsometry and Polarized Light* (North Holland, Amsterdam), 1977.
15. De Feijter, J.A.; Benjamins, J.; Veer, F.A. *Biopolymers* **1978**, *17*, 1759.
16. Schaaf, P.; Déjardin, Ph.; Schmitt A. *Langmuir* **1987**, *3*, 1131.
17. Quash, G.A.; Rodwell, J.D.; in *Covalently Modified Antigens and Antibodies in Diagnosis and Therapy*; Marcel Dekker, Inc.: New York, Basel, 1989; p. 156.
18. Feder, J. *J. Theor. Biol.* **1980**, *87*, 237.
19. Doolittle, R.F. *Sci. Amer.* **1981**, *245*, 126.
20. Cuypers, P.A.; Hermens, W.T.; Hemker, H.C. *Ann. N. Y. Acad. Sci.* **1977**, *283*, 77.
21. Wojciechowski, P.; Brash, J.L. *J. Biomater. Sci. Polymer Edn* **1991**, *2*, 203.
22. De Baillou, N. *Thesis*, University of Strasbourg, 1983.
23. Pressman, D.; Sternberger, L.A. *J. Immunol.* **1951**, *66*, 609.
24. Koshland, M.E.; Englberger, F.M.; Erwin, M.J.; Gaddone, S.M. *J. Biol. Chem.* **1963**, *238*, 1343.
25. Koshland, M.E.; Englberger, F.M.; Gaddone, S.M. *J. Biol. Chem.* **1963**, *238*, 1349.
26. Ball, V.; Huetz, Ph.; Elaïssari, A.; Cazenave, J.-P.; Voegel, J.-C.; Schaaf, P. *Proc. Natl. Acad. Sci. USA*, in press.
27. Schmitt, A.; Varoqui, R.; Uniyal, S.; Brash, J.L.; Pusineri, C. *J. Colloid and Interface Sci.* **1983**, *92*, 25.
28. Yan, F.; Déjardin, Ph. *Langmuir* **1991**, *7*, 2230.
29. Lutanie, E.; Voegel, J.-C.; Schaaf, P.; Freund, M.; Casenave, J.-P.; Schmitt, A. *Proc. Natl. Acad. Sci. USA* **1992**, *89*, 9890.

RECEIVED February 10, 1995

EFFECTS OF SURFACE CHEMISTRY ON PROTEIN ADSORPTION

Chapter 25

Interactions of Hydrolytic Enzymes at an Aqueous–Polyurethane Interface

J. Paul Santerre[1], Daniel G. Duguay[2], Rosalind S. Labow[2], and John L. Brash[3]

[1]Department of Biomaterials, Faculty of Dentistry, University of Toronto, 124 Edward Street, Toronto, Ontario M5G 1G6, Canada
[2]Cardiovascular Devices Division, University of Ottawa Heart Institute, Ottawa Civic Hospital, 1053 Carling Avenue, Ottawa, Ontario K1Y 4E9, Canada
[3]Department of Chemical Engineering, McMaster University, Hamilton, Ontario L8S 4L7, Canada

Within seconds of exposure to biological fluids and tissues, the materials making up the surface of medical device implants rapidly become "hybrid biomembranes" composed of the synthetic matrix itself and adsorbed proteins, lipids and glycoproteins. The long-term stability of these altered biomaterial interfaces has been a question of concern for the past decade. Key to the degradation processes is the role of hydrolytic enzymes and their interactions at a polymer surface. This papers will present data describing the degradation of a polyester-urethane and two polyether-urethanes by cholesterol esterase. A mechanistic model for the degradation process is presented which highlights the importance of enzyme adsorption and surface chemistry. Data on the adsorption rates of the enzymes and on their activity in the adsorbed state are presented and discussed.

Finding appropriate "biomaterials" has been one of the most challenging problems in the development of biomedical devices, particularly for long-term implantation where the biostability of the materials is a major concern (1) Blood proteins, electrolytes, coagulation factors and white blood cells activated by the body's immune system can chemically react with the biomaterials and lead to degradation and failure of the implant (2-4) Furthermore, there is the concern that the degradation products themselves may be toxic (5). Tissue injury resulting from the surgical procedures used to implant devices automatically triggers the processes of wound healing and tissue repair (6). Involved in this response are phagocytic white blood cells, such as monocyte-derived macrophages. These cells have the ability to migrate towards and strongly adhere to the surface of biomedical implants. Phagocytic cells can undergo release of constituents, including various hydrolytic (including cholesterol esterase) and oxidative enzymes in response to the surfaces (7). It has been proposed that enzymatic attack is one of the principal contributors to degradation, given the ability of enzymes to significantly lower the activation energy of reactions (8). Enzymatic degradation of many polymers has been demonstrated by different groups and Table I lists several of these studies (9-17).

The specific chemistry of the biomaterial will in part define the biochemistry of the implant site with respect to the type and rate of interactions occurring between the material and its biological environment. Of special interest as implant materials are the"polyurethanes". The medical research literature on these materials has

0097–6156/95/0602–0352$12.00/0

grown exponentially over the past decade primarily because of their wide range of application throughout the medical field *(1)*. Among the more extensive studies of polyurethane degradation by enzymes to date, are those of Smith et al. *(18)* who investigated the enzymatic degradation of several polyurethanes by hydrolytic enzymes such as trypsin, chymotrypsin, esterase, papain, bromelain, ficin, cathepsin C and collagenase, as well as oxidative enzymes such as xanthine oxidase and cytochrome oxidase. In addition, a rabbit liver homogenate was used to expose a mixture of enzymes to the materials.   Using a series of ^{14}C-labelled polyurethanes, degradation was monitored up to 15 days after incubation in the various enzyme solutions. Their method assumed that the extent of degradation would be represented by the release of the radiolabelled compounds, containing a ^{14}C-ethylene diamine monomer, into the incubation solution. They observed that a polyether-urethane, prepared from polytetramethylene oxide (molecular weight

Table I. Polymer Degradation by Enzymes

Polymer (s)	Enzymes	Reference
polyurethanes	cholesterol esterase, xanthine oxidase, cathepsin B, collagenase	9
polyglycolic acid	esterase, chymotrypsin, trypsin	10
polyester	esterase	11
polyester-urea/phenylalanine	urease, pepsin, chymotrypsin, elastase, papain, acid protease	12
lignin	peroxidase compound I	13
2,2-bis[4-(2-hydroxy-3-methacryloyloxy-propoxy)phenyl] propane/triethylene-glycoldimethacrylate copolymer	esterase	14
poly(3-hyroxybutyrate)/poly (ethylene oxide) blend	PHB depolymerase	15
nylon 66, poly(ethylene terephtahlate),poly (methyl methacrylate)	esterase, papain, trypsin chymotrypsin	16
copoly(l-lactic acid/ε-caprolactone)	carboxytic esterase, lipases	17

1000),was affected by esterase, cathepsin C and rabbit liver homogenate, but not by the other enzymes. However, polyether-urethanes prepared from polyethers of molecular weight 650 and 2000 were not affected by any of the enzymes. Since interactions of enzymes with their natural substrates are highly specific, it would seem probable that enzyme/polyurethane interactions would also be specific. While hydrolytically cleavable groups are present in all polyurethanes, they must be accessible to the enzyme and the enzyme must adsorb in a state that would permit cleavage to occur. The data from Williams' *(18)* showing the dependence of enzyme degradation on both the distribution of chemical groups within the polyurethane and enzyme type, would support this.

Polyurethane chemistry can be complex since the polymers may have several chemical domains. Depending on the exact formulation, polyurethanes can have one or several types of hydrolysable chemical groups available for cleavage by enzymes, including urea, urethane, ester and carbonates. In addition, it has been hypothesized that oxidation of other polyurethane chemical groups (such as the ether linkages) may generate chemical groups susceptible to hydrolytic cleavage (19). Recent work in our laboratory (20,21) confirmed observations by Williams and further showed that the degradation of polyurethanes with physiologically relevant enzymes could be quantitatively monitored in-vitro. Using several ^{14}C radiolabelled monomer components, a number of different products were isolated. The data showed that release rates of radiolabel, for samples incubated with enzyme solution, were initially high and then slowed down but remained significant compared to the "buffer control" solutions (21). In addition, it was shown that the specific inhibition of the enzyme activity reduced the amount of radiolabelled product released (21).

The exact mechanism by which hydrolytic enzymes degrade polyurethanes has not been completely resolved. However, recent modelling work by our research group has led to the development of a mechanistic model comprised of seven principal steps (22). 1) Water/electrolytes contact and begin to interact with the polyurethane at the surface; 2) Polyurethane is restructured as the surface changes to adapt to an aqueous environment; 3) Enzyme adsorption and desorption is established; 4) Adsorbed enzyme reacts with one or several susceptible bonds at or near the surface; 5) Resulting products are released into solution; 6) Water/electrolytes/enzymes migrate within the polymer to degrade the polyurethane; 7) The degradation process is propagated by solution components over the longterm. The model requires the definition of initial polyurethane surface properties which will be defined by both material chemistry and sample processing. While these steps are listed sequentially, several of them are likely to be proceeding simultaneously. Figure 1 describes steps 1 through 5. One of the pivotal steps in the mechanism is the enzyme adsorption/desorption (step 3). The ability of the enzyme to adsorb onto the surface in a conformational state that will permit cleavage of the polymer is important in determining the relation between biomaterial chemistry and the stability of the material in the biological environment. This paper presents and discusses experimental data specifically related to the enzyme interaction steps of the above model and defines relative time frames in which these processes occur.

Experimental

Preparation of Radiolabelled Monomers. [14C] 2,4-toluene diisocyanate (custom synthesis from NEN DuPont) was received in an amber glass ampoule dissolved in dried toluene (2.0 mCi dissolved in 5 mL). In a nitrogen atmosphere, the 0.5 mL solution was divided into 50 µL aliquots and stored in the dark at -20°C in amber glass vials.

Polyurethane Synthesis. A total of three different radiolabelled polyurethanes were synthesized for this study. The materials used for the synthesis included 2,4-toluene diisocyanate (TDI) (Eastman Kodak), polycaprolactone diol of average molecular weight 1250 (PCL) (Aldrich), poly(tetramethylene) oxide of average molecular weight 1000 (PTMO) (DuPont Canada), polyethylene glycol of molecular weight 1000 (PEO) (Aldrich Chemical Company) and ethylene diamine (ED) (Aldrich).

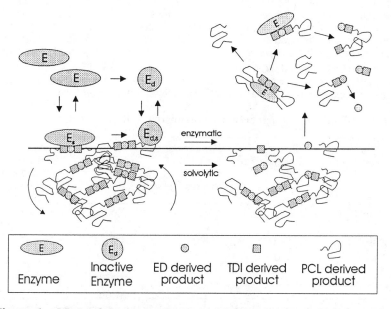

Figure 1. Model for enzymatic degradation of a polyurethane. E is the active enzyme in solution, E_s is the enzyme adsorbed to the surface, $E_{d,s}$ is the adsorbed deactivated enzyme, E_d is the deactivated enzyme in solution. ED is ethylene diamine, TDI is 2,4-toluene diisocyanate and PCL is polycaprolactone.

Prior to use, TDI was vacuum distilled at 0.025 mmHg. PCL, PTMO and PEO were degassed for 24 hours at 0.5 mm Hg. ED was distilled at atmospheric pressure. Radiolabelled TDI was mixed with non-radiolabelled analogs to provide the specific activity required for the polyurethane synthesis. The polymers were synthesized using a standard two step solution polymerization. The polymerization was carried out in a controlled atmosphere glove box containing dried nitrogen gas. The selected solvent was distilled within 24 hours of the synthesis. Details of this type of synthesis were previously provided (9). A complete listing of polymer nomenclature, reaction stoichiometry, solvent, temperature and specific radioactivity are provided in Table II. After reactions were complete, the polymers were precipitated in distilled water, finely chopped and washed in 1 L of distilled water three additional times for periods of 12 hours with vigorous stirring. The washing procedure was required to remove low molecular weight radiolabelled material and unreacted monomers. The polymers were then re-dissolved in their reaction solvent and the resulting solution was centrifuged at 1500 x g for 10 minutes, in order to remove any non-soluble polymer gels. This solution was then reprecipitated in distilled water and the polymer chopped and washed three times. The polymer was then dried in an oven at 50^0C for 48 hours. The specific activity for each polymer was obtained by dissolving 100 mg of the purified polymer in dimethylacetamide (DMAC) and counting in a liquid scintillation counter.

Table II. Polyurethane Chemistry and Nomenclature

[a]Polymer Nomenclature	[b]Reaction Stoichiometry	Prepolymer Reaction Temperature (degree C)	[c]Reaction Solvent	Specific Radioactivity (counts/min) per 100 mg
[14]TDI/PTMO/ED	2/1/1	60-70	DMSO	6.8×10^5
[14]TDI/PTMO-PEO/ED	2/.5-.5/1	60-70	DMAC	8.5×10^5
[14]TDI/PCL/ED	2.2/1/1.2	60-70	DMSO	7.4×10^5

[a]Polymer nomenclature has three alpha symbols xxx/yyy/zzz, where xxx is the diisocyanate, yyy is the diol and zzz is the chain extender. The superscript "14" indicates the radiolabelled component.

[b]The indicated stoichiometry corresponds to the xxx/yyy/zzz nomenclature.

[c]Dimethylsulfoxide (DMSO), dimethylacetamide (DMAC).

Molecular Weight Determination. Molecular weights were determined by size exclusion chromatography (SEC). The system consisted of a solvent mobile phase reservoir, a high pressure solvent delivery system (Waters 510 pump), a Rheodyne 7125 injector with 200 μL sample loop, three Waters Ultrastyragel columns, a thermostatted oven for the columns with temperature controller set at 80^0C and a refractive index detector (set at 40^0C), and data processing station. The solvent phase was dimethylformamide (DMF) containing 0.05 M LiBr, the sample size was 200 μL and the polymer concentration was approximately 0.2 g/100 mL. The flow rate was set at 1 mL/min and a typical chromatogram showed a retention volume of 40 mL. All molecular weight data are reported as calibration grade polystyrene molecular weight equivalents.

Differential Scanning Calorimetry (DSC). DSC studies of thermal transition behaviour were carried out for all polyurethane samples. Polymer films were prepared from a 20% wt/vol solution of polymer in DMAC. Solutions were centrifuged at 1500 x g for 10 minutes as before, in order to remove any particulate forms and then filtered. The films were cast on Teflon sheets and placed in an oven at 50^0C for 48 hours, then further dried in a vacuum oven for 24 hours at 50^0C. Films were gently lifted off the Teflon sheets and stored in a refrigerator until required. The film thickness was 0.3 ± 0.02 mm and the sample size was approximately 10 mg. The specimens were scanned from -100^0C to 220^0C at a rate of 5^0C per minute.

X-ray Photoelectron Spectroscopy (XPS). Data were obtained using a Kratos Axis HS spectrometer system. A monochromatic Al Ka radiation was used for excitation and the spectrometer was operated in Fixed Analyzer Transmission (FAT) mode throughout the study, using electrostatic magnification. Charge compensation was achieved by using a thermionic emission electron flood gun. In order to verify that the surfaces were not damaged during variable take-off angle studies, the surface was scanned at zero degree take-off angle both before and after the angle dependence measurements. A low resolution/high sensitivity spectrum was obtained over the range of take-off angles from 10 to 80 degrees.

Preparation of Tubes for Enzyme Adsorption Experiments. The polyurethanes were coated onto 5 mm long tubes, 3 mm I.D. and 4 mm O.D. using a 10 % w/w polymer solution in DMAC. The tubes were dipped into solution and residual polymer in the tube was blown out using a Pasteur pipette (blowing repeated 10-15 times). The glass tubes were supported at one end on a Teflon plate and placed in a 50^0C oven for overnight drying. The tubes were coated four times using the above procedure, alternating ends, supported on the Teflon plate, during the drying process. The tubes were then allowed to dry an additional 48 hours in an air oven at 50^0C and then dried overnight in a vacuum oven for 24 hours. They were stored at 4^0C in 2 mL Eppendorf tubes until required for experiments.

Radioiodination of Cholesterol Esterase and Adsorption Experiments. Cholesterol esterase was labelled with ^{125}I (Amersham, England) by a lactoperoxidase method using the reagent enzymobeads (BioRad, Richmond, CA) *(23)*. Solutions of radiolabelled enzyme were prepared in 0.05 M phosphate buffer, pH = 7.0. Polymer coated tubes were equilibrated overnight in 0.05 M phosphate buffer solution at 4^0C and then allowed to warm up to room temperature on the day of the experiment. Samples were placed in a microtitre plate (96 wells) and 195 μL of radiolabelled enzyme solution was added to the wells. Experiments were run in triplicate. For determination of isotherms, the adsorption time was fixed at 3 hrs and the concentration range was 0 to 1 mg/mL. For the kinetic adsorption experiments, the selected enzyme concentrations were 0.08 mg/mL and 0.240 mg/mL and the time ranged from 0 to 180 minutes.

Preparation of Coated Glass Tubes for Biodegradation Experiments. The polyurethanes were coated onto hollow glass tubes (I.D. was 2 mm and O.D. was 3 mm) using a 10 % w/w polymer solution in DMAC. These coated tubes were dried in a vacuum oven for 48 hours at 50^0C after dip coating 4 times with intermittent overnight drying. The resulting film thickness was approximately 10 μm. The coated tubing was sectioned into 10 pieces of 2.55 mm length (approximate total surface = 36 cm^2) and placed into a sterile 15 mL vacutainer

(Becton-Dickinson). All incubation experiments were prepared in a laminar flow hood, employing sterile techniques.

Biodegradation Experiment. Cholesterol esterase (CE) was selected as the hydrolytic enzyme for this study since it is present in the intracellular granules of liver cells, aortic intima and leukocytes *(24)*. The probability of CE being released from the lysosomes is very high when these cells are stimulated by tissue injury at the site of an implantation. Vacutainers were set up with 0.05 M phosphate buffer, pH 7.0, either with bovine pancreas CE or without (control solution). The enzyme concentration was 0.1 unit mL^{-1}. CE (#C 3766, Sigma) solutions were prepared by dissolving the powder in 0.05 M phosphate buffer, pH 7.0 at a concentration of 1 unit mL^{-1} and stored frozen at -40°C until required. The required units per mL were estimated based on the specifications of the supplier (800-1600 units gram^{-1} protein; a unit hydrolyzes 1 μmol of cholesteryl oleate to cholesterol and oleic acid per min at pH 7.0 and 37°C in the presence of taurocholate). The actual activity data generated in this study are based on p-nitrophenylacetate as a substrate, at pH 7.0 and 25°C, with a unit defined as a change in absorbance of 0.01 min^{-1} at 410 nm. All solutions were sterile filtered using a 0.22 μm filter. Aliquots were removed from the polymer incubation solutions and counted in a liquid scintillation counter for radioactivity. Samples were removed twice daily for a week, and daily for the second and third weeks. The activity lost between sample times was calculated and when the samples were removed for counting, fresh enzyme was added in order to maintain the incubation volume and the enzyme activity (enzyme half-life was previously reported *(9)*). Bacterial cultures were run on samples at the conclusion of all incubation periods in order to determine if sterility was maintained throughout the experiment. Duplicate samples at each reaction condition were run over three weeks and the experiments were repeated twice.

Results and Discussion

Material Characterization. The molecular weight data for the three polymers used in the study are given in Table III. All three polymers showed molecular weight values and polydispersities representative of commercial polyurethane materials *(25)*. The bulk structure of the materials was assessed by DSC. Thermograms for the three polymers are shown in Figure 2. The polyester-urethane, TDI/PCL/ED, shows one thermal transition near 45°C associated with the melting of the soft segment PCL domains. There is no transition at higher temperatures where hard segment domain related transitions could be expected and it can therefore be concluded that there is little or no structured hard segment domain within the polymer matrix. The two polyether based polyurethanes show distinct glass transition temperatures for the soft segment and transitions at higher

Table III Molecular Weight Data

Polymer	Weight Avg. MW	Number Avg. MW	Polydispersity
[14]TDI/PTMO/ED	2.6×10^5	1.1×10^5	2.3
[14]TDI/PTMO-PEO/ED	7.3×10^4	4.1×10^4	1.8
[14]TDI/PCL/ED	1.1×10^5	5.0×10^4	2.2

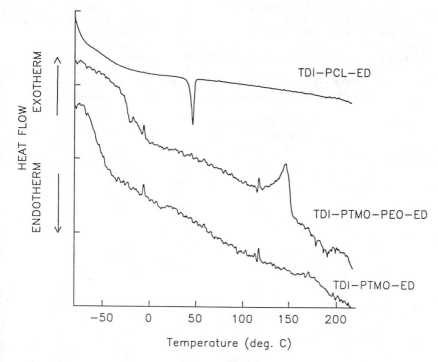

Figure 2. DSC thermogram data

temperatures, indicating good phase separation. In addition, the PEO containing material shows a transition near 150^{0}C which is probably related to enhanced hydrogen bonding between PEO and urethane/urea linkages. A good understanding of domain structure in polyurethanes is essential since the lack or presence of phase separation will influence the mobility and accessibility of chemical groups at the surface of the polyurethane and therefore influence interactions with the enzymes *(20)*.

XPS data for the three polymers are given in Table IV. All three materials show a reduced nitrogen concentration in the immediate surface region (i.e. 80^{0}C take off angle). The two polyether-based polyurethanes show a continuous decrease of nitrogen concentration as the surface is approached. Elemental nitrogen may be associated with the hard segment domains since nitrogen is contained in both components making up the hard segment (i.e. ED and TDI). Therefore the data suggest a depletion of hard segment domain in the near-surface regions of these materials. The TDI/PCL/ED polymer shows different nitrogen concentration values, alternating between high and low values throughout the profile.

Table IV XPS Analysis of Polyurethanes: Elemental Composition at Various Take-off Angles (relative accuracy = 10%)

Polymer	Takeoff Angle (degree)	C	O	N
			(atomic percent)	
TDI/PTMO/ED	10	78.2	17.3	3.5
	30	80.0	16.2	3.8
	40	80.2	16.5	3.3
	60	81.0	16.2	2.9
	80	85.5	13.3	1.2
TDI/PTMO-PEO/ED	10	80.7	16.0	3.3
	30	81.5	15.5	3.0
	40	82.0	15.4	2.6
	60	84.3	14.2	1.5
	80	86.0	14.0	0.0
TDI/PCL/ED	10	77.5	21.7	0.8
	30	77.8	20.6	1.6
	40	78.5	20.6	0.9
	60	79.4	18.9	1.7
	80	81.0	18.7	0.3

Effect of Soft Segment Chemistry on Enzyme Interactions. Figure 3 shows three hour adsorption data for CE onto the three polyurethanes and a glass control surface. It was shown in preliminary work that there was no preferential adsorption of the radiolabelled enzyme (unpublished data). Several features of the data presented in Figure 3 should be noted. All three polyurethanes adsorb significantly greater amounts of enzyme than does the glass surface, thereby indicating a relatively high capacity for enzyme interaction with polyurethanes. All three polymers adsorb similar amounts at low concentration whereas at higher concentrations the PEO-based material adsorbs slightly more than the TDI/PCL/ED and the TDI/PTMO/ED polymers. At this time there are not sufficient data available on the nature of the specific interactions between CE and the three polymers to explain these observations. However, the data do show that there is sufficient

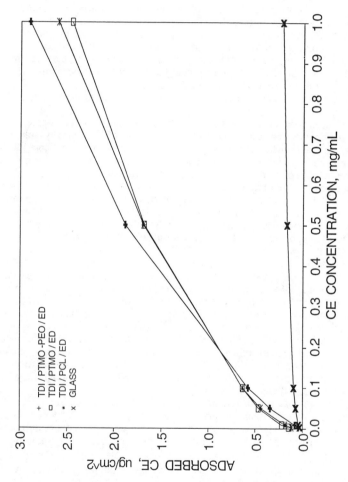

Figure 3. Cholesterol esterase adsorption isotherms. Each adsorption value was determined in triplicate. The standard deviations were of the order of ± 10 % of the mean values.

enzyme adsorbed at the interface of the materials to explain the enzyme related degradation observed in Figure 4.

Biodegradation data for the three polymers are presented in Figure 4 at different time points and are plotted as cumulative radioactivity (counts/min, CPM) released versus time of incubation. The presence of enzyme causes a higher radiolabel release for all three polymers as compared to buffer only solutions (controls). However, the greatest radiolabel release is observed with TDI/PCL/ED. It is intuitive to suggest that the esterase is degrading all ester bonds in the polymer with equal specificity. However our previous studies [9] have indicated that not all ester bonds are equally susceptible to cleavage by the enzyme since bulk analysis of weight loss data were not correlated with hard segment radiolabelled release for TDI/PCL/ED. Weight loss was significantly lower than release of the radioactivity would have predicted. Since for those studies the radiolabelled carbon was contained within the ethylene diamine component, it was concluded that preferential cleavage of the ester bonds was occurring near the hard segment domains. If ester bonds remote from the rigid aromatic hard segment domains were not cleaved, weight loss would be reduced, since the PCL component makes up the bulk of the polymer.

The significant radiolabel release for TDI/PCL/ED (Figure 4) as compared to the other two polymers is primarily due to the availability of the hydrolysable ester bonds at the surface (indicated by the high surface oxygen content in Table IV) but as well is enhanced by the fact that the high degree of phase mixing for this polymer increases the number of hard segment domains in the proximity of the ester bonds. The high degree of phase mixing is supported by the DSC data in Figure 2. While the urea and urethane sites in the two polyether-based materials are susceptible to hydrolysis, a lower degree of cleavage would exist for the well phase separated materials since diffusion and access of the relatively large CE molecule to the cleavage sites would be hindered by the structured state of the hard segment.

Mechanism of Enzyme Interactions. The data in Figures 3 and 4 indicate that while CE may be present in similar amounts at the surface of all three polyurethanes, not all of it is actively involved in biodegradation. This is supported by the data showing that the amount of radiolabel released from the polyetherurethane is significantly different from the polyester-urethane. Despite the high levels of enzyme adsorbed onto the TDI/PCL/ED surface, which corresponded to at least monolayer coverage (Figure 3), a previous study (9) in which the surface of the same polymer was examined, using scanning electron microscopy, showed no gross surface degradation. This result would support the suggestion that not all ester sites on the polymer are equally susceptible to cleavage, eventhough they have enzyme adsorbed near them.

Figure 5 shows dose response data for the effect of CE on TDI/PCL/ED. For this polymer/enzyme system the saturation value was approximately 0.2 U/mL, equivalent to about 0.08 mg/mL. Based on the enzyme adsorption data in Figure 3, 0.08 mg/mL corresponds to an enzyme surface concentration of roughly 0.5 $\mu g/cm^2$, which is well below the adsorption capacity of greater than 2.0 $\mu g/cm^2$. These data indicate that all adsorption sites on the enzyme are not necessarily related to cleavage sites on the polymer. In regard to the development of a model for the process of polyurethane degradation, the mechanism of enzyme adsorption and desorption should be defined separately from the mechanism of enzyme related cleavage, specifically with respect to "active sites", i.e. binding sites in the case of adsorption and cleavage sites in the case of hydrolytic enzyme activity. This would rationalize the separation of these two steps in the proposed model, shown in Figure 1.

In order to obtain an estimate of the time frame for the various processes illustrated in Figure 1, kinetic data were obtained for enzyme adsorption and the

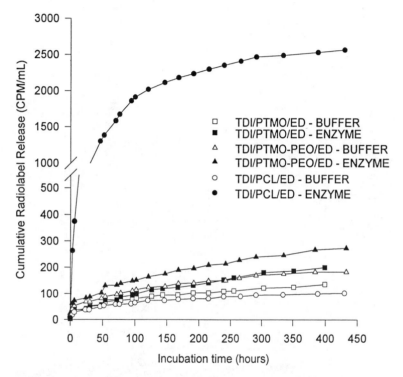

Figure 4.   Effect of CE on radiolabel release from different polyurethanes.   Enzyme concentration is 0.1 U/mL in phosphate buffer, pH 7.0, incubated at 37°C. The standard deviations were of the order of $\pm$ 10 % of the mean values for TDI/PCL/ED and $\pm$ 20% of the mean values for TDI/PTMO/ED and TDI/PTMO-PEO/ED.

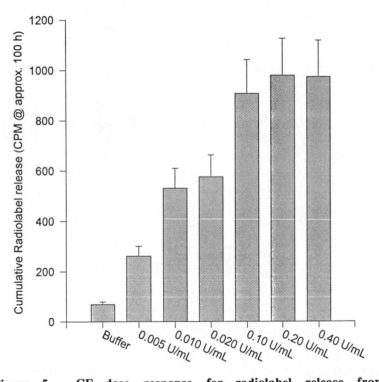

Figure 5. CE dose response for radiolabel release from TDI/PCL/ED.

release of degradation products from the polymer. As shown in Figure 6, enzyme adsorption onto the surface of TDI/PCL/ED occurred within minutes. Within one hour of incubation with CE, the adsorption of enzyme required for saturation levels of degradation products to be released was achieved. In fact, within minutes enough enzyme was adsorbed to the surface of the material to begin the degradation process. As shown in Figure 7 differences in the amounts of radiolabelled degradation products for enzyme versus buffer solutions were observed within one hour of the incubation start time. The actual cleavage process would probably occur before this time since there would be some delay in time between enzymatic cleavage and the release of product(s) into the solution. The mathematical relationships that define these processes have been described in recent work *(22)*.

The four different enzyme states shown in Figure 1 suggest two distinct enzyme deactivation processes that may occur in the in-vitro test system, i.e. the deactivation of non-adsorbed enzyme and deactivation of adsorbed enzyme. The data presented in Figure 8 would suggest that the kinetics of both processes are different and that enzyme in the adsorbed state is deactivated more slowly than enzyme in solution. Simultaneous monitoring of enzyme activity in solution and release of radiolabelled products into solution was done over a 200 hour time period. While the enzyme activity in solution was no longer measurable after 30 hours of incubation with the polymer (Figure 8b), radiolabelled products were still being released at significantly higher rates than in the buffer control (Figure 8a). The slopes of the radiolabel release data after 30 hours for the 0.01 U/mL and the 0.04 U/mL solutions were 0.97 CPM/hr and 0.69 CPM/hr respectively, whereas the slope for the buffer control was 0.09 CPM/hr. There was approximately a ten fold difference in the rates of degradation.

Since it is known that product release occurred within an hour of initiating the incubation of enzyme with polymer (see previous paragraph), the subsequent release of radiolabel (after 30 hours) following loss of enzyme activity in solution could not be related to the delay of product release. Another possible explanation for prolonged release of products following the loss of enzyme activity in solution, may be related to the polymer degradation products themselves. A previous study showed that contact with TDI/PCL/ED did not influence the halflife of cholesterol esterase *(9)* and therefore it is unlikely that degradation products released into solution deactivate the enzyme. However, it is conceivable that the degradation products could be implicated in a degradation reaction with the polymer. Work by Pitt has described how degradation products of poly(e-caprolactone) catalyze the hydrolytic degradation of the polyester chains *(26)*. If this were occurring, degradation could be prolonged beyond the life of the enzyme.

The actual process of degradation product release appears to occur in two distinct phases (see Figure 7). An early phase (first 50 hrs) where product release is very high followed by a second phase which proceeds at a much slower rate but remains significant compared to the buffer- only control. It is hypothesized that early degradation results from immediate enzyme interactions with available cleavage sites at the surface. Following the consumption of these sites, the feeding of subsequent sites to the surface becomes dependent on the polymer chain mobilily and the rate at which minimal surface free energy will be re-established. At this time the kinetics of such processes are essentially unknown. However, based on contact angle kinetic data reported by Andrade and co-workers *(27)* it has been suggested that these processes occur on a time scale in the order of hours. These data would indicate that a significant drop in product release rates would follow with a dependence on surface reorganization and orientation of new cleavage sites to the surface adsorbed enzyme. Another mechanism that may in part explain the two phase profile for Figure 7 is that an accumulation of adsorbed/deactivated enzyme could possibly impede the release process. However, when it is considered that the surfaces of the polymer adsorb very large amounts of enzyme within

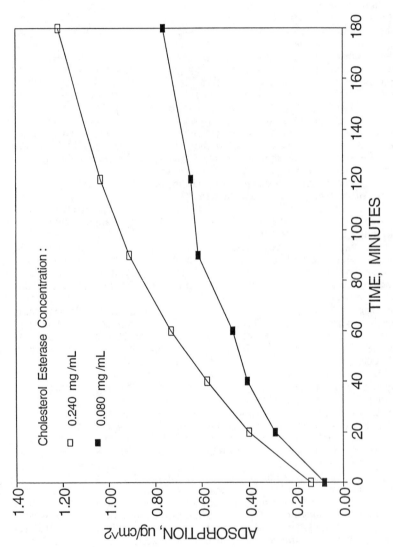

Figure 6. Kinetics of CE adsorption onto TDI/PCL/ED. Each adsorption value was determined in triplicate. The standard deviations were of the order of ± 10 % of the mean values.

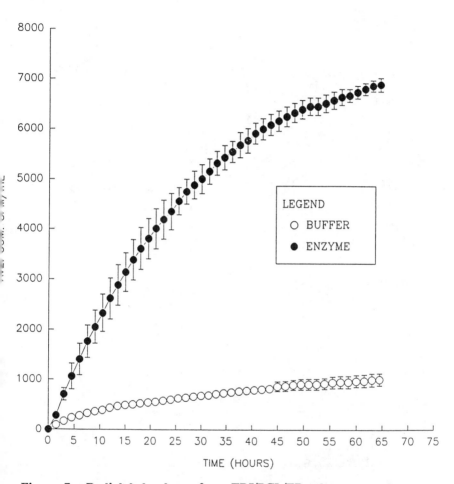

Figure 7. Radiolabel release from TDI/PCL/ED when exposed to CE. Enzyme concentration is 0.1 U/mL in phosphate buffer, pH 7.0, incubated at 37°C. The standard deviations were of the order of ± 10 % of the mean values.

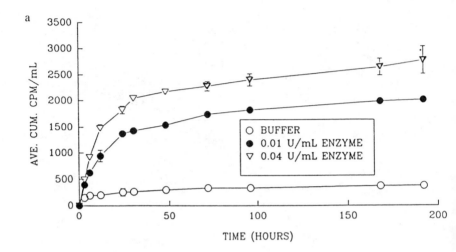

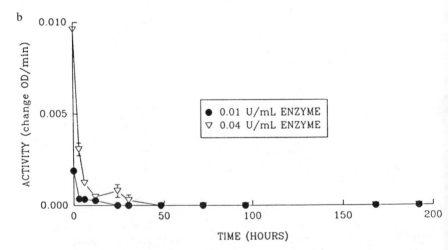

Figure 8.   Simultaneous a) radiolabel release from TDI/PCL/ED and b) deactivation of enzyme.

minutes of incubation, yet degradation proceeds at high rates for several hours, it is not likely that protein denaturation or inactivation on the surface is completely responsible for the observed decrease in radiolabel release rates. It has been proposed *(18,28)* that the degradation products themselves remain adsorbed and inhibit further degradation by occupying available cleavage sites. This explanation conflicts with the previous hypothesis that degradation products may enhance the generation of radiolabelled components at the surface.

In conclusion, the experimental data reported support the conceptual model of polyurethane interactions with hydrolytic enzymes that was previously proposed and mathematically developed *(22)*. However, the data also indicate that the effect of degradation products themselves is inadequately defined and that experiments are needed to investigate their role in the biodegradation process. Degradation of polyurethanes begins to occur within minutes of exposure to enzymes. Thus it will be of interest in future work with more relevant tissue-material systems (e.g. the response of cells to material degradation) to look for events occurring within this relatively short time frame.

Acknowledgements

The authors would like to acknowledge funding from the Ontario Centre for Materials Research, Health and Welfare Canada (NHRDP #6606-5053-52) and the Ontario Ministry of Health Career Scientist Program. As well, we are grateful for the use of the DSC facilities located at the National Research Council of Canada, Institute for Environmental Chemistry, Ottawa, Ontario, Canada (Dr. Michael Day's laboratory). The authors acknowledge the assistance of Dr. Yves Deslande at the National Research Council of Canada for the use of XPS facilities.

Literature Cited

1.   Stokes, K.B. *Cardiovasc. Pathol.* **1993**, 2, 111S-119S.
2.   Stokes, K.B. *J. Biomat. Appl.* **1988**, 3, 228-259.
3.   Szycher, M. *J. Biomat. Appl.* **1988** , 3, 297-402.
4.   Hennig, E.; Zartnack, J.F.; Lemm, W.; Bucherl, E.S.; Wick, G.; Gerlach, K. *Int. J. Artif. Org.* **1988**, 11(6), 416-427.
5.   Boyes, D.C.; Adey, C.K.; Bailor, J.; Baines, C.; Kerrigan, C.; Langlois, P.; Miller, N.; Osterman, J. *Can. Med. Assoc. J.* **1991**, 145, 1125-1132.
6.   Remes, A.; Williams, D.F. *Biomaterials* **1992,** 13, 731-743.
7.   Williams, D.F., *ASTM STP 684*, **1979**, 61-75.
8.   Royer, G.P. *Fundamentals of Enzymology: Rate Enhancement, Specificity, Control and Applications* ; Wiley Interscience: New York, N.Y., 1982, pp 1-232.
9.   Santerre, J.P.; Labow, R.S.; Adams, G.A. *J. Biomed. Mater. Res.* **1993**, 27, 97-109.
10.   Chu, C.C.; Williams, D.F.; *J. Biomed. Mater. Res.* **1983**, 17, 1029-1040.
11.   Smith, R.; Williams, D.F. *J. Mater. Sci. Letters* **1985**, 4, 547-549.
12.   Huang, S.J.; Bansleben, D.A.; Knox, J.R. *J. Appl. Polym. Sci.* **1979**, 23, 429-437.
13.   Schoemaker, H.E.; Harvey, P.J.; Palmer, J.M. *Febs Letters* **1985**, 183(1), 7-12.
14.   Freund, M.; Munksgaard, E.C. Scand. *J. Dent. Res.* **1990**, 98, 351-355.
15.   Jumagai, Y.; Doi, Y. *Polym. Deg. stab.* **1992**, 35, 87-93.
16.   Smith, R.; Oliver, C.; Williams, D.F. *J. Biomed. Mater. Res.* **1987**, 21, 1149-1165.
17.   Fukuzaki, H.; Yoshida, M.; Asano, M.; Kumakura, M.; Mashimo, T.; Yuasa, H.; Imai, K.; Yananaka, H. *Polymer* **1990**, 31, 2006-2014.

18. Smith, R.; Williams, D.F.; Oliver, C. *J. Biomed. Mater. Res.* **1987**, 21, 1149-1165.
19. Anderson, J.M.; Hiltner, A.; Zhao, Q.H.; Wu, Y.; Renier, M.; Schuber, M.; Brunstedt, M.; Lodoen, G.A.; Payet, C.R. In *Biodegradable Polymers and Plastic* ; Editors, Vert, M.; Feijen, J.; Albertsson, A.; Scott, G.; Chiellini, E., Eds.; The Royal Soc. of Chem.: Cambridge, 1992; pp 122-136.
20. Santerre, J.P.; Labow, R.S.; Duguay, D.G.; Erfle, D.E.; Adams, G.A. *J. Biomed. Mater. Res.* **1994**, 28, 1187-1199.
21. Labow, R.S.; Duguay, D.G.; Santerre, J.P. *J. Biomat. Sci.* **1994**, 6, 169-179.
22. Duguay, D.G.; Labow, R.S.; Santerre, J.P; McLean, D.D. *Polym. Deg. Stab.* **1995**, in press.
23. Yu, X.J.; Brash, J.L. In *Test Procedures for Blood Compatibility of Biomaterials*; Editor, Davids, S., Ed.; Kuwer Aacademic Publishers: 1993, pp 287-330.
24. Labow, R.S.; Adams, A.H.; Lynn, K.R. *Biochim. Biophys. Acta* **1983**, 749, 32-41.
25. Lelah, M.D.; Cooper, S.L. *Polyurethanes in Medicine*, CRC Press, Boca Raton, USA, 1986.
26. Pitt, G. In *Biodegradable Polymers and Plastic* ; Editors, Vert, M.; Feijen, J.; Albertsson, A.; Scott, G.; Chiellini, E., Eds.; The Royal Soc. of Chem.: Cambridge, 1992; pp 7-19.
27. Andrade, J.D. *Polymer Surface Dynamics* ; Plenum Press: New-York, N.Y., 1988, pp 1-182.
28. Szycher, M.; Siciliano, A.A. *J. Biomat. Appl.* **1991**, 5, 323-336.

RECEIVED April 14, 1995

Chapter 26

Adsorption of Human Low-Density Lipoprotein onto a Silica–Octadecyldimethylsilyl (C$_{18}$) Gradient Surface

Chih-Hu Ho and Vladimir Hlady

Department of Bioengineering, Center for Biopolymers at Interfaces, University of Utah, Salt Lake City, UT 84112

Adsorption kinetics of human low density lipoprotein (LDL) onto a silica-octadecyldimethylsilyl (C18) gradient surface was studied using Total Internal Reflection Fluorescence (TIRF) and autoradiography. The fluorescein-labeled LDL adsorption rate onto the negatively charged silica surface was transport-limited. On the hydrophobic C18 silica end of the gradient surface the adsorption rate was slower than the transport-limited rate. A simple adsorption model was used to determine the adsorption and desorption rate constants. The LDL adsorption was equal to a sum of adsorption processes on available hydrophilic and hydrophobic adsorption sites on either end of the C18 gradient region. The middle part of the C18 gradient displayed a retardation of the adsorption rates. The lipid-labeled LDL adsorption experiments resulted in a fluorescence adsorption pattern that resembled the protein-labeled LDL adsorption, indicating that initially both protein and lipid components of LDL remain adsorbed on the surface.

Human lipoproteins are lipid-protein complexes responsible for the transport of water insoluble lipids in the circulation. The presently accepted structure of lipoproteins depicts them as an apolar core surrounded by polar and amphiphilic components (1,2). Interest in lipoproteins arises primarily from their association with coronary artery disease. The popular interpretation of low density lipoprotein (LDL) as "bad" and high density lipoprotein (HDL) as "good" lipoprotein is based on their use as risk markers for atherogenesis and coronary artery disease. An interrelation between an atherosclerotic plaque formation and arterial thrombosis has been shown to be due to the action of lipoprotein (a), Lp(a), a variant of LDL which in its structure contains an additional protein, apolipoprotein(a), homologous to plasminogen (3,4,5). It has been shown recently that *adsorbed* LDL or Lp(a) may promote a procoagulant state (6).

0097–6156/95/0602–0371$12.00/0
© 1995 American Chemical Society

The aim of this work was to investigate the adsorption behavior of LDL in relation to the hydrophobicity of the adsorbing surface. Rather then using a series of partially hydrophobic surfaces we have utilized hydrophobicity gradient surface, originally described by Elwing (7). This linear gradient surface prepared on flat silica can be used in combination with the Total Internal Fluorescence Reflection (TIRF) technique as a convenient tool for characterizing protein surface adsorption (8). Here, we have used a surface density gradient of octadecyldimethylsilyl groups (silica-C18 gradient surface) which had a several millimeters long gradient region of increasing surface density of octadecyldimethyl-silyl groups between a clean silica end and a self-assembled C18 monolayer end. The advantage of a silica-C18 gradient surface is that one can evaluate protein adsorption and desorption rates from a flowing solution as a function of an average surface density of silica bound-C18 chains under otherwise identical experimental conditions. Our results showed that the apparent affinity of LDL decreases with increasing surface hydrophobicity. By performing two identical TIRF adsorption experiments, one with protein-labeled LDL, the other with lipid-labeled LDL, we confirmed that both protein and lipid components of LDL became adsorbed on the silica-C18 gradient surface.

Experimental

Isolation of Lipoprotein. LDL was isolated by the method of ultra-centrifugation (9). Blood was drawn from ten healthy human donors in 5 ml Vacutainer™ evacuated blood collector tubes which contained 0.05 ml of 15% (75 mg) EDTA solution (Becton Dickinson) and centrifuged at 3,000 rpm and 4°C for 30 minutes. Plasma from ten different donors was pooled. In the first step, VLDL was separated from plasma. The density of the remaining plasma was adjusted to 1.063 g/ml by adding 4.778 M NaBr solution (d = 1.3199 g/ml at 20°C). Three ml of this solution was loaded into a centrifuge tube and 2 ml of 0.844 M NaBr solution (d = 1.063 g/ml at 20°C) was floated on the top. Eighteen tubes were put in an ultracentrifuge rotor (50.3Ti, Beckman) and spun at 40,000 rpm, 20°C and vacuum for 20 hours. After 20 hours, LDL floated to the top of the tube was removed by the tube slice method and pooled. The purity of LDL was checked by one dimensional polyacrylamide gel electrophoresis (Phast, Pharmacia). Half milliliter aliquots of LDL stock solution were placed in 0.65 ml vials and frozen at -20°C. For each adsorption experiment, a desired amount of LDL was thawed. A modified Lowry's method (10,11) was used to determine the concentration of the LDL apoprotein. The LDL concentration was estimated by assuming 25% of protein and 75% of lipid in each LDL particle (12).

Preparation of C18 Gradient Surfaces. The silica plates (2.54 x 7.62 x 0.1 cm, ESCO Products, Inc.) and all glassware were cleaned in a hot chromic acid at 80°C for 30 minutes, thoroughly rinsed in purified deionized water and dried in an oven at 120°C for more than 2 hours. The silica contact angle was measured by the Wilhelmy plate technique to check the cleanliness of the surface (13). The silica-C18 gradient surface was prepared by a two phases diffusion method in which a high density solvent with a silane reagent is layered below a low density solvent in a container with clean silica plates. A solution mixture of 125 ml dichloromethane (DCM, density, d = 1.325 g/cm³, EM Science), 1 ml pyridine (J.T. Baker) and 2 ml octadecyldimethyl-chlorosilane (ODS, Aldrich) was layered below 150 ml p-xylene (d = 0.86 g/cm^3, Fluka). During the surface modification process, the silane diffuses into the low density solvent forming a concentration gradient between the two phases. The silanization reaction proceeded for 4 hours. The silanized silica plates were rinsed by DCM, ethanol and deionized water. The contact angles of silica-C18 gradient surface were measured by the Wilhelmy plate technique.

Labeling of Lipoprotein. Fluorescein isothiocyanate (FITC, Aldrich) was covalently bound to the LDL particle following the method of Coons *et al.* (*14*). Three mg of apolipoprotein (12 mg of LDL) were thawed and diluted in 3 ml of Dulbecco phosphate buffer (DPBS, 0.05 M phosphate, 0.145 M NaCl, 0.90 mM CaCl$_2$, 0.88 mM MgCl$_2$, 2.7 mM KCl, pH 7.4) (*15*). 0.6 ml of 1 mg/ml FITC in carbonate bicarbonate buffer (CBB, pH 9.2) was added to 3 ml of the LDL solution (1 mg apolipoprotein/ml) and incubated at room temperature for 3 hours. Separation of FITC-labeled LDL (FITC-LDL) from free FITC was performed on a PD-10 column (Pharmacia). 3,3'-dioctadecyloxacarbocyanine perchlorate (DiO, Molecular Probes Inc.) was used to label the LDL lipids following the procedure described by Stephan and Yurachek (*16*). 1.5 ml of DiO solution (3 mg/ml DiO in dimethylsulfoxide) was added into 2 ml LDL solution (1 mg apolipoprotein/ml) and incubated at 37°C for 15 hours. The DiO labeled LDL (DiO-LDL) was then separated from the free DiO by filtration (0.8 μm filter, Millipore) (*16*). The technique of ^{125}I-iodination of LDL was a modified ICl iodination protocol (*17*) performed at pH 10 to minimize the iodination of lipids. The free ^{125}I was removed from ^{125}I-labeled LDL (^{125}I-LDL) by ion exchange chromatography on the QAE-Sephadex A-25 column (Pharmacia). The ^{125}I-LDL was stored in the refrigerator at 4°C and used within a week.

Protein TIRF Adsorption Experiments. All details of the custom-built TIRF apparatus and the flow cell are described elsewhere (*8*). TIRF protein adsorption experiments were performed in a flow cell containing two identical rectangular flow channels so that two experiments could be performed on a same silica-C18 gradient surface. The concentration of LDL were equal to 1/100 of its respective concentrations in normal plasma: apolipoprotein concentration was 0.01 mg/ml, the total concentration of LDL was 0.04 mg/ml. In the adsorption segment of the TIRF experiment, LDL solution flowed for 11 minutes through the channel with the rate of 0.84 ml/min initially displacing the DPBS buffer. In the 11 minutes desorption segment the flow was by switched back to the buffer solution. The fluorescence from the adsorbed LDL (FITC-LDL or DiO-LDL) along the silica-C18 gradient surface was excited by a spatially-filtered and expanded Ar$^+$-ion laser beam (10 mW, @ 488 nm). The emitted fluorescence was passed through a monochromator (1681C, Spex Inc.) and recorded along the silica-C18 gradient surface every second by a charged couple device camera (Photometrics Inc.). All TIRF experiments were performed at room temperature.

Protein Autoradiography Adsorption Experiments. In this experiment, ^{125}I-LDL was measured instead of FITC-LDL (or DiO-LDL). ^{125}I-LDL was adsorbed onto the silica-C18 gradient surface from DPBS buffer. The concentration of protein solution, the flow rate and the adsorption-desorption cycle were the same as in the

TIRF experiments. The LDL solution contained a 1 : 4 ratio of ^{125}I-LDL and unlabeled LDL. After the adsorption-desorption cycle, the adsorbed LDL on the silica-C18 gradient surface was fixed with 3 ml of 0.6% glutaraldehyde solution in DPBS buffer. A calibration plate with a set of known amounts of ^{125}I-LDL was prepared separately as an autoradiography standard for the quantification of adsorbed protein. The silica-C18 gradient silica plate with the adsorbed and fixed ^{125}I-LDL and the calibration plate were placed in a polyethylene bag and brought in contact with an autoradiography film (X-OMAT AR, Kodak) in a light tight cassette. The film was exposed at low temperature (-70°C) for 21 days. The exposed film was processed in an automated developer system and its optical density was recorded by a custom-built densitometer (18). The adsorbed amount of LDL was calculated as a function of C-18 gradient position (19).

Results

Characterization of Silica-C18 Gradient Surface. The advancing, θ_{adv}, and receding, θ_{rec}, water contact angles of the silica-C18 gradient surface are shown in Figure 1a as a function of the gradient surface position. The maximum angles, $\theta_{adv} = 104°$, $\theta_{rec} = 85°$ were found at positions greater than 4.6 cm indicating the hydrophobic C18-silica surface is at the end of the gradient region. The contact angle hysteresis, i.e. the difference between θ_{adv} and θ_{rec}, $\Delta\theta = 19°$, indicated that the hydrophobic C18-silica was not a defect free C18 monolayer. The contact angles decreased smoothly towards the silica end of the gradient. At the positions smaller that 3.45 cm from the silica end the angles remained unchanged at $\theta_{adv} = 10°$, $\theta_{rec} = 0°$, indicating that the surface is a hydrophilic silica. The fractional surface coverage of the C18 chains along the gradient, $\Theta/\Theta_{max(C18)}$, was calculated from the advancing contact angles using the Cassie equation (20).

$$\cos\theta_{adv} = (\Theta/\Theta_{max(C18)})\cdot\cos\theta_{C18} + (1 - \Theta/\Theta_{max(C18)})\cdot\cos\theta_{silica} \qquad (1)$$

where the contact angle of a fully packed monolayer of C18 chains, $\theta_{C18} = 112°$ (21) and clean silica $\theta_{silica} = 0°$. Figure 1b shows the C18 fractional surface coverage as a function of the gradient surface position. The arrows indicate the four positions along the gradient region where the fractional surface coverage was 0.12, 0.22, 0.33 and 0.72 respectively.

Adsorption of LDL onto the Silica-C18 Gradient Surface. The LDL TIRF adsorption-desorption patterns are shown in Figure 2. The two markers indicate the C18 gradient region. The flow of the protein solution was from the hydrophilic end towards the hydrophobic end of the gradient. The FITC-LDL adsorption results are shown in Fig 2a. The initial fluorescence increase was linear in the region of the hydrophilic silica surface. After this initial increase the fluorescence intensity reached a steady-state level. A very small decrease of fluorescence intensity in the desorption segment of the experiment indicated a slow desorption. The initial fluorescence in the C18-gradient region and at the C18 end of the gradient increased slowly with time. The DiO-LDL adsorption results are shown in Fig 2b. Although the fluorescence intensity of the adsorbed DiO-LDL and the signal-to-noise ratio was lower than in the

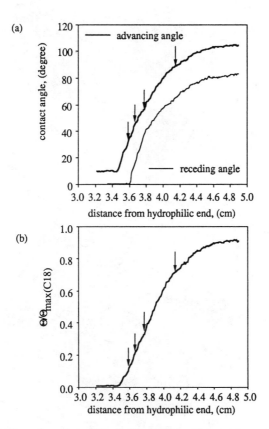

Figure 1. (a) Water contact angles of silica-C18 gradient surface and (b) fractional C18 surface coverage, $\Theta/\Theta_{max(C18)}$, shown as a function of gradient position.

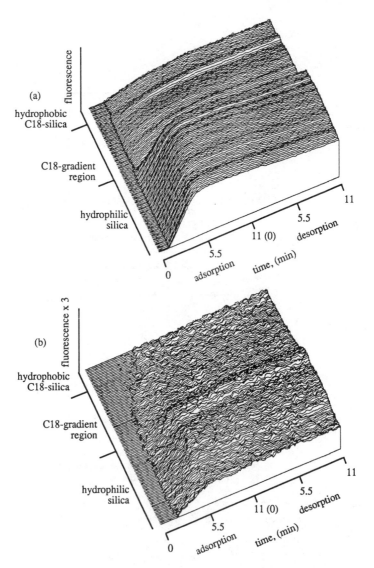

Figure 2. Spatially-resolved TIRF adsorption fluorescence pattern showing the fluorescence intensity vs. gradient position vs. time. (a) FITC-LDL, (b) DiO-LDL.

case of the FITC-LDL, the overall fluorescence pattern resembled the results of the FITC-LDL adsorption experiment.

Quantification of Lipoprotein Adsorption. A quantitative analysis of the LDL adsorption kinetics required that the amount of adsorbed LDL is known as a function of time. Two independent quantification schemes were performed and the results compared. In first, the initial FITC-LDL adsorption (Fig 2a) on the hydrophilic silica was assumed to be limited by transport, i.e. by the availability of protein molecules at the adsorbing surface and the flux of protein molecules to the surface, dA/dt, was compared with the initial increase of adsorbed protein fluorescence, dF/dt (22,23). As described previously (8,20), when the desorption rate is very small or zero, this comparison can be used to calculate the adsorbed amount from the initial linear fluorescence increase since the flux of protein molecules to the surface, dA/dt, in rectangular flow channel can be computed from the Leveque equation:

$$dA/dt = (\Gamma(4/3))^{-1} \cdot 9^{-1/3} \cdot (6q/b^2 \, w \, l \, D_{LDL})^{1/3} \cdot D_{LDL} \cdot c_{LDL} \qquad (2)$$

where Γ is the gamma function, q is the experimental volumetric flow rate (0.84 ml/min), b is the thickness of the TIRF flow cell (0.05 cm), w is the width of the TIRF flow cell (0.5 cm), l is the distance from the entrance of the flow chamber (2.8 cm), D_{LDL} is the diffusion coefficient of LDL (1.8×10^{-7} cm^2 s^{-1}) (15), and c_{LDL} is the bulk concentration of LDL. In the LDL adsorption experiment (Fig 2a) dA/dt amounted 1.84×10^{-2} µg cm^{-2} s^{-1} and the factor which relates the protein flux and the fluorescence increase, $Z = (dA/dt)/(dF/dt)_{silica}$ (µg cm^{-2} count^{-1}) was 1.01×10^{-4} µg cm^{-2} count^{-1}. All experimental FITC-LDL fluorescence results were converted to the surface density of LDL, Γ_{LDL}, using the same conversion factor Z.

In the second quantification scheme, autoradiography was used to measure the adsorbed amount of ^{125}I-LDL along the silica-C18 gradient surface. Since in the autoradiography experiment the signal from the surface adsorbed ^{125}I-LDL can not be recorded independently from the solution ^{125}I-LDL, only the final adsorbed amount of ^{125}I-LDL was measured after the unbound protein was washed out of the flow cell. The comparison between the two LDL adsorption quantification schemes is shown in Figure 3. Given the fact that the protein label was different, the agreement between the two quantification schemes was remarkable, especially at the both ends of the C18 gradient, but worsened in the middle of the gradient. Accordingly, the analysis of LDL adsorption kinetics was limited to several positions close to the ends of the silica-C18 gradient surface (indicated with arrows in Fig 1b). Figure 4a shows the respective experimental adsorption-desorption kinetics. On the hydrophilic silica surface the LDL adsorption reached a maximum of $\Gamma_{LDL} = 0.40$ µg cm^{-2} after four minutes. The initial rate of LDL adsorption decreased and the linearity of the initial slope disappeared as the hydrophobicity of the surface increased.

Modeling kinetics of the lipoprotein adsorption. A simple protein adsorption model, to which the experimental data were fitted, comprised of two opposing processes: adsorption and desorption (24):

$$d\Gamma(t)/dt = k_{on} \cdot (1 - \Gamma(t)/\Gamma_{max}) \cdot c(0,t) - k_{off} \cdot (\Gamma(t)/\Gamma_{max}) \qquad (3a)$$

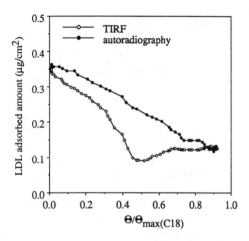

Figure 3. Comparison between the FITC-LDL adsorption quantified by using Eq 2 and the ^{125}I-LDL adsorption measured by autoradiography, both shown as a function of fractional C18 surface coverage.

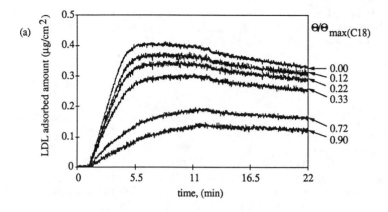

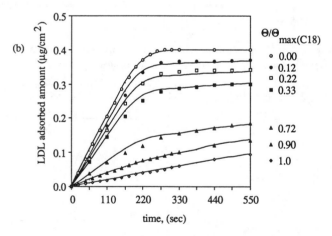

Figure 4. (a) Experimental FITC-LDL adsorption-desorption kinetics. (b) Comparison between the experimental FITC-LDL adsorption (symbols) and the "adsorption sum" model (Eq 5, solid lines) for several positions along the silica-C18 gradient surface. Fractional C18 surface coverage, $\Theta/\Theta_{max(C18)}$, is indicated.

where k_{on} and k_{off} are the intrinsic adsorption and desorption rate constants, Γ_{max} is the maximum adsorption, $\Gamma(t)$ is the adsorbed amount per unit surface at time t, $(1 - \Gamma(t)/\Gamma_{max})$ is the fraction of unoccupied adsorption sites and $c(0,t)$ is the protein concentration right next to the adsorbing surface. A numerical computer routine was used to calculate $c(0,t)$ from the Fick's diffusion law across the unstirred layer close to the surface and to model the adsorption using Eq 3a. The model also allowed that the intrinsic adsorption and desorption rate constants are defined as exponential functions of protein surface concentration, Γ:

$$k_{on} = k_1 \exp(-\alpha\Gamma) \qquad\qquad (3b)$$
$$k_{off} = k_{-1} \exp(\beta\Gamma) \qquad\qquad (3c)$$

where k_1 is the initial intrinsic adsorption rate constant, k_{-1} is the initial intrinsic desorption rate constant and α and β are the "cooperativity" adsorption and desorption constants, respectively. Fitting of the experimental results of LDL adsorption was carried out for the hydrophilic ($\theta_{adv} = 0^o$, $\Theta/\Theta_{max(C18)} = 0$) and hydrophobic C18-silica surface ($\theta_{adv} = 104^o$, $\Theta/\Theta_{max(C18)} = 0.90$). k_{-1} was found first by fitting the desorption segments (t > 11 minutes, Fig 4) to the Eq 3 using the condition: $c(0,t) = 0$. Γ_{max} was assumed to be 0.4 µg cm^{-2} at the hydrophilic surface and 0.3 µg cm^{-2} at the hydrophobic C18-silica surface. These two Γ_{max} values agreed very well with a previous study (15). The comparison between the model (solid line) and the experimental results (• symbols) is shown in Figure 5. The parameters used to achieve the fits shown in Fig 5 are listed in Table I.

Table I. The LDL adsorption parameters obtained by fitting the experimental results to the model given by Eq 3

	hydrophilic silica	hydrophobic C18-silica	hypothetical C18 monolayer
θ_{adv}	0°	104°	121°
$\Theta/\Theta_{max(C18)}$	0	0.9	1.0
α	-2	2.5	2
β	0	0	0
Γ_{max} (µg cm^{-2})	0.4	0.3	0.26
k_1 (cm^{3}µg^{-1} s^{-1})	$7.5 \cdot 10^{-4}$	$4.5 \cdot 10^{-5}$	$2.2 \cdot 10^{-5}$
k_{-1} (s^{-1})	$2.9 \cdot 10^{-4}$	$2.3 \cdot 10^{-4}$	$2.3 \cdot 10^{-4}$
K (M^{-1})	$6.5 \cdot 10^9$	$4.9 \cdot 10^8$	$2.4 \cdot 10^8$

The apparent affinity constant, K, (in M^{-1}) for each surface was calculated from the ratio k_1/k_{-1}, assuming that the $M_{wLDL} = 2.6 \cdot 10^6$ Da. The apparent affinity of LDL for the negatively charged silica surface (K = $6.5 \cdot 10^9$ M^{-1}) was larger by an order of magnitude compared to the hydrophobic C18-silica surface (K = $4.9 \cdot 10^8$ M^{-1}). This difference was due to the intrinsic adsorption rates differences since the respective desorption rates were quite similar (Table I).

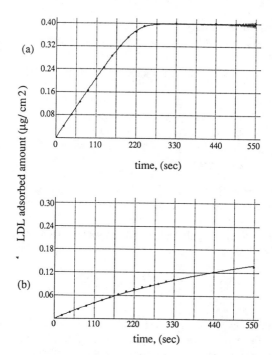

Figure 5. Comparison between the experimental FITC-LDL adsorption (•) and the model (Eq 3, solid line) for the hydrophilic silica (a) and hydrophobic C18-silica surface (b), respectively.

The LDL adsorption kinetics indicated that the adsorption onto the hydrophilic silica surface takes place via a mechanism that is different from the mechanism of adsorption onto the hydrophobic C18 silica. To answer the question whether the LDL adsorption on a mixed hydrophobic-hydrophilic surface behaves as a simple addition of two independent adsorption processes or the hydrophobic-hydrophilic neighboring sites affect each other, the experimental LDL adsorption to the silica-C18 gradient region was fitted to an "adsorption sum" model. The model adds the adsorption on the hydrophobic (subscript $(C18)$) and hydrophilic (subscript (sil)) surface sites, respectively, and weights each adsorption process according to the probability of finding a given adsorption site:

$$d\Gamma(t)/dt =$$
$$(1 - \Theta/\Theta_{max(C18)})\cdot\{k_{on(sil)}\cdot(1 - \Gamma(t)/\Gamma_{max(sil)})\cdot c(0,t) - k_{off(sil)}\cdot(\Gamma(t)/\Gamma_{max(sil)})\} +$$
$$(\Theta/\Theta_{max(C18)})\cdot\{k_{on(C18)}\cdot(1 - \Gamma(t)/\Gamma_{max(C18)})\cdot c(0,t) - k_{off(C18)}\cdot(\Gamma(t)/\Gamma_{max(C18)})\}$$
$$(5)$$

In order to be able to use the "adsorption sum" model (Eq 5), it was necessary to know the adsorption parameters for LDL adsorption onto a fully covered C18 surface. Since the LDL adsorption kinetics at $\Theta/\Theta_{max(C18)} = 0$ and $\Theta/\Theta_{max(C18)} = 0.90$ were experimentally determined, the hypothetical adsorption of LDL to a fully covered C18 surface, where $\Theta/\Theta_{max(C18)} = 1.0$, was estimated from the two experimental kinetics:

$$\Gamma(t)_{(\Theta/\Theta max(C18)=1.0)} = \Gamma(t)_{(\Theta/\Theta max(C18)=0.90)} - 0.1 \cdot \Gamma(t)_{(\Theta/\Theta max(C18)=0)} \quad (6)$$

and the resulting adsorption vs. time data set was subsequently fitted to the Eq 3 to obtain the hypothetical k_1, k_{-1} and Γ_{max} (data shown in Table I). Eq 5 was used next to fit the adsorption kinetics measured in the C18 gradient region. The comparison between the experimental adsorption kinetics and the fits to the "adsorption sum" model is shown in Figure 4b. Included in the same figure is the hypothetical LDL adsorption fit for a fully covered C18 surface. The inspection of Fig 4b shows that the "adsorption sum" model (Eq 5) fits very well the experimental LDL adsorption for the C18 surface coverage close to 1 or to 0. However, the fit worsens as one moves towards the middle of the gradient where $\Theta/\Theta_{max(C18)} ---> 0.5$, especially in the initial adsorption stage. Notice that the adsorbed amount of LDL calculated from the fluorescence (Fig 3) showed an unexplained minimum at approximately half C18 surface coverage.

Discussion

We have shown previously that silica-C18 gradient surfaces can be used to study how protein adsorption depends on surface hydrophobicity (20). The objective of this study was to determine how the surface density of C18 chains affects the LDL adsorption. The adsorption of LDL was monitored using three different LDL labels: FITC-LDL, DiO-LDL and ^{125}I-LDL. The TIRF measurements (Fig 2) suggested that FITC-LDL and DiO-LDL followed similar adsorption kinetics laws. Since these two labels tag different parts of LDL particle, one can tentatively conclude that both protein and lipid components of LDL particle adsorb together to the surface. It remains to be determined whether the structural integrity of adsorbed LDL particle is affected by a longer residence time at a given surface or not.

The comparison between the two quantification schemes indicated that the LDL adsorption on the hydrophilic silica is indeed transport-limited. The adsorption rates, k_1, (Table 1) were calculated using a simple adsorption model (Eq 3). In order to obtain the best fit between the model and the experiment, a small positive cooperativity ($\alpha = -2$) had to be assumed for hydrophilic silica and a negative cooperativity ($\alpha = 2.5$) for C18 silica, respectively. The physical meaning of the adsorption cooperativity also remains to be determined. The adsorbed amount of LDL decreased with increasing coverage of C18 chains (Figs 2 and 3). It is known that LDL binds strongly to negatively charged adsorbents; some of the commercially available LDL-apheresis devices use negatively charged dextran sulfate adsorbents (25, 26). The electrostatic interactions can explain the order-of-magnitude larger affinity of LDL for a negatively charged silica surface ($K = 6.5 \cdot 10^9$ M^{-1}) than for a hydrophobic C18-silica surface ($K = 4.9 \cdot 10^8$ M^{-1}). The LDL adsorption took place from a buffer containing Ca^{++} ions which might act as bridges between local negative charges of LDL and negatively charged silanol groups on the silica surface.

The mechanism of LDL adsorption onto the hydrophobic C18-surface is less obvious. The adsorption kinetics suggests an energy barrier to adsorption, probably due to the orientation of LDL particle during the collision with the surface. The energy barrier could also have an electrostatic origin. When a charged particle arrives into the close proximity of a low dielectric surface, electrostatic image forces are created opposing further approach and contact of the particle with the surface (27).

The "adsorption sum" model fitted the experimental results very well at either end of the C18-gradient region. One can expect this since a smaller fraction of given surface sites can not dramatically influence the adsorption process occurring at other surface sites present in excess. In the middle part of the C18-gradient region the "adsorption sum" model failed to fit the experimental fluorescence results. Local retardation of the fluorescence-derived adsorption rates at approximately half C18 surface coverage has been found in a number of repeated TIRF FITC-LDL adsorption experiments. The origin of this local fluorescence decrease is not known and has not been further investigated.

Summary

Adsorption kinetics of human low density lipoprotein (LDL) onto a silica-octadecyldimethylsilyl (C18) gradient surface was studied using the Total Internal Reflection Fluorescence (TIRF) and autoradiography techniques. The silica-C18 gradient surface was prepared by a two phases silanization reaction. The advancing water contact angles were used to calculate the fractional surface coverage of C18 chains, which increased from zero (clean silica) to 0.9 on the hydrophobic end of the gradient surface. The FITC-LDL adsorption rate was transport-limited on the negatively charged silica but significantly slower on the hydrophobic C18 silica. The lipid labeled-LDL adsorption experiments resulted in a fluorescence adsorption pattern that resembled the protein-labeled LDL adsorption. A simple adsorption model was used to calculate the adsorption and desorption rate constants. The LDL adsorption was equal to a sum of adsorption processes on available hydrophilic and hydrophobic adsorption sites at either end of the C18 gradient region. The middle part of the C18 gradient displayed an unexpected retardation of the LDL adsorption rates.

Acknowledgments

The authors are indebted to Dr. Lilly Wu (University of Utah) for teaching them how to isolate human lipoproteins. We thank all blood donors for providing us with their low density lipoproteins and J. D. Andrade and H. P. Jennissen for helpful discussions. This work was financially supported by the Whitaker Foundation, NIH grant (R01 NIH-44538) and the Center for Biopolymers at Interfaces, University of Utah.

References

1. Smelie, R. M. S., Ed.; *Plasma Lipoproteins*; Biochemical Society Symposium No 33; Academic Press: New York, NY, 1971.
2. Gotto, A. M. Jr., Ed.; *Plasma Lipoproteins*; Elsevier: Amsterdam, 1987.
3. Utermann, G. *Science* **1989**, *246,* 904-910.

4. Miles, L. A.; Fless, G. M.; Levin, E. G.; Scanu, A. M.; Plow, E. F. Nature **1989,** *39,* 301-303.
5. Hajjar, K. A.; Gavish, D.; Breslow, J. L.; Nachman, R. L. *Nature* **1989,** *339,* 303-305.
6. Simon, D. I.; Fless, G. M.; Scanu, A. M.; Loscalzo J. *Biochemistry* **1991,** *30,* 6671-6677.
7. Elwing, H.; Nilsson, B.; Svensson, K.-E.; Askendahl, A.; Nilsson, U. R.; Lundström, I. *J. Colloid Interface Sci.* **1988,** *125,* 139-145.
8. Hlady, V. *Appl. Spectroscopy* **1991,** *45,* 246-252.
9. Mills, G. L.; Cane, P. A.; Weech, P. K. in *A guide book to lipoprotein technique, Laboratory techniques in biochemistry and molecular biology, Vol. 14.;* Burdon, R.H.; van Knippenberg, P. H., Eds.; Elsevier: Amsterdam, 1976; pp 18-78.
10. Lowry, O. H.; Rosebrough, N. J.; Farr A. L.; Randall, R. J. *J. Biol. Chem.* **1951,** *193,* 265-275.
11. Peterson, G. L. *Anal. Biochem.* **1977,** *83,* 346-356.
12. Soutar, A. K.; Myant,N. B. In *International Review of Biochemistry, Chemistry of Macromolecules IIB, Vol. 25;* Offed, R. E., Ed.; University Park Press: Baltimore, MD, 1987, pp 55-119.
13. Andrade, J. D.; Smith, L. M.; Gregonis, D. E. In *Surface and Interfacial Aspects of Biomedical Polymers, Vol. 1;* Andrade, J. D., Ed.; Plenum: New York, NY, 1982, pp 262-290.
14. Coons, A. M.; Crech, H. J.; Jones R. N., Berliner, E. J. *J. Immunology* **1942,** *45,* 159-170.
15. Hlady, V.; Rickel J.; Andrade, J. D. *Colloids and Surfaces,* **1988,** *34,* 171-183.
16. Stephan, Z. F.; Yurachek, E. C. *J. Lipid Res.* **1993,** *34,* 325-330.
17. Shepherd, J.; Bedford, D. K.; Morgan, H. G. *Clin. Chem. Acta,* **1976,** *66,* 97-109.
18. Ho, C.-H.; Hlady, V.; Nyquist, G.; Andrade, J. D.; Caldwell, K. D. *J. Biomed. Mater. Res.* **1991,** *25,* 423-441.
19. Lin, Y. S.; Hlady V.; Janatova, J. *Biomaterials* **1992,** *13,* 497-504.
20. Lin, Y. S.; Hlady V.; Gölander, C.-G. *Colloids and Surfaces,* **1994,** in press.
21. Ulman, A. *J. Mat. Ed.* **1989,** *11,* 205-280.
22. Lok, B. K.; Cheng, Y.-L.; Robertson, C. R. *J. Colloid Interface Sci.* **1983,** *9,* 87-103.
23. Lok, B. K.; Cheng, Y.-L.; Robertson, C. R. *J. Colloid Interface Sci.* **1983,** *9,* 104-116.
24. Corsel, J. W.; Willems, G. M.; Kop, J. M. M.; Cuypers, P. A.; Hermens, W. T. *J. Colloid Interface Sci.* **1986,** *111,* 544-554.
25. Yokoyama, S.; Hayashi, R.; Kikkawa, T.; Tani, N.; Takada, S., Hatanaka, K.; Yamamoto, A. *Arteriosclerosis* **1984,** *4,* 276-282.
26. Ikonomov, V.; Samtleben, W.; Schmidt, B.; Blumenstein, M.; Gurland, H. J. *Int. J. Artif. Organs* **1992,** *15,* 312-319.
27. Israelachvili, J. *Intermolecular and Surface Forces, 2nd ed.,* Academic Press, London, 1922.

RECEIVED December 22, 1994

Chapter 27

Reduced Protein Adsorption on Polymer Surface Covered with a Self-Assembled Biomimetic Membrane

Kazuhiko Ishihara and Nobuo Nakabayashi

Institute for Medical and Dental Engineering, Tokyo Medical and Dental University, 2–3–10 Kanda-surugadai, Chiyoda-ku, Tokyo 101, Japan

We have synthesized phospholipid polymers containing 2-methacryloyloxyethyl phosphorylcholine(MPC) moieties as new blood compatible polymers and have evaluated their interactions with blood components. It was found that in the absence of anticoagulants blood clotting was delayed, and blood cell adhesion and activation were effectively prevented on the MPC copolymer surface. Protein adsorption on the MPC copolymer from human plasma was also reduced with increasing MPC fraction. The MPC copolymers are able to collect phospholipid molecules from plasma due to MPC moiety-phospholipid interactions and the accumulated phospholipids appear to create a self-assembled biomimetic membrane structure on the MPC copolymers which minimizes protein adsorption. When the MPC copolymer was treated with dipalmitoylphosphatidylcholine(DPPC) liposomal solution, DPPC molecules having the organized liposomal structure were adsorbed on the MPC copolymer surface. On the other hand, DPPC molecules were randomly adsorbed on conventional polymers. The amounts of plasma proteins adsorbed on the polymer surfaces decreased on pretreatment with DPPC liposomal solution for all polymers case, but the smallest amounts of adsorbed proteins were found on the MPC copolymer. The differences in protein adsorption on the different polymers probably reflect the differences in the state of organization of adsorbed phospholipid molecules on the polymer surfaces.

In recent years, many polymer materials have been used for the fabrication of biomedical devices that contact blood, body fluids and tissues(1). Upon blood contact, biocomponents in blood such as lipids and proteins interact with these surfaces and are adsorbed on them. Well designed polymers are required to regulate their interactions. The adsorption state of these biocomponents strongly affects blood cell adhesion and activation which induce thrombus formation.

The effect of proteins adsorbed onto polymer surfaces on nonthrombogenicity has been investigated by many researchers (2). Lyman *et al.* and Kim *et al.* claimed the importance of albumin adsorption for the nonthrombogenicity of a polymer surface (3,4). They proposed that predominant adsorption of albumin

0097–6156/95/0602–0385$12.00/0

with a high adsorption rate   and   large   equilibrium   amount   will   make a polymer surface nonthrombogenic. Kim *et al.* also proposed that a polymer surface which can adsorb protein reversibly and/or minimally (or possibly no protein at all), will show minimal thrombogenicity (*4*). To minimize the amount of adsorbed protein, hydrophilic polymers, especially hydrogels have been prepared (*5*). The general results concerning protein adsorption on hydrogels clearly indicate that the amount and degree of denaturation of adsorbed protein decreases with increase in the water content of the hydrogel. Anderson *et al.* reported the adsorption and conformational change of proteins on hydrogels (*6*). We have also found that proteins are adsorbed from plasma to hydrogels such as acrylamide or vinylpyrrolidone copolymers even when their fractional water contents are higher than 0.4 (*7*). Moreover, some experimental results strongly suggested serious problems such as calcification, platelet deposition and complement activation when blood comes in contact with hydrogel surfaces (*8*).

We have proposed the preparation of new nonthrombogenic polymers based on a different approach (*9-18*). We have noted that biomembrane structures which consist mainly of phospholipids and proteins show good biocompatibility since such surfaces interact only minimally with biocomponents. A similar approach was reported by Chapman *et al.*(*19*). They found that polymeric lipids and polyesters having a phosphorylcholine group are relatively nonthrombogenic based on thrombelastographic studies. It has also been reported that polyamide microcapsules coated with a lipid membrane reduced the adhesion of platelets(*20*). These studies indicated that the surface adsorbed with phospholipids interacted only minimally with blood cells. However, the effects of the lipid adsorption state on biocompatibility were not investigated. We synthesized copolymers having a phospholipid polar group, the 2-methacryloyloxyethyl phosphorylcholine(MPC) moiety, as a new type of biomaterial (*11,15*). We hypothesized that if a biomembrane-like structure can be constructed on a polymer surface by spontaneously adsorbing natural lipids from a living organism, it will prevent the adsorption and activation of proteins and cells (*14,16*). It has already been reported that the MPC copolymers have an affinity for phospholipids (*21,22*) which were adsorbed on the surface in significant amounts (*16,17*). Moreover, the copolymers effectively suppressed thrombus formation (12,16,18). It was also found that the amount of adsorbed protein decreased with an increase in MPC content of the copolymers (*13,16,18*).

In this review, the physical and protein adsorption properties of the MPC copolymers will be discussed with attention given to the probable creation of a biomembrane-like structure by organization of phospholipid molecules accumulated on the MPC copolymer surface.

Materials and Methods

Materials.   MPC was synthesized by a previously reported procedure(*11*). The structure of MPC is shown in Figure 1. Poly[MPC-*co-n*-butyl methacrylate(BMA)], poly[2-hydroxyethyl methacrylate(HEMA)], and poly(BMA) were prepared by a conventional radical polymerization technique using 2,2'-azobisisobutyronitrile as initiator. The MPC mole fraction in the copolymer was determined by phosphorus analysis. In this study, we used two poly(MPC-*co*-BMA)s with 0.25 and 0.16 MPC mole fractions. Cast films of each polymer were prepared by a solvent evaporation method. Dipalmitoylphosphatidylcholine(DPPC), bovine serum albumin(BSA), and bovine serum γ-globulin(BSG) were obtained from Sigma Chemical Co. St Louis, MO, USA and used without further purification. Acrylic beads (250 - 600 μm diameter) or slabs (9 x 50 x 1 mm) were used as substrates onto which the polymers were coated. Polymer coating on these substrates was carried out by a solvent evaporation technique using a 0.5 wt% polymer solution (*12, 14*). The MPC mole fraction at the surface on the poly(MPC-*co*-BMA) membrane was determined using X-ray photoelectron spectrometry (XPS, Shimadzu ESCA-750).

Evaluation of Organization of Phospholipid Adsorbed on Polymers Treated with Phospholipid Liposome. DPPC liposomes were prepared by the sonication method (*14,21,22*). Polymer membranes were incubated in the DPPC liposomal solution at 45 °C for 10 min. After washing with distilled water, thermal analysis was carried out by differential scanning calorimetry(DSC) to determine the gel-liquid crystalline phase transition temperature(Tc) of the DPPC liposome. Furthermore, by using a quartz resonator with gold electrodes coated with the poly(MPC-*co*-BMA), the phase transition phenomena of the DPPC liposome adsorbed on the surface were observed (*23*).

Determination of Amount and Conformational Change in Proteins Adsorbed on Polymers. The BSA and BSG were dissolved in PBS at a concentration of 4.5 g/dL and 1.6 g/dL, respectively, that is, the same concentration as in plasma. Acrylic slabs coated with polymer were immersed in 100 mL of protein solution at 30 °C. To avoid denaturation of the proteins in PBS, the maximum adsorption time was 30 min. The slabs were rinsed with PBS for 1 min, and then immersed in PBS in a quartz cell for spectroscopy. The amount of protein adsorbed on the surface was determined by ultraviolet(UV) spectrometry as described previously (*14*). For pre-adsorption of phospholipids on the polymer surface, the polymer-coated acrylic slabs were immersed in the DPPC liposomal solution(0.5wt%) and incubated at 45 °C for 10 min. The amount of protein adsorbed on the polymer surface pretreated with DPPC was also measured by UV spectroscopy (*14*).

Circular dichroism(CD) spectroscopic measurements on proteins adsorbed on the polymer surface was carried out to evaluate conformational change. A quartz slab was used as a substrate for the polymer coating and the polymer-coated quartz slab was immersed in the BSA solution for 60 min at 30 °C. The mean residual ellipticity of adsorbed BSA was calculated based on the method proposed by Akaike *et al.* (*24*).

Protein Adsorption on MPC Copolymer Surface

When polymer materials contact blood, adsorption of lipids and proteins precedes cell adhesion. Thus protein adsorption not only on the polymer surface but also on the lipid adsorbed surface need to be clarified.

Figure 2 shows the adsorption of BSA and BSG on three polymer surfaces. It is seen that adsorption increases with time in all cases. Baszkin and Lyman reported the maximum amount of adsorbed proteins calculated for a monolayer in an end-on type orientation, that is 0.90 μg/cm^2 for BSA and 1.85 μg/cm^2 for BSG (*25*). On hydrophobic poly(BMA), the adsorption rate of BSA was high and the adsorbed amount at 30 min was about 16 times higher than that for an end-on monolayer. Suggesting that multi-layer adsorption occurred. Similar adsorption behavior was observed for poly(HEMA). Compared with these polymers, the amount of BSA adsorbed on poly(MPC-*co*-BMA) was quite low. Adsorption of BSG was also suppressed on poly(MPC-*co*-BMA) compared with poly(BMA) and poly(HEMA), and was lower than the calculated value for a monolayer.

Concerned with protein adsorption on the MPC copolymers, Sugiyama *et al.* also reported that the amount of BSA adsorbed on poly(MPC-*co*-alkyl methacrylate) microspheres was lower than that on polystyrene microspheres and decreased with an increase in the hydrophilicity of the comonomer used for the preparation of the microspheres(*26*).

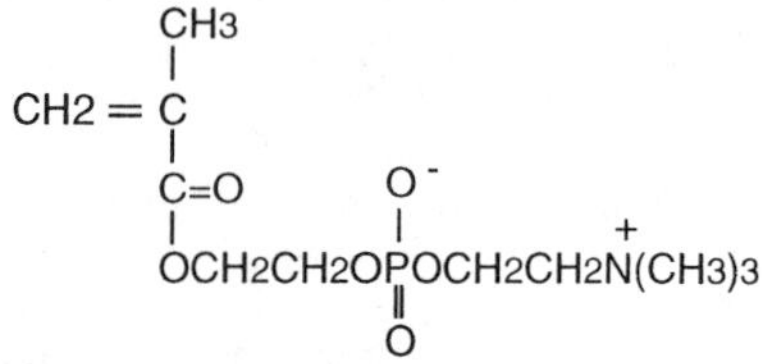

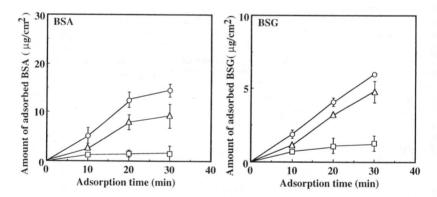

Figure 1. Structure of 2-methacryloyloxyethyl phosphorylcholine (MPC).

Figure 2. Adsorption of BSA and BSG onto polymer surfaces at 30 °C. (O)
Poly(BMA), (△) Poly(HEMA), (□) Poly(MPC-*co*-BMA), MPC mole fraction;
0.25. Initial concentrations of BSA and BSG were 4.5 g/dL and 1.6 g/dL,
respectively (Reproduced with permission ref. 14).

Figure 3 demonstrates the CD spectra of BSA in PBS and adsorbed on the polymer surfaces (*27*). The molar ellipticity of native BSA dissolved in PBS was negative between 210 nm and 230 nm. In the case of BSA adsorbed on poly(HEMA), this value was slightly higher. A more significant change in the spectrum was found in the case of BSA adsorbed on poly(BMA), that is, the molar ellipticity was close to zero in the observed range. On the other hand, BSA adsorbed on poly(MPC-*co*-BMA) had the same molar ellipticity as that of BSA in PBS. The ellipticity reflects the secondary structure of the protein, so it appears that no significant conformational change of BSA occurred during adsorption on poly(MPC-*co*-BMA).

It is considered that ζ-potential is an important variable in relation to protein adsorption. We found that the ζ-potential of a poly(ethylene terephthalate)(PET) membrane was -40.5 mV (Ueda, T.; Ishihara, K.; Nakabayashi, N. *J. Biomed. Mater. Res.*, in press.), and coating of poly(HEMA) on the PET membrane reduced this value to -15.7 mV. Moreover, a poly(MPC-*co*-BMA)-coated PET membrane showed a potential of -0.4 mV. In general, hydrophobic polymer surfaces have a negative ζ-potential, whereas the potential approaches zero with increasing hydrophilicity. However, the fractional equilibrium water contents of membranes of poly(HEMA) and the poly(MPC- *co*-BMA) were very close to 0.38 and 0.39 respectively. Park and coworkers suggested that when proteins are adsorbed on a polymer surface via hydrophobic interactions, an exchange of bound water between the protein and the surface must take place (*28*). Therefore, we considered that the amount of bound water may be the key parameter for understanding the exceptionally mild interaction between proteins and the MPC copolymer surface. When DSC measurements of the hydrated polymer membranes were carried out to evaluate water structure, the fraction of free water(not bound water) in the poly(MPC-*co*-BMA) membrane was 0.74 and it found to be significantly higher than that in conventional hydrogels such as poly(HEMA)(0.41) (K.Ishihara, Tokyo Medical and Dental University, unpublished data.).

Organized Adsorption of DPPC on MPC Copolymer

We have previously reported protein and phospholipid adsorption on a poly(MPC-*co*-BMA) surface from human plasma (*16*). The amount of proteins adsorbed on the poly(MPC-*co*-BMA) membrane decreased, while adsorption of phospholipid increased, with increasing MPC mole fraction of the copolymer. This result strongly suggested that preferential adsorption of phospholipid relative to protein occurred when the poly(MPC-*co*-BMA) came in contact with plasma. Moreover, it was also considered that the poly(MPC-*co*-BMA) might have specific affinity for the phospholipid in plasma. Therefore, we tried to clarify the interactions between phospholipids and poly(MPC-*co*-BMA) using DPPC as a model phospholipid (*14, 17, 21, 22*).

Figure 4 indicates the relation between the amount of DPPC adsorbed on the poly(MPC-*co*-BMA) and MPC mole fraction in the copolymer(*17*). The amount of adsorbed DPPC on poly(BMA) was 2.14 μg/cm^2, and was the same as on poly(HEMA) even though the hydrophilicity of these materials are quite different. However, the adsorbed amount of DPPC on poly(MPC-*co*-BMA) is seen to depend on the MPC mole fraction, that is, the adsorbed amount of DPPC increases with increasing MPC mole fraction. This same tendency was observed on poly[MPC-*co*-styrene(St)] (*21*). The hydrophilicity of the MPC polymers increases with an increase in the MPC mole fraction (*11, 15*). These results suggest that the hydrophilic nature of the substrate is not the main factor determining the adsorption behavior of phospholipids. It is well known that phospholipid molecules adopt an organized structure in aqueous medium, presumably reflecting specific interactions

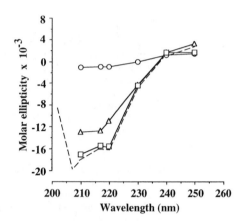

Figure 3. CD spectra of BSA in PBS and adsorbed on the polymer surfaces after contact with BSA solution for 60 min. (-----) in PBS, adsorbed on (O) poly(BMA), (△) poly(HEMA), and (□) poly(MPC-*co*-BMA). MPC mole fraction : 0.16. Initial concentration of BSA: 0.45 g/dL (Reproduced with permission with ref. 27).

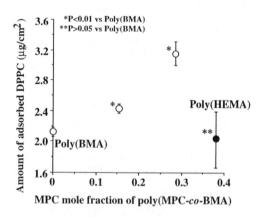

Figure 4. Relationship between the amount of DPPC adsorbed on the poly(MPC-*co*-BMA) and the mole fraction of MPC of the copolymer. The initial concentration of DPPC in the liposomal solution was 1.0 wt % which was prepared by sonication method.

between phospholipid molecules.   The increase in the amount of adsorbed DPPC may be due to the affinity of the MPC moiety for the DPPC molecule based on the "self-assembling" properties of the phospholipids since the MPC moiety has the same polar group as DPPC.

When   phospholipid   molecules associate in an aqueous medium due to their amphiphilic nature and form a so-called multi- or bi-layered membrane structure, a gel-liquid crystalline transition of the long alkyl chain in the molecule is observed (29).   Therefore, the adsorption state of DPPC molecules on polymer surfaces can be examined using DSC.   Figure 5 shows DSC curves of polymer membranes treated with the DPPC liposomal solution (14).   The DSC curve of the DPPC liposomal solution shows an endothermic peak at 41.8°C, corresponding to the Tc of the DPPC liposomes (29).   If the bilayer membrane structure is destroyed, the endothermic peak will disappear.   In the case of poly(BMA) and poly(HEMA) which were treated with the DPPC liposomal solution, there was no endothermic peak in the range of 20 °C to 50 °C (14).   On the other hand, an endothermic peak was observed at 42 °C in poly(MPC-*co*-BMA) treated with the DPPC liposomal solution.   There were no DSC peaks in that temperature range for any of the "conventional" polymer membranes which did not have MPC units.   This result clearly indicates that the multi- and bi-layer membrane structure, as in the DPPC liposomes, was maintained on poly(MPC-*co*-BMA) whereas it was destroyed an the poly(BMA) and poly(HEMA) surfaces.   Thus, it is considered that MPC moieties in the poly(MPC-*co*-BMA) play an important role in maintaining the bilayer membrane structure of adsorbed DPPC molecules.   Very recently, we determined the phase transitions of DPPC liposomes adsorbed on the poly(MPC-*co*-BMA) which was coated on the quartz crystal(QC) microbalance as function of frequency change of the QC (23).   Similar results were found for another MPC copolymer system: an endothermic transition was seen after addition of poly(MPC-*block*-St) to a DPPC liposomal solution. On the other hand, it disappeared with the addition of poly(St) latex (21).

Protein Adsorption on Poly(MPC-*co*-BMA) Treated with DPPC Liposome

We considered that the interaction between proteins and the phospholipid adsorption layer depends on the orientation of the lipid molecules.   Figure 6 shows the amount of proteins adsorbed on polymer surfaces which were pretreated with DPPC liposome. By comparison with Figure 2, it can be seen that the treatment of a polymer surface with a DPPC liposomal solution reduced BSA adsorption in the poly(BMA) and poly(HEMA) cases.   For the case of poly(MPC-*co*-BMA), though the effect of DPPC treatment on protein adsorption seems very small, the amount of proteins adsorbed was reduced by about one-half compared to the bare polymer surface(first few minutes). The same effect of DPPC treatment on BSG adsorption was observed.

The surface of the polymers becomes hydrophilic by adsorption of DPPC even if adsorption is random and the hydrophobic interaction between proteins and the surface is weakened.   Moreover, when the polar groups of DPPC are structured on the surface, other interactions such as hydrogen bonding and/or electrostatic interactions seem to decrease.   As previously mentioned, poly(MPC-*co*-BMA) appears to organize DPPC molecules adsorbed on the surface.   Therefore, it is considered that the difference in protein adsorption between poly(MPC-*co*-BMA) and the other two polymers treated with DPPC is due to the difference in the organized state of DPPC molecules adsorbed on the surface.

In conclusion, MPC copolymers adsorbed very little protein from plasma, possible due to the preferential adsorption of phospholipid molecules and the formation of a biomembrane-like organized adsorption layer of the phospholipids. MPC copolymers

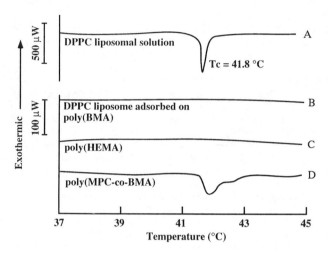

Figure 5. DSC curves of polymer membranes treated with DPPC liposomal solution. A: DPPC liposomal solution(2 wt%), B: poly(BMA), C: poly(HEMA), D: poly(MPC-*co*-BMA). MPC mole fraction, 0.25 (Reproduced with permission from ref. 14).

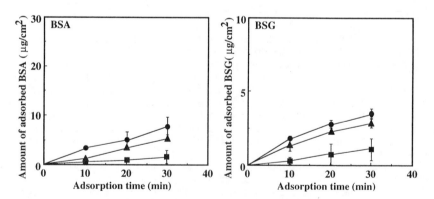

Figure 6. Adsorption of BSA and BSG onto polymer surfaces pretreated with a DPPC liposomal solution at 30 °C. (●) Poly(BMA), (▲) Poly(HEMA), (■) Poly(MPC-*co*-BMA). MPC mole fraction, 0.25. Initial concentrations of BSA and BSG were 4.5 g/dL and 1.6 g/dL, respectively (Reproduced with permission from ref. 14).

may thus minimize adsorption of proteins from plasma and blood and thus afford a good nonthrombogenic surface. Such a surface does not offer ligands which are recognized by receptors on cell membranes and it may protect from adsorption and activation of platelets. These phenomena may be explained by creation of a "self-assembled biomimetic membrane" structure by adsorbed phospholipids on the MPC copolymer surfaces.

Literature Cited

1. *Biomedical Applications of Polymeric Materials;* Tsuruta, T.; Hayashi, T.; Kataoka, K.; Ishihara, K.; Kimura, Y., Eds.; CRC Press: Boca Raton, FL, 1993.
2. *Proteins at Interfaces : Physicochemical and Biochemical Studies;* Brash, J.L.; Horbett, T.A., Eds.; ACS Symposium Series 343; American Chemical Society: Washington, D.C., 1987.
3. Lyman, D.J.; Metcalf, L.C.; Albo, Jr., D.; Richards, K.F.; Lamb, J. *Trans.Amer.Soc.Artif.Int.Organs* **1974,** *20,* 474.
4. Kim, S.W.; Lee, R.G.; Oster, H.; Coleman, D.; Andrade, J.D.; Lentz, D.J.; Olsen, D. *Trans.Amer.Soc.Artif.Int.Organs* **1974,** *20,* 449.
5. *Hydrogels for Medical and Related Application;* Andrade, J.D., Ed.; ACS Symposium Series 31; American Chemical Society: Washington, D.C., 1976.
6. Castillo, E.J.; Koenig, J.L.; Anderson, J.M.; Lo, J. *Biomaterials* **1984,** *5,* 319.
7. Ueda, T.; Ishihara, K.; Nakabayashi, N. *Seitai Zairyo (J.Jpn.Soc.Biomat.)* **1991,** *9,* 288.
8. Horbett, T.A.; Payne, M.S. *J.Biomed.Mater.Res.* **1987,** *21,* 843.
9. Ishihara, K. In *New Functionality Polymers, Volume B;* Tsuruta, T.; Dojima, M.; Seno, M.; Imanishi, Y., Eds.; Elsevier Science Publishers B.V.: Amsterdam, 1993, pp233-238.
10. Ishihara, K.; Nakabayashi, N.; Nishida, K.; Sakakida, M.; Shichiri, M. In *Diagnostic Biosensor Polymers;* Usmani, A.M.; Akmal, N., Eds.; ACS Symposium Series 556, American Chemical Society: Washington, DC, 1994, Chapter 16; pp194-210.
11. Ishihara, K.; Ueda, T.; Nakabayashi, N. *Polym.J.* **1990,** *22,* 355.
12. Ishihara, K.; Aragaki, R.; Ueda, T.; Watanabe, A.; Nakabayashi, N. *J. Biomed. Mater. Res.* **1990,** *24,* 1069.
13. Ishihara, K.; Ziats, N.P.; Tierney, B.; Nakabayashi, N.; Anderson, J.M. *J. Biomed. Mater. Res.* **1991,** *25,* 1397.
14. Ueda, T.; Watanabe, A.; Ishihara, K.; Nakabayashi, N. *J. Biomater. Sci. Polymer Edn.* **1991,** *3,* 185.
15. Ueda, T.; Oshida, H.; Kurita, K.; Ishihara, K.; Nakabayashi, N. *Polym. J.* **1992,** *24,* 1259.
16. Ishihara, K.; Oshida, H.; Ueda, T.; Endo, Y.; Watanabe, A.; Nakabayashi, N. *J. Biomed. Mater. Res.* **1992,** *26,* 1543.
17. Ishihara, K.; Oshida, H.; Endo, Y.; Watanabe, A.; Ueda, T.; Nakabayashi, N. *J. Biomed. Mater. Res.* **1993,** *27,* 1309.
18. Ishihara, K.; Tsuji, T.; Kurosaki, T.; Nakabayashi, N. *J. Biomed. Mater. Res.* **1994,** *28,* 225.
19. Hall, B.; Bird, R.R.; Kojima, M.; Chapman, D. *Biomaterials* **1989,** *10,* 219.
20. Kono, K.; Ito, Y.; Kimura, S.; Imanishi, Y. *Biomaterials* **1989,** *10,* 455.
21. Kojima, M.; Ishihara, K.; Watanabe, A.; Nakabayashi, N. *Biomaterials* **1991,** *12,* 121.
22. Ishihara, K.; Nakabayashi, N. *J. Polym. Sci., Part A: Polym. Chem.* **1991,** *29,* 831.
23. Tanaka, S.; Iwasaki, Y.; Ishihara, K.; Nakabayashi, N. *Makromol. Rapid Commun.* **1994,** *15,* 319.

24. Akaike, T.; Sakurai, Y.; Kosuge, K.; Senba, Y.; Kuwana, K.; Miyata, S.; Kataoka, K.; Tsuruta, T. *Kobunshi Ronbunshu* **1979,** *36,* 217.
25. Baszkin A.; Lyman, D.J. *J. Biomed. Mater. Res.* **1980,** *14,* 393.
26. Sugiyama, K.; Aoki, H. *Polym. J.* **1994,** *26,* 561.
27. Ishihara, K.; Ueda, T.; Saito, N.; Kurita, K.; Nakabayashi, N. *Seitai Zairyo(J. Jpn. Soc. Biomat.)* **1991,** *9,* 243.
28. Lu, D.R.; Lee, S.J.; Park, K. *J. Biomater. Sci., Polymer Edn.* **1991,** *3,* 127.
29. *The Liposome;* Nojima, S.; Sunamoto, J.; Inoue, K., Eds.; Nankodo: Tokyo, 1988.

RECEIVED February 10, 1995

Chapter 28

Analysis of the Prevention of Protein Adsorption by Steric Repulsion Theory

Timothy B. McPherson, Samuel J. Lee, and Kinam Park

School of Pharmacy, Purdue University, West Lafayette, IN 47907

It is known that adsorption of proteins to surfaces can be prevented by modification of the surfaces with hydrophilic polymers. The grafted hydrophilic polymers are thought act by exerting steric repulsion against the adsorbing proteins. We have examined the prevention of fibrinogen and lysozyme adsorption by covalent grafting of Pluronics, which are triblock copolymers of ethylene oxide (EO) and propylene oxide (PO), to hydrophobically modified glass surfaces. The adsorption of both proteins decreased to negligible levels as the surface concentration of the grafted Pluronic molecules increased. The decrease in protein adsorption by the grafted polymer chains was also theoretically examined by calculating the attractive and repulsive interaction energies between lysozyme and the surface by computer simulation. The calculation showed that the surface density of the grafted chains must be high enough to cover most of the surface to exert effective steric repulsion. This study suggests that one of the important factors in the surface modification by hydrophilic polymers is the surface concentration of the covalently grafted polymers.

It is known that when biomaterials come into contact with blood, plasma proteins adsorb to the surface very rapidly. Depending on the type and quantity of the adsorbed protein, the protein layer may cause platelet adhesion and activation, leading to mural thrombus formation. It is often thrombus formation and subsequent embolization that result in failure of implanted biomaterials. Two main approaches have been taken to improve biocompatibility of the implanted materials: synthesis of new polymer materials exhibiting less inherent thrombogenicity; and surface modification of existing materials. The latter method is appealing because the surface properties of a material can be modified to suit the particular biomaterial application, while retaining its desirable bulk properties.

Biomaterial surfaces have been modified with various water-soluble polymers, such as albumin, heparin, and poly(ethylene oxide). In general, covalent grafting of these water-soluble polymers results in significant decreases in blood protein adsorption and platelet adhesion. While surface modification has been successfully

0097–6156/95/0602–0395$12.00/0

exploited to enhance blood compatibility, the underlying mechanism has not been fully understood. Recent studies in our laboratory and others suggest that steric repulsion is the operative mechanism in surface passivation by the surface-grafted water-soluble polymers (1-5). Since the steric repulsion theory can explain many of the observed experimental results, whether positive or negative, it is beneficial to describe the steric repulsion theory and the conditions under which it applies.

Steric Repulsion Theory

When an adsorbate molecule approaches a surface grafted with terminally bound water-soluble polymers, the surface polymer molecules become compressed. Compression results in decreased conformational entropy for the compressed polymer chains, which is an unfavorable thermodynamic state. The system attempts to move back to the higher entropy state, repelling the adsorbate molecule. In addition, when a polymer chain is compressed by an adsorbate molecule, the local concentration of monomer units increases relative to the surrounding areas, increasing local osmotic pressure. Water molecules then diffuse in to equalize the osmotic pressure, thereby repelling the adsorbate. Fig. 1-A describes how steric repulsion by the surface-grafted flexible molecules prevents protein adsorption. Steric repulsion is referred to as an osmotic-entropic process, due to the synergy of these two complementary components providing the observed phenomenon. The same is true with globular proteins such as albumin as shown in Fig. 1-B. It should be noted that albumin is a flexible molecule, rather than a hard ball, and thus can exert steric repulsion.

For the surface-bound polymers to exert effective steric repulsion, they must satisfy the following conditions: (i) tight anchoring to the surface; (ii) complete surface coverage; and (iii) flexibility of the grafted molecules. If polymer molecules are not tightly bound to the surface, they may be displaced by the blood proteins or cells which have a greater affinity for the substrate. Covalent bonding to the surface is most preferred. Complete surface coverage is also important, since proteins and cells can directly interact with the bare surface between polymer chains if the surface coverage is inadequate. Flexibility of the grafted molecules provides the steric repulsion as described above. If the grafting of polymer molecules does not conform to the above three conditions, enhancement in blood compatibility will not be fully realized. In most of the literature data, the surface concentrations of the grafted polymer molecules are not determined mainly due to the difficulty in measuring it. Thus, if the grafting of hydrophilic polymers does not prevent protein adsorption or cell adhesion, it is likely that the surface concentration of the grafted polymers is not high enough.

The purpose of this study was to examine the ability of the covalently grafted PEO chains to prevent protein adsorption, and to analyze the steric repulsion property of the PEO chains by the scaling analysis (3,6-9) and computer simulation (10-12). The steric repulsion property was measured from the ability to prevent adsorption of fibrinogen and lysozyme to the surface. For computer simulation, lysozyme was used because its crystal structure is available.

Experimental

Protein and Pluronic Adsorption Experiments

Glass Tubing Preparation. Glass tubes (6 inch-long, inner diameter of 2.45 mm and outer diameter of 4 mm) or capillaries (1.8 mm inside diameter x 50 mm long) were cleaned by immersion in chromic acid solution at room temperature overnight, followed by rinsing with deionized distilled water (DDW) and drying at 60°C. Cleaned glass tubes were then hydrophobically modified for the covalent grafting of

Pluronics. The details of the hydrophobic modification of glass will be described in a future publication. The modified glass tubes were rinsed sequentially in fresh chloroform, absolute ethanol, and DDW, and dried at 60°C overnight.

Pluronic Grafting. A variety of Pluronics were kindly provided by BASF Corp. They were dissolved in phosphate buffered saline solution (PBS, pH 7.2) to make a 100 mg/ml stock solution. The modified glass tubes were filled with Pluronic solutions of desired concentrations without introducing air bubbles and left undisturbed for one hour. Then the tubes were flushed with PBS to remove non-adsorbed Pluronics. The tubes, filled with fresh PBS, were then γ-irradiated for 6 hours (0.08 Mrad/h) to covalently graft the polymer to the surface. Following irradiation the tubes were evacuated with DDW and dried at 60°C. The concentration of Pluronic solutions was varied from 0.01 mg/ml to 25 mg/ml. Two control samples used in this study were modified glass with and without exposure to γ-irradiation. The prepared tubes were used in the protein adsorption experiment within 1 week.

Protein Adsorption. Human fibrinogen and chicken egg lysozyme (Sigma) were radiolabeled with [125]I (Amersham) using Enzymobead reagent (Bio-Rad). Free ^{125}I and the Enzymobeads were removed from the radiolabeled protein solution by gel filtration over a Sephadex G-10 column, previously calibrated with blue dextran. Radiolabeled lysozyme was mixed in a mass ratio of 1:9 with unlabeled lysozyme to make a solution containing 0.15 mg/ml protein, while fibrinogen was mixed in a ratio of 1:39 with unlabeled fibrinogen to a concentration of 0.10 mg/ml protein. The Pluronic grafted glass tubes were filled with PBS and allowed to hydrate for 2 hours at room temperature prior to exposure to the protein solution. The radiolabeled protein solution was introduced into the tubing, displacing PBS. After 1 hour of protein adsorption, the protein solution was similarly displaced from the tubes with fresh PBS. The tubes were cut into one inch sections, and the middle four sections were assayed on a gamma counter (Beckman) to measure the surface adsorbed radioactivity. These data were converted to $\mu g/cm^2$ by comparison to the specific activity of the standard protein sample.

Pluronic Adsorption. p-Methoxyphenyl-Pluronic F127 (MP-PF127) was synthesized by a previously described method (13) with azeotropically dried benzene as the solvent. The MP-PF127 was radiolabeled using the Enzymobead reagent and purified as above. The radiolabeled polymer was diluted with PBS to concentrations from 0.01 mg/ml to 25 mg/ml. The modified glass capillary tubes were filled with the radiolabled MP-PF127 solutions and left for 1 hour at room temperature. Radiolabeled polymer was not mixed with non-labeled polymer in these experiments. The lowest concentrations exhibited relatively low γ counts to begin with due to the 1000-fold dilution of the starting solution, and further reductions by mixing with non-labeled polymer resulted in immeasurable surface concentrations for these samples. The polymer solution was displaced with PBS, and the tubes assayed for surface-adsorbed radioactivity. The data were converted to $\mu g/cm^2$. Values for the scaling theory parameters D and L, used in the computer simulation, were then calculated.

Computer Simulation

The surface was generated using Quanta polymer generator. The energy minimized surface consisted of three 65 x 65 Å layers. The atom coordinates of lysozyme were obtained from the Brookhaven Protein Data Bank. Atomic coordinates are necessary because the intermolecular interaction is dependent on the distance between each atom pair. Lysozyme was placed on the surface of the polymer and the distance dependent calculation consisting of van der Waals, electrostatic, and solvation energy was performed (10-12,14). The repulsive energy due to the grafted chains were calculated using the scaling analysis (3,6-9).

Fig. 2-A shows the interaction between brushes, a model used in the scaling theory. The grafted Pluronic molecules are assumed to exist as a brush on the surface. Thus, a similar model can be used to study protein adsorption on to the brush as depicted in Fig. 2-B. This model shows that the repulsive energy per cm^2 against the protein by the grafted polymer chains is:

$$E(H) = \frac{kT}{D^3}\left[\left(\frac{L^{9/4}}{1.25H^{5/4}} + \frac{H^{7/4}}{1.75L^{3/4}}\right) - \left(\frac{L}{1.25}\right) + \left(\frac{L}{1.75}\right)\right] \quad (H < L) \quad \text{(Equation 1)}$$

where D is the average distance between the attached points, k is the Boltzmann constant $(1.38\times10^{-23}$ J/K), T is the temperature, L is the thickness of the grafted chains, and H is the separation distance between the surface and the protein. Equation 1 was derived as done in reference 8. Here L is related to the density of the grafted chains as follows:

$$L = N a \, \sigma^{1/3} \quad \text{or} \quad L = N a \left(\frac{a}{D}\right)^{2/3} \quad \text{(Equation 2)}$$

where N is the number of monomers per chain, and a is the monomer size. Here, σ $(=a^2/D^2)$ is equal to the surface density of the grafted chains (6,7).

Results

The adsorption of fibrinogen to the modified glass was high as expected. The surface fibrinogen concentration was 0.47 μg/cm^2. As Pluronics were grafted to the surface, the surface fibrinogen concentration decreased dramatically with the exception of Pluronic L31 as shown in Fig. 3. The number of EO residues varied from 2 (for L31) to 128 (for F108). Even Pluronic L61, which has only 3 EO residues, was effective in preventing the adsorption of fibrinogen to the surface. Except for Pluronic L31, the surface fibrinogen concentrations on the Pluronic-grafted surfaces were all smaller than 0.02 μg/cm^2, which is known to be the minimum surface fibrinogen concentration necessary to activate platelets (15).

To examine the effect of grafted Pluronic surface concentration on protein adsorption, we chose Pluronic F127 as a model Pluronic. As a model protein, lysozyme was used since it is smaller than fibrinogen, and its crystal structure is known. The radius of gyration of lysozyme calculated using QUANTA is about 14Å. Thus, lysozyme is expected to enjoy easier access to the surface than fibrinogen. The surface lysozyme concentration on the control modified glass was 0.27 μg/cm^2. The value decreased to 0.21 μg/cm^2 if the surface was exposed to γ-irradiation.

Fig. 4 shows the surface concentrations of lysozyme and Pluronic as a function of the bulk Pluronic concentration used for adsorption. The lysozyme and PF127 adsorption profiles reveal some interesting points when directly compared. At high PF127 surface concentrations, lysozyme adsorption is low. Conversely, when the PF127 surface concentration is low, lysozyme is free to adsorb to the surface. When the bulk Pluronic concentration was 0.01 mg/ml, the surface Pluronic concentration was only 0.015 μg/cm^2. At this condition, the adsorption of lysozyme was similar to that on the control surface. As the solution concentration increased up to 5.0 mg/ml, the surface Pluronic concentration steadily increases up to 0.22 μg/cm^2. Further increase in the bulk Pluronic concentration up to 25 mg/ml did not increase the surface Pluronic concentration any more. The PF127 solution concentration where the adsorption of lysozyme reaches its minimum coincides closely with the maximum surface PF127 concentration. When the surface Pluronic concentration was 0.22

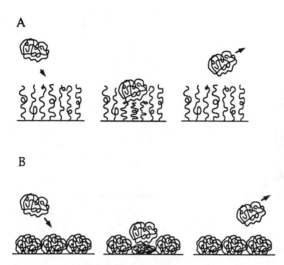

Figure 1. Schematic description of steric repulsion exerted by the grafted hydrophilic polymers such as PEO (A) and globular proteins such as albumin (B).

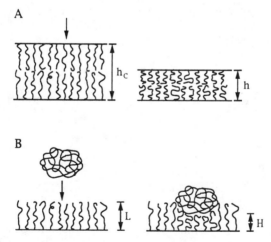

Figure 2. Comparison of the interaction between brushes (A) and the interaction of a protein with a brush (B).

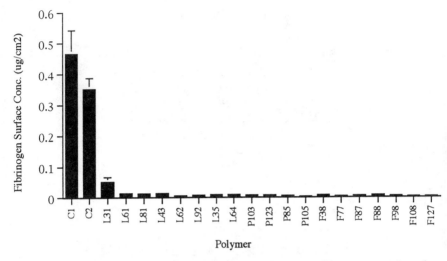

Figure 3. Fibrinogen adsorption to Pluronic-grafted glass. The concentration of Pluronics was 1 mg/ml and grafted to the surface by γ-irradiation for 6 h. The fibrinogen concentration in the adsorption solution was 0.1 mg/ml. C1 and C2 are control samples, untreated and γ-irradiated, respectively. EO/PO/EO values for Pluronics are as follows: L31 (2/16/2); L61 (3/30/3); L81 (6/39/6); L43 (7/21/7); L62 (8/30/8); L92 (10/47/10); L35 (11/16/11); L64 (13/30/13); P103 (20/54/20); P123 (21/67/21); P85 (27/39/27); P105 (38/54/38); F38 (46/16/46); F77 (52/35/52); F87 (62/39/62); F88 (97/39/97); F127 (98/67/98); F98 (122/47/122); and F108 (128/54/128).

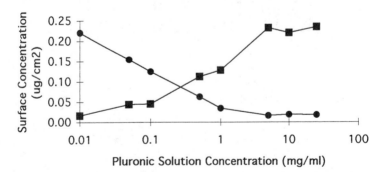

Figure 4. The surface concentrations of lysozyme (●) and Pluronic F127 (■) as a function of the bulk Pluronic concentration in the adsorption solution.

μg/cm^2, the surface lysozyme concentration was reduced to a minimum of 0.017 μg/cm^2, which corresponds to approximately 6.85 x 10^{-4} nmol/cm^2. This is more than 90% decrease in the surface lysozyme concentration by Pluronic grafting at high surface density. It is interesting to note the inverse relationship between the lysozyme adsorption profile and the PF127 isotherm.

The attractive and repulsive interaction energies between lysozyme and the modified glass were calculated as a function of the separation distance from computer simulation. The calculation requires the information on the thickness (L) of the grafted chains, which can be determined from the density of the grafted chains (σ). The average distance between PEO chains (D) and the thickness of the brush surface (L) were calculated from the Pluronic adsorption data. This was accomplished by converting the surface Pluronic concentration value (in μg/cm^2) to molecules/cm^2. The inverse of the square root of these values is considered to be D, but since Pluronic molecules contain two equivalent PEO chains, this value must be halved. Therefore, the D value calculated from the Pluronic adsorption data is:

$$D = (1/2) \text{ (molecules/cm}^2)^{-1/2}$$

The L values are then calculated using Equation 2. The steric repulsive force due to the grafted Pluronic chains was calculated using Equation 1 for each set of data. The calculated data are listed in Table I.

Table I. D and L values of the grafted PEO chains at different surface concentrations of Pluronic F127

F127 solution conc. (mg/ml)	F127 surface conc. (μg/cm^2)	D (Å)	L (Å)
0.01	0.016	82	29
0.05	0.043	49	40
0.1	0.045	48	41
0.5	0.111	31	55
1.0	0.128	29	58
5.0	0.231	21	70
10.0	0.220	22	69
25.0	0.234	21	70

The results of the computer simulation of lysozyme-brush interaction energy are shown in Fig. 5. The figure shows the total interaction energies between lysozyme and the PEO-grafted surface. Only the results with 4 different grafting conditions were plotted to make it easy to identify the curves. As shown in Fig. 5, the total interaction energy was repulsive at all separation distances if the D values were 29 Å or less (i.e., L values were 58 Å or larger). When the D values were 31 Å and 48 Å, the total interaction was attractive unless the separation distance was less than 2 Å. Only attractive interaction was observed when the D value was 82 Å. The results in Fig. 5 agrees rather well with the data in Fig. 4. When the total interaction energy was all repulsive, the adsorption of lysozyme was minimal. On the other hand, the PEO on the surface could not prevent adsorption of lysozyme when the total interaction was attractive. The results in Fig. 5 clearly show that the distance (D) between the grafted chains needs to be smaller than the thickness (L) of the brush to exert effective steric repulsion. This means that the surface is completely covered with the grafted chains, since they are highly flexible. The calculation with other D and L values also indicates that the surface density of the grafted chains can be smaller as the chain length (or L value) becomes larger. It is clear that effective steric repulsion by the grafted chains relies on both the chain length and the chain density on the surface.

Discussion

In general, computational models utilize assumptions. This occurs as a result of shortcomings in resources to completely represent the real system. For example, protein adsorption onto surfaces occurs in the solvent environment. In order to model such a system, explicit inclusion of solvent molecules would be ideal. For the most part, however, computational resources do not allow such large systems to be modeled. Dielectric constant has been used to simulate the effect of solvent on electrostatic interactions. In this study the dielectric constant used for water was 80. Since the use of dielectric constant may be inadequate, the solvation interaction energies were calculated using the fragment constant method (11). This method utilizes data from partition coefficient experiments. Therefore, the effect of solvent which the dielectric constant alone may not represent can be modeled more effectively.

We used a static computational model for protein adsorption. Proteins may undergo conformational changes during the adsorption process, although lysozyme is known to be rather inflexible. Although conformational changes are not explicitly modeled, other important aspects of the protein were considered. Electrostatic charge values were obtained from the validated Quanta/CHARm database, and ionized amino acid residues were represented accurately based on pKa values. We have previously determined that the orientation of the protein relative to the surface influences the magnitude of interaction energy calculations (10), so an orientation with high interaction energy was used in this study. The distance at which interaction becomes insignificant was not greatly affected by the protein orientation, however.

The adsorption of lysozyme decreased sharply as the surface Pluronic concentration increased, then leveled off to a minimum at high Pluronic concentrations. The profile implies three distinct surface conditions, indirectly defining the Pluronic adsorption isotherm. At the surface Pluronic concentrations of $0.016 \, \mu g/cm^2$ or less, lysozyme adsorption was similar to that on the control surface. Therefore, any Pluronic on the surface is completely ineffective in repelling protein adsorption. This condition would correspond to a very large D value in the computer simulation data, where D is much greater than L. No repulsion was seen at any separation distance between the protein and the surface. When the surface Pluronic concentration was between $0.045 \, \mu g/cm^2$ and $0.231 \, \mu g/cm^2$, the Pluronic molecules on the surface become dense enough to prevent protein adsorption. The protein no longer enjoys free access to the entire surface. The repulsion, however, is still not large enough to prevent protein adsorption. This situation is described by the calculation where D is of similar magnitude to L (e.g., D=48 Å and L=41 Å). Here, there is little repulsion except when the protein is very close to the surface. In the tail portion of the lysozyme adsorption profile in Fig. 4, lysozyme adsorption is essentially constant regardless of the surface Pluronic concentration. This region corresponds to small D values in the computer simulation data, where the distance between polymer chains is small and D is substantially less than L (e.g., D=21Å and L=70Å). Repulsive forces in this region increase quickly as a protein approaches to the surface, resulting in "complete" shielding of attractive forces.

Our calculation shows that the surface density of the grafted chains necessary for the prevention of protein adsorption varies depending on the thickness of the grafted layer, which is related to the molecular weight of the grafted chains. As the molecular weight decreases, the surface density necessary to have effective repulsive energy

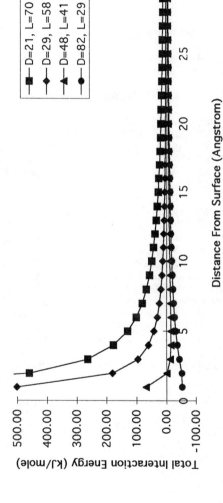

Figure 5. The total interaction energy as a function of the separation distance between lysozyme and the surface. The thickness of the brush was calculated from the surface Pluronic concentrations listed in Table I. The D and L values used to calculate the interaction energy are shown in the box.

increases. Thus, even a low molecular weight chain, such as Pluronic L61 with 3 EO residues, can have significant repulsion if the surface density is high. It is very likely that Pluronic L61 has a high surface density due to its small size. As the PEO chain length of Pluronics increases, the surface density may decrease substantially due to the large size of the molecules.

In conclusion, the covalent grafting of hydrophilic polymers in high surface density appears to be most important in successful prevention of protein adsorption to the surface. The steric repulsion theory is useful in explaining the success or failure of the grafted hydrophilic chains in the prevention of protein adsorption. The steric repulsion theory will also be useful in the design of new materials with excellent biocompatibility.

Acknowledgments

This study was supported by the National Heart, Lung, and Blood Institute of the National Institute of Health through grant HL 39081.

REFERENCES

1. Amiji, M.; Park, K. *J. Biomater. Sci. Polymer Edn.* **1993**, *4*, 217-234.
2. Amiji, M.; Park, K. *ACS Symp. Ser.* **1994**, *540*, 135-146.
3. Jeon, S. I.; Lee, J. H.; Andrade, J. D.; de Gennes, P. G.*J. Colloid Interface Sci.* **1991**, *142*, 149-158.
4. Han, D. K.; Jeong, S. Y.; Kim, Y. H.; Min, B. G.; Cho, H. I. *J. Biomed. Mater. Res.* **1991**, *25*, 561-575.
5. Bjorling, M. *Macromolecules* **1992**, *25*, 3956-3970.
6. de Gennes, P.G. *Scaling Concepts in Polymer Physics*; Cornell University Press: Ithaca, NY, 1979.
7. de Gennes, P.G. In *Physical Basis of Cell-Cell Adhesion*; Bongrand, P., Ed.; CRC Press: Boca Raton, FL, **1988**; Chapter 2.
8. Luckham, P. F.; Klein, J. *J. Chem. Soc. Faraday Trans.* **1990**, *86*, 1363-1368.
9. Patel, S.; Tirrell, M.; Hadziioannou, G. *Colloids Surf* **1988**, *31*, 157.
10. Lu, D.R.; Park, K. *J. Biomater. Sci. Polymer Edn.* **1990**, *1*, 243-260.
11. Lu, D.R.; Park, K. *J. Biomater. Sci. Polymer Edn.* **1991**, *3*, 2127-147.
12. Lee, S. J.; Park, K. *J. Vacuum Science and Technology,* in press.
13. Amiji, M. A.; Park, K. *J. Colloid Interface Sci* . **1993**, *155*, 251-255.
14. Lee, J. H.; Kopecek, J.; Andrade, J. D. *J. Biomed. Mater. Res.* **1989**, *23*, 351-368.
15. Park, K.; Mao, F. W.; Park, H. *J. Biomed. Mater. Res.* **1991**, *25*, 407-420.

RECEIVED December 22, 1994

Chapter 29

Direct Measurement of Protein Adsorption on Latex Particles by Sedimentation Field-Flow Fractionation

Yong Jiang[1], J. Calvin Giddings[1,3], and Ronald Beckett[2]

[1]Department of Chemistry, Field-Flow Fractionation Research Center, University of Utah, Salt Lake City, UT 84112
[2]Department of Chemistry, Water Studies Centre, Monash University, P.O. Box 197, Caulfield East, Victoria 3145, Australia

In previous work from this group it was shown that sedimentation field-flow fractionation (SdFFF) is capable of the direct measurement of quantities as small as a few attograms (10-18 g) of various materials, including γ-globulin, adsorbed on latex beads. Here, following a review of the principles of adsorbed mass measurements by SdFFF, the previous work is extended to include numerous measurements of immunoglobulin (IgG) adsorbed on polystyrene latex beads under diverse conditions. A few additional γ-globulin measurements are reported as well.

Adsorption of IgG on latex was quite rapid with adsorption times usually <10 min. The desorption process was generally much slower. Adsorption isotherms were obtained and fitted to the Langmuir equation. The initial slope of the isotherm was different for latex beads prepared in the presence and absence of surfactant but the plateau was similar. The isotherm plateau was also little affected by exposing the latex to a protein concentration increased in two stages rather than a single stage. However, the adsorption plateau was found to be slightly dependent on pH with a maximum near pH 7, close to the reported isoelectric point range of IgG.

Sedimentation field-flow fractionation (SdFFF), a versatile technique for separating colloidal particles, is capable also of directly measuring particle mass. The measurement of mass is extremely sensitive, sufficient to account for small amounts of adsorbed material consisting of no more than a few tens of protein molecules on a single particle. More specifically, if colloidal particles gain mass by adsorption, the incremental mass due to the presence of the adsorbate can be measured down to amounts as small as a few attograms (10^{-18} g) per particle. Thus SdFFF, through the direct and accurate measurement of incremental adsorbate masses in an experimentally convenient procedure, is a promising tool for protein adsorption studies.

In an earlier study, SdFFF was used to measure small adsorbed masses of γ-globulin, ovalbumin, and other substances on latex beads (1). The beads consisted of both polystyrene (PS) and vinyl toluene butylstyrene, which is a neutrally

[3]Corresponding author

0097–6156/95/0602–0405$12.00/0
© 1995 American Chemical Society

buoyant latex in water. It was shown that full adsorption isotherms could be constructed. The preferred conditions for measurement were described theoretically.

Caldwell and coworkers have applied both SdFFF and a related technique, flow FFF, to the study of various films adsorbed on latex particles (2-4). These studies have included measurements of the thickness of adsorbed Pluronic triblock polymeric surfactant films (2) along with adsorbed IgG-anti IgG complexes (3) and milk proteins (4).

SdFFF is an elution technique. Small samples (requiring only a few micrograms of colloidal particles) are injected at the head of a narrow channel and driven through the channel by the flow of a suitable carrier liquid. The emergence time (called the retention time) depends rigorously on particle effective mass, which is the true mass less the mass of fluid displaced by the particles. Thus particle mass can be calculated from the observed retention time of a small population of particles.

If the particle population is relatively monodisperse, the particles emerge as a single peak whose retention time can be identified by a sensitive detector (e.g., an HPLC UV detector). An eluting population of 0.215 μm polystyrene latex beads is shown as the left hand sample peak of Figure 1.

If protein or other adsorbate attaches to the particles, their mass will increase and the particle peak will shift to a different retention time. The center peak of Figure 1 has been shifted to the right of the original peak by the adsorption of IgG on the latex surface from a 0.05 mg/mL IgG solution. A further shift right is caused by the adsorption of more protein from a more concentrated (1 mg/mL) IgG solution, as illustrated by the right hand peak of Figure 1.

The shift in latex peak positions caused by the adsorption of protein on the latex surface can be related rather exactly to the mass of adsorbed protein (see Theory section). Thus observed retention time shifts yield a direct measure of adsorbed mass. As noted above, the method is sensitive to small incremental masses, which means that measurable retention time shifts are induced by minute amounts of added adsorbate.

Numerous accounts of the SdFFF mechanism have appeared in the literature (5-9). Briefly, SdFFF is one of several FFF techniques used to separate and characterize macromolecular and particulate materials. All FFF techniques utilize a driving force with a direction *perpendicular* to the flow axis of a stream of carrier liquid (into which the sample pulse is injected) being pumped through a ribbonlike channel (see Figure 2). The field acts across the thin dimension of the channel, which is typically of the order of 10^2 μm thick. For SdFFF, the driving force is generated by a centrifuge, within which the channel is incorporated.

The sedimentation force drives particles toward one of the channel walls (the accumulation wall). An equilibrium concentration distribution soon forms, representing a balance between the driving force and Brownian motion (Figure 2). The distribution at equilibrium is exponential and has a characteristic (mean) thickness of ℓ where $\ell = kT/F$, in which k is Boltzmann's constant, T is the absolute temperature, and F is the sedimentation force acting on a single particle.

Flow causes the displacement of particles through the channel. The flow profile between the major channel walls is parabolic. When F is large and ℓ is small, the particles are held near the accumulation wall at the very edge of the parabolic profile where the velocity is low. Such particles elute slowly and have long retention times. Particles with smaller forces acting on them have larger ℓ values and elute earlier. Thus the retention time depends on the applied force F (see Theory section). However, the force acting on a particle depends on the mass of the particle including the mass of material adsorbed on its surface. As adsorbate loads increase, F increases and retention time increases roughly in linear fashion.

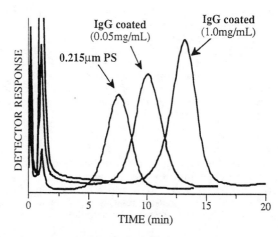

Figure 1. Comparison of SdFFF elution profiles (fractograms) of uncoated and IgG coated 0.215 μm PS latex beads. The concentration of IgG in the solution to which the latex was exposed is shown with the two right hand peaks.

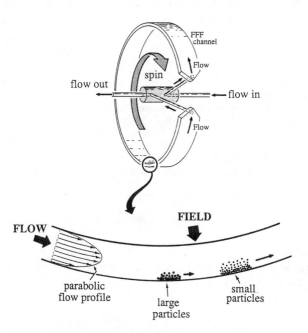

Figure 2. Schematic illustration of the SdFFF system. The enlarged edge view of the channel (bottom) shows the equilibrium concentration distribution of two particle sizes.

The theoretical equations of FFF can then be used to deduce the adsorbate mass as will be shown below.

The high sensitivity of measured retention time to small masses and mass increments is a natural outcome of the above mechanism. Since the driving force is balanced against Brownian motion, which is extremely weak but well defined, forces at very small levels, down to 10^{-16} N or even 10^{-17} N, affect the compression (hence the ℓ value and the retention time) of the particle distribution. Such forces can be generated by only a few attograms of mass at accelerations of only 1000 gravities, corresponding to the upper limit of conventional SdFFF instrumentation. Larger masses are best measured at lower accelerations. This matter is discussed further in the original publication (1).

Theory

The retention time t_r of eluting particles in SdFFF depends on the effective mass m' of the particles (including the mass of any adsorbed material) according to well established principles (1, 6-9). Two equations (and their derived forms) are needed. First is the basic FFF expression relating t_r for constant field conditions to the sedimentation force F acting on each of the particles

$$t_r = \frac{t^0 Fw / 6kT}{\coth(Fw / 2kT) - 2kT / Fw} \tag{1}$$

where t^0 is the void time (i.e., t_r for unretained particles), w is the channel thickness, k is Boltzmann's constant, and T is the absolute temperature. When $Fw \gg kT$, this equation simplifies to

$$t_r = t^0 \frac{Fw}{6kT} \tag{2}$$

While equation 1 is used here for all calculations, equation 2 is a simple linear expression that provides a more direct insight into FFF behavior. Most importantly, t_r is found approximately proportional to F. Since F is proportional to m', t_r becomes proportional to m' also. This high sensitivity of t_r to m', along with a capability of measuring forces down to 10^{-16} N, makes SdFFF ideal for the direct and accurate measurement of small quantities of added mass, such as those associated with thin adsorbed films on small particles.

The second major equation (following equation 1 and its limiting form, equation 2) expresses the physical law relating F to m' and eventually to the mass and thickness of the adsorbed film. Quite simply, F is the product of the two terms

$$F = m'G \tag{3}$$

where G is acceleration, $\omega^2 r_0$ (ω being the rotational velocity of the centrifuge and r_0 its radius). For a particle of uniform density ρ_p (lacking an adsorbed film), $m' = m_p(1 - \rho / \rho_p)$ and F becomes (1)

$$F_p = m_p G(1 - \rho / \rho_p) = \frac{\pi}{6} G d^3 \Delta \rho_p \tag{4}$$

where the particle mass, density, density increment (relative to the surrounding carrier liquid), and effective spherical diameter are given by m_p, ρ_p, $\Delta\rho_p$, and d, respectively.

When the particle adsorbs a coating of mass m_c having a density ρ_c and a density increment $\Delta\rho_c$, F becomes the sum of two terms, the first for the particle and the second an incremental term for the coatings (*1*)

$$F = F_p + Gm_c\frac{\Delta\rho_c}{\rho_c} = G\left(m_p\frac{\Delta\rho_p}{\rho_p} + m_c\frac{\Delta\rho_c}{\rho_c}\right) \qquad (5)$$

If the particle diameter is d and the mean film thickness is h, F becomes

$$F = F_p + \pi d^2 h\Delta\rho_c G = d^2\pi G\left(\frac{d\Delta\rho_p}{6} + h\Delta\rho_c\right) \qquad (6)$$

providing $h \ll d$. (A spherical particle with an idealized uniform film is shown in Figure 3.)

Since F is obtained by FFF retention measurements and is thus a known quantity (via equation 1), it remains only to invert the above expressions for F to obtain values for the coating mass and thickness, m_c and h, respectively. We obtain (*1*)

$$m_c = \frac{F\rho_c}{G\Delta\rho_c} - m_p\frac{\Delta\rho_p}{\rho_p}\frac{\rho_c}{\Delta\rho_c} \qquad (7)$$

and

$$h = \frac{F}{\pi Gd^2\Delta\rho_c} - \frac{d\Delta\rho_p}{6\Delta\rho_c} \qquad (8)$$

These expressions are valid when the particles and their external coating are both denser than the carrier medium and sink to the outer wall, as is the case here. A more general treatment accounting for negative $\Delta\rho$ values and "floating" particles is given in the original publication (*1*).

If the adsorbed mass is small compared to the bare particle mass, ($m_c \ll m_p$), then t_r may be only slightly increased by adsorption, that is, by the addition of m_c to m_p. A small error in the final terms of equations 7 or 8 would then cause a large error in m_c or h, respectively. Thus it is perhaps more accurate, and generally preferable in practice, to obtain F_p (by numerical means) from equation 1 for the bare particle rather than rely on equation 4 to get F_p from particle parameters. The latter (less accurate) approach underlies equations 7 and 8. The proposed improvement is implemented by rearranging equations 5 and 6 as follows

$$m_c = \frac{\rho_c(F - F_p)}{G\Delta\rho_c} \qquad (9)$$

and

$$h = \frac{F - F_p}{\pi d^2 G\Delta\rho_c} \qquad (10)$$

where force F on the coated particles and force F_p on the bare particles are to be obtained using equation 1 from the respective t_r values measured at the same field strength G. If different G values are used for the two t_r measurements, $(F - F_p)/G$ would be replaced by $(F/G) - (F_p/G_p)$.

Steric perturbations to retention (6, 8), although small under most circumstances, might best be compensated by adjusting G and G_p values such that $F \cong F_p$. However, if these perturbations are significant, a sterically corrected version of equation 1 should be used.

For cases in which equation 2 is a good approximation to equation 1, m_c and h can be expressed explicitly in terms of the time increment Δt_r between the elution of coated and uncoated particles. For this purpose we use equation 2 to obtain

$$F - F_p = \frac{6kT}{wt^0} \Delta t_r \qquad (11)$$

The substitution of equation 11 into equations 9 and 10 yields

$$m_c = \frac{6kT\rho_c}{wt^0 G \Delta\rho_c} \Delta t_r \qquad (12)$$

and

$$h = \frac{6kT}{wt^0 \pi d^2 G \Delta\rho_c} \Delta t_r \qquad (13)$$

Equation 12 will be used later to show that Δt_r is highly sensitive to small adsorbed masses. Specifically, even with G values as small as 544 gravities (as used here), only 10-20 IgG molecules are needed to give measurable shifts in Δt_r.

Experimental

Equipment and Materials. The sedimentation FFF system used in this work has been described in the literature (10). The device is similar to the Model S101 Colloid/Particle Fractionator from FFFractionation, Inc. (Salt Lake City, UT). The channel has dimensions of 90 cm in length, 2 cm in breadth, and 0.0127 cm in thickness. All work was done at room temperature ($22 \pm 1°C$). The spin rate used in these experiments was 1800 rpm, producing 544 gravities of acceleration.

The FFF system was equipped with a Kontron HPLC pump (Kontron, London, UK) to pump the carrier liquid into the channel. The eluted sample components were detected by a UV detector (Model 757 Absorbance Detector, Applied Biosystems, Ramsey, NJ) with the wavelength set at 280 nm. The detector signal was recorded by a chart recorder, stored in a computer, and analyzed by a special program written in this laboratory.

The particles used for the coating work were 0.215 µm polystyrene latex beads with surfaces specially cleaned for protein coating (Polyscience, Inc., Warrington, PA). Human IgG was purchased from Sigma Chemical Company (St. Louis, MO).

The carrier was a 10 mmol Tris[hydroxy methyl] aminomethane (United States Biochemical Corp., Cleveland, OH) solution. The pH of the solution was adjusted to 9.0 using HNO_3 (J. T. Baker Chemical Co., Phillipburg, NJ). Phosphate buffer solutions of various pH values were used for sample preparation in this work. They were prepared by using proper ratios of Na_2HPO_4 and NaH_2PO_4 (Mallinckrodt Specialty Chemicals Co., Paris, KY). The actual pH of the solutions were determined using a Model 630 pH meter (Fisher Scientific, Fair Lawn, NJ).

Preparation of Coated Particles for FFF Analysis. The 2.5% (*w/v*) original particle suspension was first diluted to 0.0125% with triple distilled water. The IgG sample was dissolved in 20 mM phosphate buffer solution of the required pH value to yield 20 mg/mL stock solutions. The stock solutions were then diluted to the desired concentration in the same buffer solution. A 50 μL aliquot of each protein solution was added to a 50 μL 0.0125% polystyrene latex suspension. The mixture was then allowed to stand at room temperature (22 ± 1°C) for various specified intervals before sedimentation FFF analysis.

Approximately 10 or 20 μL of sample was injected directly into the FFF channel. Following the injection, the flow was stopped for 4 minutes for relaxation, a process to allow the particles to reach an equilibrium state in the channel. After relaxation, the flow was resumed and the run proceeded under various specified conditions.

Results and Discussion

Precision in Measurement of Coating Mass. The mass of protein adsorbed m_c is calculated from the measured SdFFF retention time (t_r) for the coated bead using equations 7 and 12. In order to assess the errors in m_c and the minimum adsorption level detectable by the method, 37 runs were made over 3 days on a 0.215 μm diameter polystyrene under the same operating conditions of field strength (544 gravities, 1800 rpm) and flowrate (2.0 mL/min) as used generally in these studies. The measured t_r values and standard deviations obtained for each day's work are given in Table I. The standard deviation for the entire data set was 0.06 min. This shows that the precision in measuring t_r is less than 1%. There was a small (0.08 min) difference in the mean of the results for day 2 compared to the mean for days 1 and 3, which were identical. Equation 12 can be used to obtain an expression for the standard deviation of the calculated coating mass σ_{m_c}

Table I. Retention Time t_r and Standard Deviation σ_{t_r} (Both in Minutes) of 0.22 μm PS at 1800 RPM and a Flowrate of 2.0 mL/min

	Day 1	Day 2	Day 3
t_r (min)	7.233	7.330	7.267
	7.167	7.333	7.267
	7.267	7.300	7.267
	7.300	7.265	7.267
	7.200	7.330	7.233
	7.200	7.267	7.233
	7.300	7.330	7.233
	7.233	7.267	7.267
	7.267	7.360	7.200
	7.233	7.330	7.200
	7.253	7.400	7.233
	7.233	7.400	7.167
	7.237		
t_r (min) (average)	7.24	7.32	7.24
σ_{t_r} (min)	0.038	0.046	0.033

$$\sigma_{m_c} = \frac{6kT\rho_c}{wt^0 G\Delta\rho_c}\sigma_{t_r} \tag{14}$$

This yields a value for σ_{m_c} of 1.5×10^{-18} g for our experimental conditions, assuming $\rho_c = 1.37$ g/mL. For human IgG, which has an average molecular weight of 158,500, σ_{m_c} corresponds to ~6 molecules. If the measured mass of the coated particles is $\geq 3\,\sigma_{m_c}$ greater than the mass of the uncoated particles, then the difference in mass which we attribute to the adsorbed protein is highly significant statistically. Thus the coating mass that can be determined with relative certainty by the SdFFF technique at a field strength of 544 gravities is only 4.5×10^{-18}g, a mass that corresponds to about 17 protein molecules. This value indicates the high sensitivity of the method to the adsorbed amount.

Under conditions of adsorbate and adsorbent concentrations where the change of the protein concentration is small, the conventional depletion method for studying adsorption has limitations. In SdFFF, the adsorbed mass is determined directly from the retention time or retention time increment and not indirectly by the difference in the concentration of protein remaining in the solution. The high sensitivity of SdFFF makes the method suitable for the investigation of protein adsorption over a large range of conditions which should be particularly useful in probing the low adsorption-density end of the isotherm and for adsorption isotherms where the adsorption is strong but the equilibrium concentration in solution is below the detection limit for the analytical method being used.

Determination of Adsorption Isotherms. As the concentration of IgG added to the latex suspension increases, the peak generated by the latex is shifted to higher retention times, indicating an increase in mass of the beads. We attribute this increase to the adsorption of IgG on the particle surface. Using equation 7 or 12, the adsorbed mass m_c may be calculated either from the measured force (F) exerted on the particles in each of the experiments or from the time increment Δt_r. The adsorption density X, expressed as mass adsorbed per unit area, may also be calculated based on m_c and the estimated surface area of the particle (1.45×10^{-13} m^2). In addition, we also calculated the thickness of the adsorbed layer by means of equation 13 (or equation 8). (The calculated h is, of course, only a mean thickness for a compact coating of protein.) These data are summarized in Table II.

The adsorption isotherm (X vs C_{eq}) of IgG on the surface of PS latex beads is displayed in Figure 4a. The equilibrium concentration of IgG, C_{eq}, remaining in the solution after the adsorption is finished was obtained by subtracting the mass adsorbed to the particles in unit volume of solution from the initial concentration. In most cases, the initial concentration and the equilibrium concentration differ only slightly.

The isotherm obtained can be fitted to the Langmuir equation

$$X = \frac{aC_{eq}}{1 + abC_{eq}} \tag{15}$$

where a and b are constants. Equation 15 can be rearranged to give

$$\frac{C_{eq}}{X} = bC_{eq} + \frac{1}{a} \tag{16}$$

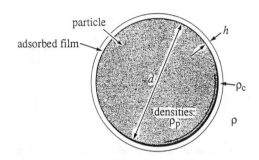

Figure 3. Illustration of an idealized particle with adsorbed coating of thickness *h*. (Reprinted with permission from *Langmuir*. Copyright 1991 American Chemical Society.)

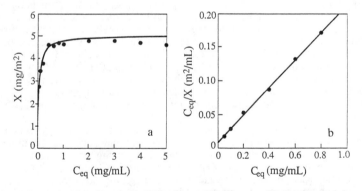

Fig 4. Adsorption isotherm plots for human IgG onto PS latex particles: (a) adsorption density X versus equilibrium solution concentration; (b) Langmuir plot C_{eq}/X versus C_{eq}. The line in Figure 4a is a Langmuir model curve plotted using the parameters a and b obtained from the regression line in Figure 4b.

Table II. Compilation of Mass m_c, Adsorption Density X, and Thickness h of Film of IgG on Polystyrene Latex Beads for Different Protein Concentrations

C_{eq}	λ	m_c (×10^{16})	X (mg/m²)	h (Å)
0.05	0.0164	3.97	2.75	19.9
0.10	0.0152	5.04	3.47	25.3
0.20	0.0147	5.54	3.81	27.9
0.40	0.0137	6.65	4.58	33.4
0.60	0.0137	6.65	4.58	33.4
0.80	0.0136	6.77	4.66	34.0
1.0	0.0136	6.77	4.66	34.0
2.0	0.0135	6.89	4.74	34.6
3.0	0.0135	6.89	4.74	34.6
4.0	0.0136	6.77	4.66	34.0
5.0	0.0137	6.65	4.58	33.4

Thus the Langmuir constants a and b can be obtained from the intercept and slope, respectively, of a plot of C_{eq}/X versus C_{eq}. The straight line plot shown in Figure 4b demonstrates that the data follow the Langmuir form. Using the constants so obtained ($a = 109$, $b = 0.20$) the solid line in Figure 4a was plotted.

The plateau in the Langmuir adsorption isotherm is $1/b$. For the case of IgG adsorbed to polystyrene at pH 7 and ionic strength 0.01 M, the maximum (plateau) adsorption level was found experimentally to be 4.75 mg/m². This is in agreement with the values found for adsorption of globular proteins, which usually have a value of a few mg/m² depending on the nature of the protein and the particle surface (*11, 12*). Assuming an average molecular weight of 158,500 Dalton, the adsorption density above corresponds to about 2600 IgG molecules per particle. The average area occupied by a protein molecule is thus about 56 nm², which is also within the range reported in the literature of about 10-100 nm² (*11*). This area is equivalent to a square with a side length of 7.4 nm. Since a spherical particle of the same mass and density as an IgG molecule will have a diameter of about 7.2 nm, we concluded that the protein molecules are rather tightly packed on the particle surface.

In our earlier work on protein adsorption (*1*), PS latex beads were used without any further clean-up procedure. For these beads, residual surfactant (sodium lauryl sulfate) is probably present on the surface due to the procedure used in the preparation of the latex particles (*13*). The adsorption behavior of γ-globulin on these particles was studied. We found that, although the isotherm is also similar to that of a Langmuir type, the plateau adsorption was reached at a much higher concentration. In the present study we have used particles prepared without surfactant. The isotherms of these two type of particles are shown in Figure 5. The figure shows that the plateau adsorption values on the surface of particles with and without the surfactant are fairly close, being 2.94 and 3.10 mg/m², respectively. However, the concentrations of proteins required to reach the plateau adsorption differ by about one order of magnitude, approximately 0.3 mg/mL for the clean surface and 4 mg/mL for the surface with surfactant. The results suggest that the protein and the surfactant compete for adsorption sites on the surface of the latex beads.

Influence of pH on the Adsorption Maximum. The maximum adsorption density for the IgG-polystyrene system was measured at different solution pH values but with the ionic strength maintained constant at 0.007 M. The results, which are plotted in Figure 6, display a small but distinct peak in the adsorption maximum occurring near pH 7. The adsorption density at the pH of this peak is at least 25% above the values obtained outside the pH range 5.5-9.5. Since the isoelectric point of various IgG proteins is reported to be in the pH range 5.8-7.3 (*14*) it is likely that the maximum in the adsorption density measured here occurs at the isoelectric point for the sample. A number of reasons could contribute to this observation. For example, intermolecular charge repulsion would be minimized at the isoelectric point which would lead to less charge repulsion between adsorbed protein molecules on the surface. It is also possible that reduced intramolecular repulsion could lead to a contraction of the protein molecule, leading to a smaller cross sectional area per molecule. In addition, protein-surface charge repulsion could also contribute to the reduction in adsorption density at any pH above the isoelectric point as the latex surface would be negatively charged under the conditions studied.

Adsorption Characteristics. The adsorption of IgG onto polystyrene is quite rapid with adsorption times generally being less than 10 minutes, sometimes much less.

With increasing IgG concentration, the adsorption density increases in a series of quasi-equilibrium steps. Figure 7 illustrates that the same adsorption density is obtained irrespective of whether the IgG is added in one step or two. In this series of experiments, 0.05 mg/mL IgG was equilibrated with polystyrene beads for 60 min and then the IgG concentration was increased to 1 mg/mL. The adsorption process was followed by making SdFFF runs after specific elapsed times. The results for this two step adsorption were then compared (Figure 7) with those from a single adsorption with 1 mg/mL IgG.

Protein adsorption can be accompanied by a relaxation process where the molecular conformation changes over time in such a way that the proteins occupy a higher area per molecule than on initial adsorption (*15*). This could possibly lead to a lower amount adsorbed in the two step adsorption experiment. The fact that the same adsorption density was achieved in two steps as in one would appear to indicate that either this surface conformational change of the IgG molecules is reversible or that the adsorbed molecules can undergo a further rearrangement which reduces the area occupied per molecule when the surface becomes crowded.

Protein Desorption. The SdFFF adsorption method will only measure relatively strongly bound protein as during elution (typically 5-20 min) the particle surface is continually washed with fresh carrier solution. Desorption of the protein over longer time periods was tested in a series of experiments where initial adsorption was carried out in solutions of 1 mg/mL IgG and then the suspension was diluted to give 0.025 mg/mL. After various periods of time at the lower concentration, SdFFF runs were performed to measure the amount of IgG remaining on the surface.

The results of these experiments are plotted in Figure 8. Slow desorption to about 20% of the original IgG adsorbed mass occurred over about 2 hours. However, negligible desorption occurred after this for periods up to 6 hours.

Conclusions

New results are presented here on the direct measurement of human IgG adsorption on polystyrene latex beads using sedimentation FFF. The results confirm the high sensitivity and overall efficacy of sedimentation FFF used as a tool for measuring adsorption on colloidal particles.

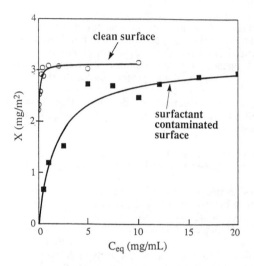

Figure 5. Comparison of isotherm of human γ-globulin on the surface of PS latex spheres with and without surfactant on the surface. The plateau adsorption is reached at a much lower protein concentration on the clean surface.

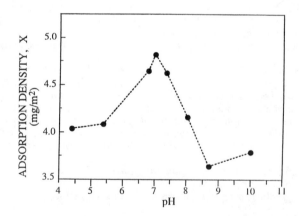

Figure 6. Maximum adsorption density X for human IgG on PS latex particles versus solution pH. Ionic strength was maintained constant at 0.007 M.

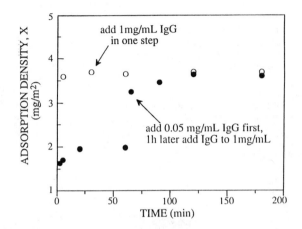

Figure 7. Adsorption density X versus time for 0.05 mg/mL and 1 mg/mL IgG additions (open symbols) and for a two-step adsorption in which the concentration was increased to 1 mg/mL after a preliminary adsorption for 1 hour at 0.05 mg/mL (closed symbols).

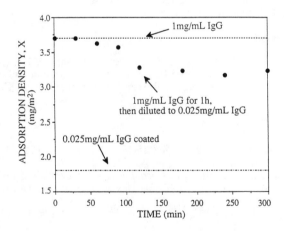

Figure 8. Adsorption density versus time for PS coated at 1 mg/mL IgG followed by dilution of the protein concentration to 0.025 mg/mL.

Glossary

a	constant in equation 15
b	constant in equation 16
C	concentration
C_{eq}	equilibrium concentration
d	particle diameter
F	force exerted on a single particle
F_p	force exerted on a bare particle
G	field strength measured as acceleration
G_p	acceleration of a bare particle
h	coating thickness
k	Boltzmann's constant
ℓ	equilibrium concentration thickness
m'	effective particle mass
m_c	mass of particle coating
m_p	particle mass
r_0	centrifuge radius
t^0	void time
t_r	retention time
T	absolute temperature
X	adsorption density
w	channel thickness

Greek

$\Delta\rho_c$	density difference between coating and carrier
$\Delta\rho_p$	density difference between particle and carrier
ρ_c	density of particle coating
ρ_p	particle density
σ_{m_c}	standard deviation of calculated mass of coating
ω	rotational velocity of centrifuge

Acknowledgments

This work was supported by Public Health Service Grant GM10851-37 from the National Institutes of Health in the U.S. and by the Australian Research Council and the Land and Water Resources Research and Development Corporation in Australia.

Literature Cited

1. Beckett, R.; Ho, J.; Jiang, Y.; Giddings, J. C. *Langmuir* **1991**, *7*, 2040-2047.
2. Li, J. T.; Caldwell, K. D. *Langmuir* **1991**, *7*, 2034-2039.
3. Langwost, B.; Caldwell, K. D. *Chromatographia* **1992**, *34*, 317-324.
4. Caldwell, K. D.; Li, J.; Li, J.-T.; Dalgleish, D. G. *J. Chromatogr.* **1992**, *604*, 63-71.
5. Giddings, J. C.; Yang, F. J. F.; Myers, M. N. *Anal. Chem.* **1974**, *46*, 1917-1924.
6. Giddings, J. C.; Caldwell, K. D. In *Physical Methods of Chemistry*; Rossiter, B. W.; Hamilton, J. F., Eds.; Wiley: New York, 1989, Vol. 3B; pp 867-938.
7. Giddings, J. C. *Science* **1993**, *260*, 1456-1465.
8. Martin, M.; Williams, P. S. In *Theoretical Advancement in Chromatography and Related Separation Techniques*; Dondi, F.; Guiochon, G., Eds.; NATO

ASI Series C: Mathematical and Physical Sciences; Kluwer: Dordrecht, 1992, Vol. 383; pp 513-580.

9. Beckett, R.; Nicholson, G.; Hotchin, D. M.; Hart, B. T. *Hydrobiologia* **1992**, *235/236*, 697-710.
10. Giddings, J. C.; Myers, M. N.; Caldwell, K. D.; Fisher, S. R. In *Methods of Biochemical Analysis*; Glick, D., Ed.; Wiley: New York, 1980, Vol. 26; pp 79-136.
11. Andrade, J. D. In *Surface and Interfacial Aspects of Biomedical Polymers*; Andrade J. D., Ed.; Plenum: New York, 1985, Vol. II; pp 1-80.
12. Kondo, A.; Oku, S.; Higashitani, K. *Biotechnol. Bioeng.* **1991**, *37*, 537-543.
13. Bangs, L. B. In *Uniform Latex Particles*; Seradyn: Indianapolis, IN, 1984.
14. *CRC Handbook of Biochemistry: Selected Data for Molecular Biology*; Sober, H. A., Ed.; The Chemical Rubber Co.: Clevelend, OH, 1968; pp c-39.
15. Lundstrom, I. *Progr. Colloid Polym. Sci.* **1985**, *70*, 76-82.

RECEIVED December 22, 1994

Chapter 30

Modification of Silica with a Covalently Attached Antigen for Use in Immunosorbent Assays

Kathryn A. Melzak[1] and Donald E. Brooks[1,2]

[1]Department of Chemistry, University of British Columbia,
2036 Main Mall, Vancouver, British Columbia V6T 1Z1, Canada
[2]Department of Pathology, University of British Columbia,
2211 Wesbrook Mall, Vancouver, British Columbia V6T 2B5, Canada

This paper describes the preparation and characterization of a silica surface with a covalently coupled antigen, and gives an example of the use of the modified silica in immunosorbent assays. Silica beads and slides were cleaned with chromic acid and modified with 3-aminopropyltriethoxysilane. The carboxyl groups of a cyclic hexapeptide antigen, ferrichrome A, were coupled to the amine-modified silica using 1-ethyl,3,3-dimethylaminopropyl cabodiimide to promote formation of a peptide bond. Modified slides and beads were characterized with a combination of X-ray photoelectron spectroscopy and particle electrophoresis measurements, with the amount of ferrichrome A adsorbed or coupled to the beads being determined by solution depletion measurements.

Solid phase immunosorbent assays are widely used in many different formats, because the solid surface provides a convenient method for separating the bound antibody-antigen complex from the unbound reagents (*1,2*). If an antigen is adsorbed or coupled to a solid surface, then antibodies towards that antigen will bind to the surface, while other antibodies can be washed away. The surface-bound antibody can be detected and used to calculate the solution concentration by comparison to a set of standards.

The antigen must be attached to the surface by a stable bond, so that the antigen does not desorb during the course of the assay. Immunosorbent assays commonly involve antigens non-specifically adsorbed to a hydrophobic surface (*3*). Larger antigens will often adsorb in an effectively irreversible manner because of multiple attachment points (*4*), but smaller molecules may not bind, or may desorb slowly in the presence of molecules competing for adsorption sites (*5*). Covalent attachment has the advantage of preventing the desorption of small antigens during the course of the assay.

Silica is a useful substrate for covalent attachment of antigens because it can be derivatised with a large variety of functional groups for covalent attachment of different antigens (*6, 7*). The silica surface can be modified by reaction with compounds of the form $RnSiX(4-n)$, where X is a hydrolyzable group and R contains the functional group to be used for covalent attachment of the antigen.

0097–6156/95/0602–0420$12.00/0

The silanols formed after hydrolysis of the silane condense with silanols on the silica surface, leaving a surface modified with the remaining group R.

This paper describes the preparation and characterization of silica surfaces with a covalently coupled antigen and gives an example of their use in an immunological assay. The silica was cleaned and modified with 3-aminopropyl-triethoxysilane to give a surface covered with amino groups. A water soluble carbodiimide, 1-ethyl-3-(3-dimethylaminopropyl carbodiimide) (EDC) was used to promote formation of a peptide bond between the surface amine groups and the carboxyl groups on an antigen, to give the antigen-modified surface that could be used for immunological assays.

The antigen used was ferrichrome A, a cyclic hexapeptide that chelates iron through three hydroxamate groups. The coupling of the antigen to the silica involved several steps: cleaning of the silica, silylation, and EDC mediated coupling of ferrichrome A to the amine-modified silica. The modified surface was characterized after the various steps using different techniques.   X-ray photoelectron spectroscopy was used to determine the efficiency of the cleaning procedure prior to modification with the silane and also to determine the extent of modification with the silane on flat silica. A ninhydrin assay was used to detect and measure the amines on silica beads after silylation.   The amount of ferrichrome A adsorbed or coupled to the silylated silica surface was determined by solution depletion measurements using the beads, since they had a large specific surface area. Particle electrophoresis measurements were used to monitor modification of the silica beads.

The antigen-modified beads and slides were used for immunosorbent assays with a monoclonal antibody against ferrichrome A.

Experimental

Silica beads and slides. Silica beads with a 0.8 μm diameter and flat silica slides were both used. They were initially intended to give complementary information about the same system, since the beads had a high specific surface area and could be used for measuring adsorption and coupling of the antigen while the slides had a flat surface for angularly resolved X-ray photoelectron spectroscopy (XPS) studies. The antibody-binding properties differed for the beads and the slides, implying that they should be treated as related but separate systems. There was also other evidence that the beads and slides were different: the beads had a density of 1.83 g cm^{-3}, and the slides a density of 2.1 g cm^{-3}. The lower density of the beads may be due to alkyl groups incorporated during the synthesis or to formation of a porous stucture.

Silica beads were prepared by condensation from tetraethyl orthosilicate (TEOS, Aldrich, Milwaukee) in a solution of ethanol, water and ammonia (*8, 9*). The reaction mixture contained ethanol, water and TEOS in proportions of 150:30:6 by volume (*8*). The ethanol/water solution was saturated with ammonia at 4° and then equilibrated at 5 °C in an ethylene glycol bath. Distilled TEOS was added through a syringe to a solution stirred at approximately 400 rpm. The addition was made as rapidly as possible, and took less than two seconds. After the reaction had been allowed to proceed overnight, the beads were collected by centrifugation.   The bead size distribution was determined using transmission electron microscope photographs of a total of 314 beads. An average diameter of 0.779 μm was calculated, with a standard deviation of 0.042 μm.

Flat silica slides were obtained from Quartz Scientific International (Fairport Harbor). The slides were cut into 8x8 mm sections, wiped clean with lens paper and placed in separate 13x100 mm test tubes. All cleaning and drying procedures were carried out in these test tubes to minimize sample handling. Some preliminary measurements were carried out with glass slides (Canlab, Mississauga, Ont.).

Cleaning the silica. The silica was cleaned before modification with the silane so that silanol groups on the silica would be accessible to the modifying reagents and not covered with a contaminating layer. Silica beads and slides were both cleaned by being heated for one hour in chromic acid at 80 °C, followed by rinses in water, concentrated hydrochloric acid and water, before being dried under vacuum. The chromic acid was prepared by slow addition of a slurry of 5 g of potassium dichromate and 5 g of water to 100 ml of concentrated sulfuric acid (BDH Assurance, BDH, Vancouver) (*10*). The solution was stirred in a water bath to dissipate the excess heat generated during the addition of the slurry to the acid. Chromic acid was used repeatedly before being disposed of (*11*).

The hydrochloric acid used was BDH AnalaR (BDH) and the water was distilled and then deionized using a Millipore milli-Q ion exchange apparatus.

Silylation of the silica. All glassware was cleaned with chromic acid before use. Glassware used for the silylations was stored separately and was rinsed immediately following use to ensure that it remained clean and that there were no deposits of organic material on the glass during the chromic acid wash. Freshly distilled 3-aminopropyltriethoxysilane (Hüls America, Bristol, PA) was added to heptane (BDH OmniSolv) at a 1% w/v ratio and stirred briefly before addition of the silica beads or slides. The reaction was carried out in a 250 ml round bottom flask stirred from the bottom with a magnetic stirrer. The silica beads dispersed readily in the silane solution. The solution was heated to reflux for one hour, after which the silica was rinsed in fresh heptane, dried under vacuum and stored overnight at room temperature (22 °C) under vacuum. The modified silica was then sonicated in water for one minute in an attempt to remove any polymerized clumps of silane adsorbed to the surface and dried under vacuum after removal of most of the water with a pipette.

Modification of silica with silanes has been carried out in a variety of solvents (*12, 13*). Heptane was used to decrease the water in solution , to try to minimize the solution polymerization of the silane. Water present in the heptane and water adsorbed to the silica surface would be available for the initial hydrolysis of the silane.

XPS measurements. X-ray photoelectron spectroscopy measurements were made using a Leybold Heraeus MAX 200 spectrometer. Silica or glass slides were mounted on the sample holder using copper strips and transferred through air to the loading chamber of the XPS machine, which was evacuated overnight to a final pressure of approximately 1×10^{-8} mbar. The samples were then transferred to the analysis chamber and analyzed at a pressure of 8×10^{-9} mbar using a 2x4 mm sample area near the centre of the slide, chosen visually to avoid the copper strips. Samples were analyzed using an Alkα 1486.6 eV source (excitation voltage 15.0 kV, emission current 25.0 mA) or with a Mgkα 1253.6 eV source (excitation voltage 15.0 kV, emission current 20.0 mA). Ejected photoelectrons were measured normal to the surface of the silica unless otherwise specified. The kinetic energy of the photoelectrons was measured using a hemispherical analyzer at a constant pass energy of 192 eV. The carbon 1s peak at 285.0 eV (*13*) was used for charge correction of the binding energies.

The relative amounts present were calculated for all elements detected with photoelectrons in the binding energy range of 100-1100 eV (Mgkα X-ray source) or 100-1300 eV (Alkα X-ray source). Areas under characteristic peaks for each element were measured with baselines determined visually. Atomic percentages for the elements were calculated using the elemental sensitivity factors provided by the instrument manufacturer.

Particle electrophoresis. The electrophoretic mobility μ of silica particles was measured in order to calculate the surface charge density (*15, 16*). Particles were suspended in 25 mM NaCl and the suspension was added to the chamber of a Rank Mark I particle electrophoresis apparatus (Rank Bros., Cambridge, U.K.). A voltage was applied across the sample (40 V, electric field 4 V/cm) and the movement of particles through a narrow cylindical chamber was viewed and measured with a microscope focussed at the stationary layer where the movement of the particles is not affected by electroosmotic flow through the chamber (*17*).

The electrophoretic mobility was calculated from the transit time of the particles across a grid in the microscope eyepiece:

$$\mu = \frac{1}{t} \frac{D \, Le}{V}$$

where t = transit time, D = size of grid division in μm, Le = electrical path length in cm, determined as described in (*16*) and V = applied voltage. The surface charge density σ can be estimated from the electrophoretic mobility using the following expressions (*15*):

$$\sigma = \frac{(2 \, \varepsilon \, k \, T \, no)^{1/2}}{\pi} \quad \sinh \frac{z \, e \, \zeta}{2kT}$$

where
σ = surface charge density (esu cm^{-2})
ζ = zeta potential = $(4 \, \pi \, \eta \, \mu) / \varepsilon$
ε = dielectric constant of suspending medium
k = Boltzmann constant
T = absolute temperature
n_0 = (number of ions)/cm^3 in the bulk solution
z = valence of the ionic species in the suspending medium
e = electron charge
η = viscosity of the suspending medium (Pa-s)

Ninhydrin assay. Ninhydrin assays were used to determine amines on the modified beads (*18*). A ninhydrin solution (0.5 ml 10 mM ninhydrin in 100 mM sodium acetate pH 5.0) was added to 10 mg of modified beads (ninhydrin from Sigma, St. Louis MO and sodium acetate from Fisher, Fairlawn, NJ). The beads were sonicated but remained in small clumps due to the high salt concentration. The beads and ninhydrin were then placed in a boiling water bath for 5 minutes, during which time the beads turned blue. When the beads were shaken, the blue colour came off into solution, leaving the beads white. The remainder of the amine-ninhydrin reaction product was washed off the beads with a 50 % v/'v solution of ethanol in water. The supernatants from the beads and the ethanol rinse were collected and diluted to one ml. The absorbance of this solution at 570 nm was then used to determine the concentration of amines on the surface, with glycine (Fisher) used as a standard.

Isolation of ferrichrome A. Ferrichrome A was isolated (*19*) from the culture supernatant of *Ustilago sphaerogena* Burril obtained from the American Type Culture Collection (cat. no. 12421). The initial freeze dried fungus was grown in 10 ml of a low iron medium (*19*) in a 125 ml Erlenmeyer flask using a reciprocal shaker at 30 °C. Larger amounts of culture used for isolation of ferrichrome A were grown in 500 ml of medium in a 2.8 litre Fernbach flask. Three or four days after innoculation of the large flask, the cells were removed by centrifugation. Ferric chloride was added to the culture supernatant to produce the coloured iron hydroxamate. The supernatant was then saturated with ammonium sulphate and extracted with benzyl alcohol. After addition of diethyl ether to the organic phase, the ferrichrome A was extracted using water and was obtained as a dark red crystalline material. The ferrichrome A was purified by recrystallization from water.

Ferrichrome A was chosen as an antigen because of several useful features: the ferric hydroxamate group absorbs at 436 nm and can be used to determine the peptide concentration, there are three carboxyl groups per molecule for covalent attachment to the amine-modified silica, iron can be distinguished by X-ray photoelectron spectroscopy and the isolation results in a high yield of peptide which can be readily purified.

EDC coupling of ferrichrome A to silica beads. The EDC was either added to ferrichrome A at a constant mole ratio of 50:1 or was added at a constant concentration of 20 mM. The EDC and ferrichrome A solution was mixed at room temperature (22 °C) for 40 s, the shortest convenient time, before being added (0.75 ml) to dry silica beads weighed out in a polyethylene centrifuge tube. Immediately thererafter the beads were sonicated and vortexed until a homogeneous suspension was obtained. The suspension was mixed at room temperature for ten minutes before the beads were separated by centrifugation. The absorbance of the supernatant was used to calculate the initial amount of ferrichrome A associated with the beads (ε_{436}= 2.778 ml mg^{-1} cm^{-1}). The amount of ferrichrome A remaining on the beads after a rinse with phosphate buffered saline (0.15 M NaCl, 10 mM phosphate, pH 7.2; PBS) was also determined, by measuring the absorbance of the PBS rinse.

Monoclonal antibodies. A monoclonal antibody (*20*) was raised against ferrichrome A using the procedure from Agriculture Canada (Vancouver) (*21, 22*). Ferrichrome A coupled to keyhole limpet haemocyanin (Calbiochem) using EDC was used as the initial immunogen. The myeloma cells used were FOX-NY cells (ATCC CRL 1732) and the mice used for the immunizations and spleen cells were BALB/c. The antibody was purified from culture supernatant of myeloma cells grown in a supplemented serum-free medium (*23*) using a protein A column (Pharmacia, Dorvall) (*24*). The antibody isotype was IgG2b with κ and λ light chains. The κ light chains presumeably come from the FOX-NY myeloma cells (*25, 26*) and the λ light chains from the immunized mouse.

Immunological assays. Silica beads and slides modified with EDC-coupled ferrichrome A were both used for immunological assays to detect either antibody or antigen in solution. The surface area of silica used for each measurement was about 74 mm^2. The beads or pieces of cut slides were placed in separate polyethylene centrifuge tubes and incubated for 30 minutes in PBS with 0.2 % bovine serum albumin (BSA) to block non-specific protein binding sites. For assays to determine antibody concentration, the silica was then rinsed in PBS, incubated at 37 °C for one hour with an appropriate dilution of antibody in PBS with 0.2% BSA, rinsed again and incubated with a second anti-mouse IgG enzyme linked antibody (horseradish peroxidase (HRP), goat anti-mouse, Cappel, Cooper Biomedical, West Chester PA). After a final set of rinses, the HRP remaining on the beads was detected using tetramethylbenzidine (TMB) as a substrate, since HRP in the presence of peroxide will act on TMB to give a coloured product (*27*). Results were presented as absorbance (of the TMB product) vs. initial solution concentration of antibody. For assays to determine the solution concentration of antigen, a fixed concentration of antibody was incubated with a serial dilution of antigen and added to the modified silica after the initial blocking step. The remainder of the assay was the same as for detection of antibody. Results are presented as absorbance vs. antigen concentration. For assays to determine the solution concentration of antigen, a fixed concentration of antibody was incubated with a serial dilution of antigen and added to the modified silica after the initial blocking step. The remainder of the assay was the same as for detection of antibody. Results are presented as absorbance vs. antigen concentration.

Results and Discussion

Cleaning of silica. The chromic acid cleaning procedure was shown to be efficient, as determined by XPS measurements showing the decrease in carbon on the surface (Table I). Since there should be no carbon in the silica or glass slides that were used, the amount of carbon left on the surface is a good indicator of the effectiveness of the cleaning procedure. Initial measurements were carried out with glass slides and showed that the amount of carbon on the surface decreased from 15.5 % on the uncleaned glass to 4 % on the glass that had been cleaned with chromic acid, rinsed with hydrochloric acid and water and dried in vacuum (Table I). The hydrochloric acid rinse was necessary to remove sulfur from the sulfuric acid used in the preparation of chromic acid. Slides dried in air instead of in a vacuum did not show a significant decrease in the amount of surface carbon, implying that the samples must be stored under vacuum to maintain cleanliness.

Table I. Surface Elemental Composition of Glass and Silica after Different Cleaning Procedures

| | Atomic % of the different elements present | | | | | | |
	Si	O	C	S	Na	Mg	Ca
SiO_2 (theoretical)...............		33.33	66.67				
Glass slides, not cleaned....	20.73	52.31	15.53		7.58	2.02	1.83
Glass, chromic acid cleaned and dried in vacuum............	22.15	54.82	15.35	3.99	1.56	1.47	0.68
Glass cleaned with chromic acid and HCl, dried in air....	26.54	57.34	13.00	-	0.39	1.65	1.09
Glass cleaned with chromic acid, HCl, dried in vacuum.	25.29	64.67	5.65	-	3.04	0.75	0.61
SiO_2 cleaned with chromic acid, HCl, dried in vacuum.	32.55	63.40	4.05	-	-	-	-
SiO_2 cleaned with chromic acid, HCl, dried in vacuum, refluxed in heptane..............	33.66	61.64	4.69	-	-	-	-
SiO_2 cleaned with HNO3, dried in vacuum...................	13.81	84.08	2.11	-	-	-	-

A nitric acid cleaning procedure was also tried, since it has been reported to give better results than cleaning with chromic acid (*28*). Two silica slides cleaned with hot concentrated nitric acid, rinsed with water and dried in vacuum gave anomalous oxygen to silicon ratios (Table I) so the procedure was discontinued.

Refluxing the chromic acid-cleaned slides in heptane, the solvent used in the silylations, did not significantly increase the amount of carbon on the slides.

Silylation of silica. The XPS measurements of the silylated silica slides showed an elemental composition of 31.5 % Si, 51.6 % O, 14.5 % C and 3.35 % N, an increase in the amount of carbon and a confirmation of the presence of nitrogen on the surface. It can be seen that the carbon and nitrogen are added in the 3:1 ratio expected for addition of an aminopropyl groups to the surface.

X-ray photoelectron spectroscopy is surface-sensitive because electrons do not travel very far through solids. Photoelectrons produced too far away from the surface are scattered inelastically by neighbouring atoms and are not ejected from the surface. The sampling depth will depend on the nature of the material being measured and can vary from 10 to 50 Å. The greatest sampling depth will be achieved when the ejected photoelectrons are measured normal to the surface. If the photoelectrons are measured at a shallow angle, θ degrees away from the normal, then electrons travelling the maximum distance d through the solid will be coming from a sampling depth $d\cos\theta$ away from the surface. Angularly resolved measurements can therefore be used to obtain depth profiles. They will also give information about the organization of overlayers on the surface (*29, 30*). If a substrate is covered by a continuous overlayer, elements from the underlying substrate will not be detected at shallow angles where the sampling depth is less than that of the adsorbed layer. If the overlayer is discontinuous, then there will be regions of substrate not covered by the overlayer, and elements from the substrate will be detectable when photoelectrons are measured at larger angles away from the normal. Angularly resolved measurements with a discontinuous or rough overlayer can be difficult to interpret (*29*).

The bonding environment of an atom will affect the binding energy of core electrons, electronegative substituents increasing the binding energy (*31*). The silicon in silica is bonded to four oxygens. The silicon in the silane layer should be bonded to three oxygens and one carbon from the alkyl subtituent; the binding energy for the silicon in the silane layer should shift to a lower value. The Si2p spectra in Figure 1a show the expected shifts in binding energies. The silicon 2p electron from the cleaned, unmodified silica had a binding energy of 103.6 eV. When the photoelectrons were measured normal to the silylated surface where the sampling depth is the greatest, there was a shift of 0.3 eV downwards and at shallower angles, when the signal from the region near the surface is enhanced, there was a shift of an additional 0.6 eV. This shows that the surface of the silica slide is in fact modified after the reaction with the silane. It also shows that there are no large regions of uncovered silica, since the the signal from regions of unmodified silica would still be seen at shallower takeoff angles.

The XPS measurements on the silica slides show that the silylation procedure used was effective in modifying a silica surface. The same modification procedure was used for the beads as for the slides. The slides would be expected to have a more homogeneous surface than the beads, so the layer of silane on the beads may have corresponding irregularities. The beads were not characterized by XPS.

Ninhydrin assays on the freshly silylated beads gave a surface concentration corresponding to an area per amine group of 20 ± 6 Å^2, similar to the area per molecule of uncharged amines at an oil/water interface (*32*). There would be sufficient silane to cover the beads with a monolayer, if the bead surface were smooth. Porosity of the beads and formation of silane multilayers would increase the amount of silane on the beads and would decrease the area measured per amine group.

Particle electrophoresis measurements showed that the unmodified beads had a negative mobility of $-2.66 \pm 0.07 \times 10^{-4}$ cm^2 V^{-1} s^{-1}, corresponding to an

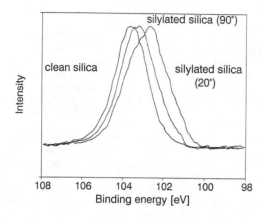

Figure 1a. The Si2p spectra of cleaned unmodified silica measured normal to the surface (90°) and of silylated silica measured at 90° and 20°.

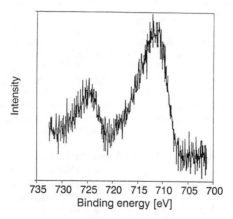

Figure 1b. The Fe2p spectra from silica slides coupled to ferrichrome A.

area per charge of 1200 Å². Silylation reversed the surface charge, and gave the beads a positive mobility of $5.41 \pm 0.25 \times 10^{-4}$ cm² V⁻¹ s⁻¹, with a corresponding area per charge of 472 Å². The surface charge density was lower than the value that would be expected from the ninhydrin assay. This may due to negatively charged contaminants adsorbing to the bead surface, to silane polymerization or to a shift in the location of the shear plane relative to the charge location.

Non-specific adsorption of ferrichrome A to silylated silica. Non-specific adsorption of ferrichrome A to the modified silica was studied to assist in selection of coupling conditions (data not shown, (33)). Ferrichrome A adsorbed readily to the silylated beads in water, with adsorption behaviour consistent with an ionic interaction. The adsorption was completely reversed by one wash in PBS and was completely inhibited by concentrations of 100 mM or greater of NaCl. Since ferrichrome A has three carboxyl groups per molecule, it woud be expected to interact electrostatically with the positively charged surface of the modified beads.

The maximum surface concentration of the non-specifically adsorbed ferrichrome A was 1.76 mg m⁻², giving an area per molecule of 99 Å². The crystal structure of ferrichrome A shows that the three carboxyl groups form a triangular face with an area of 70 Å² (34). If the ferrichrome A adsorbed with this face down and if the bead surface were smooth, as assumed in the calculations of bead surface area, the implied packing fraction of the ferrichrome A molecules on the surface is very high. The surface area of the beads was calculated assuming that they were smooth spheres. If the beads were porous, the surface area would be higher than calculated and the area per molecule would be correspondingly higher. The beads had a density of 1.83 g cm⁻³, lower than the density of fused silica (2.1 g cm⁻³) and possibly reflecting a more open structure. Although the amount of ferrichrome A on the beads will be described in terms of surface concentration throughout the remainder of this paper, it might be more realistic to give the amounts of ferrichrome A found on the beads in terms of weight of ferrichrome A per weight of beads.

EDC-mediated coupling of ferrichrome A to the silylated silica. The EDC-mediated coupling was carried out without added salt or buffer to maximize the adsorption of ferrichrome A to the beads. Since EDC itself has a net charge of +1 in solution, it acts to increase the ionic strength of the medium as its concentration is increased. The pH of the ferrichrome A-EDC solution after addition to the beads was 5.7, within the range of optimum pH for EDC coupling reactions (35). Decreasing the pH to 4.7 by addition of hydrochloric acid resulted in a decrease in the amount of ferrichrome A left on the bead surface.

The amount of ferrichrome A initially associated with the beads at different solution concentrations of ferrichrome A and the amount left on the beads after rinsing with PBS are shown in Figures 2 and 3 for a constant ratio of EDC to ferrichrome A and for a constant EDC concentration, respectively. Experiments carried out with different bead preparations resulted in plots with reproducible features but some variation in the surface concentration of ferrichrome A (data not shown). The amount bound using different bead preparations varied by about 7% at low solution concentrations of ferrichrome A and by about 15% at high solution concentrations.

As ferrichrome A bears a net charge of -3 at the pH of the reaction, it is concentrated in the electrical double layer near the positively charged surface. Reaction of each carboxyl group on ferrichrome A with EDC eliminates its charge while the group is activated, however, so the species bearing activated groups capable of forming covalent bonds with surface amines would be expected to be at a lower average concentration than those which were non-activated and fully charged. Hence, in Figure 2 there is a large discrepancy between the adsorbed and

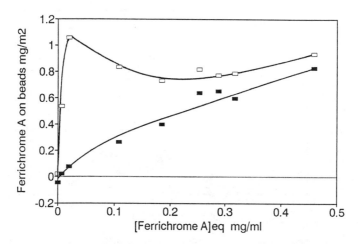

Figure 2.   Ferrichrome A adsorbed (□) and coupled (■) to silylated silica at
a constant EDC:ferrichrome A mole ratio of 50:1; surface concentration as
a function of solution concentration of ferrichrome A.

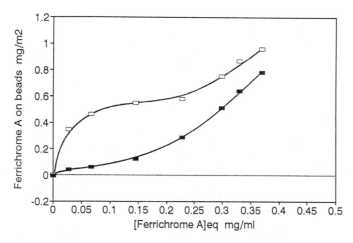

Figure 3.   Ferrichrome A adsorbed (□) and coupled (■) to silylated silica at
a constant EDC concentration of 20 mM.

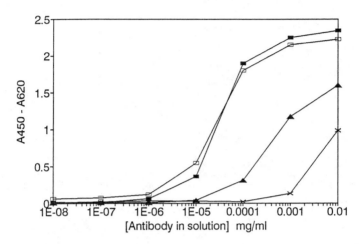

Figure 4. Immunoassays on silica beads modified with 0.83 mg m^{-2} ferrichrome A (■,□), 0.26 mg m^{-2} ferrichrome A (▲) and 0 mg m^{-2} ferrichrome A (×).

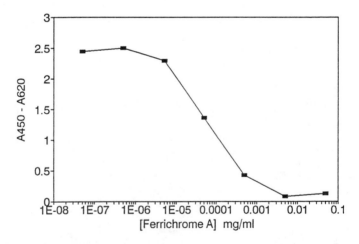

Figure 5. Immunoassays on silica beads modified with 0.83 mg m^{-2} ferrichrome A: inhibition of antibody binding with free ferrichrome A at an antibody concentration of 10^{-4} mg ml^{-1}.

coupled amounts at low ferrichrome A and EDC concentrations and therefore low ionic strengths where electrostatic effects are accentuated. The non-specific adsorption is highest at low EDC and hence low ionic strength, producing the initial peak in the isotherm. This peak was found to be reproducible in experiments carried out on different days (dataa not shown). The system in Figure 3 is at higher ionic strength at all points other than at the most concentrated solution of ferrichrome A, so the surface-associated amounts are reduced in comparison to Figure 2 and no peak is seen. The maximum surface concentration of ferrichrome A measured with a 50:1 ratio of EDC:ferrichrome A was 0.93 mg m^{-2}. After washing the beads with PBS, 89 % of the ferrichrome A was left on the beads, unlike the negligible amount that remained after the non-specific adsorption. Ferrichrome A remaining on the beads after the PBS rinse was taken to be covalently coupled. .

Other evidence for formation of a covalent bond is the decrease in amine groups on the bead surface after EDC-mediated coupling of acetic acid to the beads. The surface concentration of amines decreased from 8×10^{-6} mol -NH$_2$ m^{-2} on the silylated beads to 2.8×10^{-7} mol -NH$_2$ m^{-2} after acetylation. Acetic acid was used instead of ferrichrome A for these measurements so that the amount of product coupled to the surface would not be limited by cross-sectional area.

The iron in ferrichrome A coupled to silica slides could be detected by XPS (Figure 1b). Reaction conditions that gave a surface concentration of 0.83 mg m^{-2} ferrichrome A on the beads gave a surface composition of 0.3 % iron by XPS, using the iron 2p3/2 peak.

Immunosorbent assays. The most useful immunoassay results were obtained with beads modified with a high surface concentration of antigen (*33*). These gave a strong and reproducible response when detecting antibody in solution (Figure 4), and also permitted detection of the antigen through inhibition of antibody binding (Figure 5). Figure 4 shows the assays for two sets of beads modified with the same high surface concentration of ferrichrome A, for beads modified with a lower concentration and for beads with no ferrichrome A, to show the background results. The antibody binds to the beads with no ferrichrome A at solution antibody concentrations greater than 10^{-3} mg ml^{-1} (Figure 4).

An initial antibody concentration of 10^{-4} mg ml^{-1} was used for the inhibition assay shown in Figure 5. Ferrichrome A inhibited the assay response at solution concentrations down to 10^{-5} mg ml^{-1}, or 10^{-8} M.

A 10 mg bead sample could be modified using a total of 0.25 mg of ferrichrome A to give a surface concentration of 0.6 mg m^{-2}, as was used in the assay shown in Figure 5. This would provide enough beads for 560 data points under the conditions used.

Conclusions. All the steps in the procedure for coupling ferrichrome A to silica were shown to be efficient. The chromic acid cleaning followed by rinsing in HCl and drying in vacuum gave a clean surface with a low carbon content and no other detectable contaminants. The silylation procedure resulted in a high surface concentration of amines on the beads and the EDC-mediated coupling was simple to carry out and gave high surface concentrations of antigen. The coupled antigen formed a stable bond with the surface, as was desired for the immunosorbent assays.

Acknowledgments

The monoclonal antibody was raised with advice from Andrew Wieczorek of the Vancouver Research Branch of Agriculture Canada, whose help is greatly appreciated. We thank Dr. Johan Janzen for performing the electrophoretic mobility measurements and Dr. Philip Wong for the XPS analysis. This work was supported by Medical Research Council of Canada Grant MT 5759.

Literature Cited

1. *Manual of Clinical Laboratory Immunology;* Rose, N.R., Friedman, H.; Fahey, J.L. Eds.;4th Ed; American Society for Microbiology: Washington, D.C., 1992.
2. *Enzyme-Mediated Immunoassay;* Ngo, T.T.;Lenhoff, H.M., Eds. Plenum Press: New York, 1985.
3. Sorenson, K.; Brodbek, U.; *J. Immun. Methods*; **1986**, *95*, 291-293.
4. Andrade, J.D. In *Protein Adsorption*; Andrade, J.D., Ed.; Surface and Interfacial Aspects of Biomedical Polymers; Plenum Press: NewYork, 1985; Vol. 2, pp. 46-52.
5. Wojciechowski, P.; Ten Hove P.; Brash, J.L.; *J. Colloid Interface Sci.,* **1986**, *111*, 455-465.
6. *Silicon Compounds: Register and Review*; Petrarch Systems, Inc.; Bristol, PA, 1984.
7. Pluedderman, E.P. *Silane Coupling Agents*; Plenum Press: New York and London; 1982.
8. Tan, G.C.; Bowen, B.D.; Epstein, N.; *J. Colloid Interface Sci.*, **1987**, 118, 290-293.
9. Stöber, W.; Fink, A.; Bonn, E.; *J. Colloid Interface Sci.,* **1968**, 26, 62-69.
10. Furniss, B.S.; Hannaford, A.J.; Smith, P.W.G.; Tatchell, A.R. *Vogel's Practical Textbook of Organic Chemistry;* Longman Scientific and Technical: Harlow, England; 1989; p. 28.
11. Armour, M.A. *Hazardous Laboratory Chemicals Disposal Guide;* CRC Press, Boca Raton, FL; 1991, p. 118.
12. Arkles, B. In *Silicon Compounds: Register and Review;* Petrarch Systems, Inc.; Pristol, PA, 1984.
13. Chaimberg, M.; Cohen, Y.; *J. Colloid Interface Sci.,* **1990**, 134, 576-579.
14. Ratner, B.D. In *Spectroscopy in the Biomedical Sciences;* Gendrau, R.M., Ed.; CRC Press: Boca Raton, FL, 1986.
15. Bull, H.B. *An Introduction to Physical Biochemistry;* F.A. Davis: Philadelphia, PA; 1964, pp. 294-329.
16. Sharp, K.A.; Brooks, D.E.; *Biophys. J.,* **1985**, 47, 563-566.
17. Seaman, G.V.F. In *The Red Blood Cell;* Sturgenor, D.M., Ed.; Academic Press: New York, NY, 1975, pp. 1135-1229.
18. Hermanson, G.T.; Mallai, A.K.; Smith, P.K. *Immobilized Affinity Ligand Techniques;* Academic Press: San Diego, CA, 1992, pp. 282-284.
19. Garibaldi, J.A.; Nielands, J.B.; *J. Am. Chem. Soc.,* **1955**, 77, 2429-2430.
20. Goding, J.W. *Monoclonal Antibodies: Principle and Practice;* Academic Press: London, England, 1983.
21. Wieczorek, A.; Agricuture Canada, Vancouver Research Branch; 1989, personal communication.
22. Kannangara, T.; Wieczorek, A.; Lavender, D.P.; *Physiol. Plant.,* **1989**, 75, 369-373.
23. Stocks, S.J.; Brooks, D.E.; *Hybridoma,* **1989**, 8, 241-247.
24. Harlow, E.; Lane, D. *Antibodies: A Laboratory Manual;* Cold Spring Harbour Laboratory: New York, NY; 1988, p. 310.
25. Taggart, R.T.; Samloff, I.M.; *Science,* **1983**, 219, 1228-1230.
26. Köhler, G.; Howe, S.C.; Milstein, C.; *Eur. J. Immunnol.,* 6, 292-295.
27. Goka, A.K.J.; Farthing, M.J.G.; *J. of Immunoassay,* **1987**, 8, 29-41.
28. Pashley, R.M.; Kitchener, J.A.; *Interface Sci.,* **1979**, 713, 491-590.
29. Fadley, C.S.; *J. Electron Spectrosc. Relat. Phenom.,* **1974**, 5, 725-754.

30. Paynter, R.W.; Ratner, B.D.; Horbett, T.A.; Thomas, H.R.; *J. Colloid Interface Sci.,* **1984**, 101, 233-245.

31. Siegbahn, K.; Nordling, C.; Fahlman, A.; Nordberg, R.; Hamrin, K.; Hedman, J.; Johansson, G.; Bergmark, T.; Karlsson, S.-E.; Lindgren, I.; Lindberg, B.; *ESCA Atomic, Molecular and Solid State Structure Studied by Means of Electron Spectroscopy;* Nova Acta Societatis Scientarium Upsaliensis, Ser. IV; Almquist& Wiksells Boktryckeri AB: Uppsala, Sweden, 1965, Vol. 20, p. 76.

32. Tanford, C.; *The Hydrophobic effect: Formation of Micelles and Biologicial Membranes;* Wiley-Interscience: New York; 1980; p.101.

33. Melzak, K.A. Ph.D. Dissertation, University of British Columbia, 1993.

34. van der Helm, D.; Baker, J.R.; Loghry, R.A.; Ekstrand, J.D.; *Acta. Cryst.,* **1981**, B37, 323-330.

35. Horinishi, H.; Nakaya, K.; Tani, A.; Shibata, K.; *J. Biochem (Tokyo);* **1968**, 63, 41-42.

RECEIVED March 1, 1995

Role of Adsorbed Proteins in Cell Interactions with Solid Surfaces

Chapter 31

Mechanism of the Initial Attachment of Human Vein Endothelial Cells onto Polystyrene-Based Culture Surfaces and Surfaces Prepared by Radiofrequency Plasmas

Roles of Serum Fibronectin and Vitronectin in Cell Attachment to Surfaces Containing Amide Groups

John G. Steele[1], Thomas R. Gengenbach[2], Graham Johnson[1],
Clive McFarland[1], B. Ann Dalton[1], P. Anne Underwood[1],
Ronald C. Chatelier[2], and Hans J. Griesser[2]

Divisions of [1]Biomolecular Engineering and [2]Chemicals and Polymers,
Commonwealth Scientific and Industrial Research Organisation, P.O.
Box 184, North Ryde, New South Wales 2113, Australia

We compared Primaria and films made by plasma modification of fluoroethylenepropylene using nitrogen-containing gases for their surface composition and for the mechanism of initial colonisation with human vein endothelial cells. Our data are consistent with amide groups being a substantial constituent of the surface chemical composition, on the Primaria and the plasma-prepared samples. In addition, XPS surface analysis and contact angle measurements demonstrated considerable variability in the surface composition of the Primaria and tissue culture polystyrene materials. Attachment of endothelial cells through a fibronectin-mediated mechanism was shown to occur on the nitrogen-containing surfaces, as a result of their ability to adsorb fibronectin in competition with other serum proteins. We propose that Primaria and other nitrogen-containing surfaces differ from oxygen-containing surfaces in that both fibronectin and vitronectin can mediate initial cell attachment, whereas on tissue culture polystyrene and similar surfaces it is vitronectin that mediates initial cell attachment.

Recently it has been proposed that in order to understand the relationship between polymer surface chemistry and cell colonisation, the key issue to be determined is the adsorption of specific proteins and their mechanistic role in cell attachment. Whereas it has been put forward that cell attachment requires particular chemical species on the surface of the polymer (with various oxygen-containing groups such as hydroxyl, carboxyl and carbonyl residues having been proposed,*1-7*), other surfaces with nitrogen-containing groups have also been shown to promote cell attachment (*4,5,8*). The commonly used cell culture substratum "tissue culture grade" polystyrene, TCPS, has a surface chemistry of oxygen-containing groups whereas the culture surface Primaria has nitrogen incorporated into the polystyrene surface. The surface of Primaria differs from that of TCPS in that Primaria has nearly equal amounts of nitrogen and oxygen whereas TCPS has little or no

0097–6156/95/0602–0436$12.00/0

nitrogen (*4,5,7,9*). Furthermore, films made using plasma etching or deposition reactions support colonisation by fibroblasts and endothelial cells, and these films can be fabricated with either oxygen or nitrogen-containing chemistries (*4,5,7,10,11,12*).

When cell colonisation of surfaces occurs in the presence of serum, the surface may have adsorbed from the serum either Fibronectin (Fn) or Vitronectin (Vn). These serum glycoproteins differ in size and structure, but both can stimulate cell attachment through specific peptide sequences (including but not limited to the cell attachment motif arg-gly-asp) by cellular binding through receptors and proteoglycans on the cell surface (*13*). In previous reports we have demonstrated that the proteins on polymer surfaces involved in the attachment of human vein endothelial cells differs between two different classes of surface chemistries, when these cells are cultured in medium containing serum (*11,14*). Cell attachment to TCPS or oxygen-containing plasma films occurs as a result of the binding of Vn, but these surfaces fail to bind sufficient Fn from serum for efficient attachment of endothelial cells and fibroblasts (*4,11,14-16*). On nitrogen-containing surfaces, however, cell adhesion is dependent upon either Vn or Fn from serum, but either one of them is effective alone (*11*). This mechanistic difference was tentatively assigned to the presence of amide groups (*10,11*).

In this report we present additional data which support a putative role of amide groups in the adsorption of effective concentrations of Fn for cell attachment, including a demonstration of a role for Fn in the mechanism of cell attachment to Primaria. During this study we observed some variability in the surface composition of TCPS and of Primaria. As these materials are commonly used as reference surfaces in the evaluation of the performance of novel cell growth substrates, the surface composition of a number of specimens of both materials were characterized by X-ray photoelectron spectroscopy (XPS) in order to study putative variability in these surfaces.

Materials and Methods

Plasma modification. Plasma fims were prepared by plasma modification of fluorinated ethylene-propylene copolymer (FEP; DuPont 100 Type A, 12.7 mm wide tape) in a custom-built reactor (*17*) which enabled the modification of extended lengths of tape and hence the fabrication of numerous identical specimens for the various analyses and cell assays. The sample fabrication procedures which involved the surface treatment of FEP in an ammonia plasma (NH_3/FEP) or the deposition onto FEP of thin films from the "monomers" heptylamine (HA) and dimethylacetamide (DMAc) were identical to those previously described (*10*). For comparison with TCPS, a number of monomers containing oxygen (but not nitrogen) were also used for plasma polymerization (*10*). All plasma-treated and plasma polymerized specimens were stored for at least eight weeks prior to cell assays allowing for equilibration of the surface compositions, which can vary over initial periods of time after fabrication due both to surface restructuring and oxidative processesbut which stabilize typically within a few days to several weeks (*18-21*).

Surface composition analyses of plasma films and polystyrene surfaces. Primaria and TCPS specimens for XPS and contact angle analyses were cut out of the bottoms of multiwell tissue culture plates carrying various batch numbers. XPS analyses were performed on a VG Escalab V unit equipped with a non-monochromatic Al K_α source and a hemispherical analyser operating in the fixed analyser transmission mode. A power of 200 W (10 kV * 20 mA) was used for excitation. The pressure in the chamber during analysis was typically 2×10^{-9} mbar. The binding energy scale was calibrated with data from sputter-cleaned foils

(Ni, Ag, Au and Cu) (22). With a pass energy of 30 eV, the width (FWHM) of the Ag $3d_{5/2}$ peak was 1.5 eV. Linearity of the binding energy scale was assessed using PTFE tape as a reference material that yielded a binding energy separation of the F 1s and C 1s (CF_2 component) peaks of 397.3 ± 0.1 eV, in agreement with literature data (23). A value of 285.0 eV for the binding energy of the main C 1s component (CH_x) was used to correct for charging of the specimens under irradiation, except for NH_3/FEP, for which charge correction was done assuming a peak position of 291.6 eV for the CF_2 component. High resolution spectra were recorded on individual peaks at a pass energy of 30 eV with a step width of 0.078 eV/channel. The elemental composition of the surface was determined by a first principles approach (24); atomic number ratios were calculated with integral peak intensities, using a nonlinear Shirley type background, and published values for photoionisation cross-sections (25). The inelastic mean free path of the photoelectrons was assumed to be proportional to $E^{0.5}$, where E is the kinetic energy (26). The transmission function of the analyser had previously been determined to be proportional to $E^{-0.5}$. Individual components of the C 1s signal of oxygen-containing plasma polymers and TCPS were fitted assuming a Gaussian/Lorentzian lineshape and using damped nonlinear least squares regression (27). For the nitrogen-containing plasma surfaces and Primaria, too many components and secondary shifts were expected to warrant such fitting. For both types of samples, however, the $\pi\pi^*$ shakeup satellite to the C 1s peak was clearly separated and could be quantified. Spectra were collected at various angles of the photoelectron emission relative to the specimen surface (angle dependent XPS; ADXPS). The ADXPS intensity data were converted into depth profiles using a algorithm by Tyler, Castner and Ratner (28) and assuming a value of 3 nm for the electron attenuation length of a C 1s photoelectron in a polymeric matrix (29).

The random error associated with quantitative elemental analysis had previously been determined to be < 8 % on our unit using known, reproducible standard samples. The contribution of systematic errors is difficult to estimate, but the F/C ratio determined on PTFE tape agreed to within 3% with the theoretically expected value of 2. Also, several plasma-treated polymers and plasma polymer coatings had earlier been analyzed both on our unit and on a Surface Science Instruments SSX-100 spectrometer equipped with a monochromatic Al K_α source, located at NESAC/BIO, University of Washington, Seattle, and elemental ratios obtained on the two different instruments agreed to within 1 to 2 % (18,19). As polymer samples can be susceptible to X-ray induced decomposition under non-monochromatic irradiation, degradation-related compositional changes were minimized by exposing specimens to X-rays for less than one minute before the start of the data acquisition. Total exposure time did not exceed 30 minutes. The contact angle apparatus and method has been described elsewhere (30) and measurements were performed in quadruplicate. Standard deviations were typically 1 to 2°, with somewhat larger uncertainties at (receding) angles below 10°.

Cell culture and cell attachment assays. The culture of human vein endothelial (HUVE) cells was as previously described (11,14). The HUVE cells to be used in an attachment assay were metabolically labelled with radioactive methionine, as follows. Confluent cultures were passaged on the day before use in the attachment assay, then labelled by culture for 18 hr in methionine-free Medium 199 (Cytosystems, Sydney) containing 15% (v/v) fetal bovine serum (FBS) and supplemented with ^{35}S-methionine (Amersham, Australia) at a concentration of 5 uCi/ml. Cells were washed briefly with phosphate buffered saline (PBS) then a cell suspension prepared by incubation with 0.1% (w/v) trypsin/0.02% EDTA (in PBS) solution at 37°C to cause retraction of the cell boundaries. The cells were displaced from the TCPS by pipetting of the solution, collected and centrifuged in serum-free culture medium then resuspended in fresh serum-free medium. The cell attachment assays were performed in Primaria (24-well tray, Becton Dickinson cat. no. 3847), or TCPS (Corning 24-well tray, cat. no. 25820) dishes. In experiments where various different sera were used, the cells were seeded in 50 ul of serum-free medium into wells which had been pre-loaded with 50 ul of 2X serum concentration. FBS depleted of either Vn or Fn was prepared by methods described previously (*15,16*). Immunoassays showed that the Vn-depleted FBS retained normal Fn content and the Fn-depleted FBS retained normal Vn content. The cells were cultured in an atmosphere of 5% CO_2 at 37°C for 90 minutes then the wells were washed twice with PBS to remove nonattached cells and the attached cells were removed by overnight digestion with 0.1% (w/v) trypsin/0.02% EDTA solution. The digests were transferred to counting pots and the ^{35}S content was determined by liquid scintillation counting.

Determination of adsorption of Fn and Vn to TCPS and Primaria. The amounts of Fn and Vn which adsorbed to Primaria and TCPS from serum-free medium or from medium containing FBS, were measured using radiolabelled Fn and Vn. Fn and Vn were radiolabelled with ^{125}I using the chloramine T method (see *31*). The labelled Fn and Vn were stored for up to 4 weeks at -70°C. Following the radiolabelling reaction and subsequent storage, a proportion of the radioactivity is not able to adsorb to polymer surfaces (*31*). The proportion of the radiolabelled protein preparation which was adsorbable to Dynatech polyvinylchloride (PVC) trays was determined at various times of storage by a sequential binding method (*31*). At the time of use, the adsorbable fraction of ^{125}I-labelled Vn was between 60 - 78% of the total radiolabel content and for ^{125}I-Fn, the adsorbable fraction was 49 - 70% of the total radiolabel content. To measure the amount of Fn and Vn which adsorbed to Primaria and TCPS from serum-free medium, radiolabelled Fn or Vn was added to solutions of 5 or 10 ug Fn or Vn/ml and a 50 ul aliquot per well (containing $2 - 5 \times 10^5$ cpm) was incubated at 37°C for 90 min. To measure the amount of Fn and Vn which adsorbed to Primaria and TCPS from culture medium containing serum, FBS containing 50 ug of either Fn or Vn /ml of serum (made by the addition of Fn or Vn to depleted FBS) was used to make up solutions containing 10 or 20% (v/v) of FBS in culture medium. Radiolabelled Fn or Vn was added to the solutions and triplicate wells were incubated with 50 ul of solution for 90 min at 37°C. The wells were washed with three washes of PBS, then were cut from the dish using a hot wire and the adsorbed ^{125}I was measured. The total amount of adsorbable label was calculated from the total amount of radioactivity added per well and the proportion of this radioactivity which was determined to be adsorbable to Dynatech PVC trays. This total amount of adsorbable label, the radioactivity on the wells, the known Fn or Vn concentration and the surface area of the well in contact with solution (0.63 cm^2) were used to calculate the amounts of Fn or Vn adsorbed to Primaria or TCPS and expressed as ng/cm^2.

Results

Surface Analysis of Plasma films, Primaria and TCPS. The surface composition of the NH3/FEP, Primaria, DMAc and HA specimens was characterized by the presence of the elements C, N, and O. Two examples of XPS survey spectra are reproduced in Figure 1. The elemental ratios depended

Table I. Elemental composition determined by XPS at an emission angle of 70 degrees of representative specimens of Primaria and nitrogen-containing plasma polymers, and NH3 plasma treated FEP

Sample	N/C	O/C	F/C
Primaria	0.09	0.18	-
NH3/FEP	0.17	0.26	0.35
HA film	0.08	0.12	-
DMAc film	0.176	0.19	-

somewhat on the plasma conditions; representative values are listed in Table I. Oxygen was present on all surfaces, although the process gases NH3 and HA did not contain oxygen and the plasma system was leak-tested carefully. Detailed analyses presented elsewhere have shown that the oxygen content in these surfaces arose from post-fabrication uptake of atmospheric oxygen which reacted with trapped, carbon-centered radicals (18-20). Grazing angle FTIR analysis of HA plasma polymers revealed that the C=O stretch region contained two partially overlapping bands assigned to amide and carbonyl groups respectively. For DMAc, the C=O stretch band was centered at a position typical of amide groups, but a higher energy shoulder indicative of the presence of a significantly smaller amount of carbonyls was present.

XPS compositional data for four wells each from four Primaria multiwell plates and for two wells each from four TCPS plates are shown in Figure 2. There was significant compositional variability both between plates and within wells on one plate. Earlier analyses in which only one well from a plate was measured showed in some instances even larger variations from the mean of the values shown in Figure 2; for instance, one Primaria well was measured with a N/C ratio of 0.043 and an O/C ratio of 0.128, suggesting less extensive surface treatment than for the four plates shown. The variability of the Primaria surfaces was also manifested in contact angle measurements; representative data are shown in Figure 3 and suggest variable treatment efficiencies. In the case of TCPS likewise to that of Primaria, there were large variations observed in the O/C ratio. The data shown in Figure 2c were obtained from TCPS plates with different batch numbers, but analysed at the same time. Earlier analyses of single wells from individual plates had shown larger variability, for instance an O/C ratio of 0.14.

The depth distribution of the nitrogen-containing groups was assessed by ADXPS. The depth profiles of the Primaria and NH3/FEP surfaces are shown in Figures 4A,B. This NH3/FEP sample (B) possessed a relatively low oxygen content since it had not stabilized yet; aged samples showed analogous depth profiles but for a higher oxygen content (data not shown). HA and DMAc plasma polymers did not show significant compositional variations with depth. For the Primaria and NH3/FEP surfaces, IR intensities were insufficient to allow such characterization. Accordingly, the binding energy of the XPS N 1s peak was measured in order to determine the chemical nature of the dominant contribution to the N signal (Table II). In all cases the peak position is consistent with a predominant *amide* N contribution. This position would also be consistent with imides, but the IR identification of substantial concentrations of amides in the case of the HA and DMAc films suggests that these two types of surfaces comprise mainly amides. The slightly higher binding energy for Primaria is thought to relate

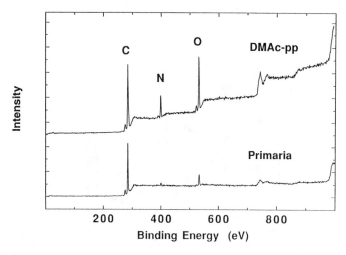

Figure 1. XPS survey spectra of Primaria and DMAc plasma polymer on FEP.

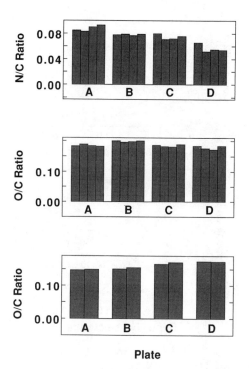

Plate

Figure 2. XPS atomic ratios determined at normal (zero degrees) photoelectron emission: (a) N/C, and (b) O/C of four wells each from four Primaria multiwell plates (A -D), and (c) O/C of two wells each from four TCPS plates (A - D).

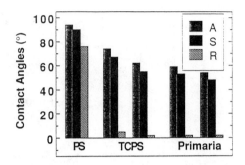

Figure 3. Advancing (A), sessile (S) and receding (R) air/water contact angles of representative untreated polystyrene, TCPS, and Primaria samples.

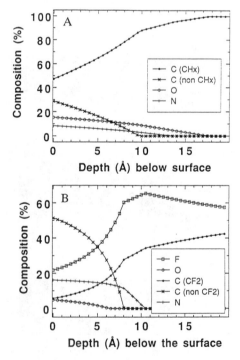

Figure 4. Compositional depth profiles by angle dependent XPS of Primaria (A) and NH3/FEP (B).

Table II. Binding energy of the XPS N 1s peak for Primaria, HA and DMAc plasma polymers, and NH_3 plasma treated FEP

Sample	N 1s B.E.
Primaria	400.67 ± 0.14[a]
NH_3/FEP	400.3[b]
HA film	400.0 ± 0.1
DMAc film	400.0 ± 0.1

[a] uncertainty may possibly be larger than for other materials; see text for discussion.
[b] from the 16 samples shown in Figure 2.

to attachment to aromatic ring structures, and the presence of a shakeup satellite to the N 1s peak is in agreement with this interpretation. For NH_3/FEP, the charge correction was done using a value for the CF_2 peak which does not take into consideration possible secondary shifts consequent upon substitutions by the plasma treatment; accordingly, there is a potentially larger and presently undefinable uncertainty in the N 1s binding energy value. Nevertheless, the value also corresponds well to that of amide groups. Several NH_3/FEP samples have been assessed elsewhere by surface derivatization by fluorescein isothiocyanate (FITC), which is a reagent probing for amine groups (*18*). After storage, the surfaces no longer were reactive, indicating disappearance of amine groups; this is consistent with oxidation of amines to amides, which are not capable of reaction with FITC.

Thus, in summary, all four nitrogen-containing surfaces appeared to be qualitatively of similar composition, with amide groups being a major component. Oxygen-containing groups also were on the surfaces as a result of post-fabrication oxidation; carbonyl groups were evident in IR. TCPS and oxygen-containing plasma polymers could not be analyzed to the same extent because there are too many possible functional groups contributing to the C 1s signal. Curve fitting enables determination only of classes of groups; derivatization reactions are needed for more detailed characterization.

Involvement of Fn in initial attachment of endothelial cells. The initial attachment of HUVE cells onto these amide-containing surfaces during the first 90 min after seeding was stimulated by the presence of FBS in the seeding medium. The attachment of HUVE cells to Primaria in serum-free medium was 40.1 ± 6.2 (mean $\pm$ SEM of 5 experiments) of that in medium containing 15% (v/v) FBS; the proportion of cells which had spread was similarly lower in serum-free medium ($87.3 \pm 3.9\%$ of cells when the medium contained FBS as compared to $51.3 \pm 20.7\%$ of cells seeded in serum-free medium). Selective depletion of Fn and Vn from the FBS prior to addition to the culture medium was conducted, in order to determine the requirements for these two adhesion factors in the stimulation of HUVE cell attachment. Table III shows that when both Fn and Vn had been removed from the FBS, HUVE cell attachment to Primaria was reduced to be less than that in serum-free medium. FBS that had been depleted of Vn alone stimulated cell attachment to equivalent levels as with intact FBS, as did FBS that had been depleted of Fn alone. These results show that the stimulation of HUVE cell attachment to Primaria by FBS was dependent upon either serum Fn, or Vn. Figure 5 shows that almost identical results were obtained for HUVE cell attachment to the ammonia plasma film surface, but that there was a higher dependence upon the

Table III. Comparison of Primaria and TCPS for the involvement of Fn and Vn in stimulation of HUVE cell attachment by 15% (v/v) serum

Surface and Culture medium	HUVE cell attachment	Fn Adsorbed	Vn Adsorbed
	Mean ± SEM	(ng/sq cm)	(ng/sq cm)
Primaria:			
	%		
FBS	(100) ± 4.0		
Depleted of Vn	77.6 ± 3.1	31.4	0
Depleted of Fn	74.1 ± 5.5	0	47.0
Depleted of Vn and Fn	14.2 ± 3.8	0	0
TCPS:			
	%		
FBS	(100) ± 6.4		
Depleted of Vn	25.6 ± 5.4	12.4	0
Depleted of Fn	80.4 ± 2.4	0	47.5
Depleted of Vn and Fn	14.6 ± 3.5	0	0

serum Fn content than the serum Vn content. These results with the amide-containing surfaces contrast with those previously described for oxygen-containing surfaces, such as TCPS and plasma films deposited with methanol, ethanol or methylmethacrylate (10,11). On these oxygen-containing surfaces, the stimulation of HUVE cell attachment by FBS was dependent upon the Vn component, whereas cell attachment was essentially independent of the serum Fn content.

The ability to selectively remove Vn from FBS enabled the involvement of Fn in HUVE cell attachment to the amide-containing surfaces to be tested by repletion experiments. Figure 6 shows the dependence of stimulation of HUVE cell attachment to Primaria and TCPS upon the serum Fn content, at several concentrations in the range of 0 - 2 ug/ml Fn. For the Primaria surface and in the absence of Vn, HUVE cell attachment correlated with the Fn concentration, whereas on TCPS no stimulation of HUVE cell attachment was demonstrated.

Adsorption of Fn and Vn in medium containing other serum proteins. The adsorption of Fn and Vn onto Primaria and TCPS was determined at two concentrations that were equivalent to the concentrations found in cell culture media. For the situation where the content of Fn or Vn in FBS is 50 ug/ml of serum and the serum content is either 10% and 20% (v/v), the concentration is 5 ug/ml and 10 ug/ml. Table IV compares the amounts of Fn and Vn that adsorbed to Primaria and TCPS at these Fn and Vn concentrations, from serum-free medium and from culture medium containing FBS.

The surface density of Fn that adsorbed from medium containing other serum proteins was reduced as compared to that from serum-free medium, the competition from other serum proteins reducing adsorption to 3 - 13% of that from serum-free medium. Furthermore, there was a difference between the surfaces in the extent to which competition from other serum proteins reduced the amount of Fn that adsorbed to Primaria and TCPS. On Primaria, 10 - 20% (v/v) serum reduced Fn adsorption to 10 - 13% whereas on TCPS, the adsorption of Fn was reduced to

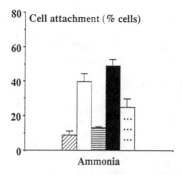

Figure 5. Requirement for serum Vn and Fn for the attachment of HUVE cells on the NH3/FEP surface. HUVE cells were seeded in serum-free medium (bars with diagonal stripes) or medium containing 15% (v/v) intact FBS (open bars), Vn-depleted FBS (shaded bars), Fn-depleted FBS (dotted bars) or FBS depleted of both Fn and Vn (bars with horizontal stripes) and cell attachment was measured after 90 min. Mean ± SEM.

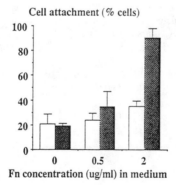

Figure 6. Comparison of Primaria with TCPS for the attachment of HUVE cells in the 90 minutes after seeding in media containing 15% (v/v) of Fn- and Vn depleted serum and replenished with the indicated final Fn concentrations. The histogram shows HUVE cell attachment to Primaria (solid bars) and TCPS (open bars). Mean ± SEM; mean of triplicate determinations.

3 - 5%. The surface densities of Fn that adsorbed onto TCPS and Primaria from serum-free medium were between 162 - 450 ng/sq cm, which compare to a calculated value of 300-5000 ng/sq cm for a close-packed monolayer of Fn, depending upon the presumed orientation of the molecule (32). Interestingly, although the amounts of Fn that adsorbed onto Primaria from medium containing other serum proteins was higher than for TCPS, the amounts that adsorbed onto Primaria from serum-free medium was less than to TCPS.

Similar surface densities of Vn adsorbed onto Primaria and TCPS, in both the case of serum-free medium and also from medium containing serum proteins (depleted of endogenous Fn and Vn). The presence of serum proteins reduced the amount of Vn that adsorbed to approximately 26 - 32% of that from serum-free medium. Whereas the Primaria surface bound more Fn in the presence of competing serum proteins than did TCPS, the surfaces showed equivalent adsorption of Vn in the presence of serum (Table IV).

Discussion

The analysis presented here of the role of amide surface groups in the mechanism of initial attachment of human endothelial cells is based upon the correlation between the presence of amide groups and the demonstrated involvement of Fn in cell attachment, shown by repletion experiments. These repletion experiments were conducted after Vn had been removed from the FBS, as for both nitrogen-containing and also other hydrophilic surfaces without amide (eg TCPS, oxygen-containing plasma films), adsorbed Vn provides one mechanism of cell attachment (11,14,16,33-35).

The surface chemistry of nitrogen-containing plasma films and Primaria. The main point of interest arising from the present study of surface compositions of Primaria and nitrogen-containing plasma films is the close analogy between the depth profiles and the surface functional groups of Primaria and NH3/FEP. Thus one may consider the four N-containing surfaces included in this and our previous studies (being DMAc, NH3/FEP, HA/FEP plasma films and

Table IV. The effect of other serum proteins upon adsorption of Fn and Vn to Primaria as compared to TCPS

Fn or Vn concentration	Surface	Adsorbed from serum-free medium (ng/sq cm)	Adsorbed from mdium with serum (ng/sq cm)
5 ug/ml Fn in 10% (v/v) FBS	TCPS	236.8	11.7
	Primaria	162.8	21.6
10ug/ml Fn in 20% (v/v) FBS	TCPS	450.7	15.4
	Primaria	293.2	29.4
5 ug/ml Vn in 10% (v/v) FBS	TCPS	154.4	40.5
	Primaria	137.8	43.6
10 ug/ml Vn in 20% (v/v) FBS	TCPS	236.6	71.4
	Primaria	229.4	70.5

Primaria) as related materials, albeit with hydrocarbon or fluorocarbon backbone structure. It is likely that polar groups, including in particular amides, will make major contributions to the surface interactions of these materials. A previous analysis of the surface chemistry of Primaria showed the nitrogen content to be similar to the oxygen content, and that the nitrogen may be bonded directly to the aromatic ring of polystyrene or one carbon from the ring (*4,5,9*). We note that our XPS data for Primaria differ somewhat from previous analyses (*4,5*) in that we found the N content to be only about half of the O content, whereas previous studies had reported nearly equal amounts of the two elements. One possible explanation rests on the observation that plasma-treated surfaces continue to incorporate O over extended periods of time (*18,19*); the Primaria specimens we analysed may simply have been older and thus have incorporated additional O by the slow, extended post-fabrication oxidative process.

This study has documented considerable variability in the elemental composition of specimens of Primaria, suggesting either variability in the commercial process used for surface modification or variability in stability during storage. Some variability of composition was also seen with TCPS, but to a lesser extent. Such variability is of concern, because Primaria and TCPS are commonly used as reference materials in cell colonisation assays. The nitrogen content of Primaria varied considerably more than did the oxygen content. Interestingly, the nitrogen contents and oxygen contents did not appear to be related, suggesting that they are not incorporated by the same primary mechanism. Furthermore, the shakeup satellite intensity appeared to be inversely proportional to the nitrogen content, although the low intensity observed made accurate quantitation difficult. The variability of Primaria surface composition was also detected in contact angle measurements. As Chinn and Ratner have found that the surface treatment effects on Primaria can be partially removed by washing (Ratner, B.D.; University of Washington, personal communication, 1993), the method of measurement of contact angles may interfere with the reliability of this method of analysis of the surface. Since the rate of such a dissolution of fragmented, modified surface polymer chains from Primaria is not known at present, it is not possible to predict whether such a dissolution process interferes with the results of either contact angle measurements or the initial cell attachment assays (which were conducted for 90 minutes culture). Furthermore, it is possible that in the cell attachment assay the adsorption of proteins onto the surface may cover it sufficiently rapidly and effectively as to block the release of low molecular weight polymeric material into the medium. The consequences of partial elutability of modified surface layers is in need of study and caution is indicated until we have an improved understanding.

Mechanism of the initial attachment of human vein endothelial cells onto Primaria and amide-containing plasma films.   The comparison between surfaces containing amide groups and oxygen-containing surfaces in the involvement of Fn in the initial attachment of endothelial cells was quite stark, when compared to the common ability of these two groups of surfaces to support cell attachment in a Vn-dependent reaction. Whereas the amide-containing surfaces showed effective cell attachment with as little as 2 ug/ml Fn in the serum component of the culture medium, the oxygen-containing surfaces did not support cell attachment in this medium (present study,*11*). These cell attachment results are supported by data comparing Primaria and TCPS for the adsorption of Fn and Vn (present study,*36*). Adsorption of higher levels of Fn was also seen previously with nitrogen-containing plasma polymers (*4*) but not with other nitrogen-containing surfaces made using a nitrogen plasma (*12*). The Primaria surface adsorbed Fn surface densities that were approximately 185-190% of those to TCPS, when the adsorption was from serum-containing medium. These results contrasted to the adsorption of Vn from serum-containing medium, which was similar on Primaria and TCPS (present study,*36*). Furthermore, the amounts of Fn that adsorbed to Primaria from serum-free medium were only 66% of those to TCPS. Taken together, these results indicate that the Primaria surface promotes the adsorption of Fn in competition with other serum proteins; This higher avidity of the Primaria surface is the result of a reaction that is somewhat specific to Fn, in that it is demonstrated under conditions of competitive binding and that it is not seen with Vn. The Fn-binding reaction of Primaria and the active conformation of the adsorbed Fn is sufficiently effective as to provide a Vn-independent mechanism for the colonisation of the amide-containing surfaces by human vein endothelial cells, not seen with oxygen-containing surfaces.

The results of the Fn adsorption experiments lead to emphasis on the *adsorption* of Fn to amide-containing surfaces as a key difference from oxygen-containing surfaces. Although the repletion experiments show that the Fn adsorbed onto the amide-containing surfaces is active, it will also be of interest to compare the conformation of Fn adsorbed onto these surfaces with that on oxygen-containing surfaces using monoclonal antibodies to the cell-binding regions of Fn *(37,38)*. In previous studies of Fn adsorbed on TCPS and the hydrophobic surface unmodified polystyrene (PS), it was shown that Fn adsorbed onto the hydrophobic surface was compromised in its cell attachment activity, including for attachment of HUVE cells *(39,40,14)*. Further evidence that the polymer surface chemistry can affect the activity of Fn was obtained with chemically-derivatized self-assembled monolayers and attachment of neuronal cells and fibroblasts*(41,42)*. Although the correlation between the presence of amide groups on the Primaria surface and the Fn-binding reaction suggests that it may involve interaction of the surface of Fn with surface amide groups, the determination of the precise binding mechanism (if indeed it is a single mechanism) will be a challenging project, particularly as it will need to be studied under competitive binding conditions.

Acknowledgements

We thank Ms S. Mayer and Mr Zoran Vasic for their expert technical assistance. This work was supported in part by the IRD program of DITAC and by AMBRI Ltd., Cyanamid Australia Pty Ltd., Telectronics Pty Ltd. and Terumo (Australia) Pty Ltd.

Literature Cited

1. Ramsey, W.S.; Hertl, W.; Nowlan, E.D.; Binkowski, N.J. *In vitro,* **1984,** *20,* 802-808.
2. Curtis, A.S.G.; Forrester, J.V.; McInnes, C.; Lawrie, F. *J. Cell Sci.* **1983,** *97,* 1500-1506.
3. Curtis, A.S.G.; Forrester, J.V.; Clark, P. *J. Cell Sci.* **1986,** *86,* 9-24.
4. Ertel, S.I.; Ratner; B.D.; Horbett,T.A. *J. Biomed. Mater. Res.* **1990,** *24,* 1637-1659.
5. Chinn, J.A.; Horbett, T.A.; Ratner, B.D.; Schway, M.B.; Haque, Y.; Hauschka, S.L. *J. Colloid Interface Sci.* **1989,** *127,* 67-87.
6. Lydon, M.J.; Minett, T.W.; Tighe, B.J. *Biomaterials* **1985,** *6,* 396-402.
7. Ertel, S.I.; Chilkoti, A.; Horbett, T.A.; Ratner, B.D. *J. Biomater. Sci. Polymer Edn.* **1991,** *3,* 163-183.
8. Klein-Soyer, C.; Hemmendinger, S.; Cazenave, J-P. *Biomaterials* **1989,** *10,* 85-90.
9. Salvati, L.; Grobe, G.L.; Moulder, J.F. *The PHI Interface* **1990,** *12,* 1-7.

10. Griesser, H.J.; Chatelier, R.C.; Gengenbach, T.R.; Johnson, G.; Steele, J.G. *J. Biomater. Sci., Polymer Edn.*, **1994**, *5*, 531-554.
11. Steele, J.G.; Johnson, G.; Griesser, H.J.; Gengenbach, T.; Chatelier, R.C.; McFarland, C.; Dalton, B.A.; Underwood, P.A. *J. Biomater. Soc., Polymer Edn* **1994**, *6*, 511-532.
12. Dekker, A.; Reitsma, K.; Beugeling, T.; Bantjes, A.; Feijen, J.; van Aken, W.G. *Biomaterials* **1991**, *12*, 130-138.
13. Hayman, E.G.; Pierschbacher, M.D.; Suzuki, S.; Ruoslahti, E. *Exp. Cell Res.* **1985**, *160*, 245-248.
14. Steele, J.G.; Dalton, B.A.; Johnson, G.; Underwood, P.A. *J. Biomed. Mater. Res.* **1993**, *27*, 927-940.
15. Underwood, P.A.; Bennett, F.A. *J. Cell Sci.* **1989**, *93*, 641-649.
16. Steele, J,G.; Johnson, G.; Underwood, P.A. *J. Biomed. Mater. Res.* **1992**, *26*, 861-884.
17. Griesser, H.J. *Vacuum* **1989**, *39*, 485-488.
18. Gengenbach, T.R.; Xie, X.; Chatelier, R.C.; Griesser, H.J. *J. Adhes. Sci. Tech.* **1994**, *8* , 305-328.
19. Gengenbach, T.R.; Vasic, Z.R.; Chatelier, R.C.; Griesser, H.J. *J. Polym. Sci., Part A: Polym. Chem.* **1994**, 32, 1399-1414.
20. Gengenbach, T.R.; Chatelier, R.C.; Vasic, R.C.; Griesser, H.J. *ACS Polym. Prepr.* **1993**, *34,* 687-688.
21. Griesser, H.J., Gengenbach, T.R.; Chatelier, R.C. *Trans. Amer. Soc. Biomater.* **1994**, *17*, 19.
22. Anthony, M. T.; Seah, M.P. *Surf. Interf. Anal.* **1984**, *6*, 95-106.
23. Briggs, D.; Beamson, G. *Appl. Surf. Sci.* **1991**, *52* , 159-161.
24. Grant, J.T. *Surf. Interf. Anal.* **1989**, *14,* 271-283.
25. Scofield, J.H. *J. Electron Spectrosc. Relat. Phenom.* **1976**, *8* , 129-137.
26. Seah, M.P.; Dench, W.A. *Surf. Interf. Anal.* **1979**, *1*, 2-11.
27. Hughes, A.E. ; Sexton, B.A. *J. Electron Spectrosc. Relat. Phenom.*, **1988**, *46,* 31-42.
28. Tyler, B.J.; Castner, D.G.; Ratner, B.D. *Surf. Interf. Anal.* **1989**, *14*, 443-450.
29. Roberts, R.F.; Allara, D.L.; Pryde, C.A.; Buchanan, D.N.E.; Hobbins, N.D. *Surf. Interf. Anal.* **1980**, *2* , 5-10.
30. Xie, X.; Gengenbach, T.R.; Griesser, H.J. *J. Adhes. Sci. Tech.*, **1992**, *6,* 1411-1431.
31. Underwood, P.A.; Steele, J.G. *J. Immunol. Methods* **1991**, *142*, 83-94.
32. Giroux, T.A.; Cooper, S.L. *J. Colloid Interf. Sci.* **1990**, *139*, 351-362.
33. Bale, M.D.; Wohlfahrt, L.A.; Mosher, D.M.; Tomasini, B.; Sutton, R.C. *Blood* **1989**, *74*, 2698-2706.
34. Preissner, K.T.; Andrews, E.; Grulich-Henn, J.; Muller-Berghaus, G. *Blood* **1988**, *71*, 1581-1589.
35. Lydon, M.J.; Foulger, C.A. *Biomaterials* **1988**, *9*, 525-527.
36. Steele, J.G.; Dalton, B.A.; Johnson, G.; Underwood, P.A. *Biomaterials* **1994**, *submitted.*
37. Underwood, P.A.; Steele, J.G.; Dalton, B.A. *J. Cell Sci.* **1993**, 793-803.
38. Pettit, D.K.; Hoffman, A.S.; Horbett, T.A. *J. Biomed. Mater. Res.* **1994**, *28*, 685-691.
39. Grinnell, F.; Feld, M.K. *J. Biomed. Mater. Res.* **1981**, *15*, 363-381.
40. Grinnell, F.; Feld, M.K. *J. Biol. Chem.* **1982,** *254,* 4888-4893.
41. Sukenik, C.N.; Balachander, N.; Culp, L.A.; Lewandowska, K.; Merritt, K. *J. Biomed. Mater. Res.* **1990**, *24*, 1307-1323.
42. Lewandowska, K.; Pergament, E.; Sukenik, C.N.; Culp, L.A. *J. Biomed. Mater. Res.* **1992**, *26*, 1343-1363.

RECEIVED March 2, 1995

Chapter 32

Platelet Adhesion to Fibrinogen Adsorbed on Glow Discharge-Deposited Polymers

David Kiaei[1,3], Allan S. Hoffman[1,2], and Thomas A. Horbett[1,2,4]

[1]Center for Bioengineering and [2]Department of Chemical Engineering, University of Washington, Seattle, WA 98195

The state of fibrinogen adsorbed on untreated and glow discharge-treated surfaces was examined by measuring platelet adhesion, the amount of fibrinogen adsorbed, and the amount of adsorbed fibrinogen which could be eluted with sodium dodecyl sulfate (SDS). Tetrachloroethylene (TCE) and Tetrafluoroethylene (TFE) glow discharge-treated polymers retain a larger fraction of adsorbed fibrinogen after SDS elution than untreated or ethylene (E) glow discharge-treated surfaces. Platelet adhesion was lowest on the TFE and TCE-treated surfaces which retain the highest amounts of fibrinogen. The tight binding of fibrinogen on the TFE and TCE-treated surfaces suggests that the fibrinogen may rearrange in order to maximize protein-surface interactions. The suggested rearrangements in fibrinogen seem to result in a state which prevents its recognition and binding by platelet receptors. The data suggest that the tight binding of fibrinogen on a surface directly affects the ability of the fibrinogen to interact with the platelet receptors, i.e. that fibrinogen must be loosely held to facilitate maximal interaction with platelet receptors.

Fibrinogen adsorption is a well-known contributor to surface induced thrombosis [1]. Chinn et al. have demonstrated that platelet adhesion on Biomer preadsorbed with baboon plasma is primarily due to adsorption of plasma fibrinogen [2]. The adhesion of platelets to Biomer coated with serum or afibrinogenemic plasma was low compared to Biomer coated with normal plasma. Furthermore, addition of fibrinogen to afibrinogenemic plasma or serum restored platelet adhesion in a dose dependent manner [2].

While the importance of fibrinogen in thrombosis is well-documented, the factors which govern fibrinogen adsorption and its biological activity after its adsorption on a surface are still poorly defined. Several investigators have radiolabeled fibrinogen and correlated platelet adhesion with the amount of fibrinogen adsorbed on a surface [3-5]. More recent work by Lindon et al. has shown that on some polyalkyl methacrylate surfaces, platelet adhesion correlates with the amount of antibody recognizable ("native") fibrinogen and not with the total amount of fibrinogen adsorbed [6, 7].

[3]Current address: Ciba Corning Diagnostics Corporation, 333 Coney Street, East Walpole, MA 02032
[4]Corresponding author

0097–6156/95/0602–0450$12.00/0

Chinn et al. have detected changes in the state of adsorbed fibrinogen when the protein adsorbed surface was stored in buffer for a period of time, the "residence time", prior to various assays [8]. They observed a decrease in platelet adhesion, polyclonal antibody binding, and sodium dodecyl sulfate (SDS) elutability of fibrinogen when the fibrinogen-coated Biomer was stored in buffer for 3 days [8]. The actual amounts of fibrinogen adsorbed on Biomer prior to and after buffer incubation were the same (i.e. no fibrinogen desorbed during the 3 days storage in buffer). When the fibrinogen-coated samples were stored in a buffer solution containing albumin, these decreases were prevented [8]. The authors have suggested that in the absence of albumin, the fibrinogen undergoes time-dependent changes which render it both less reactive towards platelets and antibodies as well as more resistant to elution by a surfactant. The albumin effect is ascribed to its occupancy of sites adjacent to the fibrinogen, thereby preventing occupancy by unfolded regions of the fibrinogen.

Fibrinogen adsorbs much more tenaciously onto tetrafluoroethylene (TFE) glow discharge-treated surfaces than on untreated surfaces [9-12]. In light of the observations by Chinn et al [8], we hypothesized that the tenacious binding of fibrinogen to TFE-treated surfaces may cause fibrinogen to assume an inactive conformation and render the surface more thromboresistant. Therefore, we investigated the state of fibrinogen adsorbed on TFE-treated surfaces [13]. In this study, we have compared the state of fibrinogen adsorbed on tetrachloroethylene (TCE) and TFE glow discharge-treated surfaces. The amounts of fibrinogen adsorbed on untreated and glow discharge-treated surfaces and the fractions elutable with SDS have been measured. In addition, *in vitro* platelet adhesion assays have been performed in order to determine the reactivity of adsorbed fibrinogen.

Materials and Methods

Materials. Poly (ethylene terephthalate) (PET) coverslips were obtained from Nunc Inc. (Naperville, IL) and cleaned using successive 15-min ultrasonic treatments in methylene chloride, acetone, and distilled water. Argon (Ar) was obtained from Air Products (Allentown, PA). Ethylene (E) was purchased from Byrne Specialty Gases (Seattle, WA). TFE was received from PCR Inc. (Gainesville, FL). TCE was purchased from Aldrich Chemical Co. (Milwaukee, WI) and used as received. SDS, ultra pure, was obtained from ICN Biomedicals Inc. (Cleveland, OH).

Glow Discharge Polymerization. The glow discharge treatment of the cleaned PET substrates with E , TFE, and TCE was carried out in a 135-cm-long, 18-mm i.d., Pyrex reactor as described in detail elsewhere [12, 13]. Briefly, PET coverslips, 11 x 16 mm, were placed horizontally in a 16-mm i.d. glass tubes to allow uniform treatment of both sides of the substrates. The substrates were initially treated with an argon plasma (2.5 W, 3 cm^3/min, 0.1 mm Hg, 3.3 mm/s). Next, the desired gas mixture was introduced and the plasma was initiated under similar conditions to the argon treatment. TCE was degassed by freeze-thawing once under vacuum, prior to the initiation of glow discharge, and was introduced in the reactor via a micrometering valve. Therefore, the flow rate of this monomer was not measured.

Surface Characterization. Surfaces were characterized by electron spectroscopy for chemical analysis (ESCA) with a Surface Science Laboratories SSX-100 ESCA spectrometer at the National ESCA and Surface Analysis Center for Biomedical Problems (NESAC/BIO) at the University of Washington. Survey scans of 0 to 1000 eV binding energy were run in order to determine the elemental composition of the surfaces. High resolution scans of the carbon 1s (C_{1s}) region (20 eV window) were also recorded and the $\underline{C}H_n$ peak was assigned to 285 eV [14]. For referencing binding energies of the fluorocarbon polymers, either the $\underline{C}F_2$ peak at 292 eV, or the fluorine 1s (F_{1s}) peak at 689 eV was used [14]. The Cl $_{2p3/2}$, at 200 eV, was used for binding energy referencing of chlorine containing glow discharge-deposited films [14].

Advancing contact angles of several test liquids were measured on surfaces using a Rame´-Hart goniometer (Rame´-Hart Inc., Mountain Lakes, NJ). The dispersion force contribution (γ^d_s) and the polar force contribution (γ^p_s) to the total surface free energy (γ_s) of each polymer were obtained from a modification of the procedure employed by Kaelble et al. [15] using the liquids and other procedures described previously [12].

Protein Preparation and Labeling. The following buffers were used for protein labeling and adsorption: CPBSz (0.01 M citric acid, 0.01 M sodium phosphate monobasic, 0.12 M sodium chloride, 0.02% sodium azide, pH=7.4), and CPBSzI (0.01 M citric acid, 0.01 sodium phosphate monobasic, 0.11 M sodium chloride, 0.01 M sodium iodide, 0.02% sodium azide, pH=7.4).

Bovine gamma globulin (BγG, cat. # G-7516) was purchased from Sigma Chemical Co. (St. Louis, MO). Baboon fibrinogen was purified from fresh citrated baboon blood (obtained from the Regional Primate Research Center at the University of Washington) by a polyethylene glycol, β-alanine precipitation protocol as described previously for bovine fibrinogen [16] with the addition of the protease inhibitor Trasylol (40 KIU/ml, Mobay Chemical Co., New York, NY) to the plasma to prevent proteolytic degradation of fibrinogen. The fibrinogen prepared by this method has been shown to be over 90% clottable by thrombin and over 95% precipitable by a 55°C heating step that is to some extent selective for fibrinogen [17].

Na^{125}I was purchased from Amersham Corp. (Arlington Heights, IL). Fibrinogen was labeled with radioactive ^{125}I by the iodine monochloride method of McFarlane [18] as modified by Helmkamp [19] and Horbett [20]. Briefly, 1 mCi of Na^{125}I was added to 0.5 mL of 0.4 M borate and 0.32 M sodium chloride at pH=7.75. This solution was mixed with 0.5 mL of cold iodine monochloride in 2 M sodium chloride. Equimolar concentrations of fibrinogen and iodine monochloride were used. The ^{125}ICl solution was added to 0.5 mL of fibrinogen solution (0.5-5 mg in CPBSz) and mixed by gentle repipetting. The mixture was allowed to react for 20-30 min at room temperature. Unincorporated ^{125}I was separated from the labeled protein by gel chromatography (Bio-Gel P-4, Bio-Rad, Richmond, CA) at room temperature using CPBSz as the mobile phase. Next, the amount of free ^{125}I in the labeled protein solution was determined by instant thin layer chromatography (ITLC) [21] to be approximately 1% of the total ^{125}I present in the protein solution. The labeled protein was collected, stored at -70°C and used within 2 weeks of preparation [21].

Protein Adsorption and Retention. All surfaces were stored overnight in CPBSzI buffer at 4°C. Prior to adsorption, the buffer was replaced with 2 mL of freshly degassed buffer and thermally equilibrated at 37°C for 2 h. Adsorption was initiated by addition of 1 mL of protein solution at three times the desired final concentration and mixed by gentle repipetting. After elapse of 2 h, samples were rinsed with 100 mL of CPBSzI by simultaneous filling and removing of the buffer, which avoids exposure of the surfaces to air [21]. Samples were subsequently placed in polystyrene tubes with 2.5 mL of SDS solution (0.01 M Tris, 0.003 M phosphoric acid, 1 % w/v SDS, pH=7.0) and the amount of radioactivity of each sample was measured in a γ counter (Tracor Analytic Model 1185R, Elk Grove, IL).   The amount of protein initially adsorbed was calculated by dividing the net sample radioactivity (after background subtraction) by the specific activity of the protein solution and planar surface area of the sample.   Following an overnight incubation in the surfactant solution, samples were dip-rinsed nine times in CPBSzI, placed in new polystyrene tubes, the retained radioactivity remeasured, and the amount of protein retained by the surface calculated.   Percent retention was calculated by dividing the amount of protein retained by the amount initially adsorbed multiplied by 100%.   The results are presented as average values ($\pm$ standard deviations) for triplicate samples from the same experiment.

In Vitro **Platelet Adhesion.**   One hundred twenty milliliters of blood from a healthy baboon were withdrawn directly into two disposable plastic syringes containing acid-citrate-dextrose anticoagulant (10% ACD, NIH formula A). Platelet rich plasma (PRP) was obtained by centrifuging blood at $200g$ for 20 min at room temperature.   The PRP was separated and centrifuged for 15 min at $1300g$ to obtain a platelet pellet.   The platelet poor plasma supernatant was removed and the platelet pellet was rinsed with 5 mL of Ringer's-citrated-dextrose (RCD, 102 mM NaCl, 2.8 mM KCl, 1.54 mM $CaCl_2 \cdot 2H_2O$, 21.2 mM $Na_2C_6H_5O_7 \cdot 2H_2O$, 0.5% w/v dextrose, pH=6.5) solution containing 0.03 mg/mL apyrase (cat. # A-9149, Sigma Chemical Co., St. Louis, MO, ADPase activity = 3.8 units/mg, ATPase activity = 4.1 units/mg).   The platelet pellet was resuspended in 9.8 mL of RCD plus 0.03 mg/mL apyrase by gently aspirating the suspension using a disposable plastic pipette.

Platelets were radiolabeled with ^{111}In-tropolone by a modification of the technique developed by Dewanjee et al. [22].   One hundred microliter of tropolone (Sigma Chemical Co., St. Louis, MO) at a concentration of 1 mg/mL was mixed with 100 µCi of [111]Indium monochloride solution (New England Nuclear, Boston, MA).   The mixture was shaken for 2 min and RCD was added to bring the final volume to 200 µL.   The ^{111}In-tropolone solution was added to the platelet suspension and incubated for 30 min at room temperature.   In order to remove any unincorporated ^{111}In, the platelet suspension was centrifuged at $1500g$ for 15 min forming a pellet and the supernatant was discarded.   The pellet was rinsed and resuspended in platelet suspending buffer (145 mM NaCl, 2.7 mM KCl, 4 mM $NaH_2PO_4 \cdot H_2O$, 1 mM $MgCl_2 \cdot 6H_2O$, 2 mM $CaCl_2 \cdot 2H_2O$, 5 mM HEPES, 5.5 mM dextrose, 0.003 mg/mL apyrase, pH 7.4).   The platelet concentration was determined using a Coulter counter (Coulter Electronics, Hialeah, FL) and adjusted with platelet suspending buffer to 1.13×10^9 platelets/mL.   The radioactivity of a 100 µL sample of this suspension was measured with a γ counter and the specific activity of the solution was calculated.

Fibrinogen-coated coverslips were placed in polystyrene cups and incubated in 2 mL of platelet-suspending buffer at 37°C.   Then, 0.25 mL of the labeled platelet suspension was added to each sample bringing the final platelet concentration to 1.25×10^8 platelets/mL.   After 2 h of incubation on an orbital shaker water bath at 37°C and 300 r.p.m., the samples were removed, and dip-rinsed in the platelet suspending buffer.   The amount of radioactivity associated with each sample was measured with a γ counter (Tracor Analytic, Elk Grove, IL).   Following correction for the background radioactivity, the number of adherent platelets was calculated based on the sample radioactivity, specific activity of the platelet suspension, and the planar surface area of each sample.   The data are presented as the mean $\pm$ standard deviations of three replicates for each type of surface.

To verify viability of the platelets following ^{111}In labeling procedure, the ability of labeled platelets to aggregate was analyzed. The platelet suspension was mixed with platelet poor plasma to achieve a platelet concentration of 3.0 x 10^8 platelets/mL. Four hundred fifty microliters of the platelet suspension and an equal volume of platelet free plasma were incubated in two siliconized glass cuvettes at 37°C with continuous stirring. After initial recording of the light transmission through the platelet suspension with a dual chamber aggregometer (Model 530, Chronolog Corp., Havertown, PA) for approximately 1 min, 50 µL of 200 mM adenosine-5-diphosphate (ADP, cat. # A-5410, Sigma Chemical Co., St. Louis, MO) in saline solution was added to the platelet suspension and changes in light transmission were recorded. All labeled platelet suspensions aggregated in response to addition of ADP.

Results

Surface Characterization. The high resolution ESCA C_{1s} spectrum of the untreated PET is composed of three peaks centered at 285, 286.5, and 289 eV (Figure 1). Consistent with the chemical composition of PET, these three peaks have been assigned to the $\underline{C}$-C, $\underline{C}$-O, and O-$\underline{C}$=O functional groups, respectively. PET is composed of ca. 72% carbon and 28% oxygen as determined by ESCA.

The C_{1s} spectrum of ethylene glow discharge-treated PET (E/PET) shows a single peak at 285 eV (Figure 2). E/PET is comprised of ca. 96% carbon and 4% oxygen. The complete coverage of the PET substrate by the ethylene glow discharge-deposited coating is indicated by the near absence of the $\underline{C}$-O and O-$\underline{C}$=O peaks of PET in the C_{1s} spectrum of E/PET. The source of the small amount of oxygen detected on E/PET is most likely post-glow discharge reaction of the surface radicals - generated by the glow discharge process - with the atmospheric oxygen.

A series of surfaces varying in their chlorine content were prepared by mixing ethylene (E) and tetrachloroethylene (TCE) in various ratios in the glow discharge treatment of PET (abbreviated as E-TCE/PET). The chlorine content of the deposited films decreased from 48% for the pure TCE treatment (TCE/PET) to 0% for the pure E treatment (E/PET) (Table I). The complete coverage of the PET substrate by the TCE glow discharge-deposited coating is indicated by the absence of oxygen in the elemental composition of TCE/PET. The oxygen content of the other glow discharge-deposited films is less than 3% which suggests complete coverage of the PET by these films as well. As the amount of TCE in the feed mixture is decreased, there is a decrease in the amount of $\underline{C}$-Cl and $\underline{C}$-Cl$_2$ groups, at 286.5 and 288 eV, as shown in the C_{1s} spectra of E-TCE/PET in Figure 2.

Table I. Surface Composition of Untreated and Glow Discharge-Treated PET

Sample	% C	% Cl	% O
PET	71.9	0	28.1
TCE/PET	52.5	47.5	0
E-TCE (25:75)/PET	53.5	46.5	0
E-TCE (50:50)/PET	59.7	39.1	1.2
E-TCE (75:25)/PET	78.3	19.1	2.6
E/PET	96.2	0	3.8

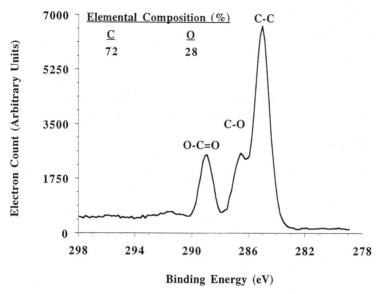

Figure 1.   High resolution ESCA C_{1s} spectrum of PET.

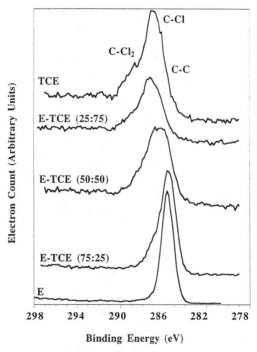

Figure 2.   High resolution ESCA C_{1s} spectra of E-TCE/PET.

The dispersion (γ^d_s) and polar (γ^p_s) components of the surface free energy of untreated PET, 44.9 and 1.6 dynes/cm, are close to the values reported by Owens and Wendt, 43.2 and 4.1 dynes/cm, respectively [23]. As the amount of E in the E-TCE mixture increases, there is a sharp decrease in the γ^p_s and slight increase in the γ^d_s (Table II). There is a negative correlation between the oxygen content of the E-TCE/PET and γ^p_s of these surfaces (Figure 3). Because of this and the fact that no polar element other than Cl was detected on these surfaces, acid-base interactions with Cl, and dipole moments between C and Cl, must be responsible for the relatively high levels of γ^p_s on E-TCE/PET.

Table II. Surface Free Energy of Untreated and Glow Discharge-Treated Polymers

Sample	γ^d_s (dynes/cm)	γ^p_s (dynes/cm)	γ_s (dynes/cm)
PET	44.9 ± 1.7	1.6 ± 0.7	46.5 ± 1.8
TCE/PET	23.5 ± 1.9	6.5 ± 1.3	30.0 ± 2.3
E-TCE(25:75)/PET	23.3 ± 2.5	5.3 ± 1.5	28.6 ± 2.9
E-TCE(50:50)/PET	26.0 ± 3.3	4.8 ± 1.8	30.8 ± 3.7
E-TCE(75:25)/PET	30.9 ± 1.2	2.9 ± 0.7	33.8 ± 1.4
E/PET	35.0 ± 0.5	0.1 ± 0.1	35.1 ± 0.5
E-TFE(75:25)/PET	32.9 ± 1.6	0.9 ± 0.6	33.8 ± 1.7
E-TFE(50:50)/PET	22.5 ± 0.9	2.0 ± 0.6	24.5 ± 1.1
E-TFE (25:75)/PET	15.6 ± 1.0	1.3 ± 0.6	16.9 ± 1.1
TFE/PET	9.5 ± 0.5	1.7 ± 0.5	11.2 ± 0.7

* E-TFE data adapted from reference 13

A series of surfaces varying in their fluorine content were prepared by mixing E and TFE in various ratios for the glow discharge treatment of PET (abbreviated as E-TFE/PET). These materials were described previously [13] but are included here for comparison to the results for TCE-treated surfaces. The fluorine content of the E-TFE deposited films decreases from 60% for the pure TFE treatment (TFE/PET) to 0% for the pure ethylene treatment. The oxygen content of all E-TFE deposited films is < 4%. As the amount of TFE in the feed mixture is decreased, there is a decrease in the amount of fluorinated groups (e.g. $\underline{C}F_3$, $\underline{C}F_2$, $\underline{C}F$-CF_n) and an increase in the amount of hydrocarbons (e.g. $\underline{C}$-CF_n, $\underline{C}$-C) based on the high resolution ESCA C_{1s} spectra [13].

Surface free energy of E-TFE/PET surfaces range from 35.1 dynes/cm for E/PET to 11.2 dynes/cm for TFE/PET [13]. The γ^d_s of E-TFE/PET surfaces decreases linearly from 35 dynes/cm to 9.5 dynes/cm with increasing concentrations of TFE in the feed mixture [13]. The γ^p_s for all surfaces in this series remains relatively low ranging between 0.1 and 2 dynes/cm [13].

Fibrinogen Adsorption. Fibrinogen, 0.1 mg/ml, was adsorbed from a binary mixture with 1 mg/ml BγG on E-TCE and E-TFE series of surfaces for 2 h at 37°C. The amounts of fibrinogen adsorbed on E-TCE surfaces are very similar (Table III). Figure 4 shows fibrinogen retention as a function of γ^p_s of surfaces. Percent

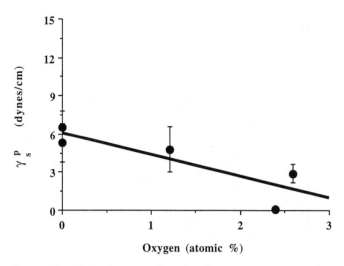

Figure 3.   Correlation between γ^p_s and surface oxygen concentration of E-TCE/PET.

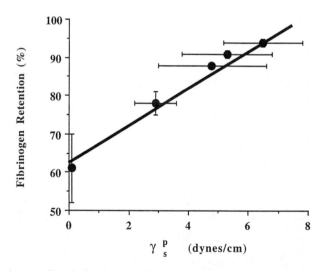

Figure 4.   Correlation between fibrinogen retention and γ^p_s for E-TCE/PET. Fibrinogen, 0.1 mg/mL, was adsorbed for 2 h at 37°C from a binary mixture with 1 mg/mL BγG. Fibrinogen was eluted overnight with 1% SDS.

Table III.   Fibrinogen Adsorption, Retention, and Platelet Adhesion on Untreated and Glow Discharge-Treated Polymers *

Sample	Fib. Adsorbed (ng/cm^2)	Fib. Retained (ng/cm^2)	Fib. Retained (%)	Plt. Adhered (x 1000/cm^2)	Norm. Plt. Adhesion (plts/ng Fib/cm^2)
E/PET	239 ± 5	160 ± 4	67 ± 1	204 ± 18	242 ± 22
E-TCE (75:25)/PET	297 ± 23	230 ± 10	78 ± 3	202 ± 18	193 ± 23
E-TCE (50:50)/PET	283 ± 12	248 ± 10	88 ± 0	119 ± 13	120 ± 14
E-TCE (25:75)/PET	275 ± 1	250 ± 2	91 ± 1	80 ± 15	83 ± 16
TCE/PET	261 ± 5	246 ± 5	94 ± 1	90 ± 24	98 ± 26
E-TFE (75:25)/PET	292 ± 7	203 ± 7	70 ± 2	179 ± 9	174 ± 10
E-TFE (50:50)/PET	269 ± 10	246 ± 17	91 ± 3	141 ± 11	149 ± 13
E-TFE (25:75)/PET	229 ± 1	223 ± 10	97 ± 4	92 ± 9	114 ± 11
TFE/PET	211 ± 7	210 ± 4	99 ± 1	75 ± 2	101 ± 4

* E-TFE data adapted from reference 13

fibrinogen retained increases linearly as the γ^p_s of surfaces increases. Fibrinogen retention also increases as the γ^d_s of E-TCE surfaces decreases (Figure 5). For the E-TFE series, fibrinogen retention increases in an approximately linear fashion as the γ_s and γ^d_s of surfaces decreases and reaches a maximum of 99% for TFE/PET (Table III, reference 13).

In Vitro **Platelet Adhesion.**   Table III presents the platelet adhesion data for E-TCE/PET and E-TFE/PET series before and after normalization by the amount of fibrinogen (determined from ^{125}I labeling) which was preadsorbed on surfaces. Normalized platelet adhesion is maximum on E/PET surface and decreases rapidly as the amount of TCE in the E-TCE mixture is increased.   Normalized platelet adhesion also decreases as the amount of TFE in the E-TFE mixture is increased and the γ^d_s of surfaces decreases (Table III, reference 13).

Discussion

The findings of this study support previous work showing that surface composition influences the state of adsorbed proteins, including fibrinogen [6-8, 12, 13].   The dependence of SDS elutability of fibrinogen on the surface composition is one manifestation of the differences in the state of adsorbed fibrinogen on the different surfaces. Elutability of fibrinogen on E-TFE/PET series of surfaces was previously found to decrease with increase of the amount of surface CF_3 groups and decrease in surface free energy [13].   The surfaces with the highest amount of surface CF_3 groups have the lowest surface energy (in air) and should have the highest interfacial energies in water, with corresponding high driving force for protein adsorption.   This driving force may also promote strong hydrophobic interactions through the formation of multiple protein-surface contact points, leading to the observed stronger binding. For E-TCE treated surfaces, protein retention appears to correlate with both γ^p_s and γ^d_s of surfaces (although the range of values is rather small) (Figures 4 and 5).   These correlations suggest that on TCE-treated surfaces, both polar, acid-base interactions (e.g. hydrogen bonding), and dispersion forces play important roles in retaining the adsorbed protein.   Irrespective of the type of protein-surface interactions, low elutability of a protein may reflect formation of multiple protein-surface bonds.

The number of adherent platelets, normalized by the amount of fibrinogen preadsorbed on surfaces, was previously found to decrease markedly as fibrinogen retention on E-TFE treated surfaces increased (Table III, reference 13). Similarly, normalized platelet adhesion on the E-TCE series of films decreases as fibrinogen retention increases   (Table III).   Thus, fibrinogen adsorbed on TFE and TCE-treated surfaces is apparently in a somewhat inactive state with respect to platelet adhesion.

It is clear from the data presented and discussed so far that platelet adhesion does not correlate with the amount of fibrinogen adsorbed on a surface.   However, platelet adhesion does appear to correlate with SDS elutability of fibrinogen (Figure 6).   The number of platelets adhered is lowest on TFE-and TCE-treated surfaces which exhibit the lowest fibrinogen elutability and highest on E/PET, which is among the highest in fibrinogen elutability.   Thus, tenacious binding of fibrinogen to surfaces may directly affect the ability of fibrinogen to interact with the platelet receptors. Alternatively, the conformation and/or orientation of the adsorbed fibrinogen may differ on the various surfaces as a result of varying degrees of unfolding or spreading of the fibrinogen on the treated surfaces as thesystem tends towards minimization of interfacial energy and maximization of protein-surface interactions.

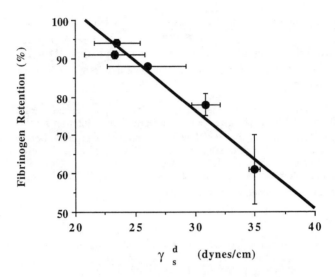

Figure 5.   Correlation between fibrinogen retention and γ^d_s for E-TCE/PET. Fibrinogen, 0.1 mg/mL, was adsorbed for 2 h at 37°C from a binary mixture with 1 mg/mL BγG. Fibrinogen was eluted overnight with 1% SDS.

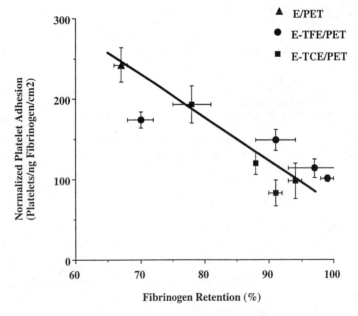

Figure 6.   Correlation between *in vitro* platelet adhesion and fibrinogen retention for E-TFE/PET and E-TCE/PET.

The C-terminus of the γ-chain of fibrinogen is one of the three sites recognized by the glycoprotein IIbIIIa (GPIIbIIIa) complex on the platelet membrane. The sequence of the C-terminus of the γ-chain is a non-RGD (Arg-Gly-Asp) containing decapeptide composed of Leu-Gly-Gly-Ala-Lys-Gln-Ala-Gly-Asp-Val (LGGAKQAGDV) [24]. The other two binding sites of GPIIbIIIa are the RGDS peptide located in the Aα chain of fibrinogen at amino acid numbers 572-575 and the RGDF sequence at Aα 95-98 [24].

Monoclonal antibody binding to the three platelet binding domains of fibrinogen is the same when fibrinogen is adsorbed on untreated or TFE-treated surfaces [13]. These results suggest that the decreases in platelet adhesion to fibrinogen adsorbed on TFE-treated surfaces are not due to changes in the availability of any of the known platelet binding domains of fibrinogen. Instead, we now believe that tight binding of fibrinogen may directly affect the ability of the fibrinogen to interact with the platelet receptors, i.e. fibrinogen must be loosely held to facilitate maximal interaction.

There are some observations from other studies which support a role for the tightness of binding of a protein on a surface in influencing protein-cell interactions. Substrate adherent platelets have been shown to undergo a centripetal redistribution of fibrinogen occupied receptors on their dorsal surface [25]. In addition, movement of fibrinogen adsorbed colloidal gold on the ventral surfaces of adherent platelets has been observed while albumin adsorbed gold particles are not moved [26]. Therefore, platelet interaction with adsorbed fibrinogen may involve movement of the fibrinogen on the substrate, and thus the tightness with which it is bound could be an important aspect. Recent results from another laboratory have shown that the changes in morphology of adherent platelets and the ability of the platelets to remove substrate bound fibrinogen are affected by the residence time of the adsorbed fibrinogen [27]. Another observation that also indicates a possible role for the strength of protein-surface interactions in affecting cell interactions is that migration rate of corneal epithelial cells correlates well with the tightness of binding of substrate adherent fibronectin [28].

Conclusions

Fibrinogen adsorbs tenaciously onto TFE and TCE-treated polymers. Fibrinogen may undergo molecular unfolding or spreading to varying extents on these treated surfaces in order to minimize interfacial free energy (in water) and maximize both hydrophobic and polar surface interactions. As a result, adsorbed fibrinogen can assume multiple states, depending on the substrate, and on TFE or TCE rich surfaces it assumes a state which reduces its recognition and binding to platelet receptors. The correlation between platelet adhesion and fibrinogen retention suggests that adsorbed fibrinogen may have to be loosely held to allow maximal interaction with platelet receptors.

Acknowledgments

Financial support for this work was provided by NIH grants HL 33229-04, HL 19419 and GM 40111-03. We would also like to thank the NESAC/BIO (NIH grant RR 01296) for the use of its ESCA instrument and the Regional Primate Research Center at the University of Washington (NIH grant RR 00166) for the baboon blood draws.

462   PROTEINS AT INTERFACES II

Literature Cited

1. Zucker, M.B.; Vroman, L. *Proc. Soc. Exp. Biol. Med.* **1969**, 131, 318.
2. Chinn, J.A.; Horbett, T.A.; Ratner, B.D.*Thromb. Haemost.*, **1991**, 65, 608.
3. Young, B.R.; Lambrecht, L.K.; Mosher, D.F.; Cooper, S.L. *Adv. Chem. Ser.*, **1982**, 199, 312.
4. Vroman, L. *Biocompatible Polymers, Metals, and Composites*; Szycher, M. Ed.; Technomic Publishing Co. Inc.: Lancaster, PA, 1983; 81.
5. Packham, M.A.; Evans, G.; Glynn, M.F.; Mustar, J.F. *J. Lab. Clin. Med.*, **1969**, 73, 686.
6. Lindon, J.N.; McManama, G.; Kushner, L.; Merrill, E.W.; Salzman, E.W. *Blood*, **1986**, 68, 355.
7. Salzman, E.W.; Lindon, J.; McManama, G.; Ware, J.A. *Ann. N.Y. Acad. Sci.*, **1987**, 516, 184.
8. Chinn, J.A.; Posso, S.E; Horbett, T.A.; Ratner, B.D. *J. Biomed. Mater. Res.*, **1991**, 25, 535.
9. Kiaei, D.; Hoffman, A.S.; Ratner, B.D.; Horbett, T.A.; Reynolds, L.O. Reynolds *J. Appl. Polym. Sci.: Appl. Polym. Symp.*, **1988**, 42, 269.
10. Bohnert, J.L.; Fowler, B.C.; Horbett, T.A.; Hoffman, A.S. Hoffman*J. Biomater. Sci. Polymer Edn.*, **1990**, 1, 279.
11. Hoffman, A.S.; Horbett, T.A.; Bohnert, J.L.; Fowler, B.C.; Kiaei, D. US patent No. 5,055,316; **1991**.
12. Kiaei, D. ; Hoffman, A.S.; Horbett, T.A. *J. Biomater. Sci. Polymer Edn.*, **1992**, 4, 35.
13. Kiaei, D.; Hoffman, A.S.; Horbett, T.A.; Lew, K.R. *J. Biomed. Mater. Res.*, in press.
14. Dilks, A. *Electron spectroscopy: Theory, techniques and applications*; Baky, A.D.; Brundle, C.R., Eds.; Academic Press: London, 1981, 4; 289.
15. Kaelble, D.H.; Moacanin, J. *Polymer*, **1977**, 18, 475.
16. Weathersby, P.K.; Horbett, T.A.; Hoffman, A. S. *Throm. Res.*, **1977**, 10, 245.
17. Horbett, T.A.; Cheng, C.M.; Ratner, B.D.; Hoffman, A.S. *J. Biomed. Mater. Res.*, **1986**, 20, 739.
18. McFarlane, A.S. *Nature*, **1958**, 182, 53.
19. Helmkamp, R.W.; Goodland, R.L.; Bale, W.F.; Spar, I.L.; Mutschler, L.E. *Cancer Res.*, **1960**, 20, 1495.
20. Horbett, T.A. *J. Biomed. Mater. Res.*, **1981**, 15, 673.
21. Horbett, T.A. *Techniques of biocompatibility testing*; D.F. Williams, Ed.; CRC Press, Inc.: Boca Raton, FL, 1986.
22. Dewanjee, M.K.; Rao, S.A.; Didisheim, P. *J. Nucl. Med.*, **1981**, 22, 981.
23. Owens, D.K.; Wendt, R.C. *J. Appl. Polym. Sci.*, **1969**, 13, 1741.
24. Andrieux, A.; Hudry-Clergeon, G.; Ryckewaert, J.-J.; Chapel, A.; Ginsberg, M.H.; Plow, E.F.; Marguerie, G. *J. Biol. Chem.*, **1989**, 264, 9258.
25. Loftus, J.C.; Albrecht, R.M. *J. Cell Biol.*, **1984**, 99, 822.
26. Goodman, S.L.; Lai, Q.J.; Park, K.; Albrecht, R.M. *Proceedings of the XII International Congress for Electron Microscopy*; San Francisco Press: San Francisco, CA, 1990; 22.
27. Sheppard, J. I.; McClung, W.G.; Feuerstein, I.A. *J. Biomed. Mater. Res.,* in press.
28. Pettit, D.K.; Horbett, T.A.; Hoffman, A.S. *J. Biomed. Mater. Res.*, **1992**, 26, 1259.

RECEIVED November 3, 1994

Chapter 33

Serum Protein Adsorption and Platelet Adhesion to Polyurethane Grafted with Methoxypoly(ethylene glycol) Methacrylate Polymers

Maria I. Ivanchenko[1], Eduard A. Kulik[2], and Yoshito Ikada

Research Center for Biomedical Engineering, Kyoto University, 53 Kawahara-cho, Shogoin, Sakyo-ku, Kyoto 606, Japan

The surface of a polyurethane film was subjected to UV-induced graft polymerization of methoxy-poly(ethylene glycol) (PEG) methacrylate monomers in the presence of L-cysteine as a chain transfer agent. The number of ethylene glycol (EG) units in the monomer side-chain was 4, 9, and 24. Adsorption of serum albumin and gamma-globulin as well as platelet adhesion to the grafted films were studied to evaluate the non-fouling property of the PEG enriched surface layer. It was found that the monomer with the smallest length of PEG chain of only 4 EG units was the most "inert" toward the blood components when the graft yield was largely reduced by the use of high concentrations of the chain transfer agent. Staining technique was employed to visualize the graft depth profile and protein penetration into the grafted films. It was concluded that extraordinarily high graft yields were not effective in preventing protein adsorption while very low graft yields were not sufficient to reduce interactions with proteins and platelets.

For many years, segmented polyurethanes (PU) have been employed to fabricate medical devices because of easy molding processes and good mechanical properties. However, problems still remain in achieving good blood compatibility. Therefore, surface modification of PU has been a target of many research groups. Recently we have reported grafting of hydrophilic chains onto the surface of PU (*1*), where methoxy poly(ethylene glycol) (PEG) methacrylate (MnG) was used as the monomer having PEG units. PEG is reported to be relatively nontoxic (*2*) and to be capable of reducing the interactions between blood components and man-made materials (*3,4*). We found that cellulose surfaces grafted with PEG chains exhibited less complement activation as the length of the PEG chain was shorter (*5*). In contrast, other research groups reported that the optimum molecular weight (MW) of PEG for minimum

[1]Current address: Bakulev Institute of Cardiovascular Surgery, Leninsky Per. 9, Moscow, Russia
[2]Current address: Institute of Transplantology and Artificial Organs, Schukinskaya 1, Moscow, Russia

0097–6156/95/0602–0463$12.00/0
© 1995 American Chemical Society

protein adsorption was in a range from 2,000 to 20,000 (6-9). It was also shown that protein adsorption decreased with increase in MW of PEG in spite of the decrease of total PEG content (10). Advantages of high molecular weight PEG in preventing protein adsorption have been calculated also theoretically (11). However, there are several reports which show no significant effect of the PEG length on the non-fouling property of the modified surface (12-14).

The purpose of this study is to gain a deeper insight into the structure of the grafted layer and its interaction with proteins and platelets. The grafted surface is obtained by photo-induced graft polymerization of MnG monomers onto a PU film. To investigate the interactions of the grafted surface with blood components, detailed surface analyses are essential, because graft polymerization occurs not only at the outermost surface but also in the bulk phase far from the surface (15), when polar polymers are used as the substrate of the graft polymerization. As a result, protein molecules may be not only adsorbed on the outermost surface but also can be sorbed into the inside of the grafted substrate polymer. A major difference between graft polymerization of monomers and chemical coupling of existing polymers for the preparation of grafted surfaces is the difficulty in controlling the molecular weight of graft chains obtained by the graft polymerization. MW of the graft chains may be a very important determinant of the microstructure, water content, and compositional depth profile of the grafted layer. In the present study, a chain transfer agent is added to the monomer solution for graft polymerization to vary the length of the graft chain and hence the graft yield, both of which should have a considerable effect on the interaction of the material with proteins and cells.

Experimental

Materials. A PU film from Pellethane (Dow Chemical, 2363-90AE) with a thickness of about 0.2 mm was used after purification by Soxhlet extraction with methanol for 24 h. MnG monomers having 4 (M4G), 9 (M9G), and 23 (M23G) ethylene glycol units were donated by Shin-Nakamura Chemical Co., Ltd., Wakayama, Japan. The chemical structure of the monomers is shown in Fig. 1. The monomers were dissolved in benzene and washed with saturated sodium chloride aqueous solution. Benzene was then evaporated to obtain the pure monomers. L-Cysteine monohydrochloride of extra pure grade and riboflavin (vitamin B2) were purchased from Wako Pure Chemical Industries, Ltd., Tokyo, Japan, and used as received. Human serum albumin (HSA, crystallized) and human gamma-globulin (IgG, crystallized) were purchased from Sigma Co., Ltd., USA, and used without further purification. Na ^{125}I for protein labeling was purchased from Dai-ichi Pure Chemical Co., Ltd., Tokyo, Japan.

Graft polymerization. UV irradiation of the PU film was performed with a high-pressure mercury lamp (75W, Toshiba SHL-100 UV type, $\lambda > 254$ nm). The density of peroxide formed on the PU film by UV irradiation was determined with the iodide method as described previously (Kulik E. A., Ivanchenko M. I., Kato K., Sano S., and Ikada Y. *J. Polym. Sci., Polym. Chem. Ed.,* in press).

The aqueous monomer solution containing riboflavin, L-cysteine, and PU film in a quartz glass tube was irradiated for 2 h at 40 °C. Unless otherwise noted, the monomer concentration was kept at 4 wt% for M4G and M9G and 10 wt% for M23G. To remove homopolymer from the grafted films, they were first washed with running tap water and then in double-distilled water at 75 °C for 24 h under stirring. The amount of grafted MnG polymers was determined by measuring the weight increase of grafted films using an electronic balance with an accuracy of 10^{-5} g. The MW of homopolymers was determined by high performance liquid chromatography with the TOSOH apparatus, equipped with an RI detector and two gel permeation columns (TOSOH PW 3.000 and PW 6.000). Polymers were eluted at a flow rate of 0.9 ml/min using double-distilled water. A set of PEG standards were used for the calibration.

Surface analysis of grafted films. The stationary contact angle of films using a water droplet of about 7 μm diameter was measured using a telescopic goniometer (M 2010-6G II type, Elma Inc., Tokyo, Japan). The measurement was done on at least five different parts of film and averaged. Dynamic contact angle analysis of hydrated films was performed at 25 °C with the Wilhelmy plate technique using equipment manufactured by Shimadzu Inc., Kyoto, Japan (*16*). Five hysteresis loops were collected to give an average value.

FT-IR analysis was done using a spectrophotometer manufactured by Shimadzu Inc. (type 8100) in the ATR mode with a KRS-5 crystal. Spectra were collected as 800 scans at 4 wave number resolution.

XPS spectra were obtained with a spectrometer using a MgK α X-ray source (ESCA 750, manufactured by Shimadzu Inc.). Emitted photoelectrons were detected at an angle of 90° with respect to the sample surface. For determination of O/C stoichiometry, a collecting factor of 2.9 was used for O1s. The XPS data were integrated with an ESCA PAC 760 analyzer.

Cross-sections of grafted films were stained with 1 wt% aqueous solution of Sky Blue 6B dye at room temperature for 2 days, followed by observation under a light microscope.

Protein adsorption. Labeling of proteins with ^{125}I was performed with the chloramine-T method(*17*). The protein concentration was adjusted with phosphate buffered saline solution (PBS, 0.1M, pH 7.4) to 0.15 and 0.2 g/L for HSA and IgG solutions, respectively. Prior to contact with the protein solution, all samples were hydrated in PBS solution for 2 h at room temperature. Protein adsorption was carried out at 37 °C for 1h. After protein adsorption, the films (1cm x1cm) were rinsed six times with 2 ml of PBS. The washing was performed for a total of about 2 h. Five species of each sample were counted and averaged.

Grafted films were subjected to adsorption of HSA and IgG labeled with fluorescein isothiocyanate (FITC). The protein labeling was made in accordance with the procedures described by Goldman(*18*). After liquid gel chromatography the labeled protein solutions were dialyzed against PBS for 8 h at room temperature. The level of labeling, determined from the adsorption spectra of the conjugates, was about one FITC molecule per molecule of HSA and two FITC molecules per IgG molecule. Protein adsorption (staining) was carried out at 37 °C for 24 h from 20 and 6 g/L HSA and IgG solutions in PBS, respectively. After staining, the films were blotted with tissue paper and the cross-section of films with a thickness of 10 μm was observed with a confocal microscope using an argon laser as a source of light. Repeated gel chromatography of protein solutions showed that less than 2 % of the bound FITC became free after the staining procedure.

Platelet adhesion. Venous blood from male rabbits was used to prepare washed platelet suspensions of 1.5×10^8 cells/mL in PBS (1 and Tamada Y., Kulik E., and Ikada Y. *Biomaterials*, in press). One ml of the platelet suspension was added to a PU film of 15 mm diameter in multidish 24 wells made of polystyrene (Corning, USA) and kept for 30 min or 1 h at 37 °C under static conditions. After incubation the film was taken out and dip-rinsed twice with PBS in order to remove the platelets which were not attached to the film surface. After washing, the lysis buffer was added to the film for the subsequent determination of adhered platelets on the same sample by the lactate dehydrogenase (LDH) method (Tamada Y., Kulik E., and Ikada Y. *Biomaterials*, in press).

Results

The density of peroxides generated by UV irradiation on the PU film increased linearly with irradiation time as shown in Fig. 2. This indicates that UV is capable of generating the active groups in PU which are necessary for the subsequent graft polymerization.

To avoid the time-consuming procedure of excluding free oxygen which inhibits graft polymerization, riboflavin was added to the monomer solution. Riboflavin functions not only as a photoinitiator, but also consumes oxygen dissolved in the aqueous medium in the course of UV irradiation to form lumichrome (*19*). In the present study, graft polymerization was carried out at a riboflavin concentration of 2.5×10^{-5} g/L because an excess of riboflavin reduced the effective UV dose on the film surface, and the maximum graft yield and the minimum static contact angle of grafted PU were observed at this initial concentration, as shown in Fig. 3. A certain weight increase was found even without the use of riboflavin, but graft polymerization did not take place unless the monomer solution containing the PU film was UV irradiated.

Effects of L-cysteine addition to the monomer mixture on the graft polymerization are shown in Figs. 4-6. L-cysteine added as a chain transfer agent for radical polymerization (*20*) effectively reduced the MW of homopolymer, as seen in Fig. 4. As shown in Fig. 5, the effect of L-cysteine addition on the weight increase of the grafted PU film was much less significant than on the MW of the homopolymer. It is evident from Fig. 5 that the monomer polymerizability decreased with an increase in PEG length in the side-chain. The grafted PU surface became more hydrophobic with increasing chain transfer agent concentration, as seen in Fig.6.

Protein adsorption to the grafted surfaces is shown in Figs. 7 and 8 for serum albumin and gamma-globulin, respectively. The adsorption dependence on the L-cysteine concentration was similar for both proteins but completely different for the monomers used for graft polymerization. When grafted with M9G and M23G polymers without L-cysteine, PU showed 5 times lower protein adsorption than the virgin PU. This low adsorption was observed for L-cysteine concentrations up to 4×10^{-2} wt %. Adsorption was enhanced by a factor of two at 4 wt% L-cysteine concentration. Protein adsorption to the PU grafted with M4G polymer was considerably reduced with an increase in L-cysteine concentration, approaching one tenth of the adsorption compared to the untreated PU. It follows that the best grafting conditions for the minimization of protein adsorption is to graft polymerize the monomer with the shortest PEG side chain (M4G) at the highest concentration of chain transfer agent. It is interesting to note that M4G was essentially ineffective in reducing albumin adsorption unless L-cysteine was used, as seen in Fig. 7.

The lowest platelet adhesion was observed when the PU films were subjected to polymerization of M4G, regardless of the L-cysteine concentration. The results are given in Fig. 9. It is interesting that the PU film grafted with M4G polymer in the absence of L-cysteine adsorbed a relatively high amount of protein, but adhered a very small number of platelets.

In the case of M4G-grafted PU the amount of adsorbed proteins markedly decreased with a decrease in the graft yield; in other words, with an increase in L-cysteine concentration. To characterize the grafted surface, analysis was done for PU grafted with M4G polymer using XPS, ATR-FTIR, and dynamic contact angle measurements. The results are given in Table I. IR spectra of the M4G homopolymer, the PU film grafted with M4G polymer without L-cysteine, and the virgin PU film are shown in Fig. 10. The absorbance ratio of $I_{1460}/(I_{1530}+I_{1460})$

Methoxy-PEG Methacrylates (MnG)

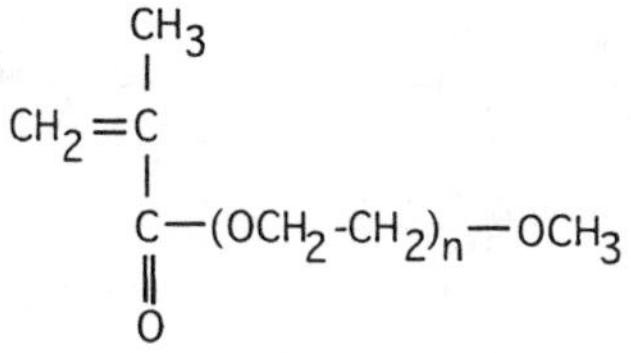

$$CH_2 = \underset{\underset{\underset{O}{\parallel}}{\underset{\displaystyle C-(OCH_2-CH_2)_n-OCH_3}{|}}}{\overset{\overset{\displaystyle CH_3}{|}}{C}}$$

monomer code	number of EG groups n	PEG in monomer, wt %
M4G	4	60.3
M9G	9	77.3
M23G	23	89.7

Figure 1. The chemical structure and EG content of monomers used.

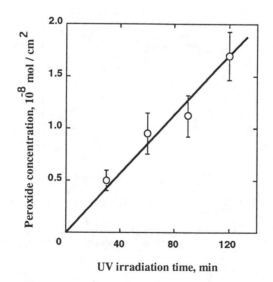

Figure 2. Peroxide formation on the UV irradiated PU as a function of irradiation time.

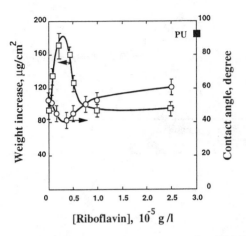

Figure 3. Weight increase and static water contact angle of the PU film grafted with M9G in the absence of L-cysteine as a function of riboflavin concentration.

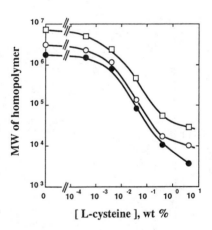

Figure 4. Molecular weight (MW) of homopolymers formed at different concentrations of L-cysteine. M4G (●), M9G (O), and M23G (□).

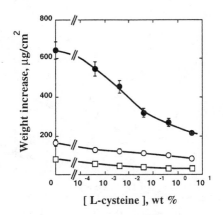

Figure 5. Weight increase of the PU film after graft polymerization. M4G (●),
M9G (O), and M23G (□).

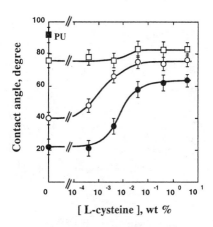

Figure 6. Contact angle of the PU film grafted at different concentrations of L-
cysteine. M4G (●), M9G (O), and M23G (□).

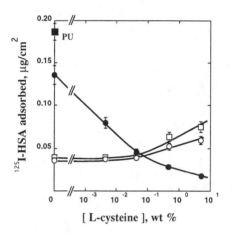

Figure 7. Albumin (HSA) adsorption onto the PU films grafted at different concentrations of chain transfer agent. Monomer: M4G (●), M9G (O), and M23G (□).

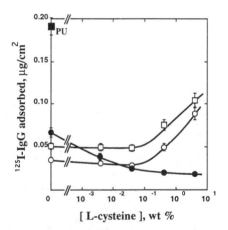

Figure 8. Gamma-globulin (IgG) adsorption onto the PU films grafted at different concentrations of chain transfer agent. Monomer: M4G (●), M9G (O), and M23G (□).

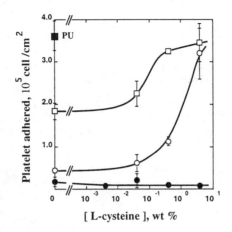

Figure 9. Platelet adhesion to the PU films grafted at different concentrations of chain transfer agent. Monomer: M4G (●), M9G (O), and M23G (□).

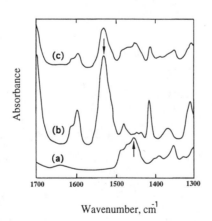

Figure 10. ATR-FTIR spectra of M4G homopolymer (a), virgin PU (b), PU grafted with M4G at 10 wt % monomer concentration in the absence of L-cysteine (c).

was used to compare the graft yield of the modified films. The absorbances at 1460 cm^{-1} and 1530 cm^{-1} correspond to methoxy groups of homopolymers and aromatic groups of the PU, respectively. As the absorbance ratio did not decrease with an increase in the depth of ATR-FTIR analysis, as shown in Table I, it is likely that the grafted polymer was distributed in the bulk phase of PU over a depth of about 100 nm. The data from XPS analysis, which are represented in Table I in terms of the atomic ratio O/(C+O), suggest that even in a very thin surface layer of the film, probably 3-5 nm depth, the content of grafted MnG polymers was likely to be less than 50 %. Both the XPS and ATR-FTIR data confirmed that the surface content of PEG decreased with an increase in L-cysteine concentration. The equilibrium water content, given in Table I, was measured after 24 h incubation in water at 25 °C. It is obvious that the water content of PU films grafted with the M4G polymer decreased with an increase in L-cysteine concentration. However, the significant weight increase seen for the film grafted with the M4G polymer was not observed for those grafted with the M9G and M23G polymers after hydration. The dynamic contact angles of hydrated films revealed that the increase in L-cysteine concentration resulted in decreased contact angle hysteresis.

Table I. Effect of L-cysteine addition to the monomer solution for graft polymerization on the surface properties of PU films modified by graft polymerization of M4G at 10 wt % monomer concentration

| | Sample | | | |
Parameter	PU	M4G (homopolymer)	PU-M4G	PU-M4G--Cysteine
L-cystein, wt %	-	-	0.0	4.0
Graft yield, wt %	-	-	34.9	29.6
ATR-FTIR, I1460/(I1530+I1460)				
Incident angle: 60°	0.04	1.00	0.32	0.14
: 45°	0.04	1.00	0.32	0.13
XPS, O/(C+O)	0.181	0.352	0.263	0.232
Water content, wt%	0.0	-	20.1	9.3
Contact angle[a]:				
advancing, Θa	78	-	45	67
receding, Θr	48	-	8	41
hysteresis, Θa- Θr	30	-	37	26

[a] Determined after hydration of samples

Fig. 11 gives the microphotographs of grafted films after staining with Sky Blue dye. Obviously, the staining is limited to the surface region for the film grafted with M9G, while the films grafted with M4G in the presence and absence of L-cysteine are stained deep into the subsurface region.

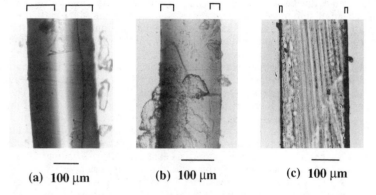

(a) 100 μm   (b) 100 μm   (c) 100 μm

Figure 11. Optical photographs of cross-section of the PU film grafted with M4G using 10 wt % monomer concentration without L-cysteine (a), M4G at 10 wt % monomer concentration and at 4 wt % L-cysteine concentration (b), and M9G at 4 wt % monomer concentration without L-cysteine (c) after staining with Sky Blue. (⌐⌐ : area stained by Sky Blue)

The fluorescent microphotographs of the films after (ad)sorption of FITC-HSA and FITC-IgG are shown in Fig. 12. It is evident from Fig. 12 that the depth of protein penetration decreases in the following order: HSA/PU-M4G(b) > HSA/PU-M4G-Cystein(c) > IgG/PU-M4G(d) > HSA/PU(a) > HSA/PU-M9G (not shown). Interference from free FITC molecules was unlikely to occur, as seen from comparison of Fig. 12(b) and 12(d).

Discussion

The following important findings of this work are helpful in clarifying the mechanism of protein interaction with PEG enriched surface layers:

1. The MnG monomer with 9 ethylene glycol units is very effective in repulsion of proteins and cells from the grafted PU surface unless the chain transfer agent is added.

2. Protein adsorption onto the PU surface is drastically decreased, when it is grafted with M4G polymer in the presence of high concentrations of L-cysteine.

3. The amount of proteins adsorbed to the grafted PU films generally does not correlate well with the number of adhered platelets. A large amount of adsorbed HSA and a small amount of adhered platelets are found for the M4G-grafted PU.

To explain the above findings we assume that the distribution profile of graft chains in the cross-sectional direction of the PU film is considerably different among the monomers used. Recently we have found that only 3 % of peroxide groups generated in the PU film by UV irradiation are located in the vicinity of the film surface and can be digested by peroxidase in aqueous solution (Kulik E. A., Ivanchenko M. I., Kato K., Sano S., and Ikada Y. *J. Polym. Sci., Polym. Chem. Ed.,* in press). Thus one can expect that the graft polymerization has proceeded also into the bulk of the PU. Indeed, as shown in Fig. 11, the M4G polymer chains are present in the deep subsurface of the PU film. The addition of L-cysteine hampered this penetration to some extent, whereas the penetration of M9G polymer into the PU film is limited to a very thin layer from the outermost surface as the stained cross-section indicates.

It is likely that penetration of proteins into the grafted samples is possible , at least when M4G is used as the monomer for graft polymerization. Staining of the cross-section of grafted films by labeled proteins, shown in Fig. 12, confirmed this assumption. Thus, it is conceivable that the main difference among the grafted samples is the variation in depth of protein penetration. From this point of view the lack of correlation between protein adsorption and platelet adhesion to the PU film grafted with the M4G monomer may be explained by the fact that the small-sized protein can penetrate into the bulk of the hydrated sample in marked contrast to the large platelet. The apparently complicated effects of L-cysteine addition on the protein adsorption may be explained in terms of the microstructure of the graft layer which should be closely related to the surface PEG content, the graft yield, and the distribution of graft chains. The chain transfer agent effects appear particularly remarkable for the PU film grafted with the shortest side-chain monomer (M4G). When the graft profile depth is much larger than the protein size, as shown in Fig. 13, the nature of the upper polymer layer is of limited importance. The protein molecules are able to pass through the upper polymer layer as easily as they penetrate into the highly swollen hydrogels which have been used for protein separation and analysis. Whether or not the protein is finally adsorbed may depend on the structure of the semipermeable layer. It is conceivable that penetration and sorption of protein often occurs when hydrophobic polymers are grafted, coupled or blended extensively with a hydrophilic polymer, because the produced hydrophilic layer will attract protein molecules less extensively, at least, at the outermost surface, but allow them to enter into the water-swollen bulk phase. Therefore, extraordinarily high graft yields are not effective in preventing protein adsorption while very low graft yields are not sufficient to reduce the interactions with proteins and platelets.

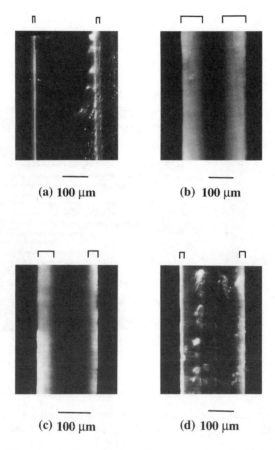

(a) 100 μm   (b) 100 μm

(c) 100 μm   (d) 100 μm

Figure 12. Fluorescent photographs of cross-section of virgin PU film (a, HSA/PU), PU film grafted with M4G at 10 wt % monomer concentration without L-cysteine (b, HSA/PU-M4G), M4G at 10 wt % monomer concentration and at 4 wt % L-cysteine concentration (c, HSA/PU-M4G-Cysteine) after staining with FITC-HSA, and PU film grafted with M4G at 10 wt % monomer concentration without L-cysteine (d, IgG/PU-M4G) after staining with FITC-IgG. (⌐⌐ : area having fluoresence)

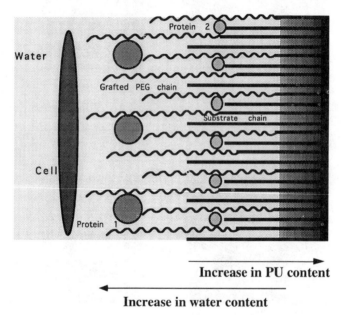

Figure 13. Schematic representation of protein and cell interaction with a modified surface having a graft depth profile much larger than the protein sizes.

Acknowledgments

One of the authors (M.I.I.) thanks the Japan Society for the Promotion of Science for financial support of this study. We would like to acknowledge Dr. E. Uchida, Kacho Junior College, Kyoto, for her help in the grafting experiments.

Literature Cited

1. Fujimoto, K.; Inoue, H.; Ikada, Y. *J. Biomed. Mater.Res.* **1993**, *27*, 347-355.
2. Chaikof,E.L.; Merrill,E.W.; Callow,A.D.; Connolly, R.J.; Verdon, S.L.; Ramberg, K. *J. Biomed. Mater.Res.* **1992**, *2*, 1163-1168.
3. Merrill, E.W.; Pekala, R.W.; Mahmud, N.A. V.3. In *Hydrogels in Medicine* ; Peppas, N.A. Ed.:New York, NY, 1987, V.3, pp..1-16.
4. Sevastianov, V.I.; Kulik, E.A.; Kim, S.W.; Eberhart, R.C. *Biomaterial-Living System Interaction* **1993**, *1*, 3-11.
5. Akizawa,T.; Kino,K.; Koshikawa, S.; Ikada,Y.; Kishida, A.; Yamashita, M.; Imamura, K. *Trans.Am.Soc.Artificial Internal Organs* **1989**, *35*, 333-335 .
6. Nagaoka, S.; Mori, Y.; Takiuchi, H.; Yokota, K.; Tanzawa, H.; Nishiiumi, S. *Polym. Preprints* **1982**, *24*, 67-68.
7. Desai, N.P.; Hubbell, J.A. *Biomaterials* **1991**, *12*, 144-153.
8. Sheu, M.S.; Hoffman, A.S.; Feijen, J. *J.Adhesion Sci.Technol.* **1992**, *6*, 995-1009.
9. Park, K.D.; Okano, T.; Nojiri, C.; Kim, S.W. *J. Biomed. Mater.Res.* **1988**, *22*, 977-992.
10. Gombotz, W.R.; Guanghui, W.; Horbett T.A.; Hoffman, A.S. *J. Biomed. Mater. Res.* **1991**, *25*, 1547-1562.
11. Jeon, S.L.; Lee, J.H.; Andrade, J.D.; De Gennes, P.G. *J. Colloid Inter. Sci.* **1989**, *142*, 149-158.
12. Pekala, R.W.; Merrill, E.W.; Lindon,J.; Kushner, L.; Salzman, E.W. *Biomaterials* **1986**, *7* , 379-385.
13. Amiji, M.; Park, K. *Biomaterials* 1992, *13*, 682-692.
14. Llanos, G.R.; .Sefton, M.V.; *J. Biomater.Sci. Polym.Ed.* **1993**, *4*, 381-400.
15. Uchida, E.; Uyama, Y.; Ikada,Y. *J.Polym..Sci. Polym.Chem.* **1990**, *28* , 2837-2844.
16. Uyama, Y.; Inoue,H.; Ito,K.; Kishida,A.; Ikada,Y. *J.Colloid Interface Sci.* **1991**, *141*, 275-279.
17. Methods in Immunology: a laboratory text for instruction and research. Garvey J.S., Cremer N.E., and Sussdorf D.H. Eds., Benjamin Inc., London, 1977.
18. *Fluorescent Antibody Techniques;* Goldman, M.,Ed.; Academic Press, New York, NY,1968.
19. Song, P.S.; Metzler , D.E.; *Photochem. and Photobiology* **1967**, *6* , 691-698 .
20. Kulik, E.A.; Kato, K.; Ivanchenko, M.I.; Ikada, Y. *Biomaterials* **1993**, *14*, 763-769.

RECEIVED February 10, 1995

Chapter 34

Selective Adsorption of Fibrinogen Domains at Artificial Surfaces and Its Effect on Endothelial Cell Spreading

Hiroko Sato

Department of Polymer Chemistry, Kyoto University, Kyoto 606-01, Japan

Adsorption behavior of fibrinogen domains and the initial attachment of vascular endothelial cells instead of platelets were studied to elucidate the mechanism of thromboresistance on artificial surfaces. For this purpose, fragment X was prepared from a late stage 2 digest of human fibrinogen by plasmin. Amounts of fragment X were much lower than those of fibrinogen adsorbed on polyether urethane nylon (PEUN). Amounts of fragment X adsorbed on PEUN were also lower than on glass and hydrophobic surfaces. Moreover, slow spreading of vascular endothelial cells was observed on fibrinogen-coated PEUN. Thus, the two thirds of the fibrinogen Aα chain from the carboxy terminus was concluded to have high affinity against thromboresistant PEUN surface. Low mobility of this pepetide region, where an amino acid sequence for cell adhesion was imvolved, was considered to result in slow cell spreading, and possibly slow spreading of platelets adhered on fibrinogen-coated PEUN.

The *in vivo* thromboresistance of artificial materials is not always found to correlate with *in vitro* test data. Artificial surfaces where a large number of platelets were adhered in contact with platelet rich plasma (PRP) did not lead always to rapid thrombus formation *in vivo*. For example, segmented polyether urethane nylon (PEUN) showed excellent thromboresistance in a canine peripheral vein for two weeks (*1*), and a copolymer of ethylene and vinyl alcohol (EVAL) was also reported to have excellent thromboresistane from results of hemodialysis (*2*). However, a larger number of adherent platelets were observed on both PEUN and EVAL than on glass and hydrophobic surfaces (*3*). Moreover, the clotting time of platelet poor plasma (PPP) in contact with PEUN or glass was much less than for hydrophobic surfaces (*4*). The adsorption of fibrinogen has also been discussed from the standpoint of thrombogenicity of artificial surfaces. Since initially adsorbed fibrinogen is known to be replaced with coagulation factors (*5*), high amounts of adsorbed fibrinogen would be expected to correlate with enhancement of thrombus

0097–6156/95/0602–0478$12.00/0

formation *in vivo*. Saturation amounts of fibrinogen adsorbed on PEUN, however, were observed to be slightly higher than on glass, as reported previously (*1*). Hence, for PEUN *in vitro* experiments such as number of platelets adhered, plasma clotting time, and amounts of fibrinogen adsorbed appear not to correspond to *in vivo* results of thromboresistance, as described above. In this work, the mechanism of thromboresistance of PEUN will be discussed mainly from the standpoint of fibrinogen adsorption.

Fibrinogen may have multiple properties, derived from its different domains, on adsorption to artificial surfaces. For the purpose of investigating the effects of fibrinogen domains on artificial surfaces, fragment X was prepared from human fibrinogen. A tri-nodular structure of the main portion of fibrinogen should be obtained from an early stage plasmic digest of fibrinogen, because plasmin starts to digest a relatively hydrophilic region of the Aα chain and the amino terminal region of the Bβ chain of fibrinogen (*6-8*). Since plasmin is a less specific protease, fragment X species have a wide range of molecular weight (M_r) from 285kD to 240kD (*9,10*). For this work, a late stage 2 digest of fibrinogen by plasmin was chosen so as to obtain fragment X preparations. Because the Aα chain of fragment X, obtained from a late stage 2 digest, should be hydrolyzed at a cleavage site as far from the carboxy terminus as possible, and thus relatively homogeneous preparations may be expected.

Fibrinogen is an adhesive protein, and plays an important role not only in platelet adhesion but in platelet aggregation (*11*). Therefore, platelets adherent on fibrinogen-coated surfaces may bring about thrombus formation directly. The number of canine platelets in PRP adherent to artificial surfaces was shown to be well correlated with that of adherent bovine aorta endothelial cells (*3*). Moreover, vascular endothelial cells have receptors for cell adhesion of the integrin family similar to platelet receptors (*12*). Thus, it is of interest to compare the adsorption behavior of fibrinogen and fragment X, and to investigate the initial attachment of vascular endothelial cells (instead of platelets) on fibrinogen-coated solid surfaces as these phenomena relate to thromboresistant surfaces.

Experimental

Polymers.   Various artificial surfaces were investigated by coating polymer solutions on glass plates (1x2x0.04 cm) or glass cover slips (1.5 cm in diameter). The quality of glass used (borosilicate glass, Matsunami, Osaka, Japan) was the same as that of cover slips used for microscopy. Glass plates and cover slips were washed with ethanol and then with distilled and deionized water. The same polymers were used as studied previously (*1, 3, 4*): polyhydroxyethyl methacrylate (PHEMA), 2% in a mixture of ethyl alcohol and water (0.95:0.05, w/w), PEUN, 0.3% in dimethyl formamide, EVAL (copolymer of ethylene:vinyl alcohol=0.31:0.69), 2% in dimethyl sulfoxide, poly-γ-benzyl-L-glutamate (PBLG), 1% in chloroform, and polyvinyl chloride (PVC) 5% in methyl ethyl ketone. The PVC used was roughly fractionated in methanol after dissolving in tetrahydrofuran, and its low molecular weight component was removed.

Proteins and fragment.   Human fibrinogen (Grade L, Kabi, Stockholm, Sweden) and plasminogen free human fibrinogen (Imco, Stockholm, Sweden) were used. Fibrinogen-Imco (10 mg/ml) was treated with thrombin (0.2 units/mg fibrinogen, Park Davis, NJ) for two hours in 0.06 M phosphate buffer at pH6.8. Fibrin gel was collected with a spatula, washed with the same phosphate buffer, and dissolved in 0.02 M acetic acid. Then, fibrin monomer, abbreviated as f, in 0.02 M acetic acid was converted to fibrin gel by adjusting above the neutral pH with 0.2 M KOH. Fibrinogen-derived fragment X was obtained from a late stage 2 digest by plasmin, as described previously (13). Streptokinase (10,000 units, Behring, Germany) was added to fibrinogen-Kabi (1 g) to form a complex with and hydrolyze plasminogen (14) bound to fibrinogen. Plasmic digestion of fibrinogen by plasmin was carried out in the presence of calcium ions (27 mM) for 60 min at 37°C. Aprotinin (Trasylol, 211 KIU/ml, Bayer, Leverkusen, Germany) was added to terminate digestion. From the digestion mixture, fragment X was isolated through columns packed with Sepharose CL6B (Pharmacia) and then Ultrogel AcA34 (LKB Instruments, Sweden). The molecular weight of fragment X was estimated to be 242kD from polyacrylamide gel electrophoresis and immunoblotting (15). SDS-PAGE patterns for the fragment X obtained were presented in previous reports for both the non-reduced (13) and reduced (15) states. Polymerization of fragment X was initiated by the release of fibrinopeptide A by thrombin. A low degree of polymerization was obtained owing to the structural impairment. Thrombin-treated fragment X, abbreviated as x, formed no gel state, and 83% of x went into solution at pH 7.4 after overnight incubation(13).

The Aα chain of human fibrinogen (American Diagnostica, New York, NY) was prepared by Dr. Shigeru Hayashi (Thrombosis Chemical Inst., Tokyo, Japan) (16). The Aα chain was isolated by passing fibrinogen through a reverse phase HPLC column packed with TSK gel, phenyl-5PW (Tosoh, Tokyo, Japan) after reduction with dithiothritol. Concentrations of proteins were determined spectrophotometrically from absorbance at 280 nm using specific extinction coefficients; 1.506 for fibrinogen and f; 1.42 for fragment X and x (17); and 0.667 for bovine serum albumin (BSA, Sigma). Amounts of Aα chain in solution were estimated from densitometric scans of SDS-PAGE patterns and from the peak area obtained by HPLC.

Adsorption experiments. Radioiodine labeling using ^{125}I (Du Pont, Wilmington, DE) of fibrinogen, x, and the Aα chain was carried out using Iodogen (Pierce, Rockford, IL). Radiolabeled products were separated through PD-10 columns (Pharmacia LKB). The buffer used for adsorption experiments was phosphate buffered saline (PBS) containing aprotinin (10 KIU/ml) and 0.02% NaN_3 to avoid biodegradation of fibrinogen, fragment X, and the Aα chain. Polymer-coated plates were immersed in solutions of radiolabeled products and rinsed lightly with PBS. The radioactivity of samples was measured with an Auto-well gammacounter ARC 605 (Aloka, Tokyo, Japan).

Initial attachment of vascular endothelial cells. Human umbilical vein endothelial cells (HUVEC) were collected according to the method, reported by Jaffe et al. (18). HUVEC were cultured in medium 199 containing 15% fetal bovine serum (Bioproducts, Walkersville, MD), porcine heparin (90 µg/ml, Opocrin, Modena, Italy), L-glutamine, (8 mM, Wako Pure Chem., Osaka, Japan), endothelial cell growth suppliment (ECGS, Collaborative Biomedical Products, Bedford, MA) and antibiotics such as penicillin (170 units/ml, Meiji Seika Kaisha, Tokyo, Japan), streptomycin (85 µg/ml, Meiji Seika Kaisha), and gentamycin (Schering-Plough, 34 µg/ml, Osaka, Japan). Cultures were carried out in plastic flasks (Iwaki Glass, Tokyo, Japan) coated with porcine gelatin (1%, Nitta Gelatin, Yao, Japan). HUVEC were incubated at 37°C under 5% CO_2, and the medium for cell culture was changed three times a week. HUVEC at confluence were detatched by treatment with 0.02% EDTA (Dojin Lab., Kumamoto, Japan) and 0.25% trypsin (Wako Pure Chem.) in

PBS for 2 min after washing twice with PBS. Cells collected and cultured were identified as vascular endothelial cells from incorporation of acetylated low density lipoprotein labelled with 1,1-dioctadecyl-3,3,3',3'-tetramethyl-indocarbocyanine perchlorate (DiI-Ac-LDL, Biomed Technol., Stoughton, MA) (*19*). The initial attachment of HUVEC on artificial surfaces was observed in serum free medium for 60 min at 37°C. Various artificial surfaces were coated with 56.6 $\mu g/cm^2$ fibrinogen-Imco or f, and rinsed with PBS before seeding HUVEC. Fibrinogen-Imco was used in order to avoid the addition of fibrinolysis inhibitors during storage. HUVEC of passage 11 were seeded in amounts of 15000 cells per well in a 24 well-plate (Iwaki Glass). HUVEC attached to artificial surfaces were washed in PBS, fixed with glutaraldehyde (2.5%) in PBS for 20 min, and washed twice in PBS. The number of HUVEC in an area of 0.636 mm^2 was counted under a Nikon inverted phase-contrast microscope TMD (Nikon, Tokyo, Japan). In addition, the shape of HUVEC was observed using a Nikon fluorescence microscope EFD at an excitation wave length of 540 nm. Cells fixed with glutaraldehyde were treated with 0.2% Triton X-100 (Nakarai Tesque, Kyoto Japan) in PBS. After washing, the above-treated cells were stained with rhodamine-conjugated phalloidin (150 nM, Molecular Probes, Eugene, OR) for 60 min at 37°C and washed three times with PBS (*20*).

Results

Amounts of human fibrinogen adsorbed on various surfaces increased with time and became constant after passing through maxima in most cases (Figure 1a). In general, fibrinogen adsorption did not show a monotonic increase as expected for a Langmuir-type mechanism (1). Fibrinogen adsorption appears to attain equilibrium at about 20 min, and the amounts of fibrinogen adsorbed are almost the same on PEUN and glass. This is in contrast to previous data using glass beads where higher amounts of fibrinogen were adsorbed on PEUN than on glass (*1,3*). It is resemble that the amounts adsorbed on PBLG, a hydrophobic surface, were 3.5 times as high as amounts adsorbed on a hydrophilic surface, PHEMA (Figure 1a). The time dependence of amounts of x adsorbed on various surfaces is shown in Figure 1b. The amounts of x adsorbed at equilibrium on PEUN and EVAL, 0.226 and 0.223 $\mu g/cm^2$ respectively, are less than the amounts of fibrinogen adsorbed at equilibrium. On the other hand, the amounts of x adsorbed at equilibrium are similar for glass, and PBLG and PVC, both of which are hydrophobic.

On the basis of the results as seen in Figure 1, the fraction (F_{ads} - x_{ads})/F_{ads} for each surface at equilibrium, as well as the amounts of fibrinogen and x adsorbed at equilibrium are depicted in Figure 2. This fractional difference (solid bar on Figure 2) in adsorption between fibrinogen and x may be taken as an indicator of the binding of the missing peptide regions (removed from fibrinogen by plasmin) to the various surfaces. The regions removed in transforming fibrinogen to x (*15*) are summarized as follows: fibrinopeptide A (1.6kD), the amino terminal region (5.7kD at longest) of the Bβ chain, and a peptide region (42kD) near the carboxy terminus in the Aα chain. Since the 42kD peptide region of the Aα chain has much higher M_r than the other digested peptides, this region probably contributes most to the difference in adsorption between fibrinogen and x on PEUN.

The adsorption behavior of the Aα chain was investigated next (Figure 3). The Aα chain was studied at low concentration (2 μg/ml) since only small amounts were available. It is noted that adsorption of the Aα chain was high on hydrophobic

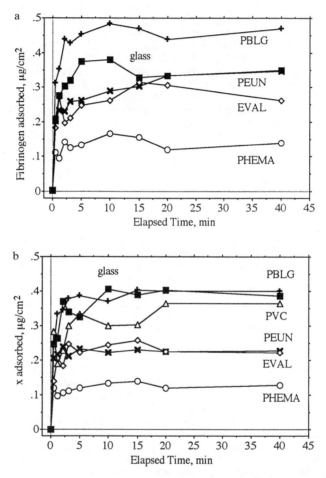

Figure 1. Adsorption kinetics of human fibrinogen (0.1 mg/ml in PBS) on various surfaces (a) and of thrombin-treated fragment X (x, 0.1 mg/ml) on various surfaces (b). Data represents average of two experiments.

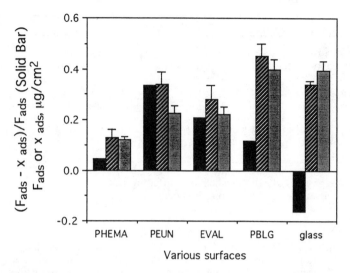

Figure 2. Adsorbed amounts of proteins on various surfaces at equilibrium. F_{ads}, fibrinogen (hatched bar); x_{ads}, x (shaded bar); $(F_{ads}-x_{ads})/F_{ads}$ (solid bar). Data represents average of four experiments.

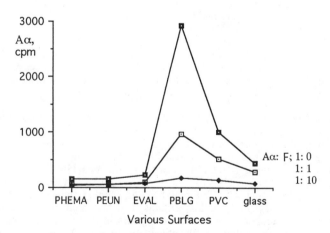

Figure 3. Adsorption of the Aα chain of human fibrinogen (2 μg/ml) on various surfaces in the presence of fibrinogen (F). Data represents average of duplicate experiments. Time of adsorption was 20 min.

Table I. Number of HUVEC/mm^2 Attached on Various Surfaces, Coated with f and fibrinogen (F) or rinsed with PBS

	PHEMA	PEUN	EVAL	PBLG	PVC	glass
f	44.7 ±13.5	11.5 ±4.1	13.8 ±3.9	22.0 ±5.5	109.4 ±41.4	27.5 ±11.9
	(6.3 ±2.5)	(6.0 ±2.4)	(6.3 ±4.7)	(11.3 ±5.2)	(81.8 ±36.5)	(5.5 ±4.2)
F	14.2 ±4.7	5.8 ±0.9	11.5 ±1.9	12.6 ±2.8	28.3 ±4.1	15.4 ±4.6
	(7.4 ±2.4)	(0)	(3.6 ±1.9)	(1.2 ±1.6)	(23.1±7.9)	(6.0 ±7.9)
PBS	3.1 ±3.0	158.8 ±13.7	61.0 ±41.2	13.7 ±6.0	46.1 ±18.2	83.8 ±8.0
	(0)	(0.6 ±0.8)	(0.3 ±0.6)	(0)	(0)	(16.2 ±6.3)

Data is average ± SD of cells attached on 3 different sites.
Brackets indicate the number of spreading cells.

surfaces, particularly PBLG. The Aα chain adsorbed, however, was observed to be desorbed by adding fibrinogen. The fractional desorption of the Aα chain in the presence of ibrinogen at a concentration equal to that of the Aα chain used for adsorption is shown in Figure 4. As is obvious from Figure 4, the Aα chain is easily desorbed from PHEMA and EVAL as well as from PEUN. The Aα chain is not expected to have affinity for hydrophobic surfaces such as PBLG and PVC. Possibly, characteristics of the whole Aα chain work as a hydrophobic peptide, and different from 42kD peptide region from the carboxy terminus.

Fibrinogen is known as one of proteins which promotes cell adhesion. In Table I, the effects of fibrinogen- and f- coating on the various surfaces on the number of cells attached in the early stages (60 min) are summarized. HUVEC were attached in higher numbers on f-coated artificial surfaces than on fibrinogen-coated ones. A large number of HUVEC were attached to PBS-rinsed PEUN just as a high number of platelets were adhered on PEUN (3). It is noted that the number of spread cells is much higher on fibrinogen or f-coated surfaces than on PBS-rinsed surfaces, where most of the initially attached HUVEC were round and unspread in the initial attachment.

Experiments on HUVEC attachment were carried out a number of times. It was found that the number of HUVEC attached changed subtly according to experimental conditions as well as passage number and treatment method. Spread cells were difficult to count without staining with a fluorescent dye. As the result of counting the number (data not shown) of attached HUVEC after fixing with glutaraldehyde, some of HUVEC attached on artificial surfaces seem to be detached through the treatment with Triton X-100 and staining steps even after fixing with glutaraldehyde. Despite possible artifacts of detachment in these experiments on HUVEC, the number of spread cells appears higher on fibrinogen- or f-coated hydrophobic surfaces than on fibrinogen- or f-coated PEUN and EVAL surfaces. Thus, the number of spread cells rather than the number of attached cells seems to be correlated with thrombogenicity of fibrinogen-adsorbed artificial surfaces such as PBLG, PVC, and glass.

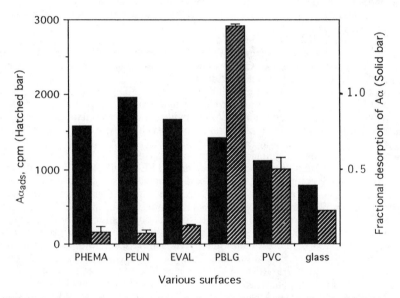

Figure 4. Adsorbed amounts of Aα chain on various surfaces (hatched bars). Fractional desorption of Aα chain in the presence of fibrinogen at a concentration the same as for Aα chain during adsorption (solid bars). Data represents average of duplicate experiments.

Discussion

Fibrinogen has two amino acid sequences for the motif of cell adhesion. One of them, RGDS, is located on the 42kD peptide region near the carboxy terminus, i.e., Aα572-575, as indicated in Figure 5. The Aα chain of fragment X, obtained from a late stage 2 digest, corresponds to the peptide Aα1-206 from which the amino acid sequence Aα79-104 was digested partially by 42% (*15*). The hydrophobicity of Aα1-206 and Aα207-610 was have been plotted using hydrophobicity factors for each of the amino acid residues taken from by Kidera *et al.* (*21*). They assigned hydrophobicity factors to each amino acid residue in different peptide structures on the basis of 188 physical properties. For example, the hydrophobicity of an amino acid residue is different in inside of a protein, i.e., buried amino acid residues, and the surface of the protein in the aqueous environment. In addition, tryptophan, tyrosine, and proline residues have strongly decreased values of hydrophobicity in the buried state, while cysteine residues have a strongly increased value of hydrophobicity on the surface.

The peptide Aα1-206 is considered to be incorporated in the tri-nodular structure of fibrinogen, and hence values of hydrophobicity in the buried state are plotted for this peptide region. On the other hand, values of hydrophobicity assigned to the surface are plotted for the peptide Aα207-610 (Figure 5). Hydrophobicity of the peptide Aα1-206 appears higher than that of the 42 KD peptide, Aα207-610. The region of peptide Aα207-610 seems relatively hydrophilic, and is abundant in serine, threonine, and glycine. The peptide regions, indicated with broken lines in the figure, represent regions of few hydrophobic side chains such as tryptophan, phenylalanine, tyrosine, leucine, and isoleucine residues in an overall hydrophilic region. This suggests that there are an amphiphilic peptide regions.

On the other hand, the structure of segmented PEUN is considered to consist of domains structure composed of soft and hard segments, based on the crystallinity of PEUN segments. Moreover, the EVAL used is a random copolymer composed of hydrophilic (=69%) and hydrophobic (=31%) monomeric units. EVAL is also a crystallizable polymer (*22*), and appears to form a domain structure of crystallites in an amorphous matrix. Thus, the above-described hydrophobic groups in the hydrophilic region of the Aα chain may have affinity to the amphiphilic domain structure on PEUN and EVAL surfaces. Therefore, the affinity of the 42 kD peptide region of the Aα chain for artificial surfaces may lead to less mobility of the motif sequence for cell adhesion, leading to slower and less extensive binding to adhesion receptors on cells. Hence, lower mobility of a ligand region of fibrinogen should lead to slow formation of focal contacts for cell adhesion (*23*), slow organization of microfilament bundles in cytoskelton (*24*), and then slow spreading of cells. In other words, spreading of vascular endothelial cells may be considered to be relatively rapid on fibrinogen-coated glass and hydrophobic surfaces, while on fibrinogen-coated amphiphilic domain structure the cells may spread slowly owing to high affinity of the peptide region containing the cell adhesion ligand for the artificial surface (Figure 6).

In conclusion, fibrinogen adsorbed on artificial surfaces is considered to affect not so much the number attached but the spreading of vascular endothelial cells in the initial attachment phase. A fibrinogen domain, the 42kD peptide region of the Aα chain appears to be adsorbed selectively on PEUN and EVAL surfaces. Slow spreading of HUVEC attached to fibrinogen-coated PEUN and EVAL may be

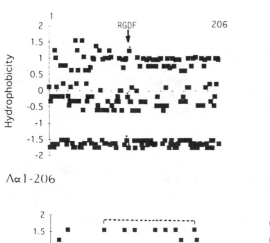

Aα1-206

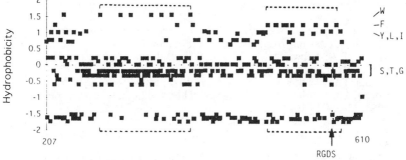

Aα207-610

Figure 5.   Hydrophobicity (*21*) of peptides Aα1-206 and Aα207-610 in the Aα chain of human fibrinogen.

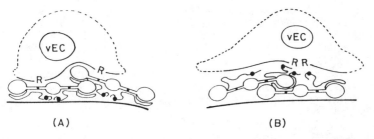

(A)   (B)

Figure 6. Schematic model for vascular endothelial cells attached on fibrinogen-coated amphiphilic domain surfaces (A) and on glass or hydrophobic surfaces (B).

correlated to low platelet spreading and aggregation (observed previously), and ultimately good thromboresistance of PEUN and EVAL.

Acknowledgement

The author thanks Prof. A. Z. Budzynski for helpful discussion and preparation of fragment X, Profs. J. Sunamoto and Y. Imanishi, Radioisotope Research Center and Research Center for Biomedical Engineering, Kyoto University. This work was supported by the 10th Esso Research Grants for Women.

References

1. Sato, H.; Morimoto, H.; Nakajima, A.; Noishiki, Y. *Polymer J.* **1984**, 16, 1-8.
2. Naito, H.; Miyazaki, T.; Shimizu, M.; Obata, S.; Haruna, K.; Kubotsu, A.; Takashima, S.; Takakura, K.; Yamane, T.; Kawahashi, M. *Artif Org.***1984**, 11, 3-6.
3. Sato, H.; Kojima, J.; Nakajima, A. *J. Dispersion Sci. Technol.* **1993**, 14, 117-128.
4. Sato, H.; Kojima, J.; Noishiki, Y. *Koubunshi Ronbunshu* **1991**, 48, 295-302.
5. Vroman, L.; Adams, A.L.; Fischer, G.C.; Munoz, P.C. *Blood* **1980**, 55, 156-159.
6. McKee, P.A.; Schwartz, M.L.; Pizzo, S.V.; Hill, R.L. *Ann. N. Y. Acad. Sci.* **1972**, 202, 127-148
7. Takagi, T.; Doolittle, R.F. *Biochemistry* **1975**, 14, 940-946.
8. Koehn, J.A.; Hurlet-Jensen, A.; Nossel, H.L.; Canfield, R.E. *Anal. Biochem.* **1983**, 133, 502-510.
9. Weisel, J.W.; Papsun, D. *Thromb. Res.* **1987**, 47, 155-163.
10. Marder, V.J.; Shulman, N.R.; Carroll, W.R. *Trans. Assoc. Am. Phys.* **1967**, 53, 156-167.
11. Kloczewiak, M.; Timmons, S.; Lukas, T.J.; Hawiger, J. *Biochemistry* **1984**, 23, 1767-1774.
12. Arnaut, M.A. *Blood,* **1990**, 75, 1037-1050.
13. Sato, H.; Weisel, J.W. *Thromb. Res.* **1990**, 58, 205-212.
14. Robbins, K.C.; Markus, G. in *Fibrinolysis*; Gaffney, P.J. and Balkuv-Ulutin, S., Ed.: Academic Press: London, **1978**, pp.61-75.
15. Sato, H.; Swadesh, J.K. *Int. J. Biol. Macromol.* **1993**, 15, 323-327.
16. Hayashi, S. *Blood Coagulation and Fibrinolysis* **1993**, 4, 921-926.
17. Marder, V.J.; Shulman, N.R. *J. Biol. Chem.* **1969**, 244, 2111-2119.
18. Jaffe, E.A.; Hoyer, L.W.; Nachman, R.L. *J. Clin. Invest.* **1973**, 52, 2757-2764.
19. Voyta, J.C.; Via, D.P.; Butterfield, C.E.; Zetter, B.R. *J. Cell. Biol.* **1984**, 99, 2034-2040.
20. Füchtbauer, A.; Jockusch, B.M.; Maruta, H.; Kilimann, M.W.; Isenberg, G. *Nature* **1983**, 304, 361-364.
21. Kidera, A.; Konishi, Y.; Oka, M.; Ooi, T.; Scheraga, H.A. *J. Protein Chem.* **1985**, 4, 23-55.
22. Matsumoto, T.; Nakamae, K.; Ogoshi, N.; Kawazoe, M.; Oka, H. *Kobunshi Kagaku* **1971**, 28, 610-617.
23. LaFlamme, S.E.; Akiyama, S.K.; Yamada, K.M. *J. Cell. Biol.* **1992**, 117, 437-447.
24. Dejana, E.; Colella, S.; Languino, L.R.; Balconi, G.; Corbascio, G.C.; Marchisio, P.C. *ibid.* **1987**, 104, 1403-1411.

RECEIVED February 10, 1995

PROTEIN BEHAVIOR AT FLUID–FLUID INTERFACES

Chapter 35

Phospholipase A_2 Interactions with Model Lipid Monolayer Membranes at the Air–Water Interface

K. M. Maloney[1], M. Grandbois[2], C. Salesse[2], D. W. Grainger[3,5], and A. Reichert[4]

[1]Department of Chemistry, Biochemistry, and Molecular Biology, Oregon Graduate Institute of Science and Technology, Portland, OR 97291–1000
[2]Centre de Recherche en Photobiophysique, Université du Québec à Trois-Rivières, 3351 boulevard des Forges, C.P. 500, Trois-Rivières, Québec G9A 5H7, Canada
[3]Department of Chemistry, Colorado State University, Fort Collins, CO 80523
[4]Materials Science Division, Lawrence Berkeley Laboratory, Berkeley, CA 94720

Mechanisms of domain formation by the membrane-active enzyme, phospholipase A2 (PLA_2), in hydrolyzed phospholipid monolayers at the air-water interface have been investigated. PLA_2-catalyzed hydrolysis of phospholipid membrane substrates yields both free fatty acid and lyso-lipid reaction products. After a certain extent of monolayer hydrolysis, PLA_2 forms large, regular two-dimensional aggregates at the monolayer interface. PLA_2-catalyzed monolayer hydrolysis has been investigated using monolayer isotherm techniques, surface potential, and dual-label fluorescence microscopy at the air-water interface, as well as electron microscopy. Our results indicate that PLA_2-produced fatty acid reaction products laterally phase separate from remaining enzyme substrate and lyso-lipid, forming two-dimensional, anionic domains. Results support our hypothesis that enzyme domains are formed by PLA_2 binding electrostatically to phase separated fatty acid through basic amino acid residues located on the enzyme's interfacial binding surface.

Monolayers at the air-water interface (*1*) have been used as organized molecular scaffolds for two-dimensional protein crystallization (*2*), and as the basic structural unit for biosensing devices (*3*). As model membrane systems, monolayers have also been used to investigate specific protein-membrane interactions in natural and synthetic systems (*4*). Membrane lateral density and phase behavior are

[5]Corresponding author

experimentally accessible parameters, allowing the determination of macromolecule-membrane interactions as functions of monolayer membrane surface density and phase behavior.

Application of surface characterization techniques such as scanning probe microscopy, quartz crystal microbalance, optical spectroscopy, and scattering methods, as well as traditional monolayer isothermal compression and fluorescence microscopy allow in-depth characterization of monolayer and monolayer-protein assemblies (see ref. *5* for a review).

Naturally, a cascading series of biological events often results from protein membrane recognition, binding and subsequent biochemical events. Biological signal transduction from phospholipase A2 (PLA$_2$) is one example (*6*). Calcium-dependent, secreted PLA$_2$'s are small (M$_r$ $\approx$ 14 kD), water-soluble enzymes derived from many sources, most common of which are mammalian pancreas and snake venoms. PLA$_2$ catalyzes the stereospecific hydrolysis of sn-2 phospholipid acyl chains to produce a membrane-resident fatty acid and lyso-lipid species (*7*). Frequently, the fatty acid released is arachidonic acid, a precursor in the potent eicosanoid pathway (*8*).

Extensive efforts have been made to elucidate the interfacial mechanism of PLA$_2$ hydrolytic activity (*9*). It is well known that the enzyme displays maximal activity towards organized lipid interfaces such as vesicles, micelles, and monolayers (*10*). The influence of the physical state of the lipid interface on enzyme activity has been the focus of our investigations. We and coworkers have focused on elucidating the mechanism of PLA$_2$ domain formation following lipid membrane hydrolysis at the air-water interface (*11–14*).

We present here a mini-review of our work relating to PLA$_2$ domain formation in these systems. We discuss our initial enzyme self-assembly results, an experimental system aimed at testing our domain formation hypothesis, and finally, our attempts to mimic PLA$_2$ domain formation using a model mixed monolayer system with a water-soluble cationic dye protein analog.

Experimental

Materials. Rhodamine-labeled phosphatidylethanolamine or 1-palmitoyl-2-[12-(7-nitro-2-1,3-benzoxadiazol-4-yl)amino]dodecanoyl-phosphatidylcholine (Rhod-DPPE and C12-NBD-PC, respectively) and cationic dye 1,1′,3,3,3′,3′-hexamethylindocarbo-cyanine iodide (H-379), were purchased from Molecular Probes (Eugene, OR) and used as received. Nonfluorescent lipids and fatty acids were from Avanti Polar Lipids (Alabaster, AL) and Fluka AG, respectively. The asymmetric lipids 6,16-PC and 16,6-PC used in the hydrolysis experiments were synthesized via the fatty acid imidazole method (*15,16*), and were generously synthesized, purified, and supplied by Prof. Mary Roberts (Boston College). *Naja N.N.* PLA$_2$ was purchased from Sigma and used as received.

Enzyme Labeling. PLA$_2$-fluorescein (PLA$_2$-FITC) conjugates were prepared by the general method of Nargessi and Smith (*17*). Labeling efficiencies were statistical, with [FITC]/[PLA$_2$] ratios typically less than 0.75. FITC concentrations were determined by UV absorbance at 490 nm and protein concentration determined by the method of Smith and coworkers (*18*).

Fluorescence Microscopy. Fluorescence microscopy of monolayers at the air-water interface was conducted on home-built, microprocessor-controlled Teflon troughs mounted on a stage of a Zeiss ACM epifluorescence microscope (*19*). Image contrast of lipid monolayers by fluorescence microscopy is based on differential solubilities of fluorescently-labeled lipid probes between monolayer phases (*20–22*). In phase-separated monolayer systems exhibiting a coexistence of fluid and condensed phases, Rhod-PE and C12-NBD-PC, preferentially partition into fluid monolayer phases. Thus, fluid phases appear bright (enriched with fluorescent lipid probe) and condensed phases appear dark (probe depleted regions). Monolayer probe concentrations were between 0.5 and 1 mol%. Using excitation and emission specific fluorescence filters integrated with the fluorescence microscope, it is possible to conduct dual-label imaging experiments with rhodamine-labeled lipid and fluorescein-labeled PLA$_2$ (*11,12*). Specific details concerning fluorescence microscopy of lipid monolayers at the air-water interface are found elsewhere (*19,22*).

Dye Binding to Monolayers. The water soluble, cationic dye H-379 was used to probe electrostatic characteristics of phase separated monolayers. H-379 was injected beneath the monolayer and into a Teflon mask positioned in the trough (*11,14*). The mask facilitated monolayer imaging by reducing monolayer surface flow. Monolayer images from inside and outside the mask were frequently compared to ensure the mask did not create any monolayer artifacts. Lipid monolayer imaging was achieved with C12-NBD-PC probe and was directly visualized with a fluorescein specific fluorescence filter. Switching between fluorescein and rhodamine specific filters allowed tandem imaging of lipid monolayer physical state and H-379 dye location, respectively.

Surface Potential. Monolayer surface potential was determined using the ionizing electrode method (*1*). An [241]Am electrode was placed 1-2 mm above a monolayer covered interface. A Pt electrode in the monolayer subphase was used as a reference electrode. Surface potential and molecular area at constant surface pressure (determined with a Wilhelmy plate) were measured simultaneously as a function of time. Surface potential, ΔV (mV), is defined as the difference in potential of a monolayer film versus a clean interface. During hydrolysis experiments, the monolayer subphase was stirred.

Electron Microscopy. L-α-DPPC monolayers were hydrolyzed by PLA$_2$ at 22 mN/m on a Tris-buffered subphase (30°C). PLA$_2$-hydrolyzed DPPC monolayers were transferred to carbon-coated EM grids by the Langmuir-Schaefer technique (*23*). Protein staining was achieved using a 2% uranyl acetate solution for 30 seconds.

Results

Dipalmitoylphosphatidylcholine (DPPC) monolayers at the air-liquid interface exhibit several phase transitions while undergoing isothermal compression. While great debate has taken place over whether or not any of the observed transitions are truly first-order (*4*), fluorescence and Brewster angle microscopy at the air-liquid interface have unequivocally shown a coexistence of liquid-expanded and solid monolayer phases in the near-horizontal region of the DPPC isotherm. A fluorescence image of a DPPC monolayer compressed beyond the phase transition point into this phase coexistence region (*19–22*) is shown in Figure 1a, as viewed through a rhodamine-specific (monolayer) fluorescence filter (*11,12*). Clearly visible are dark regions depleted of monolayer fluorescent probe solid domains surrounded by fluid LE lipid phase (fluorescent probe rich). FITC-labeled PLA$_2$ was then injected into the DPPC monolayer subphase beneath the monolayer (*11*). As viewed through a fluorescein-specific fluorescence filter, Figure 1b shows a nearly homogeneous fluorescence signal from FITC-PLA$_2$ at the monolayer interface. As time after PLA$_2$ injection increases, solid domains are hydrolyzed by the enzyme, as evidenced in Figures 1c and e. The interior of solid DPPC domains appear to be slowly degraded as the enzyme hydrolyzes at the interface between the solid DPPC domains and LE phase DPPC. However, as shown in Figure 1d (25 minutes after PLA$_2$ injection), the FITC-PLA$_2$ fluorescence signal is no longer homogeneous. Small areas of concentrated fluorescence are visible in the monolayer plane. These small areas represent aggregated FITC-PLA$_2$ at the hydrolyzed monolayer interface. As hydrolysis time increases, the enzyme domains increase in size (Figure 1f, 40 minutes after PLA$_2$ injection) and number. Moreover, the fluorescent PLA$_2$ domains in Figure 1f are also seen when the lipid monolayer is imaged with a rhodamine filter (non-fluorescent gray regions in Figure 1e corresponding to bright regions in Figure 1f). Finally, 60 minutes after PLA$_2$ introduction to the monolayer subphase, fluorescent PLA$_2$ domains are quite large (Figure 1h) with the fluorescence signal corresponding exactly to the dark bean-shaped regions in Figure 1g. PLA$_2$ domains depicted in Figure 1 resist lateral compression and dissolution at high and low surface pressures, respectively (*12*). This is consistent with tight enzyme packing and lateral enzyme association in the interfacial plane.

In an attempt to mimic PLA$_2$ domain formation, we characterized binary and ternary mixed monolayers containing DPPC, C16Lyso and palmitic acid (PA). Ternary mixed monolayers represent the composition of PLA$_2$ hydrolyzed DPPC monolayers at various extents of hydrolysis. Initially, the mole fraction of DPPC is unity. As a function of increasing PLA$_2$ hydrolysis time, mole fraction of DPPC approaches zero, while mole fractions of C16Lyso and PA (enzymatically produced in equimolar amounts) each approach 0.5. Representing various extents of hydrolysis, mixed monolayers of C16Lyso:PA (1:1; mol:mol) and DPPC:C16Lyso:PA (0.2:1:1, 1:1:1, 2:1:1, up to 10:1:1; mol:mol:mol) were investigated (*14*). C16Lyso:PA (1:1) binary mixed monolayers represent a totally hydrolyzed lipid monolayer while DPPC:C16Lyso:PA ternary mixed monolayers represent partially hydrolyzed monolayer compositions. Shown in Figure 2a is a fluorescence micrograph of ternary mixed DPPC:C16Lyso:PA (0.2:1:1) monolayers (no enzyme). Compression of the monolayer yields the phase separated microstructure shown in Figure 2. Injection of cationic, water-soluble H-379 dye results in rapid dye binding to the phase separated microstructure (Figure 2b), showing the domain region is enriched in anionic components (fatty acid). As mole fraction of DPPC increases, the size of phase separated anionic domains decreases, and is absent on the micrometer scale once a 3:1:1 (DPPC:C16Lyso:PA) ternary mixed monolayer ratio is obtained. Phase separation of anionic domains such as that shown in Figures 2a and b are *only observed in the presence of Ca^{2+} and on an alkaline monolayer subphase*. This is consistent with a Ca^{2+}-mediated fatty acid

Rhodamine Filter   Fluorescein Filter
(DPPC Monolayer)   (Phospholipase)

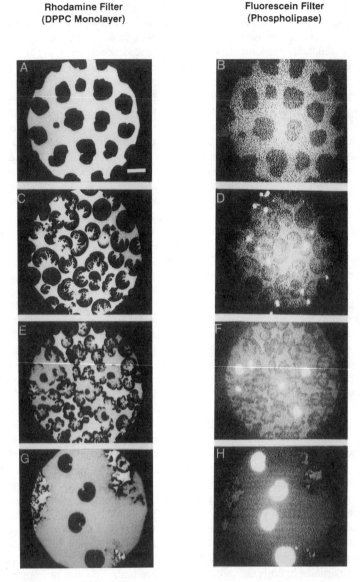

Figure 1. Fluorescence micrographs of PLA_2 hydrolyzed DPPC monolayers.
Figures **a, c, e,** and **g** were imaged through a filter specific for rhodamine
fluorescence (monolayer lipid probe). Figures **b, d, f,** and **h** were imaged
through a fluorescein specific fluorescence filter. Image pairs **a-b, c-d, e-f,**
and **g-h** were taken 0, 25, 40, and 60 minutes after PLA_2 injection into the
monolayer subphase, respectively. Subphase is 100 mM NaCl, 10 mM Tris,
5 mM $CaCl_2$, pH 8.9, T = 30°C. Surface pressure (22 mN/m) remained
constant throughout hydrolysis. Scale bar in **a** is 20 μm. Reprinted with
permission from Elsevier Scientific Press.

a

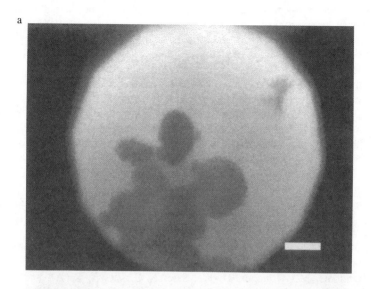

b

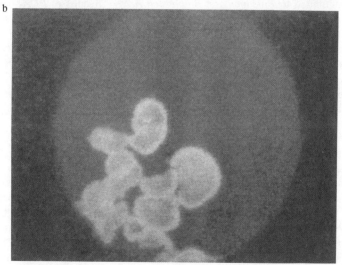

Figure 2.   Fluorescence micrographs of phase separated, ternary mixed monolayers (DPPC:C16Lyso:PA, 0.2:1:1).   Images in **a** and **b** as viewed through a fluorescein (monolayer filter) and rhodamine (H-379) specific fluorescence filter, respectively, at 12 mN/m.   Monolayer subphase is 100 mM NaCl, 10 mM Tris, 5 mM $CaCl_2$, pH 8.9, T = 20°C.   Scale bar = 25 μm.

chelating event. Alkaline subphase pH is required to keep fatty acid carboxylate head-groups ionized (pK_a ca. 5-6). By contrast, replacing palmitic acid in ternary mixed monolayer systems with its unsaturated analog, palmitoleic acid, does not yield phase separation of fatty acid enriched microstructures upon monolayer compression (unpublished data). Palmitoleic acid is unable to pack into condensed phases due to its acyl chain unsaturation. Additionally, phase separated polymerized diacetylenic fatty acid (DIAC-FA) surrounded by fluid D-DPPC lipid matrix does not lead to PLA_2 adsorption (13). However, cationic dye adsorption does occur under these phase separated DIAC-FA monolayer domains, but presumably during DIAC-FA phase separation and polymerization other monolayer species important for PLA_2 binding (i.e., PC) are expelled from phase separated DIAC-FA (13). This shows that fatty acid chemistry and the presence of substrate PC both influence protein interfacial domain formation.

To account for PLA_2 self-assembly at hydrolyzed phospholipid monolayers at the air-liquid interface, we and coworkers proposed that PLA_2-produced fatty acid reaction products were responsible for inducing interfacial enzyme domain formation (11,12). To test this hypothesis, the importance of fatty acids, as well as lyso-lipid and lipid substrate, on PLA_2 domain formation was investigated using asymmetric phospholipid substrates. Hydrolysis of these substrates results in formation of selectively water-soluble fatty acids and water-insoluble lyso-lipids (Maloney, K. M., Grandbois, M., Grainger, D. W., Salesse, C., Lewis, K. A., and Roberts, M. F., submitted to *Biochimica et Biophysica Acta*). Results of PLA_2 hydrolysis of 1-palmitoyl,2-caproyl-phosphatidylcholine (16,6-PC) monolayers are shown in Figure 3a. PLA_2 hydrolysis of 16,6-PC produces water-soluble caproic acid and water-insoluble palmitoyl-lysophosphatidylcholine (C16Lyso). Total hydrolysis of a 16,6-PC monolayer by PLA_2 thus results in formation of a pure C16Lyso monolayer. Plotted in Figure 3a is monolayer molecular area and surface potential as a function of PLA_2 hydrolysis time. The horizontal regions of the surface potential and monolayer area curves (segments AB and A'B', respectively) reflect the stability of 16,6-PC monolayers in the absence of enzyme. At point B, PLA_2 was injected into the monolayer subphase. Immediately following PLA_2 injection, surface potential and monolayer molecular area start to decrease. This decrease is linear as a function of time and continues to point C. At point C, surface potential and monolayer area continue to decrease, but at a slower rate. Point C is taken as the end of 16,6-PC monolayer hydrolysis, and the remaining slow decrease in surface potential and area reflecting the instability of the remaining C16Lyso monolayer. At the end of hydrolysis (point C), monolayer molecular area is 49.5 $Å^2$/molecule and the surface potential is 248 mV, corresponding closely to molecular area and surface potential expected for a pure C16Lyso monolayer under these conditions (50 $Å^2$/molecule and 220 mV, respectively).

Shown in Figure 3b is surface potential and molecular area versus PLA_2 hydrolysis time of 1-caproyl,2-palmitoyl-phosphatidylcholine (6,16-PC), the positional isomer of 16,6-PC. PLA_2 hydrolysis of 6,16-PC produces water-soluble caproyl-lysophosphatidylcholine and water-insoluble palmitic acid. PLA_2 hydrolysis of 6,16-PC monolayers results in the formation of a pure palmitic acid monolayer. Similar to Figure 3a, surface potential and molecular area are stable prior to enzyme introduction (segments AB and A'B', respectively). After PLA_2 injection beneath

a)

b)

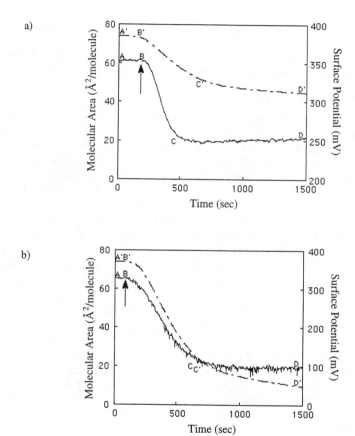

Figure 3. Surface potential (—) and monolayer molecular area (– – –) versus PLA$_2$ hydrolysis time at constant surface pressure (15 mN/m). **a** and **b** represent 16,6-PC and 6,16-PC monolayers respectively. Monolayer subphase is 100 mM NaCl, 10 mM Tris, 5 mM CaCl$_2$, pH 8.9, T = 20°C, compression rate = 2 Å^2/molecule/minute. PLA$_2$ injected at point B.

the monolayer (point B), surface potential and monolayer area immediately decrease until point C (taken as end of hydrolysis) is reached. The molecular area at point C is 19 Å^2/molecule and the surface potential is 100 mV. The area closely corresponds to the molecular area for a fatty acid molecule under these conditions, (ca. 20 Å^2/molecule) while the surface potential value does not correspond to the value expected for a palmitic acid monolayer (-30 mV) under these conditions.

As shown in Figures 4a and b, PLA_2 was observed to form domains in hydrolyzed 6,16-PC monolayers. Figure 4a shows a gray, phase separated microstructure in a rhodamine specific fluorescence filter. Imaging the same field of view with a fluorescein filter shows that FITC-PLA_2 fluorescence corresponds exactly to the gray region in Figure 4a. By contrast, PLA_2 hydrolysis of 16,6-PC monolayers results in the formation of a C16Lyso monolayer (fatty acid products are solubilized in the subphase) *with little PLA_2 domain formation*. Moreover, addition of fatty-acid binding albumin to the monolayer subphase during PLA_2 hydrolysis of 16,6-PC monolayers totally suppressed PLA_2 interfacial aggregation.

Surface potential and molecular area hydrolysis kinetics (Figure 3) show that PLA_2 does not contribute to the monolayer molecular area. Enzyme domains shown in Figure 4 are located beneath the monolayer and remain stable for hours. Similarly, PLA_2 domains shown in Figure 1 also lie directly beneath the hydrolyzed monolayer. Control experiments consisting of PLA_2 injection beneath both pure palmitic acid (unpublished data) and palmitoyl-lysophosphatidyl-choline (*12*) monolayers were also conducted. PLA_2 does not form domains beneath pure lyso-lipid monolayers. PLA_2 does, however, form domains beneath pure palmitic acid monolayers (Ca^{2+}-containing pH 8.9 subphase), though exhibiting highly irregular domain morphologies distinct from those formed in the presence of PC substrate.

To further assess the microstructure of PLA_2 domains, hydrolyzed monolayers were transferred to EM grids. Figure 5 shows an electron micrograph of a PLA_2-hydrolyzed L-DPPC monolayer (*24*). Small regions showing hexagonally ordered protein are visible. Experimental difficulties (transfer of fluid phase monolayers) prevented full transfer of intact PLA_2 domains (ca. 25 μm in diameter). Hexagonally arranged single PLA_2 crystals in Figure 5 most likely represent fragments of 2-D larger PLA_2 domains.

Discussion

Figure 6 depicts our PLA_2 domain formation hypothesis (adapted from ref. *11*). PLA_2 catalyzes the chemical conversion of a single component lipid monolayer to a ternary mixed, heterogeneous interface. We propose that after a certain extent of enzyme hydrolysis, fatty acid reaction products phase separate from remaining substrate and lyso-lipid. The result is the formation of anionic fatty acid enriched microstructures within the monolayer which prompt PLA_2 interfacial binding. Figure 1 shows, as a function of hydrolysis time, the formation of these PLA_2 microstructures. Previous accounts concerning analogous membrane microstructuring from PLA_2 action on vesicle bilayers have been reported by Jain and coworkers (*25,26*). They showed ternary mixed vesicle systems undergo lateral phase segregation yielding bilayer regions enriched in fatty acids as monitored using

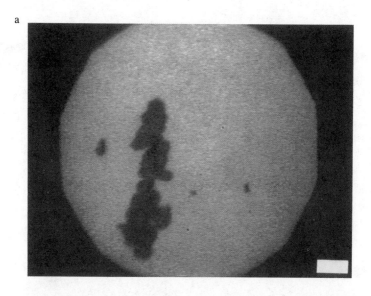

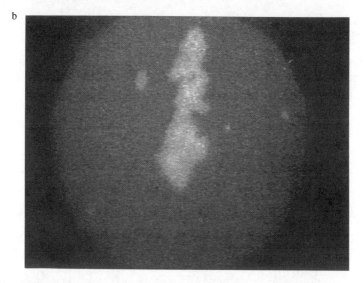

Figure 4.   Fluorescence images of PLA$_2$ hydrolyzed 6,16-PC monolayers at 15 mN/m.   Images in **a** and **b** are viewed through rhodamine and fluorescein filters, respectively.   Monolayer subphase is 100 mM NaCl, 10 mM Tris, 5 mM CaCl$_2$, 25 nM BSA pH 8.9, T = 20°C.   Scale bar in **a** is 25 μm.

Figure 5.   Electron micrograph of PLA$_2$-hydrolyzed L-DPPC monolayers. Hydrolyzed monolayers were transferred to EM grids via the Langmuir-Schaefer technique at 22 mN/m.   Scale bar is 40 nm.

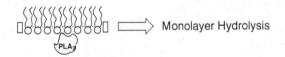

Figure 6.   Protein domain formation hypothesis for PLA$_2$-hydrolyzed phospholipid monolayers at the air-water interface.   See text for details.

cationic dye binding and fluorescence techniques and PLA_2 hydrolysis (25,26). Though PLA_2 domain formation has not been shown in vesicle systems, acceleration of enzyme hydrolysis kinetics has been correlated with vesicle fatty acid lateral segregation (26). Both monolayers and vesicles share similar physical membrane characteristics after limited PLA2 hydrolysis: laterally phase separated membrane regions enriched in fatty acid. More recently, further evidence supporting lateral phase separation in ternary vesicle systems has also been presented by Biltonen and coworkers using fluorescence techniques (27).

To test the enzyme domain formation hypothesis in monolayers, we employed asymmetric lipid substrates: enzymatic hydrolysis of 6,16-PC substrate leads to the formation of water-soluble lyso-lipid and water-insoluble fatty acids. PLA_2 domain formation in this system was observed, indicating that lyso-lipid does not play a critical role in PLA_2 interfacial aggregation. Hydrolysis of 16,6-PC, on the other hand, leads to the formation of a water insoluble lyso-lipid, and a water-soluble fatty acid. PLA_2 domain formation in this system was greatly reduced, consistent with our hypothesis that fatty acids are necessary for enzyme aggregation. In addition, the time needed to observe PLA_2 domain in 16,6-PC monolayer systems was typically 8-10 hours after PLA_2 injection, whereas PLA_2 domain formation in hydrolyzed 6,16-PC systems was usually less than one hour.

Since small PLA_2 domains were observed in hydrolyzed 16,6-PC systems, bovine serum albumin (BSA) was added to the monolayer subphase during hydrolysis to sequester fatty acid. With BSA present in the subphase, PLA_2 domain formation was completely suppressed. Though all caproic acid reaction products were probably solubilized after their hydrolytic release, aqueous self-acylation of PLA_2 is known to occur (28). Perhaps the long induction times needed for PLA_2 domain formation with the 16,6-PC substrate result from this acylation mechanism (and resulting increase in surface activity of PLA_2).

Surface potential and molecular area versus hydrolysis time (Figure 3) also support our claim that fatty acids and lyso-lipids are solubilized upon PLA_2 hydrolysis for 16,6-PC and 6,16-PC, respectively. If lyso-lipid were important for PLA_2 domain formation, PLA_2 domain formation would have either been reduced or totally suppressed in hydrolyzed 6,16-PC systems. As shown in Figure 4, this was not the case.

Recently solved high-resolution PLA_2 crystal (29–31) structures show that several cationic amino acid residues (Lys, Arg) located near the enzyme's interfacial recognition site are solvent accessible. It is reasonable to expect that these residues possess the ability to interact with negative charges concentrated within phase separated fatty acid domains. In fact, PLA_2 from porcine pancreas is well-known to prefer substrate present in a negatively charged interface (9). Additional support for electrostatic interactions between PLA_2 and fatty acids is given by surface potential data for PLA_2 hydrolyzed 6,16-PC monolayers. The final surface potential obtained after completion of 6,16-PC hydrolysis was 100 mV, which does not correspond to that of a pure palmitic acid monolayer under enzyme-free conditions (–30 mV). As shown in Figure 4, PLA_2 domain formation occurred during 6,16-PC hydrolysis. Therefore, PLA_2's dipoles and surface charges also contribute to the surface potential. Though we cannot separate PLA_2's and palmitic acid's contribution to the surface potential, this suggests PLA_2 electrostatically interacts with fatty acids. Results from ternary mixed monolayer studies also show that fatty acids play an important role in the phase behavior of resultant PLA_2 hydrolyzed lipid monolayers. Using H-379 as a cationic protein "analog," Figure 2 shows that under appropriate conditions (high pH, presence of Ca^{2+}), ternary mixed monolayers representing lipid membrane compositions at various extents of hydrolysis undergo fatty acid phase separation (13,14).

Portions of PLA_2 domains transferred to electron microscopy grids apparently exhibit 2-D hexagonal order. Collectively, these observations support our hypothesis that PLA_2 domain formation critically depends on the presence of fatty acids at the air-water interface. The shapes of PLA_2 domains in hydrolyzed DPPC and 6,16-PC films (Figures 1 and 4, respectively) are dissimilar. It is therefore possible that two different PLA_2 aggregation mechanisms are occurring in hydrolyzed DPPC and 6,16-PC systems. As shown in Figure 1, PLA_2 domains from hydrolyzed isomerically pure lipid monolayers are typically bean-shaped. Similar bean-shaped domains are also observed in PLA_2-free, phase separated ternary mixed monolayers containing enantiomerically pure DPPC, C16Lyso and palmitic acid (0.2:1:1, Figure 2). The well-characterized property of DPPC enantiomers to induce chiral lipid domains during lipid monolayer phase transitions (19–22) may also play a role in templating or structuring PLA_2 interfacial aggregation into specific morphologies. This property would also extend to DPPC's chiral hydrolysis product, C16Lyso lipid, resident after PLA_2 hydrolysis even though it lacks an observable phase transition and does not form monolayer domains (14). Nevertheless, hydrolysis of asymmetric phospholipid substrates (Figure 4) results in PLA_2 domains that are irregularly shaped and often comprise multiple domains clustered together. To obtain the regular bean-shaped PLA_2 domains after hydrolysis, C16Lyso and/or DPPC residual in phase separated fatty acid regions may play an important role in monolayer phase separation, enzyme domain formation and its resulting morphology. In 6,16-PC hydrolyzed monolayers, however, lyso-lipid products are solubilized during hydrolysis; the remaining monolayer is pure fatty acid. In this case, PLA_2 domain formation is still observed but with distinctly different morphology (Figure 4). It is our assertion in this case that PLA_2 adsorbs interfacially to fatty acid without the morphology imposed by chiral lipids. This is supported by control experiments showing PLA_2 adsorbs to pure palmitic acid monolayers resulting in highly irregular protein domain morphologies. We therefore alter our original enzyme domain formation hypothesis to include chiral PC in phase separated monolayer regions and its influence on resulting domani morphologies. PLA_2 adsorbs to pure palmitic acid monolayers, but to observe chiral protein domains (12), enzyme substrate is necessary at some point during the enzyme domain formation time course. Further work is aimed at elucidating these points.

Acknowledgments

We acknowledge the inspiration and creativity of Prof. Helmut Ringsdorf (University of Mainz) and his continued support of our work. We also thank Prof. Mary Roberts (Boston College) for synthesizing, purifying, and supplying the asymmetric phospholipids. Financial support from National Science Foundation grant DMR-9357439 (DWG) and grant INT-9303588 (KMM), Natural Sciences and Engineering Research Council of Canada, The Fonds FCAR, and the FRSQ (CS), a NATO Travel Grant (DWG and CS), and scholarships from the Fonds FCAR (MG) is greatly appreciated.

Literature Cited

1. Gaines, G. L., Jr. In *Insoluble Monolayers at Liquid-Gas Interfaces*; Prigogine, I., Ed.; John Wiley & Sons, Inc.: New York, NY, 1966; pp 73-79, 83-89.
2. Uzgiris, E. E. and Kornberg, R. D. *Nature* **1983**, *301*, 125-129.
3. Okahata, Y., Tsuruta, T., Ijiro, K., and Arigu, K. *Thin Solid Films* **1989**, *180*, 65-72.
4. Mohwald, H. *Ann. Rev. Phys. Chem.* **1990**, *41*, 441-476.
5. Ulman, A. In *An Introduction to Ultrathin Films: From Langmuir-Blodgett to Self-Assembly*; Harcourt, Brace, Jovanovich: Boston, MA, 1991; pp 1-83.
6. Mato, J. M. In *Phospholipid Metabolism in Cellular Signaling*; Mato, J. M., Ed.; CRC Press: Boca Raton, FL, 1990, pp 43-60.
7. Waite, M. In *Handbook of Lipid Research*, Hanahan, D. J., Ed.; Plenum Press: New York, NY, 1987, Vol. 5; pp 155-241.
8. Smith, W. L. *Biochem. J.* **1989**, *259*, 315-324.
9. Jain, M. K. and Berg, O. G. *Biochim. Biophys. Acta.* **1989**, *1002*, 127-156.
10. Pieterson, W. A.; Vidal, J. C.; Volwerk, J. J.; deHaas, G. H. *Biochemistry* **1974**, *13*, 1455-1460.
11. Grainger, D. W.; Reichert, A.; Ringsdorf, H.; Salesse, C. *Febs. Lett.* **1989**, *252*, 79-85.
12. Grainger, D. W.; Reichert, A.; Ringsdorf, H.; Salesse, C.*Biochim. Biophys. Acta.* **1990**, *1023*, 365-379.
13. Reichert, A.; Wagenknecht, A.; Ringsdorf, H. *Biochim. Biophys. Acta.* **1992**, *1106*, 178-188.
14. Maloney, K. M.; Grainger, D. W. *Chem. Phys. Lipids.* **1993**, *65*, 31-42.
15. Burns, R. A.; Roberts, M. F. *Biochemistry* **1980**, *19*, 3100-3106.
16. Burns, R. A.; Donovan, J. H.; Roberts, M. F. *Biochemistry* **1983**, *11*, 964-973.
17. Nargessi, R. D.; Smith, D. S. *Method. Enzymol.* **1986**, *122*, 67-72.
18. Smith, P. K.; Krohn, R. I.; Hermanson, G. T.; Mallia, A. K.; Gartner, F. H.; Provenzano, M. D.; Fujimoto, E. K.; Goeke, N. M.; Olson, B. J.; Klenk, D. C. *Anal. Biochem.* **1985**, *150*, 76-85.
19. Meller, P. *Rev. Sci. Instrum.* **1988**, *59*, 2225-2231.
20. McConnell, H. M.; Weiss, R. M.; Tamm, L. K. *Proc. Natl. Acad. Sci. USA* **1984**, *81*, 3249-3253.
21. Peters, R.; Beck, K. *Proc. Natl. Acad. Sci. USA* **1983**, *81*, 3249-3253.
22. Losche, M.; Sackmann, E.; Mohwald, H. *Ber. Bunsenges. Phys. Chem.* **1983**, *87*, 848-852.
23. Langmuir, I.; Schaefer, V. J.; Wrinch, D. M. *Science* **1937**, *85*, 76-80.
24. Reichert, A. Ph.D. Dissertation, University of Mainz, Germany, 1993.

25. Yu, B.-Z.; and Jain, M. K. *Biochim. Biophys. Acta* **1989**, *980*, 15–22.
26. Jain, M.K.; Yu, B.-Z.; Kozubek, A. *Biochim. Biophys. Acta* **1989**, *980*, 23–32.
27. Burack, W. R.; Yuan, Q.; and Biltonen, R. L. *Biochemistry* **1993**, *32*, 583–589.
28. Cho, W.; Tomasselli, A. G.; Heinrikson, R. L.; Kezdy, F. J. *J. Biol. Chem.* **1988**, *263*(23), 11237–11241.
29. White, S. P.; Scott, D. L.; Otwinowski, Z.; Gelb, M. H.; Sigler, P. B. *Science*, **1990**, *250*, 1560–1563.
30. Scott, D. L.; Otwinowski, Z.; Gelb, M. H.; and Sigler, P. B. *Science* **1990**, *250*, 1563–1566.
31. Dijkstra, B. W.; Kolk, K. H.; Hol, W. G. J.; Drenth, J. *J. Mol. Biol.* **1981**, *147*, 97–123.

RECEIVED December 22, 1994

Chapter 36

Fibronectin in a Surface-Adsorbed State

Insolubilization and Self-Assembly

Viola Vogel

Center for Bioengineering, University of Washington, Seattle, WA 98195

Human plasma fibronectin undergoes a transition from a soluble dimeric form in solution to an insoluble fibrillar form if incorporated into extracellular matrix. These initial and final states of fibronectin are well characterized; yet the molecular pathway is unknown by which fibronectin is insolubilized, self-assembled and crosslinked. A study of fibronectin adsorbed from physiological salt solution to the nonpolar air/water interface reveals that fibronectin undergoes a transition to an insoluble state at the interface. The insolubilization does not originate from covalent crosslinking, nor is crosslinking initiated. Under certain conditions, fibronectin further assembles spontaneously into fibrillar networks in contact with L-α-dipalmitoyl phosphatidylcholine (DPPC) monolayers at the air/water interface. Fibronectin self-assembly is dependent on the physical state of the DPPC monolayer and the history by which fibronectin and DPPC first come into contact.

Fibronectin is a major component of the extracellular matrix (1-3). Plasma fibronectin is soluble while circulating in blood and other body fluids, and insoluble if incorporated in the extracellular matrix (ECM). In a surface adsorbed state, fibronectin is active and regulates a wide variety of cellular processes including cell adhesion, differentiation, proliferation, and migration. On a supra cellular scale, the structural organization of fibronectin is further important in embryogenesis, wound healing and metastasis. The characteristics of fibronectin in different conformational states are under intense investigations (1-3). Fibronectin in the plasma is dimeric, soluble and assumes a compact form (4-6). After adsorption to synthetic surfaces, fibronectin expresses at least some of its biological activities (2,3,7-11), whereby its ability to promote cell adhesion to synthetic surfaces is most widely used for cell cultures (12). On cell surfaces, fibronectin is often found in a fibrillar form (1-3,13-18). The ECM-fibrils consist of disulfide-crosslinked fibronectin multimers (13) which are insoluble and biologically active.

The molecular mechanisms remain unclear by which the expression of the diverse biological functions of fibronectin are regulated through interaction of fibronectin with interfaces. It has been demonstrated that the chemical nature of synthetic surfaces and the presence of co-adsorbing proteins have distinct impact on the functions expressed by fibronectin (8-11,19,20). The most sophisticated of all surface induced conversion processes occurs when fibronectin is assembled into fibrils on cell surfaces. Several sequential events are implied to be important for matrix assembly *in vivo* (1): (a) binding of fibronectin (FN) to cell surface receptors, a step which then initiates (b) FN-FN self-assembly, and (c) FN-FN crosslinking. The cell surface receptors potentially involved in anchoring fibronectin to surfaces remain unknown, but several candidates have been proposed. They include membrane-bound receptors (integrins) for the RGDS-sequence (14) and other "matrix assembly receptors" (21). Regarding the not clearly identified role of matrix assembly receptors, it is of interest that two distinct regions in modules III_8 and III_9 in addition to the RGDS sequence are required for fibronectin to exhibit full cell adhesiveness (22). Fibronectin further contains binding sites to cell surface proteoglycans (23), and to gangliosides (24,25). More detailed assembly pathways have been proposed (26,27).

Formulation of molecular pathways by which the nature of interfaces regulates the biological activity of fibronectin is hampered by the "soft" nature of fibronectin which readily assumes different conformational states in response to environmental stimuli. Fibronectin may undergo various intermediate states during a surface induced conversion process and the roles played by specific and nonspecific surface forces in stabilizing one or the other state are unknown. Determination of the molecular pathway of matrix assembly by which fibronectin is converted from a soluble form to insoluble fibrils is a most challenging future task. It is of advantage, however, that the initial and the final states of fibronectin as incorporated into the extracellular matrix are well investigated, whereas the final states are poorly characterized for fibronectin on solid surfaces.

The focus of this article is to review our knowledge of the structure of fibronectin and the factors which contribute to fibronectin matrix assembly. In this context we will discuss (a) the structure and structural transitions of fibronectin in solution (Section I), (b) the transition of fibronectin into a water insoluble state as observed at the nonpolar air/water interface (Section II), and (c) conditions under which fibronectin assembles into fibrillar structures at DPPC interfaces (Section III). Since it is not known in which sequence fibronectin changes its characteristic attributes which distinguish the soluble initial and its final state as matrix fibril, it is of great interest to define conditions under which a partial or complete conversion of fibronectin can be initiated. Finding such conditions will shed light onto the driving forces of potential molecular pathways and the role played by either biological or synthetic interfaces in the conversion process. The questions investigated here are whether insolubilization of fibronectin at interfaces necessarily results from crosslinking or vice versa, and whether fibronectin is capable of self-assembly into fibrillar structures at interfaces in the absence of any surface receptor molecules.

I. Fibronectin in Solution

Structure.   Fibronectin is a large dimeric glycoprotein (28). Two similar but not identical polypeptides, as obtained by alternative splicing, are linked by two disulfide bonds at their carboxyl terminals (Figure 1). Each monomer strand contains a series of homologous repeating units, referred to as type I, type II and type III modules. The modules fold independently into well defined structural motifs. Clusters of several modules organize into functional domains that exhibit specific binding activities to a large variety of biomolecules. This includes binding to fibrin, heparin, collagen and DNA. The peptide sequence RGDS on the tenth type III module, III_{10}, binds to membrane bound integrins and is the most prominent (29) of several cell binding sites (22, 30). At least one of the cell binding sites acts by an integrin independent mechanism (23).

In addition to its size, the structural flexibility of fibronectin impedes the determination of its three dimensional structure by current techniques. The three-dimensional structures of the much smaller type I, II and III modules, however, have recently been resolved. These modules are not unique to fibronectin but serve as functional domains in a large number of other proteins. The type I module (Figure 2a) consists of two anti-parallel β-sheets and a wide loop with considerable structural flexibility (31). The residues buried in the core are highly conserved, whereas those exposed to the solvent are both variable and predominantly hydrophilic. Two adjacent type I modules may lock their relative orientations by formation of a common β-strand (31).

The type II module (Figure 2b) adopts a globular conformation consisting mostly of conformationally flexible loops and turns. The type II structure is stabilized by two disulfide bonds and by short sections of antiparallel β-sheets (32). The surface of the type II module exhibits a solvent exposed hydrophobic depression where several aromatic residues are clustered. This hydrophobic area may promote inter-domain aggregation or act as a potential ligand binding site (32).

The largest module, the type III module with about 90 amino acids, consists of seven β-strands (Figure 2c). They are arranged into two β-sheets lying face-to-face, one of four and one of three β-strands (33-35). The structure of the β-strands among different type III modules are predicted to be conserved, as the hydrophobic residues that stabilize the hydrophobic cores show considerable homology (35). Structural variability is found among the loops, the only exception being the highly conserved EF-loop which connects the two opposing β-sheets (see Figure 2c). The structural variability in sequence and length of the loops is likely to be responsible for the distinct binding activities of type III modules. The RGDS-motif of module III_{10}, for example, is part of the loop that connects the β-strands F and G. It is the longest and most flexible of the six loops. All type III modules lack disulfide bonding and α-helices are essentially non-existent in all of the modules of fibronectin.

Structural Transitions. The molecular shape of dimeric fibronectin in solution can vary from a compact to a more extended conformation under extreme conditions. The compact form, a rigid oblate structure (4), is favored in physiological or near-physiological buffers and has an estimated diameter and thickness of 30 nm and 2 nm, respectively (36). At temperatures above 40°C, more extreme pHs or increased ionic strengths, fibronectin undergoes a gradual transition into a more extended form (5,6,37,38). This gradual extension of fibronectin does not result from partial denaturation, as one may expect for other proteins. Extension proceeds with only

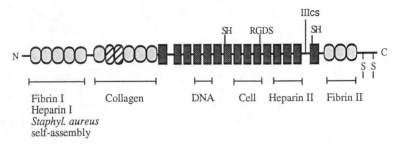

Figure 1.   Schematic drawing of the monomeric A-chain of human plasma fibronectin. Dimeric fibronectin is obtained through disulfide crosslinking of two monomers at their C-terminals. The homologous type I, II and III modules are symbolized by ◐, ◐, ▮, respectively. The specific binding activities of individual domains are indicated, as well as the two buried sulfhydryl groups and the RGDS sequence. The heparin II domain contains several additional sites that promote cell adhesion, one of them acts in an integrin independent fashion (23). Plasma fibronectin as synthesized by hepatocytes contains only the connecting segment IIIcs but not the other two alternatively spliced segments, i. e. EIIIA and EIIIB (28). IIIcs is only spliced into the A- but not the B-chain of the fibronectin dimer.

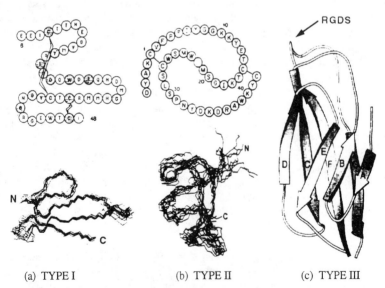

(a) TYPE I   (b) TYPE II   (c) TYPE III

Figure 2.   Schematic diagrams of the type I, type II, and type III modules. (a) Module I₇ from human fibronectin in which the consensus residues with the other type I modules are shaded (Adapted from ref. (31)). The structure is stabilized by two disulfide bonds as indicated.   (b) Module type II from a bovine seminal plasma protein (PDS-109) which is homologous to the type II module of fibronectin (Adapted from ref. (32)).   (c) Type III₁₀ module of human fibronectin in a ribbon representation (Adapted from ref. (35)). The RGDS sequence is located in the loop connecting the β-strands F and G.

minor changes in secondary structure. 90% of the secondary structure of human plasma fibronectin, consisting of 79% β-sheet and 21% β–turn, is preserved at temperatures as high as 70ºC, or at pH values as low as 3.0 (37). This indicates that the molecular extension originates from an unfolding of the chains, in analogy to the unfolding of a chain of beads (38), while the three-dimensional structure of the modules is largely preserved. The compact form is thought to be stabilized mostly by electrostatic inter-domain interactions (5). Such a gradual unfolding scheme allows exposure of binding sites in response to environmental stimuli which are otherwise buried in a compact state.

Disulfide-Crosslinking. The final state of fibronectin, if incorporated into the extracellular matrix, is a disulfide-crosslinked multimer. Disulfide-crosslinking of fibronectin can be initiated in solution in the presence of chaotropic agents. Each monomer strand contains two sulfhydryl groups which are buried in physiological salt solution. Partial unfolding of fibronectin is necessary for exposure of the sulfhydryl groups and requires a concentration of at least 2-3 M guanidine chloride (13). Oxidation of the free sulfhydryl groups leads to multimer formation *in vitro*.

Alternatively, matrix assembly *in vivo* largely involves disulfide exchange (13). Disulfide exchange requires that two disulfide bonds of different modules come close enough as to exchange their binding partners. Disulfide exchange can only occur among type I and II modules as the type III modules do not contain any disulfide bonds. It is proposed that one disulfide pair of the C-terminal module I_{12} recombines with a disulfide pair on a type II module of the collagen binding site (26). Disulfide exchange *in vivo* may not occur spontaneously, but is catalyzed by the activated factor XIII (26). Disulfide crosslinking has also been induced by a recombinant fragment of module III_1 (18).

II. Insolubilization of Fibronectin at the Nonpolar Air/Water Interface

Surface Adsorption. One of the main characteristics of fibronectin in the fibrillar state is that it has a significantly reduced water solubility. Little is known about the sequence in which fibronectin changes its characteristics during the conversion process from a dimeric soluble, to a multimeric insoluble form. It remains an open question whether the insolubilization of fibronectin necessarily results from crosslinking, or vice versa. A study of fibronectin adsorbed from a salt solution to the nonpolar air/water interface in a Langmuir trough (39) provides some insight (Figure 3). The surface pressure rises slowly over the course of several hours (Figure 3a). Once fibronectin is adsorbed to the air/water interface, the surface monolayer shows all characteristics of an insoluble monolayer. Compression of the fibronectin surface film by movable barriers leads to a steep rise of the surface pressure as shown in Figure 3b. The rise in surface pressure on film compression is unexpected, if we assume that the surface population is solely a function of the bulk concentration, as typically found for "soluble" molecules. Deviations from such a two-state model are expected, however, if the overall polarity of a molecule undergoes changes in the surface adsorbed state. Polarity changes can originate from surface induced conformational transitions, aggregation, or crosslinking, and result in an energy barrier which limits molecular desorption.

The surface pressure/area isotherms (Figure 3b) provide no indication that the transition to an insoluble form at the air/water interface originates from intermolecular crosslinking as concluded from the following observations: The fibronectin monolayer remains fluid, and the compression/expansion cycle is reversible. If extensive irreversible aggregation or crosslinking occurred as a consequence of surface adsorption or film compression, the surface monolayer would rigidify. This would lead to a rapid surface pressure drop during monolayer expansion and a shift of the recompression curve toward smaller mean areas, which is not observed. The hysteresis between monolayer compression and expansion is dependent on the barrier velocity implying that surface pressure induced inter- or intra-molecular relaxation processes occur on time scales of minutes to hours. All these observations lead us to conclude that conformational changes are responsible for the conversion of fibronectin into an insoluble form at the air/water interface. Conformational changes have also been observed for fibronectin adsorbed to a variety of solid surfaces (11,19,40-43).

Spreading of a Fibronectin Monolayer. Molecules which form "insoluble" monolayers have the advantage that they can be spread at the air/water interface and therefore allow quantification of the mean area per surface molecules as a function of the surface pressure. Quantitative spreading of proteins from buffer solutions is possible, if the solution is sufficiently diluted and slowly spread (44). A buffer solution of fibronectin (100 μl, 0.05 mg/ml) was spread directly at the air/water interface (39). The surface pressure/area isotherm, as obtained after 20 minutes of equilibration, is given in Figure 4. The surface pressure rises gradually on monolayer compression and builds up a considerable surface pressure for areas smaller than 250 nm^2. Somewhat smaller areas were found in an earlier monolayer study of human plasma fibronectin (45). It is interesting to note that a surface area of 244 - 308 nm^2 is observed by electron microscopy for dimeric fibronectin in the extended form (46). Fibronectin, if sprayed from solution to a solid surface, exhibits two arms each approximately 60 ± 7 nm in length with a diameter of 2.3 nm (46).

In summary, no indication was found from surface pressure/area isotherms or fluorescence microscopy that the transition of fibronectin to an insoluble state at the air/water interface results from disulfide crosslinking. The results suggest that insolubilization results from partial unfolding of the chains. Future studies have to show whether insolubilization also proceeds disulfide crosslinking *in vivo*, and whether it is a prerequisite for crosslinking to occur.

III.   Self-Assembly of Fibronectin

Self-Assembly Sites.   Self-assembly of fibronectin is a central element in extra cellular matrix formation and it remains unclear whether surface-bound receptor molecules are necessary for fibronectin to assemble into fibrils at biological interfaces. Fibronectin contains several binding sites that are influential in the self-assembly process. The first five type I modules at the amino termini of dimeric fibronectin, I_1-

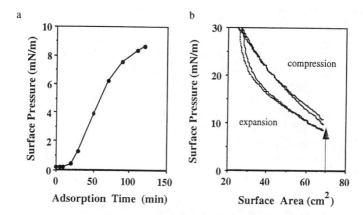

Figure 3. Fibronectin adsorption from solution to the air/water interface. (a) The surface pressure is given as a function of time for a fibronectin bulk concentration of 4 μg/ml. (b) The surface pressure is given as function of the total surface area for a fibronectin monolayer which has been adsorbed from solution for two hours (gray arrow). Two compression/expansion cycles are shown. The film was expanded immediately after compression. An equilibration of 20 minutes preceded the recompression (FN source: GIBCO BRL; subphase: 0.15 M NaCl; pH 5.6; 20°C).

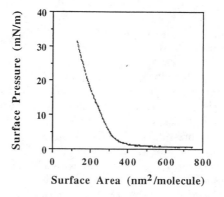

Figure 4. Surface pressure/area isotherm of fibronectin spread at the air/water interface. 100 μl of fibronectin dissolved in aqueous buffer (0.05 mg/ml) were spread on a salt solution (0.15 M NaCl). The film was equilibrated for 20 minutes prior compression.

I_5, are required for matrix assembly by fibroblasts (14,15), and are demonstrated to exhibit FN-FN binding activity (16,17). Several studies identify a second matrix assembly site involving the modules I_9-III_1 (47). Whereas recombinant FN polypeptides which lack module III_1 show matrix assembly (16), the presence of fragments of module III_1 significantly alters FN-FN binding. A recombinant fragment modeled after the C-terminal section of the module III_1 binds to fibronectin, but rather than inhibiting FN-FN assembly, it promotes spontaneous disulfide crosslinking of fibronectin into multimers (18); this recombinant III_1 fragment altogether contains 14 amino acids in the terminal positions that are not normally found in fibronectin. The functional roles of the two FN-FN assembly sites, I_1-I_5 versus III_1, are likely to be different. Whereas I_1-I_5 sites are essential for matrix assembly (14-17,21), the FN-FN binding site of module III_1 is partially buried in intact soluble fibronectin and seems to exhibit FN-FN binding activity only subsequent to surface binding (17). Cell surface molecules have been recently described with affinity to the first five type I modules (48-50).

The disulfide-bonded dimer structure is crucial for the *de novo* assembly of fibronectin into fibrils (16). In contrast to a *de novo* assembly of extracellular matrix by cells that are otherwise defective in the expression of extracellular matrix proteins, exogenously added FN monomers have been successfully incorporated into preexisting matrices of fibroblast cultures (51). Recombinant fibronectin lacking the C-terminal type I modules (52) or the type III modules, including the cell recognition site, was fully capable of matrix incorporation (16,53). It is unclear for the above cases whether covalent crosslinking occurs when the recombinant fibronectin is incorporated into extracellular matrix.

Fibronectin and DPPC-Monolayers. Considering the complexity of matrix assembly and the large number of potential contributors, it is of interest to examine whether and under which conditions fibronectin self-assembles into fibrous structures at membrane mimetic interfaces that do not contain any surface bound receptor molecules. The major lipid fraction in the outer leaflet of most cell membranes, including erythrocytes and the apical plasma membrane of aortic endothelial cells, consists of phosphatidylcholine (PC) headgroups (54,55). Membrane mimetic interfaces are prepared here by spreading L-α-dipalmitoyl phosphatidylcholine (DPPC) at the air/water interface in a Langmuir trough. The physical state of the lipid monolayer is adjusted by the use of movable barriers. The DPPC monolayer is initially compressed to a surface pressure of 25 mN/m. Fibronectin is injected underneath the monolayer to reach a final bulk concentration of 4 µg/ml. No change of surface pressure is observed over a time period of several hours. This indicates that fibronectin, if adsorbed from solution to the polar interface of a close-packed DPPC monolayer, remains underneath the PC head group plane.

Fluorescence microscopy images from photolabeled fibronectin adsorbed to a close-packed DPPC monolayer reveal that the protein film assumes a "grainy" microscopic structure and that the protein film is inhomogenous (Figures 5a and 5b). Bright areas rich in fibronectin surround dark areas which have a significantly reduced protein content (39). Most striking is that the dark areas are essentially circular in shape, a little larger in some spots and smaller in others.

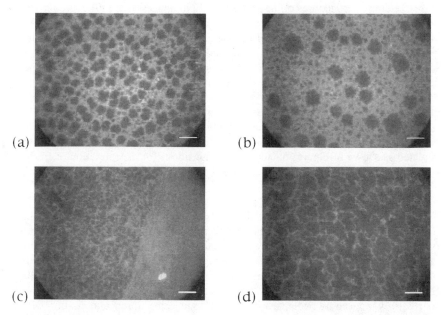

(a)   (b)

(c)   (d)

Figure 5. Fluorescence microscopy images of photolabeled fibronectin (FITC) adsorbed to an unlabeled L-α-DPPC monolayer at the air/water interface. (a,b) The DPPC monolayer was compressed to a surface pressure of 20 mN/m prior to fibronectin injection. The area per DPPC molecule was kept constant during protein injection and equilibration. Two representative surface spots are shown as the surface morphology was not uniform. (c,d) Two representative surface morphologies obtained after monolayer expansion and recompression (0.15 M NaCl; pH 5.6; 20°C; bars = 25 μm).

Patchy surface coverage has also been observed for fibronectin adsorbed to polar and nonpolar solid surfaces (56). A heterogeneous protein distribution at interfaces may have several origins, one of them being induced aggregation via surface roughness. In contrast to solid surfaces, surface steps and edges do not exist at fluid/vapor interfaces. If heterogeneous surface coverage is still found at a monolayer covered air/water interface, it can either be attributed to protein/protein aggregation (57,58), two-dimensional protein crystallization (59), or to the physical state of the monolayer. A further study in which fibronectin and the DPPC monolayer where both photolabeled, with rhodamine-DPPE and fluorescein-fibronectin respectively, revealed corresponding microtextures of the protein and lipid images (see (39)). At a surface pressure of 25 mN/m, the DPPC monolayer is in the liquid condensed state. Well ordered regions with positional and orientational order, the former DPPC domains which had formed in the liquid expanded/condensed coexistence region, are surrounded by grain boundaries or defect zones with a reduced degree of order. These defect zones are rich in rhodamine-DPPE, if the lipid monolayer is photolabeled. Adsorption of FITC-labeled fibronectin to a close-packed photolabeled DPPC monolayer shows that fibronectin has an enhanced affinity for these defect zones and evidence suggests that the defect zones initially serve as anchoring sites for fibronectin aggregation. This explains the formation of dark round areas which are protein depleted as shown in Figures 5a and b. No indications for the formation of fibrillar structures have been found by fluorescence light microscopy in the studies described above.

Fibrillar Assemblies. Prerequisite for the assembly of fibronectin into multimeric fibrils *in vivo* requires that dimeric fibronectin molecules line up (or are lined up) with respect to each other in such a way that disulfide-crosslinking can occur in a non-random fashion. We were interested in finding conditions under which the assembly of fibronectin into fibrillar structures can be induced under *in vitro* conditions. The formation of microscopic fibrillar structures underneath DPPC monolayers was successfully induced by two approaches, (a) by squeezing fibronectin out of a DPPC monolayer by successive expansion of an equilibrated DPPC/fibronectin film and its recompression, or (b) by premixing of fibronectin and DPPC under partially denaturing conditions prior spreading at the air/water interface. Figures 5c and d show the protein film morphology after expansion and recompression. Creation of lines of high protein density on monolayer recompression is not unexpected, as fibronectin was already unevenly distributed after adsorption to the close-packed DPPC monolayer (Figures 5 a and b). While the monolayer is expanded, fibronectin preferentially dissolves in the liquid expanded state of DPPC. Fibronectin is condensed and squeezed out of the monolayer on recompression, creating fine lines of fibronectin underneath the DPPC monolayer as seen by fluorescence microscopy. The observation of fibrillar structures by fluorescence microscopy does not necessarily indicate that they are disulfide-crosslinked.

Co-spreading of fibronectin and DPPC from premixed solutions has been the most successful approach, so far, in inducing spontaneous FN-FN self-assembly into fibrillar networks (Figure 6). Hereby, fibronectin has been exposed to partially denaturing conditions. The spreading solutions were prepared from aqueous buffer

A   B

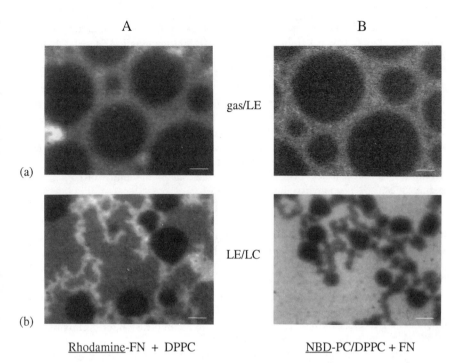

gas/LE

(a)

LE/LC

(b)

Rhodamine-FN + DPPC   NBD-PC/DPPC + FN

Figure 6.   Fluorescence microscopy images from mixed monolayers of fibronectin and L-α-DPPC in the fluid/gas (row a) and the liquid expanded/liquid condensed (row b) phase coexistence regions. Rhodamine-labeled fibronectin is co-spread with unlabeled DPPC in column A, and unlabeled fibronectin is co-spread with DPPC and 1 mol% of NBD-PC in column B at the plain air/water interface (bars = 5 μm, pH 5.6; 20°C). The molar ratio of fibronectin to DPPC was 1:340 in both experiments, and the spreading solutions were made of aqueous buffer, alcohol, and chloroform at a ratio of 12.5:75:12.5 vol%, respectively.

containing 75 vol% of alcohol and 12.5 vol% of chloroform. Two independent experiments have been conducted (39): Rhodamine-labeled fibronectin is co-spread with unlabeled DPPC (column A), and unlabeled fibronectin is co-spread with DPPC and 1 mol% of NBD-PC (column B). The formation of fibronectin fibrils is not initiated in the spreading solution as confirmed from Figure 6, row (a), where the images are shown for the mixed monolayers in the liquid/gas phase coexistence region. Fibronectin is enriched in the two-dimensional liquid phase of DPPC (bright in both columns A and B). The liquid phase surrounds the two-dimensional gas phase which is depleted of DPPC and fibronectin (dark areas in both columns A and B). Film compression reduces the diameter of the dark areas until a homogenous fluorescence is seen. This is typically observed for lipids undergoing a transition from the gas/liquid phase coexistence region to the liquid expanded state (60).

Spontaneous assembly of fibronectin into networks is observed when entering the liquid expanded/liquid condensed phase coexistence region by film compression as shown in Figure 6, row (b). Fibronectin fibrils (bright in column A) coexist with less concentrated dimeric fibronectin associated with the liquid expanded DPPC phase (gray in column A). Column B shows the distribution of the photolabeled NBD-PC. It is enriched in the liquid expanded DPPC phase (bright) and largely excluded from the liquid condensed DPPC domains (dark areas). The fibronectin networks appear gray (column B) indicating that those regions are NBD-PC depleted. The liquid condensed DPPC domains are dark in columns A and B which shows that fibronectin is exclusively associated with the liquid expanded phase of DPPC. The liquid condensed domains, however, are attached or entrapped by the fibronectin network. The self-assembled fibronectin networks cover large fractions of the entire surface and the average mesh size is on micron dimensions (39).

These results of fibronectin interacting with DPPC monolayers show that the microscopic distribution and assembly of fibronectin at the interface is dependent on various parameters which have found very little attention in previous fibronectin studies: (a) the history by which fibronectin and DPPC first come into contact, (b) the physical state of the DPPC monolayer, and (c) the mechanical treatment of the surface film. Self-assembly of fibronectin into fibrillar structures does not occur spontaneously at a close-packed DPPC monolayer, but can be initiated by either mechanical treatment through film expansion and recompression, or premixing with DPPC using partially denaturing conditions. We can only hypothesize at this point, that the different microscopic patterns of fibronectin assembly seen in contact with interfacial DPPC monolayers are distinct in their expressed biological functions.

Acknowledgments. The discussions and many research contributions of Jing Ping Chen to this paper are gratefully acknowledged as well as financial support from NIH (1R29GM49063-01A1, First Award) and The Whitaker Foundation.

Literature Cited

1.   Ruoslahti, E., *Ann. Rev. Biochem.* **1988,** *57,* pp. 375-413.

2.   Mosher, D. F., Ed., *Fibronectin* (Academic Press, 1989).

3.   Hynes, R. O., *Fibronectins.* (Springer Verlag, Heidelberg, 1990).

4.   Sjöberg, B., Erikson, M., Österlund, E., Pap, S., Österlund, K., *Eur. Biophys. J.* **1989,** *17,* pp. 5-11.

5.   Khan, M. Y., Medow, M. S., Newman, S. A., *Biochem. J.* **1990,** *270,* pp. 33-38.

6.   Benecky, M. J., Wine, R. W., Kolvenbach, C. G., Mosesson, M. W., *Biochemistry* **1991,** *30,* pp. 4298-4306.

7.   Grinnell, F., Feld, M. K., *J. Biomed. Mater. Res.* **1981,** *15,* pp. 363.

8.   Lewandowska, K., Balachander, N., Sukenik, C. N., Culp, L. A., *J. Cell. Physiol.* **1989,** *141,* pp. 334-345.

9.   Underwood, P. A., Steele, J. G., Dalton, B. A., *J. Cell Sci.* **1993,** *104,* pp. 793-803.

10. Pettit, D., Horbett, T., Hoffman, A., *J. Biomed. Mater. Res.* **1992**, *26*, pp. 1343.
11. Cheng, S.-S., Chittur, K. K., Sukenik, C. N., Culp, L. A., Lewandowska, K., *J. Colloid Interface Sci.* **1994**, *162*, pp. 135-143.
12. Horbett, T. A., *Colloids and Interface Sci. B: Biointerfaces* **1994**, *2*, pp. 225-240.
13. Mosher, D. F., Johnson, R. B., *J. Biol. Chem.* **1983**, *258*, pp. 6595-6601.
14. McDonald, J. A., Quade, B. J., Broekelmann, T. J., LaChance, R., Forsman, K., Hasegawa, E., Akiyama, S., *J. Biol. Chem.* **1987**, *262*, pp. 2957-2967.
15. Sottile, J., Schwarzbauer, J., Selegue, D., Mosher, D. F., *J. Biol. Chem.* **1991**, *266*, pp. 12840-12843.
16. Schwarzbauer, J. E., *J. Cell Biol.* **1991**, *113*, pp. 1463-1473.
17. Morla, A., Ruoslahti, E., *J. Cell Biol.* **1992**, *118*, pp. 421-429.
18. Morla, A., Zhang, Z., Rouslahti, E., *Nature* **1994**, *367*, pp. 193-196.
19. Grinnell, F., Feld, M. K., *J. Biol. Chem.* **1982**, *257*, pp. 4888-4893.
20. Lewandowska, K., Pergament, E., Sukenik, C. N., Culp, L. A., *J. Biomed. Mater. Res.* **1992**, *26*, pp. 1343-1363.
21. McKeown-Longo, P. J., Mosher, D. F., *J. Cell Biol.* **1985**, *100*, pp. 364-374.
22. Aota, S.-I., Nagai, T., Yamada, K. M., *J. Biol. Chem.* **1991**, *266*, pp. 15938-15943.
23. McCarthy, J. B., Skubitz, A. P. N., Furcht, L. T., Wayer, E. A., Iida, J., in *Cell Adhesion Molecules*, Hemler, M. E., Mihich, E., Eds. (Plenum Press, New York, 1993) pp. 127-141.
24. Spiegel, S., Schlessinger, J., Fishman, P., *J. Cell Biol.* **1984**, *99*, pp. 699-704.
25. Morley, P., Armstrong, D. T., Gore-Langton, R. E., *J. Cell Sci.* **1987**, *88*, pp. 205-217.
26. Mosher, D. F., Fogerty, F. J., Chernousov, M. A., Barry, E. L. R., *Annals. NY Acad. Sci.* **1991**, *614*, pp. 167-180.
27. Mosher, D. F., Sottile, J., Wu, C., McDonald, J. A., *Curr. Opinion in Cell Biol.* **1992**, *4*, pp. 810-818.
28. Kornblihtt, A. R., Umezawa, K., Vibe-Pedersen, K., Baralle, F. E., *EMBO J.* **1985**, *4*, pp. 1755-1759.
29. Pierschbacher, M. D., Ruoslahti, E., *Nature* **1984**, *309*, pp. 30-33.
30. Obara, M., Kang, M. S., Yamada, K. M., *Cell* **1988**, *53*, pp. 649-657.
31. Baron, M., Norman, D., Willis, A., Campell, I. D., *Nature* **1990**, *345*, pp. 642-646.
32. Constantine, K. L., Ramesh, V., Banyai, L., Trexler, M., Patthy, L., Llinas, M., *Biochemistry* **1991**, *30*, pp. 1663-1672.
33. Leahy, D. J., Hendrickson, W. A., Aukhil, I., Erickson, H. P., *Science* **1992**, *258*, pp. 987-991.
34. Baron, M., Main, A. L., Driscoll, P. C., Mardon, H. J., Boyd, J., Campbell, I. D., *Biochemistry* **1992**, *31*, pp. 2068-2073.
35. Dickinson, C. D., Veerapandian, B., Dai, X.-P., Hamlin, R. C., Xuong, N.-H., Ruoslahti, E., Ely, K. R., *J. Mol. Biol.* **1994**, *236*, pp. 1079-1092.
36. Benecky, M. J., Kolvenbach, C. G., Wine, R. W., DiOrio, J. P., Mosesson, M. W., *Biochemistry* **1990**, *29*, pp. 3082-3091.

37. Österlund, E., *Biochim. Biophys. Acta* **1988**, *955*, pp. 330-336.
38. Rocco, M., Infusini, E., Daga, M. G., Gogioso, L., Cuniberti, C., *EMBO Journal* **1987**, *6*, pp. 2343-2349.
39. Chen, J. P., Vogel, V., manuscript in preparation.
40. Pitt, W. G., Spiegelberg, S. H., Cooper, S. L., in *Proteins at Interfaces*, Brash, J. L., Horbett, T. A., Eds., (American Chemical Society, 1987) pp. 324-338.
41. Wolff, C., Lai, C. S., *Arch. Biochem. Biophys.* **1989**, *268*, pp. 536-545.
42. Wolff, C. E., Lai, C.-S., *Biochemistry* **1990**, *29*, pp. 3354-3361.
43. Narasimhan, C., *Biochemistry* **1989**, *28*, pp. 5041-5046.
44. MacRitchie, F., *Adv. Colloid Interface Sci.* **1986**, *25*, pp. 341-385.
45. Zhou, N. F., Pethica, B. A., *Langmuir* **1986**, *2*, pp. 47-50.
46. Odermatt, E., Engel, J., in *Fibronectin*, Mosher, D. F., Ed., (Academic Press, 1989) pp. 25-45
47. Chernousov, M. A., Fogerty, F. J., Koteliansky, V. E., Mosher, D. F., *J. Biol. Chem.* **1991**, *266*, pp. 10851-10858.
48. Limper, A. H., Quade, B. J., LaChance, R., Birkenmeier, T. M., Rangwala, T. S., McDonald, J. A., *J. Biol. Chem.* **1991**, *266*, pp. 9697-9702.
49. Blystone, S. D., Kaplan, J. E., *J. Biol. Chem.* **1992**, *267*, pp. 3968-3975.
50. Moon, K.-Y., Shin, K. S., Song, W. K., Chung, C. H., Ha, D. B., Kang, M.-S., *J. Biol. Chem.* **1994**, *269*, pp. 7651-7657.
51. Chernousov, M. A., Metsis, M. L., Koteliansky, V. E., *FEBS Lett.* **1985**, *183*, pp. 365-369.
52. Sottile, J., Mosher, D. F., *Biochemistry* **1993**, *32*, pp. 1641-1647.
53. Ichihara-Tanaka, K., Maeda, T., Titani, K., Sekiguchi, K., *FEBS* **1992**, *299*, pp. 155-158.
54. Chapman, D., *Langmuir* **1993**, *9*, pp. 39-45.
55. Bird, R., Hall, B., Hobbs, K. E., Chapman, D., *J. Biomed. Eng.* **1989**, *11*, pp. 231-234.
56. Schakenraad, J. M., Stokroos, I., Busscher, H. J., *Biofouling* **1991**, *4*, pp. 61-70.
57. Stenberg, M., Nygren, H., *Biophys. Chem.* **1991**, *41*, pp. 131-141.
58. Ahlers, M., Grainger, D. W., Herron, J. N., Lim, K., Ringsdorf, H., Salesse, C., *Biophys. J.* **1992**, *63*, pp. 823-838.
59. Darst, S. A., Ahlers, M., Meller, P. H., Kubalek, E. W., Blankenburg, R., Ribi, H. O., Ringsdorf, H., et al., *Biophys. J.* **1991**, *59*, pp. 387-396.
60. Knobler, C. M., *Adv. Chem. Phys.* **1990**, *77*, pp. 397-449.

RECEIVED July 3, 1995

Chapter 37

Threshold Effect for Penetration of Prothrombin into Phospholipid Monolayers

M. F. Lecompte and H. Duplan

Laboratoire de Pharmacologie et de Toxicologie Fondamentales, Centre National de la Recherche Scientifique, 118 route de Narbonne, 31062 Toulouse Cedex, France

Since the enzymatic conversion of prothrombin (II) into thrombin occurs at a membrane-solution interface, the mode of interaction of II with condensed monolayers containing phosphatidylserine (PS) was studied using two complementary methods of direct measurement at surfaces. The adsorption isotherms were determined by measuring surface radioactivity of bound tritiated proteins. Penetration of II through the monolayers could be detected using alternating-current polarography. The interactions of prothrombin with monolayers containing 25% PS, are more complex than those with 100% PS, since they are dependent on coverage. There is a threshold concentration, separating two distinct binding states, correlated with two distinct dissociation constants (Kd), both in the presence and in the absence of calcium. At low coverage, the adsorption process, mainly electrostatic, is consistent with a PS- dependent Kd. The higher affinity obtained at higher coverage is correlated with penetration of II into monolayer, by hydrophobic forces, since in this case the Kd is PS-independent.

There is a general problem in biology: how can certain soluble proteins bind to a cell membrane?

A good example is the blood coagulation cascade, where some protein factors, soluble in plasma, are converted into active enzymes through complexes located at the surface of cell membranes. Membranes play a catalytic role in these conversions, which occur at an interface.

Thus, in the last complex of the cascade, prothrombin (II) is cleaved by factor Xa to form thrombin, which plays a key role in thrombosis and hemostasis. This conversion occurs only when particular phospholipids, such as phosphatidylserine (PS), located on the inside leaflet, appear at the outer leaflet of activated platelets (*1*), while phosphatidylcholine (PC) is already at this outer surface in resting cells. Furthermore, the kinetics of this conversion are drastically enhanced in the presence of factor Va (*2*).

0097–6156/95/0602–0519$12.00/0

Since the cohesion and function of the prothrombinase complex depend on the assembly of the components, it is necessary to study the manner in which they bind to each other and, separately, to the membrane at the molecular level.

In order to study the modes of interaction, membrane models must be used, since they allow the influence of each component in these multimolecular complexes to be followed. The validity of these models was tested by determining the biological activity of the system, for example prothrombin conversion. Thus it was found that the kinetic parameters are similar for platelets and vesicles (3), and also for vesicles and condensed monolayers (4).

Concerning the binding of the individual proteins to model membranes containing negatively charged phospholipids, different modes of interaction have been invoked by several authors using different techniques. In the case of the two vitamin-K dependent proteins, II and factor Xa, which contain γ -carboxyglutamic acid residues (Gla), the interaction is generally thought to involve the Gla- containing protein domain and the polar head group of phospholipids (mainly PS), either by direct binding with Gla (5, 6, 7) through ionic interactions or indirectly through a calcium ion stabilized conformer of II (8, 9, 10).

We found through independent approaches and with different membrane models that human prothrombin penetrates into PS-containing monolayers (11) or vesicles (12). Furthermore, in both cases, it was observed that this type of interaction could occur even in the absence of calcium (11, 12, 13); this was confirmed by other workers for the interaction of human prothrombin with vesicles (14, 15), as well as for bovine prothrombin with planar membranes (16) and with positively charged vesicles (17). Thus hydrophobic interactions as well as ionic interactions needed to be considered (18), especially as calcium-induced conformational change of II expresses a hydrophobic region (19).

More recently, it was shown that factor Va engages in both electrostatic and hydrophobic interactions, depending on phospholipid composition and also on protein concentration (20). In order to determine if this is also the case for prothrombin, we investigated in this study different modes of interaction of bovine prothrombin with phospholipid monolayers at two phospholipid compositions and as a function of protein concentration.

We emphasize that, since the phospholipids in biological membranes are in a condensed state, it is necessary when studying the interaction of proteins with membrane to use methods which can give information on condensed layers. This is beyond the capacity of the current method for measuring surface pressure.

The aim of this paper is to illustrate the application of alternative techniques to measurements on condensed monolayers. This has allowed us to measure at equilibrium direct binding of prothrombin at the membrane-solution interface.

Experimental

All reagents were of the highest grade commercially available. The different inorganic salts and acids were of analytical grades. Mercury was purified and doubly distilled under vacuum. Ultrapure water was obtained from a Millipore Super Q system.

Bovine prothrombin was purified by previously described procedures (21) and its purity was checked by SDS polyacrylamide gel electrophoresis. The molecular weight of prothrombin (72,000 g) and the extinction coefficient, used to determine protein concentrations by ultraviolet absorbance, were taken from (*21*).

Radioactive ^{3}H-labeled prothrombin was prepared by oxidizing the sialic acid with sodium metaperiodate and then by reducing the obtained aldehyde with sodium ^{3}H-borohydride, as already reported (*22*). The labeled prothrombin was electrophoretically indistinguishable from the untreated prothrombin (*22*). The radioactive ^{3}H-borohydride was purchased   from the Radiochemical Center, Amersham, England.

Chromatographically pure egg lecithin (PC) and ox brain phosphatidylserine (PS), grade I, were purchased from Lipid Products (Nutfield, United Kingdom) and supplied in chloroform-methanol solution. For monolayer spreading, samples in the right proportion were evaporated in a stream of nitrogen; the lipid content was determined by weight and then samples were dissolved in hexane. A two-fold excess of the lipid was spread over an aqueous solution (subphase) containing 0.15 M NaCl and 25 mM Tris at pH 7.4 (*23*). The excess ensures a fully compressed monolayer in equilibrium with the collapsed excess lipid layers , which has a negligible contribution to protein adsorption and none at all to the electrochemical measurements. Experiments were performed at room temperature, where the lipids are in liquid-crystalline phase (*24*).

For determination of binding onto the lipid monolayer, a solution containing radioactive protein and $CaCl_2$ or EDTA was injected underneath the spread monolayer. To achieve effective stirring of the subphase, one third of it was carefully retrieved and reinjected several times with a syringe, without disrupting the monolayer. Surface radioactivity was   then counted with an ultrathin (200 nm) supported end-window gas flow counter, as previously described (*25, 26*).

The surface concentration of the prothrombin Γ_p is given by:

$$\Gamma p = \frac{(cpm)_p^{\sigma}}{A} \frac{N_p}{(cpm)_p^{b}} \frac{(cpm)_{O.A}^{b}}{(cpm)_{O.A}^{\sigma}}$$

where $(cpm)_{O.A}^{\sigma}$ and $(cpm)_{O.A}^{b}$ are the surface and the bulk (scintillation) counts per minute of an equal sample of oleic acid, $(cpm)_p^{\sigma}$ is the surface count of the protein from the total area A, and $(cpm)_p^{b}$ is the scintillation count of Np protein molecules.

Once equilibrium has been reached, as determined from the surface radioactive measurements, a hanging Hg drop was positioned in contact with the phospholipid monolayer; an Ag / AgCl sat KCl electrode was the reference, and a platinum gauge was the auxiliary electrode. Alternative current (a.c.) polarography was carried out as described previously (*11*), with the polarographic instrument already mentioned (*27*) and in a Metrohm polarographic cell. The potential was scanned at a rate of 50 mV/s, and the frequency of a.c. modulation (10 mV peak to peak) was 80 Hz. The starting potential was chosen so as to be in the stable region of the monolayer, namely between - 200 and - 900 mV relative to the Ag / AgCl sat. KCl electrode.

Results

Adsorption and Desorption Kinetics. Figure 1 shows the kinetics of prothrombin adsorption at the initial bulk concentration of 4 μg/ml, in the presence or absence of calcium, onto a monolayer containing 100 % PS. The equilibrium surface concentration is not reached instantaneously but after about one hour, when a plateau is reached. For the construction of the adsorption isotherms we tried to approach these equilibrium surface concentrations, which were assumed to be reached when doubling the time did not result in a significant change in surface radioactivity. At this point, EDTA could be added to initiate the desorption process.

Effect of Dilution of Prothrombin on the Binding. Figure 2 represents the kinetics of adsorption of ^{3}H-prothrombin of differing specific radioactivities, onto a monolayer containing 25% PS, in the presence of calcium. The equilibrium surface concentrations are Γ_{p1} = 3.1 x 10^{-12} mol/ cm^2 for 100 % of the initial mixture of ^{3}H-protein and Γ_{p2} = 1.5 x 10^{-12} mol/cm^2 for 50% of the initial mixture of ^{3}H-prothrombin mixed with 50% of cold prothrombin. This result shows that a two-fold reduction in the specific radioactivity of ^{3}H-protein dilution decreases the equilibrium surface concentration by a factor of two. The hot and cold proteins thus bind to the monolayer in a similar fashion and justify the use of this approach for binding studies.

Adsorption onto and Desorption from Pure Phosphatidylserine Monolayers. Figure 3A represents the adsorption isotherms of ^{3}H-prothrombin on condensed phosphatidylserine monolayers, in the presence of 2mM calcium or in its absence. Equilibrium was reached after about 90 to 120 min , depending on bulk protein concentration. The adsorption isotherms are dependent on calcium concentration, as was already found in the case of human prothrombin (*13*). While calcium-independent binding is significant, calcium-dependent binding is higher.

When EDTA was added, as indicated by the arrows, it lowers the surface radioactivity to values observed during binding from the same bulk protein concentration in the absence of calcium.

In Figure 3B the Scatchard plot (Γ_p / C) vs. Γ_p is presented for binding, both in the presence and in the absence of calcium, where C is the true bulk prothrombin concentration in equilibrium with the surface. As can be seen from this Figure, the slopes and the intercepts of the straight lines obtained in both cases are significantly different. Thus the dissociation constants Kd, given by the slopes, are respectively 2.2 x 10^{-8} M and 4.5 x 10^{-8} M in the presence and in the absence of calcium (Table I). Concerning the maximal surface concentrations, they are 7.2 x 10^{-12} mol/cm^2 in the presence of calcium, and 5.9 x 10^{-12} mol/cm^2 in its absence. The results obtained with calcium are consistent with what we observed in the case of human prothrombin (*26*).

Adsorption onto and Desorption from Monolayers Containing 25% PS-75% PC. Figure 4A represents the adsorption isotherms of ^{3}H-prothrombin on a condensed monolayer composed of 25% PS- 75%PC, in the presence of 2mM calcium or in its absence. Equilibrium was reached after about 60 to 90 min.. The

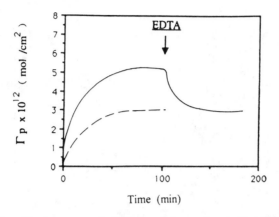

Figure 1. Effect of calcium on adsorption and desorption (with EDTA) kinetics of the binding of 4 µg/ml of prothrombin, to a phosphatidylserine monolayer: 2 mM calcium (▬) or no calcium (▬ ▬). Addition of EDTA is indicated.

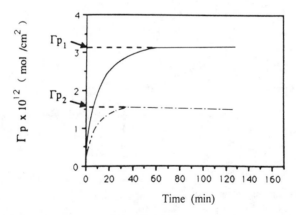

Figure 2. Binding of prothrombin of different specific radioactivities to a monolayer containing 25% PS, in the presence of calcium, as a function of time: (▬) initial mixture of ^{3}H-prothrombin (2.5 µg/ml) and after dilution by a factor of two with cold prothrombin (▬●▬).

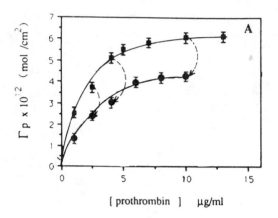

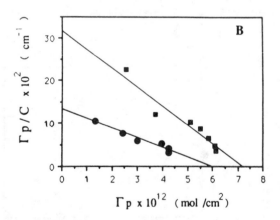

Figure 3. Binding of ^{3}H-prothrombin on condensed phosphatidylserine monolayers, in the presence of 2 mM calcium (■) or in its absence (●). Desorption by EDTA is shown by the arrows.

A) Surface concentration of bound prothrombin as a function of its concentration in the subphase, immediately after its injection.

B) Scatchard plot. Ratio of the equilibrium values of prothrombin surface concentration to its free concentration, .

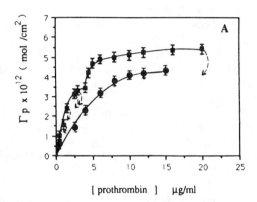

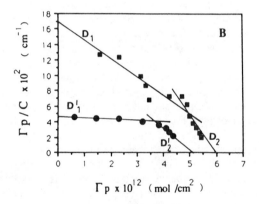

Figure 4. Binding of ^{3}H-prothrombin on condensed monolayers containing 25 % PS - 75 % PC, in the presence of 2 mM calcium (■) or in its absence (●). Desorption by EDTA is shown by the arrows.

A) Surface concentration of bound prothrombin as a function of its concentration in the subphase, immediately after its injection.

B) Scatchard plot. Ratio of the equilibrium values of prothrombin surface concentration to its free concentration, at low coverage (D1, D'1) and higher coverage (D2, D'2), adsorption   isotherms   are   dependent   on   calcium   concentration.   While   calcium-independent binding is significant, calcium-dependent binding is higher.

When EDTA was added, as indicated by the arrows, it lowers the surface radioactivity to values observed during binding from the same bulk protein concentration in the absence of calcium, at higher protein concentration; this was not the case at lower protein concentration.

In Figure 4B the Scatchard plot ($\Gamma p / C$) vs. Γp is presented for binding, both in the presence and in the absence of calcium, where C is the true bulk prothrombin concentration in equilibrium with the surface. As can be seen from this Figure, the slopes and intercepts of the straight lines obtained in both cases are significantly different. Moreover, the binding is dependent on the coverage. Thus the dissociation constants Kd, given by the slopes, are, at low and higher coverage respectively, 4×10^{-8} M and 2×10^{-8} M in the presence of calcium and 5×10^{-7}M and 3.6×10^{-8}M in the absence of calcium (Table I). Concerning the maximal surface concentrations, they are, at low and high coverage respectively, 7×10^{-12} mol/cm^2 and 6×10^{-12} in the presence of calcium, and 23×10^{-12} mol/cm^2 and 5×10^{-12} in its absence.

Effect of Prothrombin on the Differential Capacity of Monolayers Containing 25% PS- 75% PC. One way of studying the perturbations which can occur in a membrane consists of measuring the differential capacity variations of a mercury electrode in direct contact with a condensed phospholipid monolayer while macromolecules, like proteins, are introduced into the solution beneath (20). In fact, at a potential imposed between working and reference electrodes, the differential capacity depends on what is in contact with the electrode. The highest differential capacity corresponds to the pure electrolyte, so that when a macromolecule adsorbs at the electrode, the differential capacity decreases. These are the cases for condensed monolayers of proteins or of phospholipids adsorbed on a mercury electrode located above. In the case of phospholipids, the differential capacity is low around the nul charge potential (-0.5 V), where the lipids are adsorbed at the electrode by the hydrophobic part at any phospholipid composition. In the case of proteins, the differential capacity is higher. Thus, if introduction of a protein in the solution underneath a phospholipid monolayer leads to an increase in differential capacity, a perturbation of the dense phospholipid layer by the interacting proteins is indicated. This can be interpreted in terms of a penetration model.

The condensed state of the monolayer was verified by determining the differential capacity-potential curve itself, as described previously (11, Fig.1) and (20). The differential capacity variation at - 0.5 V, where the monolayer is the most stable, was selected to evaluate the effect of the protein concentration upon circuit differential capacity, as measured by a.c. polarography. In Figure 5 are represented plots of the variation of differential capacity (at equilibrium) as a function of prothrombin concentration for phospholipid monolayers composed of 25% PS- 75% PC, in the presence or in the absence of calcium. In both cases, no differential capacity change is observed at low protein concentration till about 3 µg/ml, where a cooperative interaction appears to be involved, as indicated by the sigmoîdal nature of the plots obtained in the presence of calcium. Above a certain protein concentration, saturation of the process occurs as there is almost no further change in differential

Table I Dissociation constants for prothrombin binding to monolayers of different ratios of PS to PC

		Low coverage	High coverage
100 % PS	+ Calcium	2.2×10^{-8} M	
	− Calcium	4.5×10^{-8} M	
25 % PS	+ Calcium	4×10^{-8} M	2×10^{-8} M
	− Calcium	5×10^{-7} M	3.6×10^{-8} M

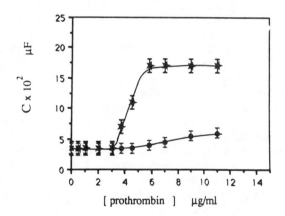

Figure 5. Effect of calcium on the differential capacity of condensed monolayers containing 25 % PS - 75 % PC, at - 0.5 V relative to sat Ag/AgCl electrode, as a function of the prothrombin concentration added: in the presence of 2 mM calcium (★) or in its absence (●).

capacity with increasing protein concentration, and a plateau is reached. This is not the case in the absence of calcium, where the increase in differential capacity is much smaller. Consequently, the penetration of prothrombin into this monolayer is quantitatively less in the absence of calcium than in its presence.

Effect of Prothrombin Concentration on its Binding to Monolayers containing 25% PS-75% PC, in the Presence of Calcium. Surface radioactivity and differential capacity of a monolayer containing 25% PS-75% PC in the presence of calcium are presented in Figure 6 as a function of prothrombin concentration and compared. At low coverage, since prothrombin binds (as inferred from radioactive measurements) but does not penetrate (as inferred from electrochemical measurements), it can be concluded that it only adsorbs, whereas at high coverage it penetrates. The similar threshold effect obtained for labeled and unlabeled prothrombin by these two different methods indicates that they are complementary and that the interaction was not affected either by the labeling procedure or by the presence of mercury.

Discussion

The effects of PS on the calcium-dependent binding of human and bovine prothrombin have been previously documented (*11, 26, 28*). In both cases, binding increases with rising PS content. The results presented here show that the interaction depends on calcium concentration at both phospholipid compositions (100% PS and 25% PS), since when calcium is added the surface radioactivity increases (Fig.3, 4) and the differential capacity too (Fig.5). This is in agreement with previous results obtained with human prothrombin interacting with monolayers containing 100% PS (*23, 26, 11*) or 25% PS-75% PC (*11, 23*) or with vesicles (*12*), and with bovine prothrombin interacting with vesicles (*28*) or with planar membranes (*16*). The dissociation constants and the maximal surface concentrations for prothrombin binding in the presence of calcium have been reported for both the membrane models used here (4); they were shown to depend on the technique used, on the membrane model employed and on the phospholipid composition.

   In addition to those calcium-dependent interactions, there are significant calcium- independent interactions at both phospholipid compositions (Fig.3, 4, 5), as was shown previously on human prothrombin interacting with monolayers (*11, 26, 23*). They were confirmed on vesicles for human prothrombin (*12, 14*) where the Kd is similar (*15*), then for bovine prothrombin by an other method (*16*). They were proposed to be electrostatic or hydrophobic (*14, 15*). They could be a result of protein molecules penetrating strongly in monolayers containing 100% PS (*29*), and slightly in a monolayer containing 25% PS (Fig.5) as observed for human II (*13*) at higher coverage only. In the latter case, penetrating molecules could come from a population already adsorbed at low coverage.

   The analysis of the different association constants and maximal binding values obtained by different groups (*4, 16*) shows that , at least in the case of monolayers, these quantities are of the same order of magnitude for the same type of membrane model, as long as the composition and the nature of the lipids are the same. Indeed, at 20 % PS the results are equivalent (*30*) to those presented here as long as the

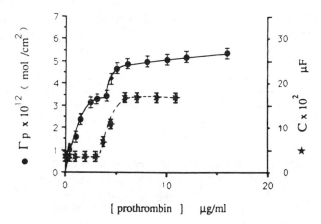

Figure 6. Binding of prothrombin on condensed monolayers containing 25 % PS - 75 % PC, in the presence of 2 mM calcium:
    Surface concentration of bound prothrombin (●) (Fig.4A) and differential capacity of the monolayer (★) (Fig.5), as a function of the prothrombin concentration in the subphase, immediately after its injection .

monolayer is not deposited on a slide (*4*), a treatment which may restrict mobility of the layer and thus may well change the binding characteristics. On the contrary, they seem to be one order of magnitude higher with monolayers than with vesicles. Why? First, in the experiments reported by Nelsestuen et al (*28*), the numerous steps necessary for vesicle preparation may have engendered a lipid composition on the outer leaflet different from that present in monolayers; this can imply a decrease in Ka , which is known to occur when % PS decreases (*28*). When albumin (*4*, ref.20) or immunoglobulin (*16*) are added, they might compete with prothrombin binding sites, mainly those involved in penetration, which would contribute to the higher affinity seen in the present study. In other words the discrepancies can come from the differences in the interaction mode which appear with the coverage, as clearly shown here (see below). Indeed it had already been observed that Kd might depend on the degree of coverage (*4*). The differences between fatty acids belonging to the phospholipids used (synthetic or natural) might have an influence, as they do in membrane fluidity and biological activity of the prothrombinase. Nevertheless, the association constants obtained in the present study are of the same order of magnitude as that obtained with vesicles (SUV), around 10^{-8}M, when equivalent conditions are used (*14-15*).

    In the present work, it was found that the dissociation constants are smaller in the presence of calcium than in its absence, for both lipid compositions studied (Table I). With 100% PS, the difference is very slight, and was not detected previously in the case of human II (*26*), whereas with a monolayer containing 25% PS, the difference is one order of magnitude at low coverage. This last result agrees with that obtained on planar membranes (*16*), and with the decrease of Kd with increasing calcium concentration obtained on vesicles (*28*).

From the desorption study with EDTA on monolayers containing 100% PS, it turns out that two populations of bound proteins can be distinguished, one calcium-dependent and one calcium- independent, as observed in the case of human prothrombin (29, 23).

Concerning the nature of the interactions of prothrombin with monolayers containing 25% PS (nearer to the maximal biological activity), they are more complex than those with 100% PS. At 25% PS, the interaction is dependent on coverage. Radioactivity measurements show that the process is biphasic, while in parallel, electrochemical measurements show an adsorption phase followed by penetration of prothrombin into the monolayer (Fig. 6). Protein molecules start penetrating only at a certain coverage of adsorbed molecules delimited by a short but significant intermediate plateau of surface radioactivity. At this level, there is a threshold concentration of about 5 µg/ml of prothrombin, separating two distinct binding states, correlated with two distinct dissociation constants, both in the presence and in the absence of calcium (Table I).

From the desorption study with EDTA, the isotherm is similar to that in the absence of calcium only at high coverage and not at low coverage. It appears as if there are two types of calcium-dependent binding: one EDTA-dependent and the other not, detected only at low coverage, since the surface radioactivity is slightly higher than in the absence of calcium. In this case, calcium should be entrapped in the protein/ membrane complex very strongly.

At high coverage, the dissociation constant is independent of PS content (above 25% PS), with or without calcium (Table I), as already observed with calcium (4); this suggests that hydrophobic forces may be involved in penetration, which occurs in this range of coverage. Furthermore, the higher affinity at higher coverage indicates that penetration should reinforce the binding which was initiated at lower coverage during the adsorption process. The forces involved in the latter case, mainly electrostatic, are consistent with a PS- dependent dissociation constant (Table I), as already observed with calcium (4). In the presence of calcium, when protein molecules start penetrating, there is a significant and abrupt increase in surface radioactivity at the threshold (Fig. 6), which could be correlated with dimerisation at the surface similar to that observed in solution, as proposed previously for human prothrombin (23), since both penetration and dimerisation are calcium- dependent. This is in agreement with Fourier transform infrared spectroscopy measurements which show that lipids induce an increase in ordered secondary structures in the prothrombin molecule, namely in prethrombin 2 (P2) domain (31). This observation favours the idea that prothrombin penetrates by means of this domain. Moreover, it was also proposed that a phospholipid binding site should be expressed on II by calcium ion-mediated stabilization of a particular conformer of II (8), where Gla should not be involved in direct binding to membrane (9).

Considering the similarities of behaviour between bovine and human prothrombin in their binding to PS- containing membranes, and particularly the threshold effect of penetration already obtained with human prothrombin (13, 32), we can propose which domains could be involved in the different processes of binding. At low coverage, it should not be prethrombin 2 which penetrates strongly (33, 34), but rather Fragment 1 (F1) which adsorbs and does not penetrate (32). Furthermore, the affinity was found to be lower for F1 than for prothrombin, with monolayers (26) as well as in planar membranes (16). It is also possible that a combination of adsorbed F1 plus prethrombin 1 (P1), ready to penetrate in a synergistic manner (*34*), probably through P2, could be involved. Indeed, it was found previously with monolayers containing 25% PS, that Gla are involved in the binding in the presence of calcium whereas it is the prethrombinic part in the absence of calcium (*35*), like on vesicles (15). We must be careful in such interpretation, since there are differences in cleavage namely between human and bovine prothrombin. Nevertheless, these interactions are consistent with what we obtained with vesicles for bovine prothrombin (*36*), with mainly P2 penetrating at high coverage.

It turns out that, as long as monolayers are in a condensed state, they can be compared to vesicles. Moreover they provide more detailed information on the mode of interaction, since a clear threshold effect could be observed for the first time with such a system. It is possible that the second layer in vesicles might stabilise the system and change the interaction. Nevertheless, the condensed monolayer gave the same threshold effect (Fig.6) whatever the nature of the second interface (helium or mercury), and the same Kd (15). Since the different model membranes can give complementary information it is prudent to compare both, as long as equivalent conditions are used. Current work is aimed at determining how well vesicles and monolayers can serve as models of membrane function.

Acknowledgments

We are very grateful to Pr.K.G.Mann for discussions and for providing purified prothrombin when M.F.L. was in his laboratory for Sabbatical. We thank D. Lane for correcting the manuscript. This research was supported in part by INSERM Grant 910306 and Grant 9007854 from Region Midi-Pyrénées.

Literature Cited

1. Bevers, E.M.; Comfurius,P.; Van Rijn, J.L.M.L.; Hemker, H.C.; Zwaal,R.F.A.*Eur.J.Biochem.* **1982**, *122*, pp.429-436.
2. Mann, K.G.; Fass D.N. *Hematology* **1983**, *2*, pp.347-374.
3. Nesheim, M.E.; Taswell, J.B.; Mann, K.G. *J.Biol.Chem.* **1979**, *254*, pp.10952-10962.
4. Kop, J.M.M.; Cuypers, P.A.; Lindhout, T.; Hemker, H.C.; Hermens, W.T. *J.Biol.Chem.* **1984**, *259*, pp.13993-13998.
5. Dombrose, F.A.; Gitel, S.N.; Zawalich, K.; Jackson, C.M. *J.Biol.Chem.* **1979**, *254*, pp.5027-5040.
6. Resnick, R.M.; Nelsestuen, G.L. *Biochemistry* **1980**, *19*, pp.3028-3033.
7. Rosing, J.; Speijer, H.; Zwaal, R.F.A. *Biochemistry* **1988**, *27*, pp.8-11.
8. Borowski, M.; Furie, B.C.; Bauminger, S.; Furie, B. *J.Biol.Chem.* **1986**, *261*, pp.14969-14976.
9. Furie, B; Furie, B.C., *Cell.* **1988**, *53*, pp.505-518.
10. Ratcliffe, J.V.; Furie, B.; Furie, B.C. *J.Biol.Chem.* **1993**, *268*, pp.24339-24345.
11. Lecompte, M.F.; Miller, I.R. *Biochemistry* **1980**, *19*, pp.3439-3446.
12. Lecompte, M.F.; Rosenberg, I.; Gitler, C. *Biochim.Biophys.Res.Commun.* **1984**, *125*, pp.381-386.

13. Lecompte, M.F. *In proteins at Interfaces;* Brash, J.L.; Horbett, T.A., Eds.; ACS Symposium Series 343; American Chemical Society: Washington, DC, **1987**, pp 103-117.
14. Prigent-Dachary, J.; Faucon, J.F.; Boisseau, M.R.; Dufourcq, J. *Eur.J.Biochem.* **1986**, *155*, pp.133-140.
15. Prigent-Dachary, J.; Lindhout, T.; Boisseau, M.R.; Dufourcq, J. *Eur.J.Biochem.* **1989**, *181*, pp.675-680.
16. Tendian, S.W.; Lentz, B.R.; Thompson, N.L. *Biochemistry* **1991**, *30*, pp.10991-10999.
17. Rosing, J.; Tans, G.; Speijer, H.; Zwaal, R.F.A. *Biochemistry* **1988**, *27*, pp.9048-9055.
18. Rhee, M.J.; Horrocks, W.De W.; Kosow, D.P. *Biochemistry* **1982**,*21*, pp.4524-4528.
19. Lundblad, R.L. *Biochem.Biophys.Res.Commun.* **1988**, *157*, pp.295-300.
20. Lecompte, M.F.; Bouix, G.; Mann, K.G.*J.Biol.Chem.* **1994**, *269*, pp.1905-1910.
21. Krishnaswamy, S.; Mann, K.G.; Nesheim, M.E. *J.Biol.Chem.***1986**, *261*, pp.8977-8984.
22. Butkowski, R.J.; Bajaj, S.P.; Mann, K.G. *J.Biol.Chem.***1974**, *249*, pp.6562-6569.
23. Lecompte, M.F.; Miller I.R. *J.Colloid Interface Science* **1988**, *123*, pp.259-266.
24. Tans, G.; van Zutphen, H.; Comfurius, P.; Hemker, H.C.; Zwaal, R.F.A. *Eur.J.Biochem.* **1979**, *95*, pp.449-457.
25. Frommer, M.A.; Miller, I.R. *J.Phys.Chem.***1968**, *72*, pp.2862-2866.
26. Lecompte, M.F.; Miller, I.R.; Elion, J.; Benarous, R. *Biochemistry* **1980**, *19*, pp.3434-3439.
27. Lecompte M.F., Clavilier J., Dode C., Elion J. Miller I.R. *J.Electroanal.Chem.***1984**, *163*, pp.345-362.
28. Nelsestuen,G.L.; Broderius,M. *Biochemistry* **1977**, *16*, pp.4172-4177.
29. Lecompte, M.F.; Miller, I.R. *Advances in Chemistry Series.***1980**, *188*, pp.117-127.
30. Mayer, L.D.; Nelsestuen, G.L., Brockman, H.L. *Biochemistry* **1983**, *22*, pp.316-321.
31. Wu, J.R.; Lentz, B.R. *Biophys.J.* **1991**, *60*, pp.70-80.
30. Lecompte, M.F.; Elion, J.; Miller, I.R. Protides Biol. Fluids Proc.Colloq. **1980**, *28*, pp.277-280.
31. Lecompte, M.F. *Biochimie*, **1984**, *66*, pp.105-109.
32. Lecompte, M.F.; Dode, C. *Bioelectrochem. and Bioenerg.* **1992**, *29*, pp.149-157.
33. Lecompte, M.F.; Dode, C. *Biochimie* **1989**, *71*, pp.175-182.
34. Lecompte, M.F.; Krishnaswamy, S.; Mann, K.G. *Thromb.Haemostasis* **1991**, *65*, pp.663.

RECEIVED March 2, 1995

Chapter 38

Formation and Properties of Surface-Active Antibodies

Shlomo Magdassi[1], Oren Sheinberg[1], and Zichria Zakay-Rones[2]

[1]Casali Institute of Applied Chemistry, School of Applied Science, and [2]Faculty of Medicine, Hebrew University of Jerusalem, 91904 Jerusalem, Israel

Surface active antibodies were formed by covalent attachment of hydrophobic groups to the IgG molecule. The modified antibodies reduced surface tension and adsorbed onto emulsion droplets at surface concentrations higher than the native antibody. The chemical modification led to a decrease in the biological activity; however, at specific conditions, surface-active antibodies, which retained their recognition ability, could be formed. By using these antibodies, a new emulsion, which has a specific recognition ability for HSV-1 infected cells, was formed.

Chemical modifications of proteins may lead to significant changes in their surface activity and functional properties. For example, when hydrophobic chains are covalently attached to ovalbumin, the modified protein becomes more surface active than the native protein: it can adsorb at higher surface concentrations at hydrophobic surfaces(1), it can reduce surface and interfacial tensions (2) and may be used as a better emulsifier than native ovalbumin (3). The modified protein is, therefore, functioning as a polymeric surfactant, which has various properties, depending on the structure of the native protein, the number of chains and chain length of the attached hydrophobic groups. Since proteins may have specific biological activity, it could be possible to form surface active proteins, which also have a specific biological activity, while the protein is spontaneously adsorbed at various interfaces.

We used such an approach for antibodies which after modification were able to reduce surface tension, adsorb onto oil droplets in oil-in-water emulsions and still retain their specific biological activity. This biological activity, which is the specific recognition, might be used for unique applications such as drug targeting and immunodiagnostics, both based on emulsions (4), clusters (5) and liposomes (6), containing a suitable probe molecule.

The use of various colloidal systems, which contain the surface active antibodies, may lead to significant advantages, such as eliminating the need for covalent attachment of the probe molecule to the antibody as suggested for drug targeting (7), or achieving a very high load of probe molecules per antibody molecule, which then can be used for enhanced immunoassays.

0097–6156/95/0602–0533$12.00/0

The present report will focus on the preparation of surface-active IgG molecules by attachment of various hydrophobic groups, and on the surface activity at oil-water interface, while retaining the recognition ability of the IgG molecules.

Experimental

The active ester was prepared by reacting N-hydroxysuccinimide with fatty acids having chain length C_8-C_{18}, by a procedure described earlier(1). The octanoic acid was covalently attached to the IgG by adding its active ester, which is dissolved in dioxane, into IgG solution (6.5 mg/ml) in phosphate buffer (pH=10), in the presence of 2% W/W sodium deoxycholate. The amount of added ester could vary according to the expected degree of modification.

The reaction mixtures were shaken in a water bath (25°C) for 3 hours, and then were filtered by 0.45μm and 0.2μm filters. The clear solutions were then dialyzed against phosphate buffer (pH=8) for 72 hours, while changing the buffer each 8 hours.

The degree of modification was evaluated by determining the concentration of free lysine groups present in the IgG before and after modification, by using the TNBS method (8, 11).

Adsorption studies were performed by mixing 2.5 ml IgG solution (native or modified) at various concentrations, with commercial oil-in-water emulsion (Intralipid, Kabi Pharmacia), which contained 10% soybean oil and was stabilized by egg phospholipids. The mixture was shaken in a water bath (25°C) for a period of time described in the text. Then, the mixture was filtered by 0.2μm filter and the concentration of protein in the clear solution was determined by the Lowry method (12). The adsorbed amount was calculated by the difference in protein concentration before and after adsorption.

The surface tension measurements were performed by the Wilhelmy plate method, using a Lauda tensiometer, which allows a continuous measurement of surface tension until a constant value is obtained within a few hours, depending on IgG concentration.

The biological activity was determined by a commercial ELISA diagnostic kit (Enzygnost anti HSV-1/IgG, Behring) suitable for detection of antibodies for Herpes Simplex 1 virus (HSV-1). The IgG prepared by suitable ammonium sulphate precipitation of human serum, had a high titer for HSV-1.

The specific recognition by emulsion was performed by incubating the various antibody solutions for 45 minutes with HSV-1 infected BSC-1 cells on a microscope slide. BSC-1 cells (originating in African green-monkey's kidneys) were infected by HSV-1 with multiplicity of infections 0.1-0.01 TCID 50 per cell. After the incubation, the cells were rinsed by phosphate buffered saline to remove the emulsion droplets, which were not adhered. The cells and presence of attached oil droplets, were viewed by a Nikon Optiphot microscope.

All modifying reagents were purchased from Sigma and were used without further purification.

Results and Discussion

Since the IgG molecule is water soluble, it was expected that the main parameter to influence its surface activity would be the number of hydrophobic groups covalently attached to the antibody molecule. The modification was performed on the lysine residues of the protein and, since each molecule contains 90 lysine groups, we could form surface-active antibodies with a very large number of hydrophobic groups. The modification was performed by various hydrophobic groups, but because of space limitations we shall concentrate on the C_8 modifications only. The modification degree, m, will define the number of hydrophobic groups which are linked to each IgG molecule.

A series of IgGs at various degrees of modification was prepared by changing the molar ratio of active ester to native protein in the reaction mixture. As shown in Fig. 1, increasing this ratio led to an increase in the number of attached hydrophobic groups. However, we could not obtain a modification degree higher than about 30, and the number of attached groups was always lower than expected, according to the initial molar ratio. This is a result of the rapid hydrolysis of the active ester in the aqueous solution, and the precipitation of the highly modified proteins, which became too hydrophobic.

One indication for the presence of surface-active antibody is its ability to reduce surface and interfacial tension, as was indeed observed for all the modified IgGs. As demonstrated in Fig. 2, both native and modified IgG can reduce surface tension; however, the modified IgG (modification degree, m=32) reduces more rapidly the surface tension to a lower equilibrium value than the native IgG. Since these measurements were performed at low protein concentrations (0.01 mg/ ml) the differences in the behaviour can be attributed to the presence of hydrophobic anchoring groups in the modified protein.

The next step was to evaluate the possibility of adsorption of the modified antibodies onto O/W emulsion droplets. In this case, the adsorption does not take place at a simple oil-water interface, but at such interface which is already covered with a phospholipid monolayer (it should be noted that we found that the modified antibodies could be used as emulsifiers by themselves).

As shown in Fig. 3, when a native IgG or modified IgG (m=9), is mixed with an O/W emulsion, both proteins are adsorbed onto the emulsion droplets within two hours, as reflected by the decrease in protein concentration in the aqueous solution with time. Since the initial concentration of the two proteins was the same, 0.75 mg/ml, it is obvious that the modified IgG is more rapidly adsorbed than the native IgG. It is important to note that the adsorption of native IgG on emulsion droplets has not been previously reported. In itself it is of great importance from the viewpoint of possible interactions of injectable emulsions with blood components, and possible applications, which could be based on these observations.

Once the adsorption onto the emulsion droplets was proved, we performed experiments which could yield adsorption isotherms. Such a typical experiment was performed by mixing IgG at various concentrations, with a constant emulsion concentration, for 12 hours, which is above the time required to reach equilibrium. Typical adsorption isotherms are presented in Fig. 4. As can be seen, the native IgG is adsorbed until it reaches a plateau value, at about 0.7 mg/m^2 (the available area for adsorption was calculated from the average droplet size in the emulsion). In comparison, the attachment of 30 hydrophobic groups leads to a significant increase of the adsorbed amount, up to 4 mg/m^2, without reaching a saturation. At lower degree of modification (m=5), a plateau value is observed, which is about twice the one obtained for the native IgG.

Comparing these results to the surface concentrations of native IgG adsorbed onto various solid surfaces, shows that the adsorption onto emulsion droplets is much lower. For example, adsorption of IgG on anionic polyrinyltoluene surface reaches a monolayer value at about 7 mg/m^2 (9). This significant lowering of adsorbed amount is probably due to the presence of the phospholipids, which are the emulsion stabilizers. In this case, the IgG molecules have to overcome the electrical repulsion caused by the negatively charged emulsifier (the IgG is also negatively charged), and penetrate through the emulsifier monolayer. This observation is in

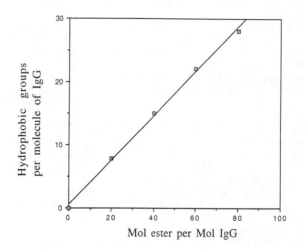

Figure 1. Number of hydrophobic groups attached to an IgG molecule, as a function of initial molar ratio of ester to antibody in the reaction mixture.

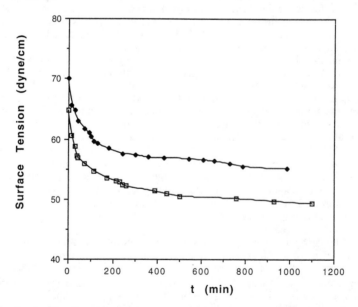

Figure 2. Surface tension reduction of aqueous solutions by native (♦).and C_8-modified IgG, m=32 (◘). Protein concentration: 0.01 mg/ml.

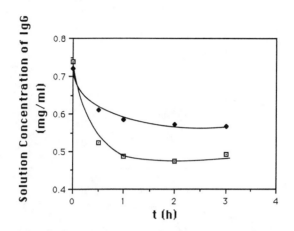

Figure 3. Kinetics of adsorption of native (♦) and C₈-modified IgG, m=9 to O/W emulsion. (□). Protein concentration is 0.75 mg/ml.

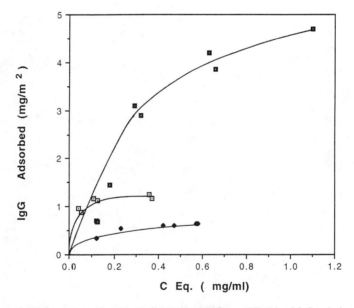

Figure 4. Adsorption of native IgG (♦), and IgG modified by 30 C₈ chains (□) and by 5 C₈ chains (▣), to O/W emulsion. Adsorption time, 12 h.

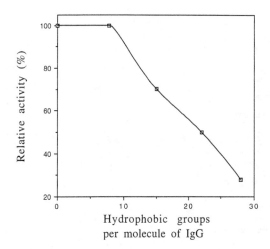

Figure 5. Activity of native and C_8 modified IgG, as determined by ELISA as a function of a number of hydrophobic groups. The activity is expressed relative to the native IgG, which is 100%.

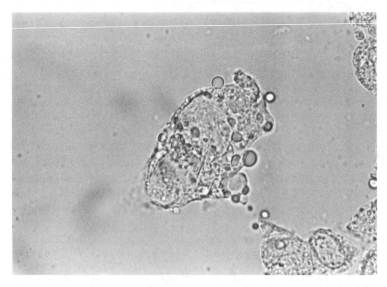

Figure 6. Photomicrograph of emulsion droplets adhered to HSV-1 infected cells. A C_8-modified IgG (m=9) was previously adsorbed on the emulsion droplets.

agreement with the results obtained by Arai et al., who measured the adsorption onto phospholipid-coated latex particles (*10*). They have found that the amount of IgG adsorbed decreased with increase of lipid coverage of the latex surface. At about 90% lipid coverage, the surface concentration of IgG was about 0.5 mg/m^2, which is similar to the present results obtained for native IgG on emulsion droplets (it is most likely that the lipid coverage in the emulsion is close to 100%). In view of these results, obtaining surface concentrations of several mg/m^2 by the modified IgG, seems to be more impressive. Furthermore, the chemical modification leads to a shift of the iso electric point of the modified IgG, towards lower pH values, meaning that the tested antibodies are more negatively charged than the native ones, at pH 7.4.

In general, it should be expected that the adsorption is decreased when the protein is below or above the iso electric point, about pH=7, as is also shown by Arai et al.(*10*). Therefore, it can be concluded, that the modification leads to the formation of antibodies which have enhanced tendency to adsorb at oil-water interface and are probably capable of replacing previously adsorbed surfactants.

The main purpose of this research was to form surface-active proteins, combining surface activity and biological activity. Therefore, the biological activity was evaluated using two methods: (a) a conventional ELISA measurement of IgG, specific to Herpes Simplex virus (HSV-1), and (b) a method designed for the evaluation of specific recognition by emulsion droplets.

The ELISA test was performed for solutions of native and modified IgG, which were tested specifically for HSV-1. As shown in Fig. 5, the activity of the modified proteins decreased with increase in the number of hydrophobic groups attached to the IgG molecule. This effect is not surprising since the chemical modification can cause denaturation of the IgG, or the presence of many hydrophobic groups can block the recognition site of the antibody (it should be emphasized that the chemical modification was not selective only toward the F$_c$ fragment of the antibody). Another possibility for the decrease in biological activity is the formation of clusters of IgG molecules, due to hydrophobic modification (*5*). However, as expected, at low degree of modification, m=8, the biological activity was similar to that of the native IgG. Therefore, the specific recognition by emulsion droplets was evaluated for low degree of modification: a sample of O/W emulsion was mixed with the modified antibody for 10 hours. This emulsion was added dropwise to a microscope slide, which contained HSV-1 infected cells, and was incubated for 45 minutes to allow adhesion and interaction between the IgG and the HSV-1 antigens present on the cell walls. Then the slide was rinsed by phosphate buffer saline to remove all droplets which were not attached. As shown in Fig. 6, we could identify, by microscope, the adhesion of emulsion droplets to the infected cells. this observation means that the emulsion droplets have the ability to recognize specific antigens.

Suitable control experiments, in which non-infected cells, emulsion without antibodies and emulsion with native antibody were used, revealed that the adhesion of emulsion droplets was indeed specific to HSV-1 infected cells.

In conclusion, we have shown that hydrophobically modified antibodies become surface-active, and that under certain conditions they can retain biological activity, while being adsorbed at an oil-water emulsion interface.

With this concept established, it should be possible now to continue with the research, aimed at both understanding the mechanism of adsorption-desorption, formation of micelle-like structures and applying the methods to development of new drug targeting and diagnostic systems.

Literature Cited

1.   Maagdassi, S.; Leibler, D. and Braun, S. *Langmuir*, **1990**, *6*, 376.
2.   Magdassi, S.; Stawski, A. and Braun, S. *Tenside*, **1991**, *28*, 264.
3.   Magdassi, S. and Stawski, A. *J. Disp. Sci. Technol.* **1989**, *10*, 213.

4. Magdassi, S.; Rones-Zakay, Z.; Lineritz, M. and Sheinberg, O. Israel Pat. Appl. No. 102718, **1992**.
5. Magdassi, S.; Rones-Zakay, Z. and Toledano, O. Israel Pat. Appl. No. 106887, **1993**.
6. Huang, L.; Huang, A. and Stephan, T. In *Liposome Technology III*; Gregoriadis, G., Ed.; 1985, pp. 52.
7. Trail, P.A.; Wilner, D.; Lasch, S.J.; Henderson, A.J.; Hofstead, S.; Casazza, A.M.; Firestone, R.A.; Hellstrom, I. and Hellstrom, K.E. *Science*, **1993**, *261*, 212.
8. Habeeb, S.A. *Anal. Biochem.* **1966**, *14*, 328.
9. Baghi, P. and Birnbaum, S.M. *J. Colloid Interface Sci.*, **1981**, *83*, 460.
10. Arai, T.; Mishiro, R. and Kitamara, H. A paper presented at 201 ACS meeting, 1991.
11. Alder-Nissen, J. *J. Agric. Food Chem.*, **1979**, *27*, 1256.
12. Peterson, G.L. *Methods Enzymology*, **1983**, *91*, 95.

RECEIVED February 10, 1995

INDEXES

Author Index

543

Affiliation Index

Subject Index

Production: Meg Marshall
Indexing: Deborah H. Steiner
Acquisition: Rhonda Bitterli
Cover design: Michele Telschow

Printed and bound by Maple Press, York, PA

Bestsellers from ACS Books

The ACS Style Guide: A Manual for Authors and Editors
Edited by Janet S. Dodd
264 pp; clothbound ISBN 0–8412–0917–0; paperback ISBN 0–8412–0943–X

Understanding Chemical Patents: A Guide for the Inventor
By John T. Maynard and Howard M. Peters
184 pp; clothbound ISBN 0–8412–1997–4; paperback ISBN 0–8412–1998–2

Chemical Activities (student and teacher editions)
By Christie L. Borgford and Lee R. Summerlin
330 pp; spiralbound ISBN 0–8412–1417–4; teacher ed. ISBN 0–8412–1416–6

Chemical Demonstrations: A Sourcebook for Teachers,
Volumes 1 and 2, Second Edition
Volume 1 by Lee R. Summerlin and James L. Ealy, Jr.;
Vol. 1, 198 pp; spiralbound ISBN 0–8412–1481–6;
Volume 2 by Lee R. Summerlin, Christie L. Borgford, and Julie B. Ealy
Vol. 2, 234 pp; spiralbound ISBN 0–8412–1535–9

Chemistry and Crime: From Sherlock Holmes to Today's Courtroom
Edited by Samuel M. Gerber
135 pp; clothbound ISBN 0–8412–0784–4; paperback ISBN 0–8412–0785–2

Writing the Laboratory Notebook
By Howard M. Kanare
145 pp; clothbound ISBN 0–8412–0906–5; paperback ISBN 0–8412–0933–2

Developing a Chemical Hygiene Plan
By Jay A. Young, Warren K. Kingsley, and George H. Wahl, Jr.
paperback ISBN 0–8412–1876–5

Introduction to Microwave Sample Preparation: Theory and Practice
Edited by H. M. Kingston and Lois B. Jassie
263 pp; clothbound ISBN 0–8412–1450–6

Principles of Environmental Sampling
Edited by Lawrence H. Keith
ACS Professional Reference Book; 458 pp;
clothbound ISBN 0–8412–1173–6; paperback ISBN 0–8412–1437–9

Biotechnology and Materials Science: Chemistry for the Future
Edited by Mary L. Good (Jacqueline K. Barton, Associate Editor)
135 pp; clothbound ISBN 0–8412–1472–7; paperback ISBN 0–8412–1473–5

For further information and a free catalog of ACS books, contact:
American Chemical Society
Product Services Office
1155 16th Street, NW, Washington, DC 20036
Telephone 800–227–5558